D0903422

PEST CONTROL

Biological, Physical,
and Selected Chemical Methods

PEST

Edited by WENDELL W. KILGORE

AGRICULTURAL TOXICOLOGY AND RESIDUE RESEARCH LABORATORY
COLLEGE OF AGRICULTURE
UNIVERSITY OF CALIFORNIA
DAVIS, CALIFORNIA

CONTROL

BIOLOGICAL, PHYSICAL, AND SELECTED CHEMICAL METHODS

and RICHARD L. DOUTT

DIVISION OF BIOLOGICAL CONTROL
COLLEGE OF AGRICULTURE
UNIVERSITY OF CALIFORNIA
BERKELEY, CALIFORNIA

ACADEMIC PRESS New York and London 1967

ACADEMIC PRESS INC.
111 Fifth Avenue, New York, New York 10003

United Kingdom Edition published by
ACADEMIC PRESS INC. (LONDON) LTD.
Berkeley Square House, London W.1

Library of Congress Catalog Card Number: 66-30122

PRINTED IN THE UNITED STATES OF AMERICA

LIST OF CONTRIBUTORS

Numbers in parentheses refer to the pages on which the authors' contributions begin.

R. C. BUSHLAND, Metabolism and Radiation Research Laboratory, Entomology Research Division, Agricultural Research Service, United States Department of Agriculture, State University Station, Fargo, North Dakota (147)

RICHARD L. DOUTT, Division of Biological Control, College of Agriculture, University of California, Berkeley, California (3)

*HUBERT FRINGS, Department of Zoology, University of Hawaii, Honolulu, Hawaii (387)

*MABLE FRINGS, Department of Zoology, University of Hawaii, Honolulu, Hawaii (387)

L. K. GASTON, University of California, Riverside, California (241)

WALTER E. HOWARD, Department of Animal Physiology, University of California, Davis, California (343)

WENDELL W. KILGORE, Agricultural Toxicology and Residue Research Laboratory, College of Agriculture, University of California, Davis, California (197)

LEO E. LACHANCE, Metabolism and Radiation Research Laboratory, Entomology Research Division, Agricultural Research Service, United States Department of Agriculture, State University Station, Fargo, North Dakota (147)

STUART O. NELSON, Electrical Conditioning, Preservation, and Protection Investigations, Farm Electrification Research Branch, Agricultural Engineering Research Division, Agricultural Research Service, United States Department of Agriculture and University of Nebraska, Lincoln, Nebraska (89)

* Present address: University of Oklahoma, Norman, Oklahoma.

RUTH R. PAINTER, Agricultural Toxicology and Residue Research Laboratory, College of Agriculture, University of California, Davis, California (267)

C. H. SCHMIDT, Metabolism and Radiation Research Laboratory, Entomology Research Division, Agricultural Research Service, United States Department of Agriculture, State University Station, Fargo, North Dakota (147)

H. H. SHOREY, University of California, Riverside, California (241)

RAY F. SMITH, Department of Entomology and Parasitology, University of California, Berkeley, California (295)

Y. TANADA, Division of Invertebrate Pathology, University of California, Berkeley, California (31)

ROBERT VAN DEN BOSCH, Department of Entomology and Parasitology, University of California, Berkeley, California (295)

DONALD P. WRIGHT, JR., Agricultural Division, American Cyanamid Co., Princeton, New Jersey (287)

PREFACE

The need to develop novel, alternative, or functional combinations of pest control techniques is emphatically a product of this decade. No reader needs to be reminded that scarcely four years have passed since pest control practices were subjected to painful public scrutiny. Disregarding the polarity of arguments on both sides of this controversy, the fact remains that responsible people continued to seek solutions to pest control problems in an imaginative fashion.

The remarkable part of the story is that developments have come so rapidly and on such a broad front. Even in this age of accelerated progress it is noteworthy that scientists have produced such a large area of information. The purpose of this book is to set forth these advances.

A characteristic feature of this work is the unmistakable recognition that pest control is not unilateral. It is basically an ecological problem of great complexity which requires the blending of talents from many contributors. This multidisciplinary approach points the way to the future, and the editors wish to thank each of the authors who have enriched this timely book.

June, 1967

WENDELL W. KILGORE
RICHARD L. DOUTT

CONTENTS

4 Radiation-Induced Sterilization

Leo E. LaChance, C. H. Schmidt, and R. C. Bushland

5 Chemosterilants

Wendell W. Kilgore

6 Pheromones

H. H. Shorey and L. K. Gaston

7 Repellents

Ruth R. Painter

8 Antifeedants

Donald P. Wright, Jr.

9 Integrated Control

Ray F. Smith and Robert van den Bosch

PART II: VERTEBRATE PESTS

10 Biocontrol and Chemosterilants

Walter E. Howard

11 Behavioral Manipulation (Visual, Mechanical, and Acoustical)

Hubert Frings and Mable Frings

PEST CONTROL

Biological, Physical,
and Selected Chemical Methods

Part I

INSECT PESTS

1 BIOLOGICAL CONTROL

Richard L. Doutt

DIVISION OF BIOLOGICAL CONTROL
COLLEGE OF AGRICULTURE
UNIVERSITY OF CALIFORNIA, BERKELEY, CALIFORNIA

Biological control of invertebrate pests and weeds has enormous and unique advantages. This splendid control measure, which is a tribute to man's knowledge of the functioning biotic world about him, is now officially designated as the preferred method of insect suppression by the Entomological Society of America, the organization to which all leading professional entomologists of the United States and Canada belong.*

Biological control is especially desirable because it is *safe, permanent,* and *economical.* It is thus unique, for no other control measure has this wonderful combination of advantages, and on logical grounds it has been strongly advocated as the first line of attack in pest suppression. Furthermore, biological control is remarkably durable. It has lasted through years of neglect by the entomological profession and through periods of appalling ignorance. It has endured criticism and has survived direct

* Statement on Pesticides. March, 1964. *Bull. Entomol. Soc. Am.* **10,** 18.

attacks by articulate, powerful, and self-serving antagonists. Finally, it has abided while one by one various pest control fads and numerous commercial products, all proposed as panaceas, have failed and faded into oblivion.

The method is not new, and perhaps it is as old as any presently known and used control measure. It is therefore especially significant that, in spite of such relative antiquity, biological control is today appropriately included among the most modern and advanced techniques for pest suppression. This is the case because its maximum usefulness is based upon sophisticated ecological knowledge, and ecology, especially population ecology, is but a comparatively recent and flourishing discipline. With this proper recognition of biological control and an increased understanding of its basic principles has come a corresponding acceleration and expansion of its use. Nevertheless, its full value and potential application still remain largely unrealized. Therefore, biological control is today virtually an unexploited resource. This is an exciting fact because the magnificent and spectacular accomplishments of biological control measures in the past are clearly destined to be matched many times in the future.

I. THE DIMENSIONS OF BIOLOGICAL CONTROL

The field of biological control has many aspects, and for this reason it easily escapes the meshes of any confining net of strict definition. Stated as a biological phenomenon it is the action of parasites, predators, or pathogens in maintaining another organism's population at a lower average density than would occur in their absence, but stated as an entomological practice it is the study and utilization of parasites, predators, and pathogens for the regulation of population densities of pests. Neither statement is entirely satisfactory, but one unifying factor can be recognized. This is simply that biological control has an ecological basis. In fact, it could correctly be considered as a special field of applied ecology, and this may well be its best definition as biological control advances into the future.

A. Ecological Dimensions

An examination of this first and fundamental dimension of biological control is therefore in order, and it is profitable to focus attention initially on the tendency of living systems to maintain an internal stability by their own regulatory devices. This concept of homeostasis was originally developed as a principle in animal physiology, but ecologists have adopted the term and have applied it to the stability of populations within ecosystems. It is nothing more than erudite terminology to avoid the unjustly maligned phrase "balance of nature."

1. *Homeostasis in the Ecosystem*

In any stable habitat, the resident species tend to maintain the same general quantitative interrelationships over long periods of time. Of course fluctuations are always occurring, but if we continued a census of the populations over a number of years in the same environment we would find that certain species would regularly be abundant, others less so, and some species would normally be comparatively rare. Barring any major change in the environment, this pattern of relative abundance would be repeated indefinitely. This inherent stability in an ecosystem is of tremendous practical importance, for biological control is basically concerned with the forces in the environment that regulate and thereby stabilize the populations of organisms.

Static and Dynamic Aspects. While these stabilizing forces on a protracted time scale impart a static quality to the average levels of resident species, nevertheless it is obvious that the population density of each species is never stationary but is continually changing. A population continually responds to the innumerable, varying, and interdependent environmental forces which influence its natality, mortality, and dispersion. As a result its numbers fluctuate up and down, but always around a relatively stable value which is called the equilibrium position. This stability in the face of constant change is no paradox. The concept is comparable to that of sea level which furnishes us a zero basis for altitude, and yet the sea is never level but is kept in constant motion by waves or tides. Both static and dynamic qualities are thus combined so that a state of dynamic equilibrium characterizes a population of organisms.

2. *Equilibrium Position*

There are entomologists who see only the dynamic aspects of population phenomena and not these static characteristics. They are so impressed by the fluctuations of insect populations, which can be of astonishing magnitude, that they are inclined to deny the existence of any equilibrium position in a species. To them its existence is contrary to facts and a denial of organic evolution. On the other hand, my colleague C. B. Huffaker (personal communication) believes that the measure of success of a species is the relative stability of its ecological position and that any adaptation which gives increased security of ecological position will tend to be perpetuated. In other words, natural selection is involved in the maintenance of balance in populations and this has survival value. Proof in either case is difficult because balanced relationships, as such, do not fossilize.

The concensus among biological control workers is that this equilibrium does exist. It is just as universal and understandable a phenomenon associated with insect populations as the concept of sea level is generally applicable to the oceans of the world. Most importantly, it is often a determinant of the pest status of a species. This is the case because the equilibrium positions are not necessarily fixed and inflexible, but instead vary in time as conditions change. The extent and manner in which such equilibrium positions are altered are important elements in determining the pest status of a species. In fact, many of the great entomological problems of the world have resulted from major displacements in the equilibrium positions of insect species. Such a displacement readily occurs when a species is transferred from its native home where natural controls maintain it at a low population density to a new area where, in the absence of effective natural enemies, its numbers may be greatly increased. Examples of this are common among the pests in North America, where the majority of such species are exotic in origin and have reached the continent without their full complement of natural enemies. Biological control in the classical sense is the procedure by which such missing natural enemies are obtained from the pest's native home and then established as a mortality component in the new ecosystem. In successful cases of biological control, these imported natural enemies depress the pest's equilibrium position to a noneconomic level, and then continue to maintain it at this reduced value.

3. *Regulation*

Such biological control agents are regulatory in nature, and it is this regulatory action which distinguishes biological control from other methods and also gives to it the advantage of permanency. The application of an insecticide will substantially reduce a pest population to a low level, but its effect is only temporary. In a distressingly short time the pest will again attain its damaging numbers, and this occurs because the insecticide is capable only of reducing the population. It lacks any regulatory action. On the other hand the agents employed in biological control are biotic in nature, attuned to the pest species, and geared to its density, so that their action becomes more intense as the pest begins to rise above its equilibrium position and conversely relaxes as the pest declines below this value. A mortality agent which functions in this manner is said to be density dependent in its action and is therefore regulatory. This fundamental concept was initially developed by entomologists engaged in biological control (see Smith, 1935) and has subsequently been intensively analyzed and generally adopted by many ecologists throughout the world. Many of the parasites, predators, and pathogens of arthropod pests act in this manner, as do the phytophagous insects utilized for control of noxious weeds. These are the tools of biological control and their use requires a thorough understanding of the principles of population dynamics. It is these principles which are employed in biological control and which emphasize its ecological basis. There is a growing awareness among professional entomologists that pest control programs will be immensely improved if field research on pest problems develops a definite ecological orientation. The Australian entomologists Geier and Clark (1961) contend that control practices are out of step with modern ecological ideas, and correctly state that ecological knowledge and an understanding of population dynamics ought to be used if lasting solutions are to be achieved.

Of course, the pest does not exist in a vacuum but is a single component in an extraordinarily complex but closely knit and integrated unit termed the ecosystem, which includes the populations of plants and animals (including man) together with the nonliving environment. It is important to understand that the populations of organisms in an ecosystem stand in a complicated functional relationship to one another, that there are essential energy flows among them, and that these populations are delicately, but dynamically, balanced. Interference with these relationships should be kept to a minimum, but this fact has been generally ignored in the development of chemical control programs. Such programs have no adequate basis for continuing effectiveness and the lack of any ecological foundation has often had disastrous results. Through experiences in using powerful insecticides, entomologists have repeatedly demonstrated that no matter whether a system be a potato patch or a national forest it is nevertheless composed of very complicated biotic communities to which one cannot with absolute impunity apply a toxic chemical. Such applications have demonstrated, sometimes in a most startling manner, the presence and the role of other invertebrate components of these communities, species of whose existence we were scarcely aware and of whose importance in the communal economy we were entirely ignorant (Doutt, 1964).

A typical case is that of certain *Lecanium* species on English walnut in California. Prior to the advent of DDT, these scales were seldom encountered and their density was so restricted that there was little to indicate that their natural enemies, particularly small hymenopterous parasites, were responsible for this condition. The importance of these parasites as controlling agents of the scales became apparent when DDT interfered with their effective action. The *Lecanium* species that previously presented no problem suddenly became major pests of walnuts. The fact that this potential existed probably would have escaped detection if it had not been for the interference of insecticides (Michelbacher, 1962).

There are many other examples of chemically induced infestations where the increase of a nontarget, phytophagous invertebrate was triggered by the application of a toxic chemical. In each case, the increase resulted from a disruptive force purposely interposed on the community. The realization of this fact has made agricultural entomologists more ecologically minded and has turned the thinking toward a more fundamental approach to the solution of pest problems. This has had the salutary effect of bringing biological control and chemical control closer together and has set the stage for the development of modern integrated control programs. There is now a strong trend in agricultural entomology to base control programs on the utilization of ecological principles. Because of the basic nature of their field, workers in biological control have long employed this approach and have always attempted to attack pest problems fundamentally and at their source. With the current expansion of this ecological approach to the solution of pest problems, it seems probable that in the future biological control will increasingly be able to assume its proper role in pest suppression (Doutt, 1965a).

B. Biological Dimensions

Until recently, the framework for the biological dimension of this field was primarily formed by the nutritional requirements of natural enemies of arthropods. Research was largely focused on the necessary interrelations, adaptations, and behaviorisms of entomophagous species in obtaining these essential sources of energy. Now, however, newer aspects of this biological dimension are being presented by the increasing use of parasites to control noxious weeds and by the recent stimulating emergence and development of biological control as an integral part of plant pathology (Baker and Snyder, 1965).

Not only is this biological dimension an extremely fascinating subject, but it also contains marvelous areas for research, especially because the efforts are so frequently rewarded with astounding discoveries. To compress this vast and complicated area into an abbreviated statement necessitates the sacrifice of enormous amounts of relevant information and, unfortunately, results in little more than a mere glimpse of some of the broader biological phenomena with which any scientist in biological control must be familiar.

1. *Antagonism*

According to Snyder (1960) biological control in the field of plant pathology is based upon an interruption of host–parasite relationships through biological means. This may be accomplished by imparting resistance to the host or by modifying the culture of the crop to avoid or reduce infection. Modification of crop culture by utilizing the action of other organisms against the pathogen invokes what plant pathologists term "antagonism." This is broadly defined as the sum total of the unfavorable influences which one organism exerts against another. Also in such a broad sense the activity of an antagonist may include physical destruction, parasitism, antibiotic excretions, and more subtle forms of attrition or competition for nutrients and space.

Although neither the exact nature nor the mechanism of action of antagonism is fully understood, there is sufficient information available to suggest that this phenomenon holds great promise for the future as an effective means of controlling soil-borne plant diseases (Snyder, 1960). It has been suggested by Patrick and Toussoun (1965) that introduced antagonists may best be used on greenhouse crops because the limited amount of soil involved can be manipulated with relative ease. They also suggest that, for practical agricultural use, biological means of such disease control should be integrated with other methods, each supplementing the other. This closely parallels modern trends in agricultural entomology.

"One is left to conclude that the total antagonism potential of soil upon which the plant pathologist may draw is enormous. The soil abounds with powerful antagonists which compete with, parasitize, or poison plant pathogens. These antagonists are selective for the pathogens they antagonize; they are selective for the kinds of organic materials upon which they thrive; and they are antagonized in turn by other elements of the flora of the soil even as they antagonize. The opportunities for playing one soil organism against another to man's advantage are there, and only await man's cleverness in dealing with antagonists" (Snyder, 1960).

2. *Parasitism*

In the field of biological control, parasitism is such a protean phenomenon that its continued usefulness as a unit concept in this work has become doubtful. The symbiotic relationship by which one species obtains its energy requirements at the expense of another is surely parasitic, and offers a system to be exploited at every opportunity in biological control. Yet the terminology presents some problems in proper communication. While a scientist will speak of entomophagous insects that develop at the expense of a single host as parasites, he actually has less justification for this than does his colleague who speaks of the phytophagous insects employed for weed control as "parasites." When communicating, these two scientists would mentally put these two types of animals into separate categories even though each was labeled "parasite." In microbial control, the problem is avoided by simply calling their useful organisms pathogens.

In actual practice, the procedures for exploiting parasitism have become so remarkably specialized that the technology required for one aspect will bear little resemblence to the skills, equipment, and techniques used by scientists engaged in another field of biological control. Even their respective philosophies may be distinguishable. It is unlikely that the same scientist could with competence manipulate species of fungi belonging to the hyphomycetous *Moniliales* to suppress plant nematodes, and with equal facility test phytophagous insects imported from Turkestan for their effectiveness in controlling a noxious weed on Nevada sheep ranges. This necessary specialization and compartmentalization of effort reflects the manifold nature of the disciplines contributing to modern biological control and demonstrates the high level of technical training required by practitioners in this field.

3. *Predation*

Definitive boundaries for predation are difficult to draw among some of the organisms which are useful in biological control. In projects involving entomophagous arthropods, a predator is considered to be an organism that requires the consumption of more than one individual (prey) to develop to maturity. At the same time the working definition for a parasite is an organism that will develop to maturity at the expense of a single (host) individual. There are many overlapping situations, so that these categories tend to lose distinctness, but the greatest amalgamation results when one considers their respective roles as parameters in the population dynamics of their host or prey. Here the parasites resemble predators because their attack results in the death of the host individual. Their action in this respect is like the predator that directly attacks and immediately devours its prey. Such parasites constitute, therefore, a rather special kind of predator and for this reason are occasionally designated as "parasitoids" instead of parasites.

These parasitoids are often more effective than predators in suppressing pests because they can function at lower host densities. Not only is the number of progeny proportional to the hosts found and attacked, but the ability to find the host is highly developed in the winged and vagile adult females.

Predation among arthropods is generally density dependent in its action and therefore is one of the key factors in regulating the population density of prey species. Several biological control projects which have been spectacularly successful and have resulted in enormous economic savings have been brought about through the action of predatory species.

4. *Behavior*

Both parasitism and predation are descriptions of the way an organism behaves in relation to its food source. It does things: it finds and parasitizes a host; or it captures and consumes a prey. Such behavior deserves far more attention than it has received as a proper field of research in entomology. The ethological work which has already been accomplished with entomophagous insects clearly demonstrates the great value and immediate application of such research. These studies have, for example, done much to explain the process of host selection and the eliminative steps that produce some degree of host specificity in nature. Entomophagous insects are especially favorable animals for such studies and so, as the importance of intensive research in insect behavior becomes more fully recognized and implemented, their use will appreciably increase. Inevitably, such research efforts will become an integral part of every major biological control laboratory that seeks proper strength through balance and diversity in its operations.

C. Taxonomic, Genetic, and Evolutionary Dimensions

At any moment, the biotic world which man perceives is the resultant of the evolutionary forces which have shaped it over geologic time. Such forces determine the appearance and survival of taxonomic units as well as the roles which the various species play in the economy of nature. Such relevant phrases as struggle for existence and the origin of species take on real meaning to the student of biological control. This is because he must constantly study, evaluate, and employ the biotic, mortality-causing agents of an environment. Furthermore, because of the close attention which must be paid to the biological characteristics of a given species of parasite or predator, he often uncovers evidence of sibling species or at least biological entities or segregates which appear to be evolving toward the species level.

In processing material through the quarantine laboratory, it has been found advisable to segregate stocks of parasites imported from various geographical localities even though these insects all appear to be conspecific. In the project to control the olive scale, *Parlatoria oleae* (Colvée), this policy led to the discovery of certain *Aphytis* which morphologically resemble the species *maculicornis* (Masi) but which are distinctly different biologically (Hafez and Doutt, 1954). These are probably sibling species, although perhaps cryptic species is better terminology.

Similarly, DeBach (1959) working with morphologically similar species of *Aphytis* attacking the California red scale, found two new species of *Aphytis* on the basis of his biological data. Although adult differences between the new species, *A. fisheri* DeBach and *A. melinus* DeBach, are very slight, distinct pupal differences exist and the two do not cross.

Hall *et al.* (1962) using differences in host specificity, cross-mating, and progeny production between a walnut aphid parasite, *Trioxys pallidus* Hal., and a spotted alfalfa aphid parasite, *T. utilis* Muesebeck, demonstrated that both of the morphologically similar and recently synonymized species are distinct sibling species. Cryptic species must therefore be numerous, and future work will undoubtedly reveal many other examples.

Considerations of the center of origin of a species and the geographical distribution of a pest and its nearest relatives are important first steps in deciding where exploration for natural enemies should be conducted. Since these considerations must necessarily be preceded by an accurate identification of the pest species, there is no applied field in entomology where the role of the taxonomist is more important in decision making than in biological control.

In theory, and in practice as well, the procedure in biological control is to return to the indigenous host population to find the most effective and presumably the best-adapted natural enemies. Accordingly, when an organism has become a pest following its introduction into a new area, great importance is immediately attached to the parasitic fauna associated with the native population of the species. Substantial expenditures of money and technical efforts are made on the basic premise that with the indigenous population will be found the most effective natural enemies. This belief seems to be fundamentally correct, for each project initiated in biological control against an exotic immigrant pest has been undertaken with the supposition that the native home of the pest is the most profitable one to explore. The soundness of this assumption is reflected in the results that have been achieved by biological control during the past three-quarters of a century throughout the world. Another way of emphasizing this point is to consider all past biological control projects merely as a series of replicated experiments to test this single idea. Viewed in this light, the results give conclusive support to the hypothesis.

A taxonomist who has mastered his subject is knowledgeable of the distribution of species in his special interest group. He uses such data in conjunction with other evidence to build a concept of the phylogenetic relationships and the general history of the organisms with which he works. The true systematist is not satisfied until he can visualize his special interest group in time and space. Accordingly we turn to such a scientist to learn where the endemic population of a pest species is probably located. There are many criteria for determining the center of origin of species, but it is not as important to locate the center of origin as it is to find the endemic population of a pest species. This may seem illogical, but actually a species may be restricted in its distribution to a particular area and yet it may not necessarily have originated there. For example, the so called relict species were formerly widespread but now exist only in much smaller areas.

A misidentification of a pest species at this point can result in years of wasted effort and a considerable amount of misspent money if it causes a search to be made in the wrong area. Such a case is recounted by LePelley (1943), which involves *Pseudococcus kenyae* LePelley. This mealybug pest became severe in Kenya and, because of misidentifications, futile searches for parasites were made in Java, Hawaii, California, and the Philippines. After the pest was correctly recognized as an undescribed species from another part of the African continent, it was a simple matter to obtain parasites from Uganda which very quickly and completely supressed the infestations.

"The crucial importance of correct specific identification of an insect is nowhere more evident than in this case, where numerous unsuccessful attempts at biological control were made, involving the tapping of beneficial insect resources in four continents, when a large number of valuable primary parasites were present all the time in a neighboring country. As soon as the insect was recognized to be quite distinct from *P. lilacinus* it seemed likely that it would prove indigenous in some part of its African range outside of Kenya" (LePelley, 1943).

For many years, *Circulifer tenellus* (Baker), the insect vector of curly top virus, was incorrectly placed in the genus *Eutettix*. In 1918 the California Commission of Horticulture sent an explorer to Australia to search for the leafhopper and its natural enemies, but the insect was not found there. In 1926–1928 the United States Department of Agriculture (USDA) surveyed the situation in Argentina and Uruguay, and still later in Mexico. Again *C. tenellus* was not located and no parasites were sent to the United States. Finally, as a result of careful taxonomic work by Oman (1936, 1948), the position of *tenellus* was clarified, for he showed it to be a representative of the Old World genus *Circulifer*. As a result, explorations were directed to Spain and North Africa where *C. tenellus* was then found in endemic populations. Large numbers of its parasites were found as well, and these were shipped to California for study. Their progeny were subsequently colonized in the western United States (Huffaker *et al.,* 1954).

Coccid specialists were consulted to pinpoint the probable center of origin of the olive scale, *Parlatoria oleae* (Colvée). They indicated that this would be in the general area of Pakistan, northern India, and Afghanistan. Subsequently, two remarkably effective parasites have been obtained from this area and have brought about economic control of this serious diaspine scale on olives in California (Huffaker and Kennett, 1966).

It is therefore obvious that biological control is dependent on good systematics, but it is also in a position to furnish valuable information to taxonomists. Of all the various fields of entomology, probably no two could mutually benefit from cooperative studies more than biological control and systematics. In 1955, Sabrosky summarized the matter as follows: "Biological-control workers are dealing with complex biological problems, with the interactions of populations of one to many parasites, with one to many species of hosts, and with each other, and with hyperparasites, and all with populations of predators, and all in turn with climatic and other environmental factors. Likewise, the taxonomy in biological-control problems can be a complex matter, involving the identification and classification of species of all those categories. All in all, the problems pose a real challenge to both taxonomists and biological-control workers, a challenge that can best be met by the fullest teamwork possible."

It has been suggested that living organisms, as hosts to parasites, form one of the three major habitats on earth, comparable to the aquatic and terrestrial habitats in which the hosts themselves dwell (Allee *et al.,* 1949). It is therefore likely that certain isolating mechanisms exist among the parasitic groups that are not encountered in other insect species. A suggestion of this is found in the work of Thorpe (1945) which contains considerable presumptive evidence of the isolating effect of host conditioning in parasitic insects. He believes that the conditioned response will give momentum to and set the direction for the selective processes tending to bring about genotypic isolation. He further believes that it is best to regard geographical, topographical, and ecological isolation as three different scales of spatial isolation. Therefore, with each host as a kind of physiological island, it is not surprising that speciation has occurred frequently among the parasitoids. (It is estimated that there are approximately 12,000 Chalcidoidea and 20,000 Ichneumonoidea.) On the other side of the coin, and of equal fascination to the naturalist, is the obvious fact that the phenotypic expression of morphological characters as well as the development of certain behavioral patterns in insects often owes to the selective pressure of natural enemies (Doutt, 1960).

The circumstances of intense competition between insects attacking the same host species has presumably brought about the specialization of natural enemies as a means of ensuring survival (Wilson, 1965). It has long been apparent to biological control workers that a widespread host will be attacked by enemies that are more restricted in habitat. In an examination of endemism, Doutt (1961) concluded that this differential distribution among parasites is an inherent characteristics of host–parasite relationships even in narrowly endemic host species.

The paper by Wilson (1965) emphasizes the genetic dimension that one encounters in the field of biological control and outlines the desirability of collaboration between geneticists and biological control workers. For example, there is the general problem of the procedures to be undertaken to ensure maximum effectiveness of natural enemies imported to a new area. As Wilson ably points out, this involves the question of genetic diversity of a species in its whole native habitat, the best means of utilizing this diversity, and the problem of the importation of the desired genetic diversity.

There are good practical reasons for being concerned about the genetic potential of an introduced species, and whether it is being properly used. To understand this concern about the introduced genetic elements, it is necessary to know that the established procedure in biological control operations is to process in quarantine the shipments of imported material collected by an entomologist in the foreign field. The parasites which emerge from this imported host material are used to start the initial cultures of the beneficial species. It is not uncommon to begin such a stock culture with only one or two females, for that may be all that emerge from a shipment. The progeny from this tiny beginning are bred for several generations and are the ancestors of the millions of individuals that are eventually colonized in the field. The implications of using the progeny from such a small stock are indeed great. The impregnated female that originates such a culture carries only the genotype of herself and her mate, not the genetic makeup of an interbreeding species population. She does not represent an average of her native population and so we may be establishing a parasite that only partially resembles the parental species in the native home. This may have potentialities in several directions, for we may be establishing an entity that is superior to the wild stock or we may be colonizing a genetic strain that is not as efficient as the parental species.

In addition there is the possibility of manipulating the gene pool to develop strains better suited to the conditions in the new habitat. This should of course be done in the laboratory before any colonization of the species is made. As Wilson (1965) suggests, this leads to the important problem of the genetic basis of ecological characteristics of species and how these may be modified. There is a great need of such studies in relation to host selection, climatic tolerances and preferences, environmental selection, host-finding capacity, mode of reproduction, and other important characteristics. The field is a wide one. Biological control workers are well aware of the value of such protracted studies on the imported species before they are colonized, but the economic realities demand the establishment of imported material as soon as is reasonably possible. There are critics of this procedure and, if biological control projects had the luxury of time, there would be merit in following their suggestions.

D. Historical and Intellectual Dimensions

Although applied biological control rests upon a foundation of ecological principles and is implemented by manipulating biological materials, it is nevertheless to a considerable degree a human enterprise. The initial achievement in biological control of an agricultural pest required a special set of circumstances that came about during the nineteenth century. It developed at that time because a pioneering spirit in a progressive new continental agriculture was a concomitant of an advanced knowledge of the biotic world outlined by intellectuals of Europe. This was a time that saw an increasing interest in population growth forms starting with the crude geometric progression suggested by Malthus and including the long overlooked work on the logistic curve by Verhulst. Darwin was influenced by Malthus and exploited the Malthusian terminology of struggle for existence. Practical men in the New World were aware of the immigrant nature of many of their serious pests and by the middle of the century had seriously suggested the alleviation of such problems by importing the natural enemies of such pests. Asa Fitch in New York and Benjamin Walsh in Illinois were early and articulate proponents of this notion, and these entomologists clearly understood the role of entomophagous species in the suppression of host or prey populations. It is significant also that these men suggested biological control of weeds by using phytophagous insects.

Modern biological control began in California with a spectacular demonstration that saved the infant citrus industry. The story is well known because nowhere in the world is the history of agriculture so well documented from its inception as it is in the State of California. This is the case because there was virtually no agriculture in the State before 1850. The Indian tribes in California were not agricultural; the Spaniards did not appear upon the scene until the last of the 18th century and they did little more than tend small mission gardens and perfunctorily run some cattle on the hillsides. Thus agriculture in California really did not begin until after the gold rush when settlers in large numbers remained to till the soil. These were inherently a pioneering breed faced with new conditions and not guided by precedent nor bound by rigid and ancient customs. They were inventive entrepreneurs and from the outset experimented with crops, irrigation systems, and new methods of pest control. They were progressive, imaginative, and fiercely competitive. These qualities were essential to economic survival because the major markets for their agricultural products were 3000 miles to the east across a nation that had no transcontinental railroad until 1869. These were the agriculturalists who were so impressed by the peregrine origins of their pests that they enacted the first quarantine legislation in the United States, fully a quarter of a century before the federal laws were passed. These were the farmers who as practical entomologists without technical assistance developed the poison bait method of grasshopper control. These were the men who invented the fumigation method for controlling pests on citrus, who experimented with sprays, and finally who enthusiastically supported the first magnificent demonstration of biological control.

The project which they endorsed was the control of the cottony cushion scale, *Icerya purchasi* (Maskell), by a colorful predaceous ladybird beetle purposely imported from Australia in 1888. This beetle, commonly called the vedalia, so drastically reduced the scale problem that the number of carloads of citrus shipped from southern California more than doubled in a single year (Doutt, 1958). This project is pointed to as the classical example of biological control for it demonstrates all the essential features of a completely successful control program. Furthermore, it established biological control as a major weapon in the arsenal of economic entomologists throughout the world, and gave the initial impetus for the vigorous and continuing support of this area of research. The project also furnishes us with a convenient time scale by which we can measure the durability of biological control, because the vedalia continues to control the cottony cushion scale throughout California as effectively as it first did in 1889.

In subsequent years, many practical projects involving biological control have been conducted with all degrees of success, and the entomologists engaged in this work have developed certain concepts that have become guiding principles. A number of men have made important theoretical contributions to the understanding of population phenomena that underlie biological control, and some of these ideas have been subjected to experimental verification. Such work continues, and has grown from the early arithmetical considerations of biological control workers to very elegant mathematical models developed by ecologists. In this computer age, theoreticians can handle complexities that were barriers to such work in the earlier days.

This area of investigation is matched by erudite studies on the taxonomy, biology, physiology, and behavior of the organisms employed in biological control endeavors. In spite of this, there are critics who contend that biological control lacks any scientific basis and is simply an empirical remedy that fortuitously succeeds. Actually, arguments on whether biological control is an art or a science seem rather pointless, for it is simply an intelligible area of study which has very great practical benefits when applied in the field and which is important as a source of challenging ideas. These are being tested in biological and ecological centers throughout the world. Inevitably there will be a feedback from the elegant models to field situations as it becomes necessary to test such computer products under natural conditions. This circular process will continue to benefit biological control and will continue to be a satisfying and rewarding experience to the scientists engaged in such studies.

E. Legal and Political Dimensions

As farmers became aware of the foreign origin of pests, they advocated laws curtailing the unrestricted transport of plants and animals between countries. These legal restrictions on the movements of commodities for the purpose of preventing, or at least delaying, the establishment of plant pests are commonly known as plant quarantine regulations. This idea of using the police power of a state to exclude pests originated with growers who had experienced staggering losses from alien insects and who were seeking ways to avoid similar problems in the future. Their basic premise was that it is better to undergo considerable inconvenience and initial expense in an effort to exclude a pest than to submit to the larger costs of suffering its damages and fighting it for an indefinite period. While this would superficially appear to be asking the nonagricultural public to bear the cost of protecting the farmers from pest control expenditures, in actuality it is a saving to the economy of a nation because increased production costs are ultimately borne by every consumer. Therefore the policy behind quarantine regulations has great merit, and the record over the years testifies to the sound manner in which the regulations have been administered. The regulations have also sensibly been kept flexible to match the rapidly changing nature of transport and the needs of modern society.

Biological control in the conventional sense must operate on the basis of importing living organisms from a foreign area with the intention of establishing these as new faunal elements in a local agricultural system. It is thus a commerce in exotics that can be maintained only as a legal exception to the general prohibitions against the introduction of foreign organisms. Therefore in order to import such organisms it is necessary first to obtain the legal clearances from plant quarantine agencies, both state and federal. Such a request is viewed very carefully and is subjected to extremely close scrutiny when the proposal involves the introduction of a phytophagous insect for weed control. Such plant parasites cannot be cleared until extensive testing of their feeding range and breeding habits has been conducted outside the country. If these tests show the insect to be restricted to the weedy species that it is to control, then the plant quarantine agencies will normally permit the introduction.

There are many different interests which must be balanced in establishing any quarantine regulation, and such conflicting interests become especially critical in the biological control of weeds. No weed control project can be initiated until the plant has been officially declared a noxious pest, and there has been some understanding of opposing views. It is not uncommon to find a species ranked as a serious pest by one segment of agriculture and at the same time considered a valuable asset by others. The yellow star thistle, *Centaurea solstitialis,* is an example. Although this weed is a scourge to cattle raisers, and is a very troublesome plant in other ways, nevertheless it is a valuable nectar source for honeybees and is therefore looked upon with some favor by beekeepers. Even poison oak and poison ivy, which each year cause misery and suffering to thousands of susceptible persons, have their defenders who oppose their removal. In each case the decision must be made upon the merits, and, if the public interest will clearly be best served by the suppression of such an organism, then the biological control project may proceed, but not otherwise.

Organizations in Biological Control

There are a number of organizations that have historically engaged in biological control, and there are newer ones appearing. It may seem odd that organizations should be discussed as a part of biological control, but this is a characteristic and necessary dimension. Biological control is by necessity a teamwork effort. It is something in the nature of a relay race, where one man carries out an initial assignment, then passes it on to the next person, and so on until the goal is attained. Such is the case with biological control, for it is both impractical and inefficient to expect one person alone to conduct a biological control project from its inception until it is finished. Conventionally the procedure is first to determine the native home of the pest, or its endemic area, and then to send some qualified explorer there to search for the host and its complement of natural enemies. This is the foreign exploration phase of biological control. Because of the increasingly efficient and rapid transportation systems in the modern world, a foreign explorer can now traverse enormous areas in a very short time, and material which he collects can be transported quickly back to the home country for processing. It is seldom now that a shipment is more than 4 or 5 days enroute from any place in the world. The foreign explorer is therefore looking primarily for the host insect. He seldom operates entirely from a fixed base or semipermanent laboratory, except in the case of weed projects. Therefore, when he finds the host species, rather than rearing parasites from them himself, he will promptly send the collected samples home for study in the quarantine laboratory.

This second phase of quarantine handling of imported material requires special facilities to prevent the possible escape of organisms and it requires special knowledge and aptitude on the part of the person who has the responsibility of rearing the parasitic species and evaluating their habits.

Imported material may consist of a very few individuals and yet hundreds of thousands may be needed for adequate colonization and establishment in the new area. Accordingly, their numbers must be increased through insectary production, and this often requires considerable ingenuity and imagination in the development of culture methods.

After the imported species finally becomes established and of value as a new beneficial element in the fauna of the agricultural area, then careful ecological studies are essential. The precise role which the imported parasite plays and the measurement of its effectiveness as a mortality factor in the pest's environment require very careful ecological measurements. This field evaluation is the final, and perhaps most difficult, step in the series of phases through which any conventional biological control project evolves.

It is obvious that a biological control project cannot hope to be undertaken, let alone succeed, unless there are certain minimal facilities and talented staff. There must be at least a quarantine laboratory and an insectary, together with the necessary equipment for handling and studying the organisms involved. There must be a staff of sufficient size and training to find the foreign material, receive and process it in quarantine, invent the necessary mass production techniques, adequately colonize the species, and then evaluate the established organism in the ecosystem. These things require substantial and continuing support, and therefore it is understandable that biological control is carried out by institutions or organizations, and that this is a necessary feature of biological control. Most laboratories specializing in biological control are supported by state or federal governments or very large agricultural organizations.

F. Economic Dimensions

One very appealing characteristic of biological control is that it is economical. Although it requires a substantial initial investment in talent and facilities, as described above, this is amply repaid by the benefits that accrue from the results of biological control. It has been conservatively estimated that in the 36 years between 1923 and 1959 the net savings from major biological control projects amounted to over $110,000,000 in the state of California alone (DeBach, 1964a).

"It would not be fair, naturally, to debit against the biological control program the cost of only those projects which were successful. The numerous projects which failed should also be charged against it. But even so it is doubted if any investments by the University of California College of Agriculture can show greater returns than those for the biological control of insect pests" (Smith, 1946). When biological control successfully solves a pest problem, the individual grower is relieved of the annually recurring expenditures of treatment for that particular pest. The permanency of the action of biological control then makes this cumulative saving an impressive figure. Even if a project is not completely successful or is occasionally disrupted by other necessary farm practices or environmental changes, it is nevertheless of real value in slowing down the rate of increase of a pest or in tending to dampen excessive fluctuations or in diminishing populations of pests in areas from which they may disperse into commercial plantings.

Biological control has always been of especial value to crops with small margins of profit and to areas where food and fiber must be produced by workers who do not have the expensive equipment or necessary training to apply safely the modern, highly toxic chemicals. Many outstanding successes of the method, however, have occurred in agricultural industries noted for their high income and high level of technology, such as citrus and sugar cane. These are the segments of agriculture which have given the greatest support to biological control in the past and are also the businesses where the costs of crop protection are most carefully and accurately scrutinized and appraised. It is therefore significant that biological control, through its obvious economic benefits, has been incorporated as a regular tactic in these economically sophisticated areas.

G. The Ethical Dimension

There is one more dimension of the biological control method that was given little, if any, attention in the earlier days, but which now figures prominently in the thinking of many influential people in policy-making positions. This is brought about by the growing public awareness of the serious nature of chemical pollutants in man's environment. It is recognized that many of these problems could be avoided or minimized if biological control could be employed instead of chemical pesticides, and every person engaged in agriculture or forestry is aware of the controversy that has raged over this fact (Carson, 1962; Rudd, 1964).

No single method of controlling pests is a panacea for all have limitations, but all have undeniable advantages. We must therefore continue to use all known control measures, and also attempt to develop new techniques. Most importantly we must strive vigorously to use these control measures as effectively as possible, either singly or in combination. Nevertheless, as human populations increase, there becomes a need not only to increase agricultural production but also a desperate need, in civilized societies, to protect and preserve natural and native values. For this reason, every possible effort should be made to use techniques for pest control that will be the least disruptive to natural environments. This returns to a consideration of conflicting interests and the need to strike a sensible balance. It involves the short-range view and use of expedient measures in opposition to a larger perspective and a more protracted time scale. Clearly there is an ethical problem here, and one which should be faced honestly by all persons engaged in any responsible aspect of pest control. It is well to remember Aldo Leopold's (1949) admonition that man has certain ethical obligations to nature and the land.

II. THE POTENTIAL OF BIOLOGICAL CONTROL

The biological control of invertebrate and weed pests has been successful in a tremendous array of diverse situations. It has worked well in temperate and tropical climates; it has succeeded on continents, subcontinents, and islands; it has prevailed in areas in northern latitudes and in regions south of the equator; it has been impressive in forests and in agricultural plantings; it has suppressed target pests of great taxonomic diversity; in short, it has worked well wherever man has persevered in its application, for it has been effective in almost direct proportion to the amount of support given to it (DeBach, 1962). It has been employed where chemical control measures were not feasible and it has been used harmoniously in conjunction with chemical measures. The tools of biological control have ranged from the original vedalia to the viruses now used to initiate epizootics. Yet the latent period in biological control may actually be the past 75 years when all of this was going on. This is the case because the true potential of this method is still to be realized, and with the advent of the new, imaginative techniques for pest suppression which may be used in conjunction with biological control, the era of enlightenment in pest control has begun. The concomitant of this is the basic ecological approach to all pest problems, which equates with biological control. The true era of biological control is thus just beginning.

A. The Challenging Possibilities

It is certain that any of the great pest problems affecting agriculture, forestry, or public health could be alleviated to some degree through biological control techniques. Even such perennially intractable pests as codling moth, which cause such direct damage to the end product of agricultural efforts, may be more amenable to biological control and the ecological approach than is generally believed. Much remains to be discovered about the codling moth, and nonchemical measures have not been put to completely adequate and long-term tests anywhere except in Nova Scotia, where they were remarkably successful and have resulted in a codling moth program that is unique in the world. Perhaps the Nova Scotia program developed by Pickett and his co-workers (Pickett and Patterson, 1953) may not be completely adoptable elsewhere, but it does show the things that can be accomplished toward integrated control if entomologists are given the encouragement to pursue such research.

The codling moth is an alien and introduced pest. Throughout the world there are approximately 120 species of parasites and 22 additional predators that are recorded as attacking this pest. It is also known to be susceptible to a fungus and to a granulosis virus. Discounting some records that may be questionable or erroneous, this is still a very impressive total. In spite of this tremendous list, only two species of parasites of the codling moth have been imported into the United States, and neither of these came from a locality in the area presumed to be the native home of this insect which, to judge from its known distribution and the distribution of wild apples, probably extends from the Caucasus through the Himalayan foothills to China. This is just one of the many examples of entomological efforts being grossly out of balance, for while millions of dollars have been spent on pesticide research and chemical control, virtually nothing has been done toward importing and testing enemy species for the biological control of the codling moth. Surely there has been ample opportunity and economic incentive to justify a major project on the biological control of this pest (DeBach, 1964a; Doutt, 1965b).

Basic studies on the behavior and ecology of the codling moth itself are by no means complete, and yet these are essential to biological control. For example, where and how do the adults actually spend the day? Is there some way that we can cause mortality of the adults during this period? We know that they have protective coloration, but is there any way that we can make them more conspicuous to vertebrate predators? Australian workers are trying to find ways to discourage the codling moth larvae from hibernating on the trees and are trying to drive them off the trees to subject them to more hazards and greater predation.

What about the exciting possibility of using a pathogen, such as the granulosis virus, to initiate epizootics? There are many other nonchemical approaches, including the use of sterilized males, interrupting the natural cycle of diapause, and so forth, but these are somewhat beyond the scope of biological control. They do show, however, that pest control without chemicals, whether it be biological control or integrated control, should never be ruled out even with such stubborn pests as codling moth.

1. *Native Pests*

The largest number of pests in the continental United States are of foreign origin and for this reason are considered to be the most appropriate subjects for biological control. This is because their full complement of natural enemies may not be present and, as a result, there is the clear opportunity to introduce biotic agents to reduce the pest's equilibrium position. On the other hand there are noxious organisms which are native and which are pests in spite of attack by indigenous natural enemies. Sometimes the native pest acquires its status by reason of an introduced food plant or the extensive cultivation of a favorable food base. A few pests are the result of man's disturbance of the environment; these species find such disturbed conditions particularly favorable for their existence and, as a result, increase in numbers. Often this is caused by a release from competition with other species.

One pest in this category is poison oak, *Toxicodendron diversiloba*. This native plant is widespread in California and is very successful in diverse habitats and is often an invader of disturbed situations. In spite of the fact that it is attacked by insects and plant pathogens, it flourishes in abundance. Herbicides are very effective, but these are expensive and cannot be applied in all the situations where poison oak is a pest. Biological control of such a plant would be highly desirable but, since it is a native species, the conventional ideas of biological control do not apply. This does not preclude the application of other ecological and biological principles for the suppression of this plant. The reasoning is based upon the known examples of destruction of native plant species by accidentally introduced organisms. The native chestnuts of the eastern United States have suffered terribly from the chestnut blight of oriental origin. The native elms are similarly affected by the Dutch elm disease. The Bermuda cedars were virtually eradicated by the attack of diaspine scales of foreign origin, and recently an insect of Australian origin has been very destructive to a species of *Leptospermum* in New Zealand. With these examples in mind, Dr. C. B. Huffaker and his staff at the Division of Biological Control, University of California, Berkeley, California, are exploring the possibility of finding some arthropod or plant pathogen of foreign origin that may be introduced to attack and suppress poison oak. This is a distinct possibility and is based upon sound biological reasoning. First, the species *diversiloba* is taxonomically distant from commercial plants raised in California but is closely related to species of *Toxicodendron* that are known to occur in the Orient. There is a very good chance that *diversiloba* may lack resistance to an enemy that attacks one of its related species in the Orient. In view of the pest status of poison oak, it is certainly worthwhile to experiment with the introduction of some potentially effective natural enemy from the Orient.

Native insects which are pests are also possible subjects for biological control in a number of ways. For example, the alfalfa butterfly, *Colias eurytheme,* is susceptible to a virus which can be stockpiled and applied at an appropriate time to initiate an epizootic (Thompson and Steinhaus, 1950). This technique may also be applied against certain defoliators of forest trees under outbreak situations.

The augmentation and encouragement of natural enemies of native pests offers considerable promise of effectiveness. The most recent example of this is the effort being made to provide overwintering refuges for *Anagrus epos,* the mymarid egg parasite of the grape leafhopper in California. Studies have shown that this egg parasite is a major mortality factor on the leafhopper, for it develops three generations to one of the leafhopper and is so effective in searching out and killing leafhopper eggs that few escape its attack after the middle of July in the vineyards where the parasite is encouraged to work. In these areas it virtually eliminates one entire brood of leafhoppers and maintains economic control throughout the season.

It was learned that the parasite, unlike its host leafhopper, does not overwinter in vineyards but instead breeds throughout the winter in eggs of a noneconomic leafhopper, *Dikrella cruentata,* that are laid on wild blackberries, *Rubus* spp. It appears that the grape leafhopper has reached pest status in some areas only because a barrier of time and space has inadvertently been created between it and its egg parasite. This is because the leafhopper spends the entire year in the vineyards whereas the parasite must recede in the winter to some blackberry thicket and often this refuge is miles from where the vineyard is planted. In the spring the grape leafhopper can therefore begin breeding in the vineyards in the absence of the egg parasites which must first bridge the distance from the overwintering refuge to the vineyard. The environment is now being manipulated to establish such refuges for the parasites adjacent to vineyards (Doutt and Nakata, 1965; Doutt, 1965a).

There are other ways to augment native parasites and predators, and one of these involves insectary production and release of such beneficial species. This has been done with *Macrocentrus ancylivorus*. The need for encouraging natural enemies is sometimes the result of agricultural practices that remove food sources for the adult parasites or predators. This is most noticeable in the areas of monoculture where a substantial part of an agricultural region is devoted to a single crop in contiguous fields with competing plants removed. Certain entomophagous species need supplementary food sources during their adult stage, and these are sometimes obtained from extrafloral nectaries of plants that are not tolerated in a monoculture situation. To offset this deficiency it may be possible to apply supplementary diets in the field. These could serve the purpose of attracting entomophagous species into the area and at the same time provide essential nutrients for the adult diets. Such experimentation is underway, and promising diets have already been devised (Hagen and Tassan, 1965).

The Canadians are investigating a long-neglected approach to augmenting natural enemies by deliberately importing and establishing alternate host species. A braconid, *Orgilus obscurator,* is the most important introduced parasite of the pine shoot moth in Ontario, but it does not reduce economic damage by its host very significantly. In Europe this braconid is the chief parasite of a tortricid moth, *Semasia hypericana* Hbn., which fortunately is also limited to the noxious Klamath weed. The introduction of *Semasia* would therefore accomplish two useful purposes, namely, it would provide a reservoir of hosts on which populations of the parasite could be increased and it would also provide an additional biological agent for the suppression of Klamath weed (Beirne, 1964).

2. *Competitive Displacement*

One of the most novel and thought provoking suggestions for an extension of biological control that has appeared in recent years is that of DeBach (1964b), who suggests the importation of an ecological homolog for the competitive displacement of a pest species. Such a homolog may or may not be a pest itself. For example, the housefly has competitors for the the larval medium and one of these, the stratiomyid fly, *Hermetia illucens* (Linnaeus), which is not itself a serious pest, eliminated the housefly through competition in certain tests (Furman *et al.*, 1959). This certainly would fit Beirne's (1964) classification of biological control as the use by man of living organisms to control pest damage. Beirne further suggests the use of agents that do not attack pests directly but which nevertheless may be employed to produce permanent control: "In relation to insect pests it might be possible, for instance, to control species of biting flies that attack man by replacing them with species that do not and that compete successfully with the harmful species for food or other common environmental requirements. In fact something like this has occurred in several parts of the world: when disease-carrying flies were reduced by various control measures they were replaced and kept down subsequently by harmless species (Gillies, 1962). Our biting-fly problem in Canada is not that we have too many flies but that we have too many of the kinds that bite. One solution might be to replace biting with non-biting kinds without necessarily reducing the total number. Moreover, this approach to pest control is relatively selective and thus, unlike extermination measures, would not harm significantly any beneficial or other useful organisms that inhabit the same environment."

The paper by DeBach also suggests the replacement of a pest by an ecological homolog which is itself a pest but which is then susceptible to eradication or to biological control. He also suggests the use of natural enemies to reduce a pest population to a low level that will then make it a more suitable subject for eradication procedures. "In the Mediterranean area, eradication attempts against the olive fly, *Dacus oleae* (Gmelin), are being considered. The importation and establishment now of effective exotic parasites could make the project much easier to accomplish. It could also, of course, make it unnecessary."

B. The Chartered Course

Biological control is now well established as a major method of pest suppression in every advanced country throughout the world, but it is not a static activity for it is rapidly expanding and new concepts are being suggested and tested. It is desirable to appraise the role of biological control in the light of the new developments in pest control and to determine the direction that biological control activities will probably take in the years to come.

The key to the development of biological control in the future is the degree of financial support that may be given to it by agricultural interests. Although the record is abundantly clear that money invested in biological control research has repaid itself many times over, it is not realistic to believe that this fact alone will persuade an increase in donations or appropriations to expand the current level of activity. Instead, any additional support of this method will be the result of the failure of current unilateral approaches to pest control and the increasing problems that result therefrom. The staggering problems of residues, resurgences, resistance, and chemically induced outbreaks are forcing a reappraisal of most pest control programs and an abandonment of the tyranny of fixed spray schedules geared arbitrarily to a calendar date and carried out irrespective of pest population levels. These considerations bring about a change in attitude toward pest control and a desire to explore other methods. In most situations the nonchemical approaches have never really been put to the test, but the trend now is to explore the possibility of integrated control and to utilize biological control as much as possible. If this continues there will certainly be a far greater use of biological control than has ever been seen in the past. In other words, there is no question but that biological control will continue and will realize expansion, but the degree of this increase will depend largely upon the amount of forced change that circumstances dictate.

It is probable that the conventional practice of biological control whereby natural enemies are sought abroad and are imported for the suppression of their alien hosts will continue unabated. In time this may be more efficiently organized through the functioning of cooperative efforts between biological control agencies or the establishment of a sensible international institute or center to assist such programs. There have been several attempts to create such agencies, but in them the American participation has been remarkably minimal or even deliberately shunned. It is unlikely that these organizations will ever perform the service of which they may be capable until there is an absolute willingness to join wholeheartedly in a cooperative venture as contributing partners.

New directions in biological control will certainly be suggested and probed with the result that the scope of the field will be considerably broadened. The theoretical contributions of population ecologists will become more frequent and will bear closer resemblance to field situations. These will then merit more careful consideration as guiding principles in the utilization of parasites and predators.

The studies on biological phenomena associated with the parasitic existence will add startling facts to the current knowledge and will probably change many of the ideas which are now accepted but which are probably erroneously based. There will be many new investigations into the behavior of such organisms and the results of this research will have immediate application to many biological control projects. Taxonomic investigations will yield badly needed revisions of higher categories and will communicate the taxa in a manner that will be clear and useful to professional entomologists. At the same time, biological studies in conjunction with taxonomic work will reveal the existence of cryptic species in groups that are now morphologically indistinguishable.

Greater attention will be paid to the attributes of a species being imported for biological control, and workers will strive to determine beforehand the comparative effectiveness of candidate species. Criteria for such an appraisal will be very complex and special techniques must be developed before they can be properly evaluated.

Attempts will be made to select genetic strains that are custom tailored to cope with particular situations, and this may include the development of insecticide-resistant natural enemies.

The ecological approach to pest control will inevitably cause research to be focused upon the management of populations and the manipulation of the environment to augment and encourage natural control agencies. This will entail intensive studies of naturally occurring (or endemic) natural enemies. Supplementary food sources will be provided for the adult entomophagous forms either artificially or by planting suitable species of plants. Supplementary food for their developmental stages will be increased by the judicious use of alternate hosts or by the inoculation of a pest population with susceptible stages of the host. The role of ecological homologs in competitive displacement will be utilized in certain limited situations.

Weed control will be enhanced by fundamental studies on the nature of weedy species. There should certainly be development of plant pathogens as controlling agents for weeds, and this may be aided by the application of such pathogens in much the same manner that is used with insect pathogens in microbial control. The possibility of mass colonizations or periodic releases of plant parasites (including arthropods) will be tested under field conditions. Certainly the studies on host selection by parasites and the bases for host specificity will be thoroughly investigated. In short, there are more fundamental questions and applied problems requiring solution than can be adequately handled by existing biological laboratories in the foreseeable future. The importance of these problems demands an enormous growth in biological control. It is evident that this will occur and that the past accomplishments are a mere prelude to what the world will witness in the future.

REFERENCES

Allee, W. D., Emerson, A. E., Park, O., Park, T., and Schmidt, K. P. (1949). "Principles of Animal Ecology." Saunders, Philadelphia, Pennsylvania.

Baker, K. F., and Snyder, W. C. (1965). "Ecology of Soil-borne Plant Pathogens, Prelude to Biological Control." Univ. of California Press, Berkeley, California.

Beirne, B. P. (1964). *Can. Entomologist* **96,** 259.

Carson, R. (1962). "Silent Spring." Houghton, Boston, Massachusetts.

DeBach, P. (1959). *Ann. Entomol. Soc. Am.* **52,** 354.

DeBach, P. (1962). *Proc. Hawaiian Entomol. Soc.* **18,** 69.

DeBach, P. (1964a). *In* "Biological Control of Insect Pests and Weeds" (P. DeBach, ed.), pp. 3–20. Chapman & Hall, London.

DeBach, P. (1964b). *Bull. Entomol. Soc. Am.* **10,** 221.

Doutt, R. L. (1958). *Bull. Entomol. Soc. Am.* **4,** 119.

Doutt, R. L. (1960). *Pan-Pacific Entomologist* **36,** 1.
Doutt, R. L. (1961). *Ann. Entomol. Soc. Am.* **54,** 46.
Doutt, R. L. (1964). *Bull. Entomol. Soc. Am.* **10,** 83.
Doutt, R. L. (1965a). *In* "Research in Pesticides" (C. O. Chichester, ed.), pp. 257–264. Academic Press, New York.
Doutt, R. L. (1965b). *Proc. Oregon Hort. Soc.* **56,** 72.
Doutt, R. L., and Nakata, J. (1965). *Calif. Agr.* **19**(4), 3.
Furman, D. P., Young, R. D., and Catts, E. P. (1959). *J. Econ. Entomol.* **52,** 917.
Geier, W., and Clark, L. R. (1961). *IUCN Symp. 8th Tech. Meeting, Warsaw, 1960* p. 10.
Gillies, H. T. (1962). *Proc. 11th Intern. Congr. Entomol. Vienna, 1960* **2,** 502.
Hafez, M., and Doutt, R. L. (1954). *Can. Entomologist* **86,** 90.
Hagen, K. S., and Tassan, R. (1965). *J. Econ. Entomol.* **58,** 999.
Hall, J. C., Schlinger, E. I., and van den Bosch, R. (1962). *Ann. Entomol. Soc. Am.* **55,** 566.
Huffaker, C. B., and Kennett, C. (1966). *Hilgardia* **37,** 283.
Huffaker, C. B., Holloway, J. K., Doutt, R. L., and Finney, G. L. (1954). *J. Econ. Entomol.* **47,** 785.
Leopold, A. (1949). "A Sand County Almanac." Oxford Univ. Press, London and New York.
LePelley, R. (1943). *Empire J. Exptl. Agr.* **11,** 78.
Michelbacher, A. E. (1962). *Proc. 11th Intern. Congr. Entomol., Vienna, 1960* **2,** 694.
Oman, P. W. (1936). *Proc. Entomol. Soc. Wash.* **38,** 164.
Oman, P. W. (1948). *J. Kans. Entomol. Soc.* **21,** 10.
Patrick, Z. A., and Toussoun, T. A. (1965). *In* "Ecology of Soil-borne Plant Pathogens" (K. F. Baker and W. C. Snyder, eds.), pp. 440–459. Univ. of California Press, Berkeley, California.
Pickett, A. D., and Patterson, N. A. (1953). *Can. Entomologist* **85,** 472.
Rudd, R. L. (1964). "Pesticides in the Living Landscape." Univ. of Wisconsin Press, Madison, Wisconsin.
Sabrosky, C. W. (1955). *J. Econ. Entomol.* **48,** 710.
Smith, H. S. (1935). *J. Econ. Entomol.* **28,** 873.
Smith, H. S. (1946). *Calif. Citrograph* **31,** 414.
Snyder, W. C. (1960). *In* "Biological and Chemical Control of Plant and Animal Pests" (L. P. Reitz, ed.), Publ. 61, pp. 127–136. Am. Assoc. Advan. Sci., Washington, D.C.
Thompson, C. G., and Steinhaus, E. A. (1950). *Hilgardia* **19,** 411.
Thorpe, W. H. (1945). *J. Animal Ecol.* **14,** 67.
Wilson, F. (1965). *In* "The Genetics of Colonizing Species" (H. G. Baker and G. L. Stebbins, eds.), pp. 307–325. Academic Press, New York.

2 MICROBIAL PESTICIDES

Y. Tanada

DIVISION OF INVERTEBRATE PATHOLOGY
UNIVERSITY OF CALIFORNIA
BERKELEY, CALIFORNIA

I. INTRODUCTION

Intensive evaluation of microbial pesticides* is taking place at present in many countries throughout the world. The interest in microbial pesti-

* Microbial pesticides are entomogenous microorganisms (pathogens) and/or their products that cause pathologies and are usually fatal to their insect hosts.

cides has expanded because of problems, such as insect pest resistance, emergence of secondary pests, and toxic residues, that have developed with the use of the broad-spectrum chemical insecticides. Moreover, recent successes with microbial pesticides have greatly encouraged their usage.

Steinhaus (1949, 1956a) was first to employ the term "microbial control" for the utilization of microbial pesticides. He defines microbial control as a phase of biological control concerned with the employment of microorganisms by man for the control and regulation of the number of animals (or plants) in a particular area or in a given population. The idea of microbial control, however, has a very early beginning (Steinhaus, 1956a). Bassi in 1835 demonstrated the infectious nature of the white muscardine fungus, *Beauveria bassiana* (Balsamo) Vuillemin, not only for the silkworm, *Bombyx mori* (Linnaeus), but also for other insects. According to Steinhaus (1956a), LeConte in 1873 presented the first clear-cut recommendation advocating the use of disease as a means of insect control to appear in the English language. Shortly thereafter, the number of advocates for microbial control increased rapidly. In 1879, Metchnikoff conducted tests with the green muscardine fungus, *Metarrhizium anisopliae* (Metchnikoff) Sorokin, and infected the scarab beetle, *Anisoplia austriaca* Herbst. Krassilstschik in 1888 mass-produced spores of the fungus and applied them in field tests against insect pests.

Up to the early twentieth century, there were only a few attempts at the microbial control of insect pests compared to the number initiated during the past decade. Both successes and failures were reported in these early attempts, and the inconsistency of the results retarded the usage and development of microbial pesticides during this period. Some of the failures were apparently due to the lack of pathogenicity of the microorganisms, e.g., *Coccobacillus acridiorum* d'Herelle (Cloaca type A) (Bucher, 1959a, 1963) and the "friendly fungi" on citrus insects (Fisher *et al.,* 1949).

From 1920 to 1950, several promising results were obtained with the use of *Bacillus thuringiensis* Berliner and several other species of bacteria for the control of the European corn borer, *Ostrinia nubilalis* (Hübner); the use of milky-disease organisms, *Bacillus popilliae* Dutky and *B. lentimorbus* Dutky, and the nematode, *Neoaplectana glaseri* Steiner, against the Japanese beetle, *Popillia japonica* Newman; and the use of nuclear-polyhedrosis viruses against the sawflies (superfamily Tenthredinoidea) and the alfalfa caterpillar, *Colias eurytheme* Boisduval.

Although limited in number, the early successes have reignited the interest in microbial control, and during the past 15 years, innumerable field tests with microbial pesticides have been successful against insect pests. The field tests with viruses have been conducted mainly with the nuclear-polyhedrosis viruses; to a lesser extent with the granulosis viruses; and to a limited extent with the cytoplasmic-polyhedrosis viruses and the noninclusion viruses. The viruses have been most effective against forest defoliating pests, such as the European pine sawfly, *Neodiprion sertifer* (Geoffroy) (Bird, 1953a), and the European spruce sawfly, *Diprion hercyniae* (Hartig) (Bird and Burk, 1961). Several field and truck-crop insects have also been controlled successfully by viruses (see Tanada, 1959a; Hall, 1961, 1963, 1964). Approximately six pathogenic rickettsiae have been found in insects, but none has been used in field tests.

Among the bacteria, the most extensive application has been with *Bacillus thuringiensis* and its varieties. Field tests with this bacillus have yielded very promising results for the control of many lepidopterous insects of field crops and forests. In the case of the milky-disease organisms, the present status indicates that the bacilli are outstandingly successful, from the biological or human community viewpoint, in preventing the development of tremendous numbers of beetles, although it is of little solace to the individual gardener and greenskeeper, since he still suffers lawn damage (Beard, 1964).

Although some of the early attempts with fungi have resulted in failures, Baird (1958) reports about 52 partially successful and successful attempts with fungi out of 88 attempts. Most of these successes have been obtained with the white muscardine fungus *Beauveria bassiana,* the green muscardine fungus *Metarrhizium anisopliae,* and species of *Entomophthora* (see Müller-Kögler, 1965). Bucher (1964), however, depreciates the reports on the successes obtained with fungi.

The protozoa and nematodes have received only limited attention for use against insect pests. Of the protozoa, primarily the microsporidia have provided promising results in field tests. Among the nematodes, several members of the family Neoaplectanidae have shown promise in the control of insects.

The large number of reviews that have appeared in recent years strongly attest to the interest in microbial control. Some of the reviews cover the general aspects of microbial control (Steinhaus, 1949, 1957a; Dutky, 1959; Tanada, 1959a; Franz, 1961a, b; Krieg, 1961a; Hall, 1961, 1963, 1964; Cameron, 1963), others involve the major groups of pathogens, such as viruses (Bergold, 1958; Huger, 1963; Bird, 1964), bacteria (Heimpel and Angus, 1960; Tanada, 1961a; Heimpel, 1964; Angus, 1965), and more specifically *B. thuringiensis* (Krieg, 1961b; Vaňková, 1964), fungi (Madelin, 1960; MacLeod, 1963; Bucher, 1964; Müller-Kögler, 1964, 1965; Pramer, 1965), protozoa (Weiser, 1961, 1963), and nematodes (Welch, 1963, 1965). Because of the extensive coverage of these reviews, duplication will be unavoidable, but I have attempted to restrict it as much as possible and to emphasize the areas that I believe are in need of special attention not only to insure the success of, but also to widen the scope of, microbial control.

In addition to the above reviews, I wish to call the readers' attention to a series of bibliographies that have been appearing at irregular intervals in *Entomophaga.* The first three numbers were prepared by Franz (1956, 1957, 1963), and the last two by Franz and Laux (1963, 1964). These bibliographies cite references to microbial control and also those pertaining to the wider field of biological control in general.

II. PROPERTIES OF THE PATHOGEN

A. Strains and Varieties

The availability of strains of variable virulence would allow for the selection of the most effective and suitable strains for microbial control of specific insect pests. Strains and varieties are known to occur in many entomogenous pathogens, especially the bacteria and fungi, and to a lesser degree in viruses and protozoa. Although I am aware of no report of strains in entomogenous nematodes and rickettsiae, this is undoubtedly due to the lack of information on these microorganisms. Thus far, the attempts to develop and utilize the most promising strains of pathogens in microbial control have been very limited.

Strains of insect viruses are presently known only in the group which forms polyhedral inclusion bodies (nuclear- and cytoplasmic-polyhedrosis viruses), whose shapes are apparently determined by the specific virus strain (Aizawa, 1961; Geršenson, 1959, 1960; Aruga *et al.,* 1961). There are few, if any, studies on the differences in the virulence of the various strains of the polyhedrosis viruses. Ossowski (1960) in his field trials has observed variations in the virulence of the nuclear-polyhedrosis virus of the wattle bagworm, *Kotochalia junodi* (Heylaerts), obtained from different localities. In the granulosis virus of the European cabbageworm, *Pieris brassicae* (Linnaeus), however, the viruses obtained from the British and Canary Islands show no differences in virulence (David and Gardiner, 1965a). Strains of granulosis viruses have not been reported thus far.* Unlike the polyhedral inclusion body, the capsule of the granulosis virus is near the visible limits of the light microscope, and other criteria, such as bioassay and serology, would be required to differentiate the strains.

Among the noninclusion viruses, the virus from *Sericesthis pruinosa*

* Stairs (1964b) has isolated from the spruce budworm, *Choristoneura fumiferana* (Clemens), a strain of a granulosis virus that forms cubic inclusion bodies rather than the oval-shaped capsules.

(Dalman) discovered by Steinhaus and Leutenegger (1963) may be a strain of the virus (TIV) of *Tipula paludosa* Meigen found by Xeros (1954). These two viruses are closely related but serologically distinct (Day and Mercer, 1964).

Strains occur among fungi, such as *Beauveria bassiana, Metarrhizium anisopliae,* and *Cephalosporium lecanii* Zimmerman (see Tanada, 1963; Müller-Kögler, 1965). Recently, virulent strains of *B. bassiana* and *Aspergillus flavus* Link have been obtained by treatment with ionizing radiation (Andreev *et al.,* 1962; Evlakhova, 1962). The use of radiation and mutagenic chemical agents should also be investigated with other entomogenous pathogens.

Among the entomogenous bacteria, the strains of *B. thuringiensis* and *B. popilliae* have been studied most thoroughly. In the early 1950's when the potential of *B. thuringiensis* as a microbial pesticide was recognized, there developed a controversial discussion concerning the relationship of *B. thuringiensis* and its varieties to *B. cereus* Frankland and Frankland. The major differences between the two species are the presence of toxic crystals in *B. thuringiensis* and its high pathogenicity for certain insects (Heimpel and Angus, 1963). At present the crystalliferous species are considered distinct from *B. cereus* (Breed *et al.,* 1957). Heimpel and Angus (1958) on the basis of biochemical tests have separated the crystalliferous bacteria as follows: *B. thuringiensis* var. *thuringiensis,* var. *sotto,* and var. *alesti, B. entomocidus* var. *entomocidus,* and *B. finitimus.* Krieg (1961b) considers these to be the five main groups of crystalliferous bacteria. After examining 24 cultures of the type *B. thuringiensis* by biochemical and serological (flagella agglutination) techniques, de Barjac and Bonnefoi (1962) have added a sixth type designated as *galleriae,* a strain isolated from *Galleria mellonella* (Linnaeus). Norris and Burges (1963) have found differences in the esterases produced by crystalliferous bacteria. Other differences among the varieties of *B. thuringiensis* are in their reactions to antibodies (Toumanoff and Lapied, 1954), to low rearing temperatures (Smirnoff, 1963b), and to the addition of urea to the culture (Smirnoff, 1963c). The varieties of crystalliferous bacilli which produce the smallest spores and the least amount of DNA also form the smallest crystals (Fitz-James, 1957), and such varieties may have a low pathogenicity since the toxic crystals play a major role in killing the insect. The grouping of the crystalliferous bacteria into the six types, however, may be an artificial rather than a natural segregation according to Lysenko (1963a).

The six types of crystalliferous bacteria vary in their pathogenicity for different insects. This is associated with the quantity and quality of the toxins produced by them and with the characteristics of the insect hosts.

The variety *sotto,* for example, is much more virulent to the silkworm than the variety *thuringiensis,* and the latter is more virulent for the European cabbageworm, *P. brassicae,* than the former.

The noncrystalliferous *B. cereus* possesses strains which are pathogenic for insects. Stephens (1952) has studied 12 strains of *B. cereus* and has found that those isolated from insects show a higher degree of virulence for the codling moth, *Carpocapsa pomonella* (Linnaeus), than strains isolated from apple. The virulence of *B. cereus* is dependent apparently on its ability to produce the enzyme, lecithinase (phospholipase C) (Heimpel, 1955a).

There are strains of the milky-disease organisms, especially *B. popilliae* (Tashiro and White, 1954; Dutky, 1963; Pridham *et al.,* 1964; Rhodes, 1965). In other entomogenous bacteria, strains have been reported for *Serratia marcescens* Bizio (Heimpel, 1955b; Steinhaus, 1959a) and *Pseudomonas aeruginosa* (Schroeter) Migula (Bucher and Stephens, 1957; Lysenko, 1963b).

B. Virulence

One of the important attributes of a pathogen as a microbial pesticide is its virulence, i.e., the disease-producing intensity or power of a pathogen. Virulence is closely associated with the ability of a pathogen to invade and injure the tissues and organs of its host. A pathogen, however, may cause disease without actually penetrating into the hemocoel. Such is the case of a pathogen which is confined to the lumen of the larval midgut, such as *Clostridium* spp. in *Malacosoma pluviale* (Dyar) (Bucher, 1957), and *Streptococcus pluton* (White) the cause of European foulbrood in honeybees (Bailey, 1963).

Differences in virulence are measured by the severity in reactions elicited in the host by the pathogens. Quantitatively, this can be determined by the bioassay of measured quantities of a pathogen applied to a homogenous host strain which is comparable in age and weight (see Section VI,B). A reliable method of determining the virulence of pathogens while under cultivation, both *in vivo* and *in vitro,* would provide a rapid and valuable tool for assaying strains and species of pathogens for use as pesticides. Sussman (1952) proposes using the respiratory gas exchange and the weight loss of insect pupae as a basis for the quantitative differentiation of virulence in fungus diseases. For certain bacteria such as *Pseudomonas* spp., Bucher (1960) suggests that proteinase production may be correlated with bacterial virulence. The proteolytic ability is based on the breakdown of gelatin. According to Lysenko (1965), however, strains of *P. aeruginosa* which exhibit identical ability to digest gelatin show differences in their ability to kill the same host.

There are several methods by which the virulence of microorganisms has been increased. Steinhaus (1949) has listed these methods as: (1) passing the pathogen successively through susceptible hosts, (2) causing it to dissociate into more virulent or less virulent strains, (3) introducing the microorganisms together with substances that may aid in increasing its invasive powers, and (4) associating it in a mutualistic relationship with other microorganisms that may render it more capable of invading tissues. In addition, the nutrition and the condition of culturing the pathogens may also affect their virulence. All of these methods have been applied to some insect pathogens.

Successful attempts at increasing the virulence of pathogens have been reported primarily with bacteria, much less frequently with fungi, and only recently with a few viruses. Serial passages of pathogens, especially of those that invade the digestive tract, should be made by feeding the pathogen to the host (*per os*), and not by inoculating the pathogen into the hemocoel since this may not select for strains capable of invading through the digestive tract. Care should also be taken that an occult or chronic strain previously present in the host would not be selected as the original strain. This is especially the case with viruses which are known to exist as chronic or latent infections.

Some bacteria (e.g., *B. cereus*) increase in virulence after serial passages through their hosts, but Bucher (1963) claims that there are no published reports of such increase for other bacteria, which he has grouped together as "potential pathogens," such as *P. aeruginosa.*

A nuclear-polyhedrosis virus has increased in virulence linearly for each passage up to the eighth passage from 10 to 88% in *Galleria mellonella* (Veber, 1964). The adaptation of an insect virus to a new host, and subsequent increase in virulence through serial passages in the new host, has been reported by Smirnoff (1963a) and Fedorintchik (1964). In *Hyphantria cunea* Drury, the increase in virulence is associated with an increase in the number of virus bundles in the polyhedral inclusion (Machay and Lovas, 1957). This suggests that some of the apparent increase may not be associated with the virulence of the individual virus particle, but with an increase in the number of virus particles per inclusion body.

Although there are several claims of an increase in the virulence of fungi after serial passages through susceptible hosts, Bucher (1964) claims that these have not been substantiated with quantitative data. Kerner (1959), however, has presented some quantitative data showing that serial passage of the fungus, *Paecilomyces farinosus* (Dickson ex Fries) Brown and Smith (*Spicaria farinosa*), through insect hosts has resulted in a decrease of the LT_{50} (lethal time).

When an alternate host is used repeatedly for propagation, a pathogen may lose its virulence for the primary host. This has occurred with the microsporidian, *Nosema infesta* Hall, (Hall, 1954) and with a strain of *B. popilliae* (Fleming, 1958).

Strains occur among the different insect pathogens and they may vary in virulence (see Section II,A). Heimpel (1964) suggests two procedures for the selection of virulent strains of bacteria: (1) separation of clones from a potentially pathogenic bacterial population, and (2) determination of what enzyme or toxin is involved in invasion. Attempts at selection for virulence by dissociating the pathogens into clones have been made mainly with bacteria and fungi, but very little with other types of pathogens.

It would be of value to investigate more extensively the use of chemicals and other nonliving substances for the purpose of increasing the virulence of pathogens. Some of the substances, such as triturated glass, cause mechanical injuries to the midgut epithelium and enable the microorganisms, such as *S. marcescens* (Steinhaus, 1958a) and *Pseudomonas noctuarum* (White) (Weiser and Lysenko, 1956), to readily invade their hosts. The basis for the increased virulence of *P. aeruginosa* on grasshoppers by mucin (Stephens, 1959), and of *B. thuringiensis* on *Porthetria dispar* (Linnaeus) by boric acid (Doane and Wallis, 1964) is still unknown. Certain chemical insecticides may increase the effectiveness of pathogens in killing their hosts (see Section VIII,A).

The association of several microorganisms may result in increased mortality of the insect host (see Section VIII,B). Such associations have been reported between viruses (Tanada, 1956a, 1959b; Shvetsova and Ts'ai Hsü-yü, 1962), virus and bacteria (Vago, 1956), among bacteria (Isakova, 1954; Steinhaus, 1959a), among bacteria and protozoa (Weiser, 1956; Jafri, 1965), among protozoa (Weiser, 1956, 1957), among fungi (Fassatiova, 1964), and between nematode and bacteria (Dutky, 1959). Isakova (1964) reports that an insect, such as *P. brassicae,* with a large intestinal microflora is more susceptible to *B. cereus* var. *galleriae* than *G. mellonella* with a less abundant flora. This implies that variation in gut microflora is responsible for the differences in resistance, but the actual resistance may be due to the innate properties of the two insect species. In microsporidian diseases, mixed infections by several microsporidian species are generally not more acute than single infections. In one case, however, a microsporidian, *Nosema lymantriae* Weiser, is made infectious for *Nygmia phaeorrhoea* (Donovan) [*Euproctis chrysorrhoea* (Linnaeus)] when it is combined with a second microsporidian, *Thelohania similis* Weiser, which is a natural pathogen of *N. phaeorrhoea* (Weiser, 1956, 1957).

The nutritional composition of the culture media may affect the virulence of pathogens. Strains of *B. popilliae* and *B. lentimorbus* vary in their infectivity for the Japanese beetle grub, and the infectivity is influenced by the method of cultivation (Pridham *et al.,* 1964; Rhodes, 1965). The toxic protein crystals are not produced by strains of *B. thuringiensis* and of *B. entomocidus* var. *entomocidus* when cultured with urea (Smirnoff, 1963c). An aqueous leaf extract of *Viburnum cassinoides* Linnaeus, when added to the culture medium, acts as a "mutagenic agent" and causes the permanent loss of the ability to form toxic crystals in crystalliferous bacilli, but does not affect sporulation and lysis (Smirnoff, 1965a).

There is ample evidence for increased growth and spore production when the nutrients of fungus culture media are altered (see Müller-Kögler, 1965), but the effect of such growth and spore production on the virulence of the fungi usually has not been investigated. According to Schaerffenberg (1957a), the addition of soluble protein (peptone) to the culture medium significantly increases the virulence of *Beauveria bassiana.*

The physical conditions under which the microorganisms are cultured may also affect their virulence. *Bacillus thuringiensis* and its varieties produce little or no toxic crystals under repeated cultivation in strong alkaline media (Toumanoff and Vago, 1952) and this affects their virulence since the crystals are primarily responsible for their pathogenicity for certain insects. Low temperature does not affect crystal formation but limits sporulation in *B. thuringiensis* (Smirnoff, 1963b).

C. Toxins

There is an increasing interest in substances formed by microorganisms that are toxic for insects. These substances not only are used for the study of the mechanisms responsible for the pathogenicity of the microorganism for insects, but also can be applied in microbial control. Such interest has been centered mainly around bacteria and fungi, but there are reports that protozoa and virus may also produce toxins.

Several comprehensive reviews have appeared on the toxins produced by the crystalliferous bacteria, *B. thuringiensis* and its varieties (Krieg, 1961b; Heimpel and Angus, 1963; Angus, 1964a; Heimpel, 1965). Krieg (1961b) reports that *B. thuringiensis* produces the following toxins: (1) thermolabile endotoxin (crystal toxin), (2) thermostable exotoxin (fly toxin), (3) bacillogenic antibiotic, (4) lecithinase, and (5) proteinase. Among the various types of toxins, the crystal endotoxin (parasporal body), which is proteinaceous, is the prominent toxin produced by this group of bacteria. Unfortunately, the proteinaceous toxin is too complex for synthesis at present.

The production of a toxin by *B. thuringiensis* var. *sotto* was first observed by several Japanese workers in 1915, but the actual association of the toxic activity to the parasporal body has only been established more recently by Angus (1954, 1956a,b). The toxic parasporal body, after dissolution by the alkaline midgut juices, affects the permeability of the midgut epithelium and enables the highly alkaline juice to permeate into the hemocoel and increase the pH of the hemolymph (Angus and Heimpel, 1956; Heimpel and Angus, 1959; Fast and Angus, 1965). The change in hemolymph pH causes general paralysis and death in 1–7 hours in certain insects, such as the silkworm. In other insect species, the syndrome is different in that the toxin causes an exfoliation of the midgut epithelium and paralyzes the gut. All the susceptible insect species have relatively alkaline midgut contents whose pH ranges from 9.0 to 10.5 (Angus, 1956a). Only lepidopterous larvae are susceptible to the toxic crystal (Heimpel, 1965).

In some insects, e.g., *P. brassicae,* the midgut juice dissociates the crystal protein into several fractions that are toxic by injection and by ingestion (Dedonder and Lecadet, 1964; Lecadet and Martouret, 1962, 1965). The toxic crystals from *B. thuringiensis* var. *thuringiensis* (Serotype I, *berliner*) lyse much faster than those from var. *anduze* (Serotype III) (Lecadet and Martouret, 1964). Apparently only one enzyme is involved in this proteolysis (Dedonder and Lecadet, 1964). This enzyme appears to be absent from the midgut of *Bombyx mori* (Angus, 1964b).

Relatively pure preparations of the toxic crystals may be obtained by several methods. Hannay and Fitz-James (1955) have separated the crystals by first destroying the spores by mechanical disruption or by germination and autolysis of the spore contents, followed by differential centrifugation. Purification has also been accomplished by sucrose density gradients (Vaňková, 1957) and by phase separation with fluorocarbon (Angus, 1956c). The crystals may also be obtained free of spores by culturing the bacteria at low temperatures (about 12°C) since such temperatures prevent normal sporulation without disturbing the crystal formation (Smirnoff, 1963b).

A second toxin of *B. thuringiensis* has been discovered by McConnell and Richards (1959). This toxin is a heat-stable, low molecular weight, water-soluble, dialyzable substance which is toxic by injection into the hemocoel, but not when ingested. The toxin is not produced by all crystalliferous bacteria but only by the variety *thuringiensis* (Cantwell *et al.,* 1964a). Since the toxin interferes with pupation of the housefly (*Musca domestica* Linnaeus), it has been called the "fly toxin" or "fly factor." Apparently when the toxin is ingested, it kills only Diptera. The toxin is also produced in the larva of *Corcyra cephalonica* (Stainton) killed by

B. thuringiensis, since the sterile filtrate from triturated larva causes 100% mortality when inoculated into susceptible larva (e.g., *G. mellonella* and *C. cephalonica*) (Tamashiro, 1960).

Contrary to the findings of McConnell and Richards (1959), others report a thermostable toxin that is active not only when injected into the hemocoel, but also when ingested by insects other than Diptera, such as Lepidoptera, Coleoptera, Hymenoptera, and Orthoptera (Burgerjon and de Barjac, 1960; Krieg and Herfs, 1963; Burgerjon *et al.,* 1964; Burgerjon and Galichet, 1965). In the case of this toxin, poisoning and death are most apparent during the period of molting. Since Krieg and Herfs (1963) maintain that the toxin acts as a "general intoxicant," Cantwell *et al.* (1964a) believe that it is different from the "fly toxin" discovered by McConnell and Richards, which has a specific action on the metamorphosis of the housefly. Moreover, they report that larvae of *Estigmene acraea* (Drury), *Plodia interpunctella* (Hübner), and *G. mellonella* are not susceptible to the fly toxin, as found by the other workers, and that the varieties *alesti, euxoae, galleriae, dendrolimus,* and *sotto* apparently do not produce the fly toxin, as claimed by Krieg and Herfs (1963).

The enzyme, lecithinase (phospholipase C), is produced by many species of bacteria, especially *B. cereus* and the crystalliferous bacteria. Heimpel (1955a) has found a significant correlation between the pathogenicity of strains of *B. cereus* for the larch sawfly, *Pristiphora erichsonii* (Hartig), and the ability of the strains to produce lecithinase. He believes that the enzyme plays an important role in the invasion and destruction of the insect by the bacteria. This enzyme is also produced by species of *Serratia* (Monsour and Colmer, 1951).

According to Vago and Kurstak (1965), the spores of *B. thuringiensis* that are transmitted by the ovipositor of *Nemeritis canescens* Gravenhorst produce a toxin upon germinating in the hemolymph of *Anagasta kühniella* (Zeller) (*Ephestia kühniella* Zeller). This toxin is present in the supernatant of cultures of this bacterium, but the nature of the toxin and its relationship to other toxins produced by the bacterium has not been established.

A thermolabile toxin has been found by Smirnoff (1964) in a commercial preparation of *B. thuringiensis.* The toxin may be a protein and possibly an exotoxin, and is effective against 12 species of sawflies.

Patel and Cutkomp (1961) report that water extracts of American foulbrood scales killed by *Bacillus larvae* White contain proteolytic fractions, one of which is toxic when fed to the larva, prepupa, and adult of the honeybee. This enzyme is harmless when fed, but toxic when inoculated into several other insect species.

Pseudomonas aeruginosa produces a toxic antigenic substance which appears to be a protein and kills *Galleria* larva when injected into the hemocoel (Lysenko, 1963b,c,d). It seems to be a specific metabolic inhibitor acting on phenoloxidases and may be an exotoxin, but this would be unusual because gram-negative bacteria are usually not true exotoxin producers (Lysenko, 1963d). Near the end of the last century, Duggar (1896) has found a very interesting bacterium which attacks the squash bug, *Anasa tristis* De Geer. This bacterium, when cultured in broth, liberates a toxin that rapidly kills the larvae of several insect species which have been immersed in the broth. The nature of the toxin is unknown.

Many species of fungi produce substances toxic to insects. Some of the fungi which have been reported to produce toxins are *Beauveria bassiana* (Dresner, 1950; Schaerffenberg, 1957b), *Myrothecium roridum* Tode (Kishaba *et al.,* 1962), *Aspergillus flavus* (Burnside, 1930; Toumanoff, 1931), *Aspergillus ochraceus* Wilhelm (Kodaira, 1962), *Spicaria prasina* (Maublanc), *Paecilomyces farinosus* (Wada, 1957), and *Metarrhizium anisopliae* (Kodaira, 1962; Roberts, 1964). The fungus, *Athrobotrys oligospora* Fresenius, produces a toxin which inactivates nematodes submerged in it (Olthof and Estey, 1963).

Toxic compounds have been isolated and purified from several species of fungi. Kishaba *et al.* (1962) have isolated two compounds ($C_{29}H_{40}O_7$ and $C_{49}H_{64}O_{11}$) from the culture filtrates of *Myrothecium roridum.* The compounds, which are active by topical application or by ingestion, markedly inhibit the feeding of the larva and adult of the Mexican bean beetle, *Epilachna varivestis* Mulsant, and subsequently cause their deaths. Kodaira (1962) has isolated two toxins, which he calls Destruxin A ($C_{29}H_{47}O_7N_5$) and Destruxin B ($C_{25}H_{42}O_6N_4$), from cultures of *Aspergillus ochraceous* and *M. anisopliae* (*Oospora destructor* Metchnikoff). Further analysis indicates that the formula of Destruxin B obtained from *M. anisopliae* is $C_{30}H_{51}O_7N_5$ (cyclo-D-α-hydroxy-γ-methylvaleryl-L-prolyl-L-isoleucyl-*N*-methyl-L-valyl-*N*-methyl-L-alanyl-β-alanyl) (Tamura *et al.,* 1964). Kuyama and Tamura (1965) have synthesized this toxin, and found it identical to the natural Destruxin B. *Streptomyces moboraensis* Nagatsu and Suzuki produces a toxin, Piericidin A, that is highly poisonous for silkworm and *Pieris rapae* (Linnaeus) (Tamura *et al.,* 1963). The formula of Piericidin A is $C_{25}H_{37-39}NO_3$ (Takahashi *et al.,* 1963). As pointed out by Roberts (1964), the mycotoxins that have been isolated are usually small molecules compared with those produced by bacteria. In general, mycotoxins cause tetanic reactions.

Some microsporidians which cause hypertrophy of the invaded cells may produce toxins (Weiser, 1961), but the toxins have not been isolated and described. The noninclusion virus of *Sericesthis pruinosa,* when cultured on ovarian cell cultures of *Antheraea eucalypti* (Scott), produces a toxic effect on the cells (Bellett and Mercer, 1964). The toxic factor is not removed by the centrifugation of the virus in a sucrose gradient or by passage in cell culture.

D. Persistence

Pathogens, when marketed as microbial pesticides, should possess a relatively long shelf-life and retain their viability and virulence in storage. The resistant spores of bacteria, fungi, and protozoa, and the polyhedral inclusion bodies of viruses are known to possess high storage life. Most of them under suitable conditions can retain their virulence for at least a year. The nuclear-polyhedrosis virus of *B. mori* has retained its infectivity for at least 15 years when suspended in the insect hemolymph and stored mostly at 4°C (Steinhaus, 1954a). In the case of the nuclear-polyhedrosis virus of the sawfly, it gradually loses its virulence and becomes completely inactivated in 12 years (Neilson and Elgee, 1960).

The pathogens with resistant stages have been applied successfully with conventional sprayers and dusters, and they have generally persisted in the field for adequate periods of time. The application of pathogens lacking suitable resistant stages, however, has been largely unsuccessful. The critical factors in the environment are desiccation, solar radiation, and temperature. Stephens (1957b) has investigated methods of increasing the survival of the nonsporeforming bacterium, *P. aeruginosa,* with the use of casein, sucrose, and mucin. The noninclusion virus of the citrus red mite, *Panonychus citri* (McGregor), can be stored by freeze drying the virus, by vacuum desiccating the diseased-mite material, and by freezing the virus in glycerin suspension (Gilmore and Munger, 1963). It is of interest that, in aqueous suspensions, the virus loses its infectivity after a few hours, but natural deposits from diseased mites remain infectious for several days. Gelatin extends the residual life of the virus.

The longevity of the conidia of entomogenous hyphomycetous fungi is affected by temperature, humidity, and light (Steinhaus, 1960a; Clerk and Madelin, 1965). Low temperature, low humidity, and darkness favor the longevity of the conidia of *Beauveria bassiana* and *Paecilomyces farinosus.* This is also the case with *M. anisopliae,* except that survival is longest at high and low humidities and least at median humidities (Clerk and Madelin, 1965).

The ability of the pathogen to persist in the host environment is especially important in the long-term or permanent control of insect pests. The pathogen persists most commonly within live primary and secondary hosts which may or may not exhibit an active infection. Persistence in alternate hosts may occur in pathogens which attack different insect species within their habitat. The spores of the microsporidian, *Thelohania hyphantria* Weiser, a pathogen of *Hyphantria cunea,* are short-lived (Weiser and Veber, 1957). Inasmuch as this microsporidian kills the overwintering host pupae, it survives in infected overwintering larvae of *Nygmia phaeorrhoea,* an alternate host.

Insect pathogens are also known to persist on foliage, in soil, and in the host's cadaver and feces. In some cases, persistence in the cadaver is an important method of survival of certain pathogens, such as *Streptococcus pluton* (White) which causes the European foulbrood (Bailey, 1959), the nematode DD-136 (Dutky, 1959), and the nuclear-polyhedrosis virus of the wattle bagworm, *Kotochalia junodi* (Ossowski, 1957). The soil is a highly favorable site for the survival of microsporidia, (Weiser, 1956, 1961; Hurpin, 1965), the milky-disease bacteria (White, 1940; Beard, 1944), spores of *Beauveria bassiana* (Evlakhova, 1953, 1954), virus polyhedra (Steinhaus, 1948; Jaques, 1964), and nematodes (Dutky, 1959).

Pathogens generally do not appear to persist for long periods on plant foliage, apparently because of the action of sunlight, rain, and wind. Simulated rain affects the crystal spore complex of *B. thuringiensis* much more than the cytoplasmic polyhedra of *Smithiavirus pityocampae* Vago (Burgerjon and Grison, 1965). There are commercial formulations of *B. thuringiensis* which contain additives that prolong the activity of the bacteria when applied to the plant foliage (Fisher, 1965). In the case of *B. cereus* spores, their persistence is conditioned by factors, other than rainfall, that affect the viability of the bacteria (Stephens, 1957a). The nuclear-polyhedrosis viruses of sawflies do not persist on plant foliage throughout the winter (Bird, 1955), but the nuclear-polyhedrosis virus of *Malacosoma fragile* (Stretch) overwinters successfully on the host plant and is a source of infection for larvae of the next generation (Clark, 1956). The nuclear-polyhedrosis virus of the alfalfa caterpillar can survive on the alfalfa stubble in the field (Steinhaus, 1948). Although entomogenous nematodes are highly susceptible to desiccation, some species persist under bark of trees where moisture is available (Dutky, 1959).

The close association of pathogens with nonsusceptible predatory animals is indicated by their persistence in the digestive tracts of such animals. Pathogens that have persisted under these conditions are the nuclear-polyhedrosis virus of *Neodiprion sertifer* through the digestive tracts of a predatory bug (*Rhinocorus annulatus* Linnaeus) and of the robin (*Erithacus rubecula* Linnaeus) (Franz and Krieg, 1957); the cytoplasmic-polyhedrosis virus of the silkworm through the domestic fowl (*Gallis gallis domesticus* Brissen) (Ishikawa and Okino, 1965a); the milky-disease organisms through birds and small mammals (Hadley, 1948); *B. thuringiensis* through birds and small animals (Smirnoff and MacLeod, 1961; Ishikawa and Okino, 1965b); and certain species of microsporidia through birds and predatory insects (Weiser, 1956, 1961; Weiser and Veber, 1957; Günther, 1959). When a suspension of *B. thuringiensis* is introduced into the rumen of cows, the counts of vegetative rods are decreased by 90% within 4 hours, but the spore viability counts do not decrease significantly for a period of 24 hours (Adams and Hartman, 1965).

E. Dispersal

Microbial pesticides have been applied with conventional dusters and sprayers, and in some cases with mist blowers and airplanes. In such applications, however, precautionary measures must be taken that the equipment does not destroy the microorganisms or inactivate the toxic principles. High temperatures and toxic solvents should be avoided. The pH of the spray solution should be maintained close to neutrality because the crystal endotoxins of *B. thuringiensis* and its varieties, and the inclusion bodies of viruses are dissolved in dilute alkali and acid solutions. The nonresistant stages, moreover, may be inactivated even with simple spray apparatus (Grison and Bourguignon, 1965).

In nature, pathogens are dispersed by the movement of their primary and secondary hosts and nonsusceptible carriers, and by physical factors such as wind, rain, and streams. The survival and movement of infected adults are very important for the dispersal of viruses, especially those of forest insects (see Bergold, 1958; Bird, 1961). The dispersal of fungi, nematodes, and protozoa by infected hosts has been known for a long time. In the case of bacteria, there is increasing evidence of their dispersal by infected hosts. *Bacillus popilliae* may be distributed by infected Japanese beetle adults (Langford *et al.,* 1942) and *B. thuringiensis* var. *dendrolimus* (var. *sotto*) by the mass migration of the Siberian silk moth, *Dendrolimus sibiricus* Tschetverikov (Talalaev, 1958).

Attempts to control insects by introducing infected individuals into healthy populations were made over 50 years ago, especially with fungi of the chinch bug and of the citrus scale insects. This method of pathogen dispersal has been applied recently. Nirula (1957) has placed grubs of *Orcytes rhinoceros* (Linnaeus) infected with *M. anisopliae* into the breeding sites of the host. Aphids infected with entomophthoraceous fungi have been dispersed in disease-free aphid populations (Hall and Dunn, 1958; Shands *et al.,* 1963). Martignoni and Milstead (1962) have demonstrated the feasibility of the transmission and dispersal of the nuclear-polyhedrosis virus by the adults of *Colias eurytheme* whose genitalia were contaminated manually with the virus. This method of virus transmission has been attempted under field conditions with adults of *Trichoplusia ni* (Hübner) contaminated with the nuclear-polyhedrosis virus, but the dispersal of the virus by this means has not been as effective as the use of a sprayer (Elmore and Howland, 1964). Nonetheless, the use of contaminated adults should be considered for the dispersal of pathogens into populations free of disease and especially in areas inaccessible to spray equipment. The noninclusion virus of the citrus red mite has been successfully introduced into healthy mite populations by liberating infected individuals (Gilmore and Munger, 1963, 1965). The virus of the mite has very short persistence when applied as a spray or dust to plant foliage.

The abnormal movements of infected insects may aid in the dispersal of pathogens. Insects infected with certain fungi and viruses climb to elevated positions just prior to their deaths, and in such positions, wind and rain are more effective in dispersing the pathogens. Other abnormal movements of infected insects that may assist in pathogen dispersal are the wandering of individuals of gregarious species and the mass movement of infected individuals (Smith *et al.,* 1956; Smirnoff, 1960, 1965b).

There are numerous reports that parasites and predators may disseminate viruses (Franz *et al.,* 1955; Smith *et al.,* 1956; Smirnoff, 1961), bacteria (White and Dutky, 1940; White, 1943; Smirnoff and MacLeod, 1961; Flanders and Hall, 1965), protozoa (Blunck, 1954; Weiser, 1956, 1957, 1961; Weiser and Veber, 1957), and fungi (see Müller-Kögler, 1965). In some cases, the parasites and predators are also susceptible to the pathogens. Birds and small mammals may serve as mechanical carriers for *B. cereus* (Smirnoff and MacLeod, 1961), *B. popilliae* (White and Dutky, 1940), and viruses (see Bergold, 1958; Tanada, 1963, 1964). In the virus diseases of the sawflies, once the epizootic is initiated, the viruses are spread very rapidly throughout a tree mainly by rain, and from tree to tree by parasites and predators and infected female sawflies (Bird, 1961).

F. Pathogenicity to Vertebrates and Plants

Insect pathogens, in general, are nonpathogenic to man, other animals, and plants (Steinhaus, 1957b, 1959b). This generalization has been confirmed by very severe and exhaustive tests in many instances. In the United States, microbial pesticides, like other pesticides, must meet the tolerance requirements set forth by the Federal Food and Drug Administration. At present, the only microbial pesticide that has been approved is spore preparations of *B. thuringiensis* (Harvey, 1960). This bacterium has undergone very rigid and controlled testing on vertebrates from three different standpoints: (1) infectivity and sensitization, (2) acute and chronic toxicity, and (3) effect on human volunteers. The experimental animals have shown no adverse or pathologic reactions (Fisher and Rosner, 1959).

The similarity between *B. thuringiensis* and its varieties to *B. cereus* and *B. anthracis* Cohn has aroused some apprehension that *B. thuringiensis* may mutate to the vertebrate pathogen, *B. anthracis*. Although Lamanna and Jones (1961) have found by agglutination and agglutinin adsorption tests that an antigenic relationship exists among the endospores of these three types of bacteria, they conclude that it is still premature to consider all three a single species. The antigenic relationship between *B. thuringiensis* and *B. anthracis* seems to be more remote than both of them to *B. cereus*. Steinhaus (1959b) has thoroughly examined the improbability of *B. thuringiensis* mutating to forms pathogenic for vertebrates and has concluded that it is highly unlikely to occur.

Some substances produced by insect pathogens are toxic to vertebrates. A highly labile substance, toxic for sawfly larvae by ingestion and for mice and toads by muscular inoculation, has been found in a commercial preparation but not in pure cultures of *B. thuringiensis* (Smirnoff, 1964). Since this labile toxin is rapidly inactivated within 24 hours in air and in 6 hours in sunlight and is toxic to vertebrates only by inoculation, there seems to be little likelihood that it will affect man and other vertebrates.

Inasmuch as lecithinase (phospholipase C) is toxic to vertebrates, adverse effects may be expected from bacteria which produce this enzyme in large quantities. However, the amount of lecithinase produced by entomogenous bacteria is apparently harmless. Both *B. thuringiensis* and *B. cereus* produce the enzyme, but they have shown no pathologic effects to vertebrates (Stephens, 1952; Fisher and Rosner, 1959). Preparations of *B. thuringiensis* have been fed, generally without harm, to hens (Briggs, 1960; Burns *et al.*, 1961), cows (Dunn, 1960; Burgerjon and Galichet, 1965), and Japanese quail (Borgatti and Guyer, 1963) for the control of the housefly breeding in feces. In one case, however, a treatment with the most effective commercial preparation has caused a reduction of feed consumption, body weight, and egg production of caged layers (Burns *et al.*, 1961). The purified exotoxin is mainly responsible for killing the housefly breeding in feces. The toxin when fed or inoculated into mice has shown no adverse effects (Krieg and Herfs, 1963).

Pseudomonas aeruginosa which has been tested experimentally in microbial control, is known to cause human and animal lesions and has also infected plants (Breed *et al.*, 1957). Ashrafi *et al.* (1965) have tested this bacterium on rats by oral feeding and subcutaneous inoculation without any harmful results. The question of strains of this bacterium, however, with variable pathogenicity for insects and for vertebrates, needs more extensive investigation.

Rickettsiella melolonthae (Krieg), a pathogen of *Melolontha melolontha* (Linnaeus), when inoculated intraperitoneally has caused pathologic symptoms and one death in white rats (Krieg, 1955). Whether other rickettsiae, which cause diseases primarily in insects, are pathogenic to vertebrates is not known at present.

None of the entomogenous viruses have caused pathologies in vertebrates. Tests have been conducted mostly with the nuclear-polyhedrosis viruses and to a limited extent with other types of viruses (Tanada, 1956b; Smirnoff and MacLeod, 1964; Ignoffo and Heimpel, 1965). Ignoffo and Heimpel (1965) have thoroughly tested the nuclear-polyhedrosis viruses of *Heliothis zea* (Boddie) and *H. virescens* (Fabricius) on white mice and guinea pigs without any adverse effects. The tests included oral feeding and inhalation of the viruses, and inoculations by intravenous, intradermal, and intracerebral routes. In addition, they found no adverse reactions in 6 vertebrates exposed to 28 different viruses from 26 different insect species.

A few entomogenous fungi have been reported to cause pathologies in vertebrates. The spores of *Beauveria bassiana* have caused nausea, headache, and allergic reactions (Hall, 1954; York, 1958). *Beauveria bassiana*, however, when fed or inoculated intradermally and intraabdominally to rats, has caused no ill effects (Dresner, 1949; Dunn and Mechalas, 1963). On the other hand, the mycotoxins, Destruxin A and B, isolated from *Oospora destructor*, are highly toxic when inoculated into mice (Kodaira, 1962). Piericidin A obtained from *Streptomyces mobaraensis* is poisonous for mice when applied intraperitoneally, orally, and cutaneously (Tamura *et al.*, 1963).

The fungus, *Entomophthora coronata* (Costantin) Kevorkian, which has a wide insect–host range, has caused phycomycosis in horses (Emmons and Bridges, 1961) and nasal granuloma in a young boy (Bras *et al.*, 1965), and is weakly parasitic on plants (Gustafsson, 1956). Gabriel (1965), however, has not been able to detect any pathology in white mice to which spores of *E. coronata* have been applied through the mouth, nasal tract, and into open wounds. For a thorough coverage on the published reports of entomogenous fungi which cause diseases in man, the readers are referred to the recent publication by Müller-Kögler (1965).

Although early tests with spore preparations of *B. thuringiensis* on the honeybee have indicated no toxic reaction, further comprehensive tests have shown that the larvae and adult bees are killed when fed very high dosages. These dosages, however, are not likely to be applied in microbial control (Krieg and Herfs, 1964; Martouret and Euverte, 1964; Cantwell *et al.,* 1964b). The component of the *B. thuringiensis* preparations most toxic to the honeybee is the thermostable toxin (Martouret and Euverte, 1964; Cantwell *et al.,* 1964b).

III. METHODS OF TRANSMISSION

In order to cause infection, a pathogen usually must penetrate into the hemocoel of the insect, although in some instances, it may just remain within the gut lumen and may, possibly through the production of toxins, cause pathology and death of the host. This is the case with the species of *Clostridium* isolated by Bucher (1957). With most pathogens, however, an invasion into the hemocoel is a necessary preliminary to the death of the host. The most common route of invasion into the hemocoel is through the digestive tract after ingestion *per os.* The viruses, bacteria, rickettsiae, protozoa, and some nematodes generally invade their hosts in this manner. When these microorganisms are used as pesticides, therefore, they must be applied to the food of the insect pests. A possible exception of a bacterium invading the host through the integument is *Micrococcus nigrofaciens* Northrup, which is reported to invade through the joints of legs, spiracles, and abdominal segments of *Lachnosterna* sp. and other insects (Northrup, 1914). There is apparently no further report concerning this interesting and promising bacterium.

The peritrophic membrane and substances in the gut juices (antifungal, antibacterial, and antiviral materials) may prevent the infection by microorganisms. On the other hand, abrasives, such as triturated glass, that injure the midgut epithelium, have enabled bacteria to gain entrance into the hemocoel.

Fungi generally invade through the integument, but some have been reported to cause infection through the digestive tract (see Müller-Kögler, 1965). Madelin (1960) concludes, apparently from the literature on fungus infections, that, "Infection commonly takes place through the walls of the digestive tract, often in addition to infection through the integument." Nevertheless, the evidence for fungus infection through the digestive tract needs further careful study and substantiation. This aspect is especially important in the use of fungi for insect control since humidity would not be a critical factor in the digestive tract for spore germination.

Invasion through the integument also occurs commonly with nematodes. Other pathogens may penetrate the integument through wounds or with the aid of insect parasites and predators, which serve as vectors. With some pathogens, however, the action of the digestive juices is needed before they can develop and multiply. Such is the case with viruses occluded in the polyhedra, which must first be dissolved by the action of digestive juices to liberate the virus particles.

In mammals, one of the major pathways of invasion by pathogens is the respiratory tract. In insects, this mode of invasion is not common. Some fungi and nematode have been reported to invade through the spiracles. No viruses, protozoa, or bacteria (except for the previously mentioned *M. nigrofaciens*) are known to cause infection through the spiracles. If the explanation for the failure to cause respiratory infection are known, methods may be developed to enable pathogens, which are highly virulent when inoculated into the hemocoel, to invade through the respiratory tract of a wide range of hosts. Insects in their dormant or nonfeeding stages and those with sucking mouth parts may become more vulnerable to infection by this means.

Adult insects, when infected chronically or latently, may transmit the pathogens either on the surface or within the egg. This is known as transovum transmission. The transovum transmission of viruses is firmly established, especially for the nuclear- and cytoplasmic-polyhedrosis viruses (see Bergold, 1958; Sager, 1960). In the microsporidia, not only are the protozoa commonly transmitted by the female through the egg (Weiser, 1961), but they are also transmitted occasionally with the sperm when an infected male mates with a healthy female (Thomson, 1958). Although the evidence is still limited, transovum transmission may occur with bacteria (Roegner-Aust, 1949; Bucher and Stephens, 1957; Toumanoff, 1960). The repeated occurrence of infection with varieties of *B. thuringiensis* of identical esterase type under conditions that have avoided methods of infection other than through the egg has been regarded by Norris and Burgess (1963) as strong evidence of the transmission of the bacillus through the egg.

Parasites, predators, and scavengers play very important roles in the transmission of pathogens by means of their ovipositors and mouth parts. They have transmitted viruses (Thompson and Steinhaus, 1950; Bird, 1953a, 1961; Smirnoff, 1959, 1961), bacteria (Metalnikov and Metalnikov, 1935; White, 1943; Toumanoff, 1959; Vago and Kurstak, 1965), microsporidia (see Weiser, 1961), and fungi (see Müller-Kögler, 1965). Virus polyhedra may contaminate the mouthparts of predators, such as *Vespa rufa consobrina* (Saussure) and *V. vulgaris* (Linnaeus), which transmit the virus to new hosts (Smirnoff, 1959). In a field test, Smirnoff (1959) has incorporated the virus polyhedra in pieces of fresh meat that served as baits for the predatory wasps. The wasps fed on the baits and apparently transmitted the virus to hosts in the vicinity. Insect parasites may inadvertently serve as vectors when their ovipositors penetrate into the gut lumen and enable the facultative (potential) bacteria, such as *Serratia marcescens* and *Proteus mirabilis* Hauser, to invade and kill the host (Bucher, 1963).

IV. INSECT HOST

An insect population in relation to the susceptibility of its individuals to disease may be composed of: (1) the typically diseased insect; (2) the atypically diseased insect; (3) the uninfected immune; (4) the uninfected susceptible; (5) the latently infected insect; and (6) the healthy carrier (Steinhaus, 1949). These different types of individuals are known to occur in insect populations, but it is still unknown whether all six types occur together in the same population. Wellington (1962) has emphasized the importance of host quality on the maintenance and dissemination of nuclear-polyhedrosis virus in populations of *Malacosoma* spp. He has found that active colonies maintain the virus at low host-population densities existing in a harsh environment, but that the sluggish colonies which are present under high host densities are responsible for the propagation of the virus. The extent of infection of potato-infesting aphids by fungi is apparently governed by the relative mobility of the several species of aphids or by variations in the susceptibility of these species to fungus infections (Shands *et al.,* 1963).

A. Host Density

Host population density has an important bearing on the expression of disease in a population. In temporary or short-term control, where the microbial pesticides are applied repeatedly, as in the case of chemical insecticides, they act as density-independent factors. Under usual field conditions, however, pathogens act as density-dependent mortality factors and may bring about permanent or long-term control. Since the threshold density of disease is generally high, the epizootics among insect populations usually occur at high host densities, but they may also occur at relatively low host densities, especially following a widespread epizootic which has distributed the pathogen extensively throughout the habitat (Bird and Elgee, 1957; Clark and Thompson, 1954; Tanada, 1961b). The initiation of an epizootic may occur at a low or high host density. The increase in the intensity of the epizootic is usually associated with the rise in the insect population, but other intrinsic and extrinsic factors may also be involved.

Under laboratory conditions, insects are more inclined to develop diseases when they are crowded together (Steinhaus, 1958b). Crowding tends to incite cannibalism and thereby increases the transmission and incidence of bacterial diseases among grasshoppers (Stevenson, 1959; Bucher, 1963) and *Scolytus multistriatus* Marsham (Doane, 1960). Bucher (1963) states that, although the crowded conditions maintained in the laboratory are conducive to epizootics in grasshopper populations, such conditions do not occur in the field and crowding per se has little effect on disease incidence. Crowding may also be relatively unimportant in virus diseases according to Bird (1964) and David and Gardiner (1965b).

An important factor in the rapid control of the European spruce sawfly, *Diprion hercyniae,* by the nuclear-polyhedrosis virus is the number of host generations per year (Bird, 1964). More rapid control is obtained in areas with three generations than in areas with two or one generation per year.

B. Host Resistance

The resistance of the host is closely associated with the virulence of the pathogen, and whether the interaction between host and pathogen is due to the resistance of the former or the virulence of the latter is not always clear. Insects, in general, possess natural immunity to microorganisms which infect plants, vertebrates, and other invertebrates. In many instances, they are not susceptible to pathogens which infect members of the same family or even the same genus. Strains or varieties within a species are known to exhibit differences in their resistance to pathogens. Such strains have been studied most thoroughly in the domesticated silkworm and honeybee. A knowledge of the fundamental basis of insect resistance to infectious diseases is of value in developing methods of overcoming the resistance and in the use of such methods in microbial control. Several excellent reviews have appeared on the general aspects of insect resistance to pathogens (Steinhaus, 1949; Stephens, 1963a; Briggs, 1964; Heimpel and Harshbarger, 1965).

Insects possess certain physical and physiological characteristics that are responsible for their resistance to pathogens. The juices in the midgut of certain insects, e.g., the silkworm, may contain antiviral, antibacterial, bacteriostatic, and antifungal substances (Duncan, 1926; Masera, 1954; Aizawa, 1962b). The identity of these substances has not been established.

Other properties of the midgut may also determine the resistance of insects to pathogens. The pH of the midgut is an important factor in insect resistance. Insects, such as sawflies, with low midgut pH (6.18–8.93) are susceptible to *B. cereus,* whereas those with high pH (8.96–10.42), such as some Lepidoptera, are resistant to infection by the bacterium. On the other hand, insects susceptible to the crystalliferous bacteria generally have a high midgut pH. According to Fisher and Sanborn (1962), the host specificity of microsporidians may be determined by the properties of the insect gut, especially the presence of a peritrophic membrane. The basis for their conclusion is that surgically introduced implants of the fat body of a *Tribolium* sp. infected with *Nosema whitei* Weiser are able to cause infections in several insect species which are insusceptible to infection *per os* by the microsporidian. The presence of antagonistic microflora in the digestive tract may also prevent pathogenic bacteria from multiplying and infecting the host (Afrikian, 1960).

Cellular and humoral immunities are known to occur in insects, but there are still some uncertainties concerning the importance of the two types of immunities (see Steinhaus, 1949; Wagner, 1961; Stephens, 1963a; Briggs, 1964). Humoral factors, not comparable to the antibodies but more analogous to the "nonspecific acquired immune factors" of vertebrates, are active in insects against bacteria (Briggs, 1958; Stephens, 1959, 1963b; Gingrich, 1964) and viruses (Aizawa, 1953, 1962a). Feir and Walz (1964) have found a rather labile agglutinating factor which occurs naturally in the hemolymph of the large milkweed bug, *Oncopeltus fasciatus* (Dallas), a shorthorned grasshopper, and the small milkweed bug, *Lygaeus kalmii* Stål, but not in *Anasa tristis* or a naucorid bug. Since the agglutinating property suggests the presence of an antibody, this factor should be thoroughly studied and its character established. Phagocytosis (cellular immunity) occurs commonly in insects and is considered an important mechanism for resistance. When the phagocytes are blocked by the inoculation of india ink, the susceptibility of *G. mellonella* to the nuclear-polyhedrosis virus is greatly increased (Stairs, 1964a).

In general, the resistance of insects to pathogens seems to increase as the insects advance in age. This has been the case especially with bacteria, viruses, fungi, and protozoa. The generalization, however, needs substantiation by quantitative analyses, such as bioassays. Recently, Stairs (1965a) has quantitatively analyzed the susceptibility of the larva of *Malacosoma disstria* (Hübner) to a nuclear-polyhedrosis virus. He has found that the susceptibility decreases markedly with an increase in larval age. Some fourth-instar larvae may survive dosages two billion times higher than those required to kill first-instar larvae.

Some diseases in insects are restricted to certain stages in the insect life cycle; e.g., the foulbrood diseases do not occur in the adult honeybee, and the *Nosema* disease does not develop in the honeybee grub. In virus diseases, the early belief that adult lepidopterous insects are immune to active virus infections may not be true for most cases. Lepidopterous adults have been found infected by the nuclear-polyhedrosis virus (Sager, 1960; Aizawa, 1963), the cytoplasmic-polyhedrosis virus (Tanada and

Chang, 1960; Smith, 1963; Neilson, 1965; Sidor, 1965), and by the non-inclusion virus of *Sericesthis pruinosa* (Day, 1965). These adults have been infected in the immature stages or through inoculation of the virus into the hemocoel (Martignoni, 1964a). At present, there appears to be no evidence of an adult insect becoming infected by ingesting the virus.

At certain periods in their development, the insect may be more or less susceptible to infection by pathogens. The embryonic regenerative cells of the sawfly and of Lepidoptera are immune to nuclear-polyhedrosis virus, but become susceptible as they mature (Bird, 1953b; Stairs, 1965b). The time of larval molt is the critical period for the invasion of bacteria through the midgut into the hemocoel (Bucher, 1963). On the other hand, the sensitivity of the silkworm to the induction of virus disease is very low during molting, but increases rapidly after ecdysis (Aruga, 1963).

Within the insect cuticle, there are antifungal substances that appear during the development of the integument after molting (Sussman, 1951; Koidsumi, 1957; Evlakhova and Chekhourina, 1964). These substances occur in the waxy epicuticular layer and are apparently medium-chain fatty acids, caprylic or capric acids (Koidsumi, 1957). Ether extracts of the cuticle of *Eurygaster integriceps* Puton suppress spore germination, mycelial growth, and spore formation of *B. bassiana* (Evlakhova and Chekhourina, 1963). Since the removal of the waxy layer with organic solvents and abrasives apparently gets rid of the fungus-inhibiting substance, fungus spore dust should include abrasives and sorptive additives that will remove the waxy layer.

There are four types of syndromes expressed by different lepidopterous insects to infection by *B. thuringiensis* and its varieties (Heimpel and Angus, 1959; Martouret, 1961). These syndromes reflect the sites of action of the toxins and bacterial spore. In the type I syndrome, there is general paralysis which terminates with death in 1–7 hours, and is associated with an increase in blood pH. Gut paralysis may occur with limited dosages of the bacteria and death may result from septicemia. The type II syndrome exhibits gut paralysis, but not general paralysis, and there is no increase in blood pH. Death results from septicemia. The type III syndrome exhibits neither gut nor general paralysis and septicemia causes death. Thus far, *Anagasta kühniella* is the only insect exhibiting this type of syndrome. The type IV syndrome occurs in noctuids, with mortality resulting mainly from the heat-stable toxin rather than from the spore and crystal toxin.

When reared at high temperature, some insects increase their resistance to virus infections. *Drosophila* spp. may become resistant to the sigma virus when reared at 30°C (L'Héritier and Sigot, 1946), *Pieris rapae* to the granulosis virus at 36°C (Tanada, 1953), *Diprion hercyniae* to the nuclear-polyhedrosis virus at 85°F (29.4°C) (Bird, 1955), *Trichoplusia ni* and *Heliothis zea* to their respective nuclear-polyhedrosis viruses at 39°C (Thompson, 1959), and *G. mellonella* at 28°C to the *Tipula* iridescent virus (Tanada and Tanabe, 1965) and the *Sericesthis* iridescent virus (Day and Mercer, 1964). The *Sericesthis* iridescent virus in a cell culture of *Antheraea eucalypti* loses its infectivity at 25°C but not at 20°C (Bellett and Mercer, 1964). The resistance to infection is apparently not associated with the failure of the virus to invade the host, but is more likely due to factors such as the destruction of the virus, a reduction or cessation of virus multiplication, or the occurrence of immune and nonimmune host responses [e.g., the febrile response (Baron, 1963), or the production of interferon (Isaacs, 1963) in the case of vertebrates]. High temperature is also known to destroy protozoa which are causing infections in insects. There is the possibility, therefore, that the occurrence of high temperatures in certain habitats may result in the recovery of the insect from chronic virus and protozoan infections and in the disappearance of latent virus infections.

Nutrition may predispose the insects to infectious diseases. The effect of the quality and quantity of food on insect resistance has been investigated especially with virus diseases (see Bergold, 1958), and only to a limited extent with fungus (see Müller-Kögler, 1965) and other diseases of insects. The incidence of virus diseases increases when unfavorable food plants are fed to *Porthetria dispar* (Vago, 1953; Schmidt, 1956; Kovačevič, 1956), *B. mori* (Vago, 1951), *Malacosoma neustria* (Linnaeus) (Kovačevič, 1956), and *Hyphantria cunea* (Kovačevič, 1958). According to Shvetsova (1950), the occurrence of virus epizootics may depend on the insect diets, especially those with a high nitrogen or high carbohydrate content. Pimentel and Shapiro (1962) have confirmed that a high protein diet increases the susceptibility of *G. mellonella* to the nuclear-polyhedrosis virus, but a high carbohydrate diet does not have any effect. In *P. brassicae,* unaccustomed food fed to the larvae increases the incidence of granulosis in the susceptible but not in the resistant host strain (David and Gardiner, 1965b). In these cases, nutrition may be acting as a stressor in activiating latent infections (see Section VII,C).

Very little is known of the existence of insect strains and varieties resistant to infectious diseases under natural conditions. There are some reports of insect populations which vary in their susceptibility, such as the wattle bagworm, *Kotochalia junodi* (Ossowski, 1960), and the California oakworm, *Phryganidia californica* Packard (Martignoni and Schmid, 1961), to the nuclear-polyhedrosis viruses, and *P. brassicae* to the cytoplasmic-polyhedrosis virus (Sidor, 1959). The continual spread of the European spruce sawfly has led to the speculation of possible increase in the resistance of the sawfly (Bird and Elgee, 1957), but this hypothesis is not confirmed by laboratory tests (Neilson and Elgee, 1960). An increase in the LD_{50} in population of the larch bud moth. *Eucosma griseana* (Hübner), which has been exposed to a granulosis, has led Martignoni (1957) to conclude that an apparent uptrend in resistance has occurred.

Resistance has developed among laboratory insects which have been continuously exposed to pathogens. A culture of the cabbageworm, *P. brassicae,* which survived a granulosis outbreak, is resistant to the virus (Rivers, 1959; David and Gardiner, 1960). After 4 years without additional exposure to the virus, this strain has remained resistant when fed the virus (David and Gardiner, 1965a). According to Smith (1959), the granulosis virus when passed through *Pieris napi* (Linnaeus) becomes infective for the resistant strain. The possibility exists, however, that the virus passed through *P. napi* may not be the original virus, but one that has been activated from the occult state. Aizawa *et al.* (1961) have developed a strain of *B. mori,* which, after 13 generations, is resistant to virus induction. Since this strain is not resistant to virus infection, the factors for induction and infection appear to be independent of each other.

The first instance of an insect becoming resistant to *B. thuringiensis* has been reported by Harvey and Howell (1965). The resistance of the housefly, after culture for 50 generations on media treated with the bacillus, has increased from eightfold to fourteenfold between generations 27 and 50 on the basis of ratios of LD_{50} values. This result is not unexpected because the action of *B. thuringiensis* on the housefly is due to the thermostable exotoxin (fly factor), and the development of resistance, as with chemical insecticides, is associated with a nonmutagenic chemical that is not a factor in the growth and development of the bacterium in the housefly. Further tests with the bacillus would probably reveal the development of the resistance of certain insects to the toxic crystal, especially in cases where the bacterium is not dependent on the toxins for survival. Where the pathogen is closely associated with the host, as an obligate pathogen, host resistance may not develop as readily because the pathogen would also be able to mutate into or select for more virulent forms, as the host becomes more resistant.

V. ENVIRONMENTAL FACTORS

Environmental conditions may affect the application of microbial pesticides, but the significance of their effects varies in degree with the type of control, short-term (temporary) or long-term (permanent) control. In general, for short-term control, the physical factors, such as rain, wind, and sunlight, are more important than the biotic factors. The effects of these factors on the application of microbial pesticides are similar to their effects on chemical insecticides. Many of the resistant stages of the pathogens are generally able to withstand the vagaries of weather as well as the chemical insecticides. In long-term control, the interaction of the physical and biotic factors with both host and pathogen population is much more complex. The environment may affect the virulence, stability, persistence, dispersal, and transmission of the pathogen, the resistance and susceptibility of the host, and the interaction between pathogen and host. The environmental factors may act not only on active infections, but also as stressors on latent infections in insect populations (see Section VII,C). There are very few quantitative data on the specific environmental factors and the mechanisms involved in their actions in short-term and long-term control.

A. Physical Factors

Among the physical factors, temperature and humidity have been most thoroughly investigated. Some workers believe that high humidity has little effect on virus diseases (Thompson and Steinhaus, 1950; Shvetsova, 1950; Bird, 1964), although rain may increase or decrease the incidence of disease by washing the virus off or by distributing the virus in a vertical direction throughout the plant (Bird, 1964). Others, however, have associated virus epizootics with wet weather, as in the case of *Porthetria dispar* (Wallis, 1957), *Trichoplusia ni* (Hofmaster, 1961), and *Pseudaletia unipuncta* (Haworth) (Marcovitch, 1958). Climatic conditions may regulate the occurrence of microsporidian species in certain areas, and seasonal variations may control the reproductive cycles of the microsporidian species (Weiser, 1965).

In the laboratory, an excess of moisture often leads to the outbreak of bacterial diseases, particularly those caused by potential pathogens (Bucher, 1963). In protozoan diseases, the effect of humidity is primarily on the survival and persistence of spores (Weiser, 1956, 1961). The spores of the microsporidian, *Thelohania californica* Kellen and Lipa, depend on an alternation of wetness and dryness before they are capable of infecting the mosquito, *Culex tarsalis* Coquillett (Kellen and Lipa, 1960).

Low humidity is generally considered a limiting factor in fungus diseases for the spore germination, infection, and subsequent sporulation of the fungus on the host. Some workers, however, claim that humidity may not be as important as generally believed because the fungus spores can germinate at a relative humidity as low as 60% (Schaefer, 1936; Schaerffenberg, 1957a, 1964). The discrepancies appearing in studies of spore germination at different relative humidities may reflect differences in measurements made at the micro- and macroclimate levels. Moreover, within a given genus of fungi (e.g., *Entomophthora*), there are species that differ in their humidity requirements (Gustafsson, 1965). Some species [e.g., *Entomophthora aphidis* (Hoffman)] are much more effective in humid, and others (e.g., *E. thaxteriana*) in semiarid areas (Rockwood, 1950).

The nematodes are known to require high humidity and, in some cases, even a surface film of water, in order to survive, persist and migrate in their environment. The failure of the nematode DD-136 to control the Colorado potato beetle, *Leptinotarsa decemlineata* (Say), has been attributed to the absence of such conditions (Welch and Briand, 1961). The nematode, when protected from desiccation by paper and burlap bands placed around trees, kills a high percentage of the larvae and pupae of the codling moth, *Carpocapsa pomonella* (Dutky and Hough, 1955).

High temperature generally accelerates the course of a disease and decreases the period of lethal infection. Under such conditions, the insect is killed rapidly by the microbial pesticides. At temperatures of about 10°C, the period of lethal infection of the nuclear-polyhedrosis virus in the cabbage looper lasts five times longer than at 22°C (Hall, 1957). Below 15°C, the period of infection by *B. thuringiensis* is not only prolonged, but also the percentage of larval mortality is reduced (Fedorintchik, 1964). Also at low temperatures (15°C), the granluosis virus predominates over the nuclear-polyhedrosis virus in a mixed infection in *Agrotis segetum* Schiffermüller, but the second virus is dominant at high temperatures (Shvetsova and Ts'ai Hsü-yü, 1962). The hibernating conditions of *A. segetum* favor the granulosis virus. The effect of high temperature on the host resistance to virus infections has been discussed in Section IV.

The effectiveness of the milky-disease organism, *B. popilliae,* is greatly reduced when the soil temperature falls below 21.1°C in the control of *Popillia japonica* (Tashiro, 1957; Fleming, 1958), and below 15°C in the control of *Melolontha melolontha* (Hurpin, 1964).

Weather and host plant, in general, have no effect on microsporidian infections (Weiser, 1956, 1961). More microsporidian-infected European cornborers, however, are killed by the low winter and the high summer temperatures in Illinois than uninfected individuals (Kramer, 1959).

There are only limited, if any, data on the effect of other physical factors on the microbial pesticides and their applications. Sunlight is known to inactivate many insect pathogens. A low soil pH may affect the spores of *B. popilliae* (Beard, 1945). The green muscardine fungus (*M. anisopliae*) and the red muscardine fungus [*Sorosporella uvella* (Krassilstschik) Giard (*Tarichium uvella*)] have different soil pH requirements: the former is favored by an acid, the latter by an alkaline, soil (Bünzli and Büttiker, 1959; Pospelov, 1940; Pyatnitzkiĭ, 1942). The application of fertilizer affects the soil pH and thereby restricts the type of fungi available to infect soil insects. Some nematodes (Mermithidae) have marked preference for calcareous soils (Théodoridès, 1952).

The flour beetles, *Tribolium castaneum* (Herbst) and *T. confusum* duVal, when infected with the schizogregarine, *Farinocystis tribolii* Weiser, are more sensitive to x-rays (Jafri, 1965). The ionizing radiation disintegrates the gut tissues and enables the bacteria in the lumen to invade the insect hemocoel.

B. Biotic Factors

Biotic environmental factors may substantially affect the application of microbial pesticides, especially for long-term or permanent control. The role of parasites, predators, and other animals in the transmission, dispersal, and persistence of pathogens has been discussed in Sections II,D and E, and III. Very intimate interrelationships occur between pathogens and biotic factors when host species closely related taxonomically have similar life habits and occur on the same host plant. This intimate association has been observed in microsporidians and their hosts (Weiser, 1961).

The effect of nutrition on the susceptibility of insects to infectious diseases has been discussed in Section IV,B. In addition to its nutritional value, the foliage of plants contains bacteriostatic and bactericidal substances, and the quantity and type of these substances vary with the plant species (Nickell, 1959). The leaf juices of conifers are more inhibitory to *B. thuringiensis* than those of other plant foliage (Kushner and Harvey, 1962; Smirnoff and Hutchison, 1965). The quality of the host plant, therefore, may affect the dosage, survival, and dissemination of bacterial pesticides. It has not been determined whether substances that are detrimental to pathogens other than bacteria also occur in plant foliage. The presence of such substances may explain the observation of Chiang and Holdaway (1960) that European cornborer larvae, collected from corn varieties resistant to the borer, are less infected by the microsporidian than those collected on susceptible corn varieties. This may also be the reason why the white muscardine fungus (*Beauveria* sp.) is most prevalent in deciduous forests, whereas a *Fusarium* sp. is predominant in pine forests (Sulzdalskaya, 1954). These fungi play important roles in the control of *Eurygaster integriceps*.

Pathogens interact with each other and other microflora mutually, synergistically, or antagonistically when present in the same host (see

Section VIII,B). The type and quantity of intestinal microflora may alter the effectiveness of microbial pesticides. *Pieris brassicae* with an abundant microflora is more susceptible to the application of *B. thuringiensis* than *G. mellonella* which has a limited flora (Isakova, 1964). This difference in susceptibility, however, may be associated with the inherent properties of the two species rather than with the differences in the microflora. Inasmuch as low dosages of *B. thuringiensis* spore preparations injure the midgut epithelium of certain insects, this enables the potential pathogens present in the lumen to invade the hemocoel. Some gregarines injure the gut wall and allow the invasion of gut bacteria in grasshoppers (Bucher, 1959b). In the mosquito, *Anopheles gambiae* Giles, the microsporidian, *Nosema stegomyiae* Marchoux, Salimbeni and Simonds, inhibits the development of the malarial parasite, *Plasmodium falciparum* (Welch), in the gut of the host (Weiser, 1961). The microsporidian has a high potential in microbial control not only because of its effect on an insect pest but also because it inhibits the development of a human pathogen.

VI. MASS PROPAGATION

The lack of adequate quantities of pathogens has been a serious impediment toward the development and application of microbial pesticides. Mass propagation methods for pathogens have progressed very slowly in view of the fact that Krassilstschik in 1888 first produced large quantities of the green muscardine fungus, *M. anisopliae,* and used it successfully. At present only two preparations, composed of the milky-disease organisms, *B. popilliae* and *B. lentimorbus,* and the crystalliferous bacterium, *B. thuringiensis,* are commercially available in the United States. Each investigator has had to produce, at times laboriously, the other types of pathogens for his own investigations.

A. Methods

The goal of commercial mass production is to produce sufficient quantities of the pathogens at competitive prices and, at the same time, maintain the virulence and infectivity of the pathogens at high levels. Martignoni (1964b) has thoroughly discussed the historical development and the problems involved in the mass production of insect pathogens. From the commercial standpoint, Briggs (1963) has compared the production methods of three insect pathogens which vary in their dependence and control by man. Fisher (1965) has discussed the industrial-scale production of *B. thuringiensis.*

The most common and usually the simplest method for mass-producing pathogens is to cultivate them on artificial media. This method, however, is suitable only for facultative pathogens. Even among the facultative pathogens, some of the more fastidious microorganisms (e.g., *B. popilliae*) must be propagated on live insects in order to produce an abundance of spores (Steinkraus and Tashiro, 1955; Steinkraus and Provvidenti, 1958; Dutky, 1963; Rhodes, 1965). At present, the nutritional requirements of *B. popilliae* are being investigated intensively, and an efficient method of mass producing the spores of this bacillus on artificial media may develop in the near future (Sylvester and Costilow, 1964; Pepper and Costilow, 1964).

For industrial-scale production, the use of surface culture becomes uneconomical and impractical for the propagation of microorganisms, and submerged culture with its reduced spatial requirements and more efficient utilization of the medium must be used. Submerged culture is successful with bacteria (presently *B. thuringiensis* is produced by this method), and recently has been found applicable to fungi (Müller-Kögler, 1964; Novák and Samšiňaková, 1964). *B. bassiana* spores produced by submerged cultures are effective against forest insects (Novák and Samšiňaková, 1964).

Obligate pathogens, such as rickettsiae, viruses, and microsporidia, are necessarily propagated on live insects. Clones of insect tissue culture, however, are presently available, and they offer a promising future for the mass propagation of obligate pathogens (Grace, 1962a). Insect viruses that have been cultured in this manner are the noninclusion virus of *Sericesthis pruinosa* (Bellett, 1965a,b), the cytoplasmic-polyhedrosis virus of *Antheraea eucalypti* (Grace, 1962b), and the nuclear-polyhedrosis viruses of the silkworm (Trager, 1935; Aizawa and Vago, 1959; Vaughn and Faulkner, 1963) and of *Peridroma saucia* (Haworth) (Martignoni and Scallion, 1961). The presence of inclusion-body protein in the medium appears necessary for the purified virus particles to infect the ovarian tissue cultures (Vaughn and Faulkner, 1963). The subject of insect tissue culture has been reviewed by Day and Grace (1959) and Martignoni (1962).

The mass propagation of the nematode DD-136 on live waxmoth larva is much more efficient than on artificial medium. A yield of up to 200,000 infective stage nematodes per larva has been obtained (Dutky *et al.*, 1964).

There are three ways in which live insects have been used as substrates for the propagation of insect pathogens: (1) healthy insects are collected in the field and infected in the laboratory, (2) insects are reared and infected in the laboratory, (3) diseased insects are collected from field populations which had succumbed to naturally or artificially induced epizootics. The insects may be primary or alternate hosts of the pathogen.

When alternate hosts are used, care should be taken that the virulence of the pathogen is not affected (see Section II,B).

The use of field-grown insects is the least expensive in time, space, and equipment, but the insects occasionally may not be available in sufficient quantities and in appropriate sizes. Inasmuch as field-collected insects often harbor other pathogens and parasites, this method cannot readily provide the desired pathogens in a pure state.

The rearing and infecting of insects in the laboratory is generally tedious and time consuming. Recent developments in artificial and semisynthetic diets, however, have greatly reduced the cost and time of feeding some species. The mass rearing of the cabbage looper, *Trichoplusia ni,* and of the corn earworm, *Heliothis zea*, on semisynthetic diets, and the production of their viruses are economically practical on a commercial scale (Ignoffo, 1964, 1965a,b). The viruses produced in this manner are as effective as naturally collected viruses (Ignoffo, 1964; Ignoffo *et al.*, 1965).

There are advantages in using the natural host for the mass production of pathogens. With the use of the primary host, there is less likelihood of a loss in the virulence of the pathogens. In using the insect host, however, the pathogens should be applied so that infection occurs through their natural mode of invasion. Introducing the pathogens into the hemocoel when the natural mode of invasion is through the mouth and digestive tract may select for strains with a lower capacity to invade through the digestive tract. In the mass propagation of the milky-disease organism, *B. popilliae,* however, inoculating the bacillus into the hemocoel has shown no loss in virulence (Dutky, 1963). The problem of selecting strains of low invasive powers is also inherent in the use of insect tissue culture.

Loss in the virulence of pathogens in mass production is more likely to occur when they are reared on artificial media. The various methods of increasing the virulence of the pathogens and the conditions that may affect their virulence during propagation have been discussed in Section II,B.

B. Dosage and Activity Determinations

Shortly after *B. thuringiensis* was commercially available in the United States, differences in the virulence of the various commercial spore preparations became evident when they were applied in the field. The necessity for standardization was obvious. Not only the pathogens had to be standardized, but also the formulations of the pathogens. Moreover, the bioassay could be used to detect contamination of the preparation by a similar strain or other types of microorganisms.

Most of the bioassay tests have been performed with bacteria and viruses, but very little, if any, with other insect pathogens. Tests have been conducted with *Bacillus thuringiensis* (Burgerjon, 1957, 1959; Menn, 1960; Aizawa *et al.*, 1962; Krieg, 1965; Mechalas and Anderson, 1964; Mechalas and Dunn, 1964), *Bacillus popilliae* (see Dutky, 1963), the granulosis virus of *Eucosma griseana* (Martignoni and Auer, 1957), the nuclear-polyhedrosis viruses of *Trichoplusia ni, Heliothis zea,* and *H. virescens* (Ignoffo, 1964, 1965c), and the noninclusion virus of *Sericesthis pruinosa* (Bellett, 1965b).

The bioassay of any pesticide must be conducted under strictly controlled conditions, such as a pure preparation of the pathogen or its formulations, a uniform test insect, and standardized test conditions. Unlike the chemical pesticides, complications may develop quickly with microbial pesticides because of the possible association of pathogenicity with not one but a multiplicity of toxins. This has been the case with *B. thuringiensis,* which produces several toxins and toxic enzymes, and thus, its bioassay is not a simple one. Not only do the various toxins differ in their modes of actions, but also the susceptibility of a given host (e.g., *P. brassicae*) to certain of the toxins depends upon the nature of the enzymes produced in the midgut. Heimpel (1965) suggests using a series of bioassays and *in vitro* tests with at least three different insect species for the standardization of *B. thuringiensis* preparations. He considers the silkworm and the housefly essential to such standardization because of their characteristic syndromes. The third test insect should also be the one being considered for control. In the United States, the commercial firms have selected *Estigmene acraea* as a standard test insect because of its widespread availability, ease of rearing and handling, freedom from catastrophic diseases, and satisfactory response to infectivity tests (Hall, 1963). Accordingly, a pathogen with diverse modes of actions would require bioassays on several insect species, each reacting to the different modes of actions.

The problem with *B. thuringiensis* is compounded by the differences in formulation by each of the commercial firms. Burgerjon (1964a,b) suggests that each commercial preparation be considered a distinct product since the occurrence of contaminants may alter the results from preparation to preparation. Contamination has been observed in commercial preparations of *B. thuringiensis* used in mosquito control (Kellen and Lewallen, 1960), and may also explain the presence of the toxic labile substance observed by Smirnoff (1964) only in a commercial product.

The mode of action of viruses is apparently less complex than those of bacteria and fungi. Insect viruses, with the present availability of insect tissue cultures, may soon be tested like the vertebrate viruses. The *Sericesthis* iridescent virus has been assayed quantitatively with fluorescent microscopy and autoradiography, but these methods are more tedious and less precise and versatile than a plaque titration (Bellett, 1965a).

An estimate of median lethal dose (LD_{50}) of an insect pathogen based on a single assay has definite limitations because of the low slope of the regression line and the low precision of estimating and administering the dose (Bucher and Morse, 1963). Replicate assays are required to investigate the lower potency ratios that are likely to occur when changes in virulence or resistance are being studied.

Certain chemical substances have been proposed as standards in place of the bioassay with pathogens. Since dipicolinic acid is formed in association with sporulation, an assay of the acid may indicate the quantity of spores formed by the bacilli (Fisher, 1963). There is a close correlation between the quantity of fly toxin and the nontoxic calcium dipicolinate produced by *B. thuringiensis,* and this suggests the possibility of using the salt in the bioassay of fly toxin-producing strains (Cantwell *et al.*, 1964a). The toxicity curves of *B. thuringiensis* and sodium arsenate obtained from tests with *P. brassicae* and *Lymantria monacha* (Linnaeus) are so similar that sodium arsenate has been suggested as a standard for the bacillus (Herfs and Krieg, 1964).

The concentration of microbial pesticides is usually expressed on the basis of the number of organisms or their products found in the preparation. The number may be given in terms of the resistant stages, such as spores, polyhedra, capsules, cysts, and dauer larvae, that are used in the preparations. At present, the commercial preparations of *B. thuringiensis* indicate their concentrations on the basis of the number of spores per gram or milliliter. Unfortunately, however, there is often no correlation between virulence and spore count. In the case of virus polyhedra, the difficulty lies in the variations in the sizes of the polyhedra and in the number of virus particles present in each polyhedron. Thus far, the variation in the sizes of virus polyhedra has not created a serious problem in applying viruses, but an intensive study in virus titration is needed. With the granulosis virus, the variation inherent in the polyhedra is negligible because each capsule rarely contains more than one virus rod.

VII. METHODS OF APPLICATION

Microbial pesticides may be applied for: (1) short-term or temporary control, and (2) long-term or permanent control. The short-term control is comparable to that achieved with most chemical insecticides and requires repeated applications for insect control. When used as a long-term control agent, the microbial pesticide is equivalent to biotic agents such as the insect parasites and predators. The principles relating to the application of the microbial pesticides vary with the two control methods.

The economic threshold of the insect pest has an important bearing on the type of control applied. If the "economic level" of the insect pest density is considerably lower than the "threshold level" of the disease, the control would be temporary and the pathogens would have to be applied repeatedly; if the level is higher than the disease threshold, then long-term control comparable to that of parasites and predators would result (Steinhaus, 1954b).

Steinhaus (1956b) has visualized four major alternatives in the ulitization of microbial pesticides: (1) introduction or colonization into an insect population, (2) application as sprays, dusts, or baits, (3) use with insecticides, both compatibly and synergistically, and (4) use with parasites, predators, and other pathogens. A fifth method is to use stressors to activate latent or chronic infections or to combine the stressors with pathogens to enhance their pathogenicity.

A. Short-Term or Temporary Control

The general methods of applying chemical insecticides are applicable to microbial pesticides (see Section II,E). Innumerable tests, especially with bacteria and viruses, in experimental field plots have resulted in the successful control of many insect pests. Control has been economically practical and feasible with susceptible foliage-feeding insects with a high economic threshold. A more stringent test of the effectiveness of microbial pesticides would be their ability to economically control insect pests of low economic threshold, where a single or few insects can cause economic damage, especially to a crop of high value.

Inasmuch as the period of lethal infection of many pathogens generally is considered to be relatively long, there has been some doubt that microbial pesticides would kill with sufficient rapidity to prevent the insect pest from causing damage. This, however, is not invariably the case. The period of lethal infection is dependent, besides various other factors, on the dosage of the pathogen and on the age of the insect. First- and second-instar larvae in many cases die within 1–3 days after ingesting virus and bacteria. Although the larva which is infected with *B. thuringiensis* may live for several days, it generally stops feeding shortly after ingesting the bacterial preparation and causes no further plant damage. Proper timing and thorough coverage where the insect pest is present on the plant are essential for successful control.

Several attempts have been made to control insects of low economic threshold. Promising results with *B. thuringiensis* and the nuclear-polyhedrosis virus have been obtained on the corn earworm, *Heliothis zea,* on sweet corn (Tanada and Reiner, 1962; Ignoffo *et al.,* 1965), and on cotton (Falcon *et al.,* 1965; Ignoffo *et al.,* 1965). Other workers

(Anderson and Reynolds, 1960; Jaques and Fox, 1960), however, have not been able to obtain adequate control of the corn earworm on sweet corn with the bacillus. On this crop, control is in part dependent upon (1) the thorough and repeated coverage of the corn silk, and (2) on the age of the larva boring into the corn ear. Young first- and second-instar larvae are killed rapidly by the bacillus and virus before they cause damage. Older larvae which migrate down from the tassel are much more resistant to the pathogens and, although infected, usually cause damage before succumbing to the pathogens. With a high population and in the presence of older larvae migrating from the tassel, therefore, the pathogens may not be able to control the corn earworm. According to Wolfenbarger (1964), the addition of conventional paraffin-base oils greatly increases the effectiveness of the nuclear-polyhedrosis virus applied to the corn silk.

Although early workers (Metalnikov *et al.,* 1930; Metalnikov and Metalnikov, 1935) have obtained promising results in the control of the European corn borer with bacteria, their results have not been duplicated in most investigations. The commercial granular formulation of *B. thuringiensis,* however, offers as much promise against the corn borer as some insecticides (Raun, 1963). There is a reduction in the population of codling moth treated with *B. cereus,* but the amount of control obtained has been inadequate (Phillips *et al.,* 1953; Stephens, 1957a; McEwen *et al.,* 1960). *Bacillus thuringiensis* spore preparations greatly reduce the rate of infestation of the artichoke plume moth, *Platyptilia carduidactyla* (Riley), on young artichoke plants (Tanada and Reiner, 1960). The plume moth is a serious and difficult pest to control because it bores into the stems, petioles, and flower buds of the artichokes. The recent commercial preparations of *B. thuringiensis* are greatly superior to the early preparations, and they may be more promising against insects of low economic threshold.

Chemical pesticides have been used for the control of flies breeding in feces by feeding the chemicals to animals, such as cows and poultry. Similar attempts at housefly control have succeeded by feeding B. *thuringiensis* to chickens (Briggs, 1960; Burns *et al.,* 1961), to Japanese quail (Borgatti and Guyer, 1963), and to cows (Dunn, 1960; Gingrich, 1965). *B. thuringiensis* spore preparations when applied to poultry droppings, however, do not control the little housefly, *Fannia canicularis* (Linnaeus) (Eversole *et al.,* 1965).

Although many toxins are produced by pathogens, only a heat-stable toxin of *B. thuringiensis* has been purified and applied in field control. This toxin appears promising for the control of *Diprion pini* (Linnaeus) and may be used as a selective pesticide in integrated control (Burgerjon and Biache, 1964).

Some apprehension has existed that the presence of a virus, in an occult or chronic form, may cause the development of resistance in an insect population to subsequent applications of viruses. Thus far, the applications of the native virus have shown no resistance in a native insect population in which the virus is already present (Steinhaus and Thompson, 1949; Franz and Niklas, 1954). Insect populations, however, may vary in their resistance to viruses (see Section IV,B).

The utilization of fungi is one of the most problematical areas in microbial control in spite of the common occurrence of fungus epizootics among insect pests in nature. The early failures in insect control with fungi, especially in the control of citrus pests and the chinch bug, *Blissus leucopterus* (Say), have discouraged the use of this group of pathogens. These failures have contributed significantly to the assumption that the application of fungus spores is impractical because the spores are already distributed extensively throughout the environment, and that they could not be effective without the proper environmental conditions, especially of temperature and humidity (Billings and Glenn, 1911; Fawcett, 1944). This generalization, although accepted by many investigators, may apply only to certain fungi, and it should be thoroughly examined and tested in the field. Fungus epizootics do not invariably follow periods of high humidity and moderate temperature. The absence of epizootics during humid conditions may indicate the lack of or insufficient quantities of spores to cause an epizootic. Under these conditions, the artificial distribution of spores may be warranted. There is also the likelihood that humid conditions may cause many of the spores (e.g., conidia) to germinate in the absence or lack of adequate hosts. Such noninfective germination may seriously reduce the number of spores available in the environment. It would be most valuable to study this aspect in the field.

Most workers agree that the environmental factors, particularly humidity and temperature, govern the development of fungus diseases, but Schaerffenberg (1964) maintains that these factors play a minor role except during sporulation, when high humidity becomes essential. He believes that the predisposition of the host and an adequate virulence are the most important prerequisites for fungus epizootics. Most evidence, however, supports the importance of humidity and temperature. Some of the confusion may result from the inadequate consideration of the microclimate surrounding the insect host.

In the use of fungi, Madelin (1966) has stressed the importance of the properties of the pathogen. He concludes that the application of fungi in microbial control will continue to be highly empirical because of the serious lack of knowledge on (1) the capacity of some fungi, yet not others to attack insects, (2) the host specificity shown by so many fungus parasites, (3) the different degrees of virulence expressed by strains of the same species, and (4) the capacity for changes in the virulence of particular strains. The problem, however, is much more complex than merely the properties of the fungi. The use of fungi may require the synchronization of a complex of conditions, such as temperature, moisture, light, air currents, host population density, host activity, antagonism, synergism, pathogen density, genetic makeup, and stress (MacLeod and Soper, 1965).

B. Long-Term or Permanent Control

When microbial pesticides are applied as a long-term control measure, the principles of the epizootiology of infectious diseases must be considered (Tanada, 1963, 1964). After a pathogen has been introduced into a host population, its control effect is indicated by the incidence of enzootics and epizootics. An efficient pathogen would regulate the host at a low density level and would be capable of reacting rapidly to changes in the host density. In general, the closer the pathogen is synchronized with and responsive to the host density, the more rapidly the host will be controlled.

Insects that have been controlled by pathogens on a long-term basis generally have a high economic threshold. Insects with low economic thresholds would be difficult to control on a long-term basis because of the lag in the spread and transmission of the pathogens throughout the host population.

The factors that regulate the effectiveness of microbial pesticides for long-term control are: (1) the properties of the pathogen population; (2) the properties of the insect host population; (3) an efficient means of transmission; (4) the physical and biotic environmental factors; and (5) the economic threshold of the insect pest. All of these factors are closely associated and interrelated with each other.

Let us examine the factors which play dominant roles in some of the successful examples of long-term control with microbial pesticides. The success obtained with the nuclear-polyhedrosis viruses of the sawflies is one of the outstanding examples of long-term microbial control. A nuclear-polyhedrosis virus of the European pine sawfly, *Neodiprion sertifer,* has been introduced from Europe and has successfully controlled the sawfly when sprayed in pine forests in Canada (Bird, 1955) and in the United States (Dowden and Girth, 1953; Benjamin *et al.,* 1955). The nuclear-polyhedrosis virus of the European spruce sawfly, *Diprion hercyniae,* apparently was accidentally introduced into Canada with insect parasites (Balch and Bird, 1944; Balch, 1958). When applied to disease-free populations of the spruce sawfly, the virus has spread rapidly and brought about long-term control (Bird and Burk, 1961).

The effectiveness of these viruses has been primarily due to their pathogenicity and efficient means of dispersal and transmission. Viruses with low virulence have given poor control. The dispersal and transmission of the viruses are brought about by infected females, which transmit the viruses through their eggs, and by the insect parasites and predators. The extent of dispersal by the infected female depends upon the ovipositional habit of the species: those laying their eggs singly distribute the virus much more widely than those laying their eggs in clusters. More rapid control has been obtained in areas with three host generations per year compared to those with one or two generations. Environmental factors, such as temperature and humidity, are relatively unimportant.

The Japanese beetle, *Popillia japonica,* has been effectively controlled in the eastern United States by the application of the milky-disease organisms, *B. popilliae* and *B. lentimorbus.* The bacteria control the beetle successfully because of their capacity to persist in the host larval environment, the soil. Infected adults and other animals naturally disperse the bacteria, but dispersal by such means has been relatively slow, and this has necessitated the application of the bacteria in disease-free populations, especially in new areas of beetle infestation. Once the bacteria are established, however, they maintain themselves in the area for many years. The persistence and propagation of the bacteria are also dependent on the immigration of susceptible individuals into the area. According to Dutky (1963), highly virulent strains of the bacteria are not preferred because such strains do not multiply and propagate themselves as well as less virulent strains. Highly virulent strains kill the larvae so rapidly that they prevent sporulation of the bacteria.

Long-term control of the spotted alfalfa aphid, *Therioaphis maculata* (Buckton), has resulted from the application of several entomopathogenic fungi (Hall and Dunn, 1958). The success in the control of the aphid is dependent mainly on the virulence, rapid dispersal, and transmission of the fungi, and on environmental conditions. The dispersal of the fungi is largely by the wind carrying the conidia and by the infected host females. The persistence of the fungi during long periods of dry weather is also important.

The reader may have noticed that there are two points of view concerning the importance of the virulence or the pathogenicity of the pathogens for long-term control. Bird (1964) maintains that a high virulence is required for viruses, whereas Dutky (1963) and Heimpel (1964) consider that high virulence may not be as important, particularly for pathogens such as the milky-disease organisms. A pathogen, obviously, should possess sufficient virulence to kill adequate numbers of the host and maintain the host population below the economic threshold. A highly virulent pathogen, on the other hand, may kill the host so rapidly that the growth and multiplication of the pathogen may be curtailed or reduced to a level which prevents further dispersal and transmission. The question of pathogenicity may vary, therefore, with the properties of the host and pathogen, methods of transmission, the environmental factors, and the economic threshold of the insect pests. For short-term control where the pathogens are applied repeatedly, a highly virulent pathogen would generally be preferable.

In some cases, the immediate death of the host is not necessarily the only indication of the effectiveness of the pathogen in long-term control. The pathogen may have debilitating effects on the pupa and adult, and may sterilize the adults. The sterilization of the adult may be just as effective for insect control as killing the host immediately. Neilson (1965) states: "In forest entomological problems the effect any biological control agent may have in influencing the magnitude and direction of population fluctuations of the host over several generations is often much more important than the immediate insecticidal effect that might be achieved a few days after the first application."

One of the aims of long-term control is the ability to predict the development of epizootics of infectious diseases in insect populations. Little has been accomplished in this area. Fungus epizootics are closely associated with high humidity, but not invariably in all cases. Some virus outbreaks appear to correlate with weather conditions, but the forecasting of epizootics on this basis is still questionable. The development of epizootics and the subsequent decline in population of the fir engraver beetle, *Scolytus ventralis* Le Conte, attacked by the internal nematode parasites, *Parasitylenchus elongatus* Massey and *P. scrutillus* Massey, can be predicted from the short galleries produced by infected beetles (Massey, 1964).

The prediction of epizootics would be facilitated if we are able to detect early stages of infection, even before the appearance of external symptoms. The detection of early stages of nuclear polyhedrosis may be made with the fluorescent antibody technique (Krywienczyk, 1963) and with the use of a hand refractometer to determine hypoproteinemia (Martignoni and Milstead, 1964). The presence of inapparent cytoplasmic-polyhedrosis-virus infection in populations of the alfalfa caterpillar can be established with the use of biotic stressors (Tanada *et al.*, 1964).

C. Activation of Infections with Stressors and Incitants

Numerous laboratory studies have been conducted with stressors and incitants, but little has been done in field studies. A "stressor" may be any stimulus, or succession of stimuli, that tend to disrupt the homeostasis of an insect (Steinhaus, 1960b). An "incitant" stimulates or enhances the activity of the pathogen. Under some conditions, differentiation between a stressor and an incitant may not be possible.

At present there is ample evidence that chronic and latent infections commonly occur in insect populations. The substantial evidence for the presence of occult virus in insects is the carbon dioxide activation of the sigma virus of *Drosophila* spp. (see L'Héritier, 1958), and the induction of nuclear polyhedrosis in tissue culture of the ovary of the white-marked tussock moth, *Hemerocampus leucostigma* (J. E. Smith) (Grace, 1958). There are reports of latent infections caused by other pathogens, such as bacteria (Vago, 1952), microsporidia (Bergold, 1951; Canning, 1953), and fungi (Pospelov, 1938), but whether the present definition of latency applies also to these pathogens may be debatable.

Extensive studies have been conducted on the activation of latent virus infections in insects, but the basic principles governing their activations are still unknown. Aruga (1963) has proposed three main factors that are involved in the induction of occult insect viruses: (1) a genetic factor responsible for the susceptibility of an insect to the action of inducing agent; (2) the inducing agent itself; and (3) the physiological conditions that allow the insect to respond to the action of the inducing agent by the production of virus. The physiological conditions of the insect may be affected by: (1) physical conditions during rearing, (2) food quality, (3) application of various chemicals, (4) infection with insect viruses from other host species, and (5) superinfection with species-specific virus (Bergold, 1958). Manipulation of any or all of these conditions has induced occult virus in the laboratory. The type of stressor effective for a virus disease in a specific insect, however, is not necessarily active on other viruses in other insects. No universal stressor has yet been found, but temperature and humidity apparently affect many occult viruses (see Bergold, 1958; Aruga, 1963; Smith, 1963).

Unlike the stressors, some substances, such as glucose and citric acid (Krieg, 1957) and zinc and cobalt (Geršenson, 1958), may inhibit the activation of a latent virus infection. Studies with inhibitors are greatly limited in number.

There are very few, if any, published reports of the application of stressors for the activation of occult viruses in the microbial control of insect pests. The application of small and sublethal quantities of DDT to larvae of *Porthetria dispar* and *Hyphantria cunea* apparently activates viral and bacterial diseases in these insects (Kovačevič, 1959). The use of Enterobactérine, a Russian formulation of *B. thuringiensis,* provokes virus epizootics (Fedorintchik, 1964). With the use of two stressors, a granulosis virus and a microsporidian of *Pseudaletia unipuncta,* we have shown that an inapparent cytoplasmic-polyhedrosis virus occurs commonly in the populations of the alfalfa caterpillar in California (Tanada *et al.*, 1964). This virus originally was found by Steinhaus and Dineen (1959) in alfalfa-caterpillar larvae stressed by crowding.

Chronic diseases may be intensified to acute infections by the action of stressors and incitants. When larvae of *Trichoplusia ni* are fed low dosages of the nuclear-polyhedrosis virus, those that have been placed under a constant physical stress of vibration have up to four times higher incidence of polyhedrosis than unstressed larvae (Jaques, 1961).

Stressors and incitants may also be applied to insect diseases other than those caused by viruses. Certain bacteria, the conditioned or potential pathogens (Lysenko, 1959; Bucher, 1960), occur commonly in the digestive tracts of insects but usually cause no infection because of their low invasive power of penetration through the midgut wall. In nature, however, they are the most frequent causes of bacterial diseases in insects (Lysenko, 1959). These bacteria are highly pathogenic when they invade the hemocoel of hosts which have been stressed by the feeding of triturated glass (Weiser and Lysenko, 1956; Steinhaus, 1958a), by the presence of gregarines in the midgut, and during the period of molting (Bucher, 1959b). Apparently no one has utilized the toxic crystals of *B. thuringiensis,* which are known to disrupt the midgut epithelium, together with these bacteria for insect control. Secondary infections by these bacteria occur frequently in laboratory insects fed *B. thuringiensis* spores.

VIII. COMPATIBILITY WITH OTHER METHODS OF CONTROL

The microbial insecticides are not only innocuous to vertebrates and plants, but they are also relatively specific in their pathogenicity to various insect species. Even the pathogens with fairly wide host ranges usually do not upset other existing host–parasite relationships and create secondary pest problems as do some wide-spectrum chemical insecticides. The pathogens are generally more tolerant to chemical insecticides than parasites and predators. The compatibilities of microbial pesticides to other biotic factors and to chemical insecticides indicate their potential for integrated control measures (Beirne, 1962; van den Bosch and Stern, 1962).

A. Chemical Control

Microbial pesticides composed of the resistant stages of pathogens can be combined with various adjuvants and surfactants. Care, however, should be insured that the pH of the suspension is maintained close to neutrality, because the inclusion bodies (polyhedra and capsules) of the insect viruses and the toxic crystals of *B. thuringiensis* and its varieties are readily dissolved by alkaline or acidic condition, and may lose their activity. Some of the failures reported with the use of viruses and crystalliferous bacteria may be associated with the loss in pathogenicity due to an acid or alkaline pH of the spray suspension.

There are an increasing number of reports on the compatibility of microbial pesticides with various chemical insecticides. Even the entomopathogenic fungi show variations in their reactions to fungicides (Fisher and Griffiths, 1950; Hall and Dunn, 1959). These variations provide a wider choice in the selection of combinations of fungi and fungicides. In some cases, the combination of microbial and chemical pesticides has not resulted in increased effectiveness, but in others, the combination not only has improved the control over that of each pesticide when used alone (Genung, 1960; Wolfenbarger, 1965), but has also resulted in effective control even with reduced or sublethal quantities of the chemical insecticides in the mixture. Most encouraging results in this area have been obtained in Russia with *B. bassiana* and *B. thuringiensis* (Telenga, 1959; see Fedorintchik, 1964; Müller-Kögler, 1965) and in Czechoslovakia with protozoan pathogens (Rosický, 1951; Weiser, 1961). The effectiveness of the chemical insecticides at sublethal dosages when combined with pathogens will not only reduce residue problems, but may also conserve the other beneficial biotic factors. This aspect has a promising future.

B. Interaction of Pathogens, Parasites, and Predators

The combined action of pathogens, entomophagous parasites, and predators in the control of an insect pest occurs commonly in nature. "The entire biological complex interacts to determine the over-all efficiency of pathogenic organisms" (Cameron, 1963). In general, the pathogens are more effective at high host densities, whereas, at low host densities, parasites and predators play a more significant role in population regulation (see Steinhaus, 1954b; Balch, 1958; Tanada, 1963, 1964). There are, however, exceptions, such as after a severe epizootic when the pathogen is so widely dispersed in the host environment that it causes an epizootic and is the major regulatory factor even at a low host density.

Although the effective joint actions of pathogens, parasites, and predators are clearly evident in the long-term control of insect pests, similar interactions may also occur when the pathogens are used for short-term control. In some cases, the pathogen alone may not destroy as many insects as some of the broad-spectrum chemical insecticides, but due to its narrow host spectrum, its action together with that of the parasites and predators may be as effective as the chemical insecticides. In our short-term control studies with field-crop insects, we have often observed an abundance of parasites and predators in the plots treated with microbial pesticides. These parasites and predators, undoubtedly, contribute to the control of the insect pests.

Parasites and predators may play important roles in the dispersal, persistence, and transmission of pathogens (see Sections II,D, E, and III). The presence of certain parasites in an insect larva may increase the predisposition of the larva to pathogens. A *Pieris* larva when infested with *Apanteles* sp. is more susceptible to bacterial infections than is an unparasitized larva (Paillot, 1925).

Some of the pathogens, in particular the protozoa, may also infect the parasites and predators which attack their common hosts. Such infections may lower the vigor and reproductive capacity, but the extent of the debilitating effect of the disease has not been established in most parasites and predators. The microsporidian, *Nosema mesnili* (Paillot) (*Nosema polyvora* Blunck), attacks not only the insect host, *P. brassicae,* but also the braconid parasite, *Apanteles glomeratus* (Linnaeus), and causes a reduction in the percentage of diapausing individuals of both the host and the parasite (Issi and Maslennikova, 1965). The microsporidian also reduces the weight, thickness, and length of the silk of the cocoons, and the reproductive potential of the infected parasite.

When both pathogen and parasite attack the same host individual, competition may occur for the host tissues. Early death of the host caused by either one of them may adversely affect the other or both of them. The survival of insect parasites in a diseased host depends primarily on the developmental stage when the host is infected by the pathogen, and on the period of lethal infection (Biliotti, 1955, 1956a,b; Franz and Krieg, 1957; Bird, 1961). In order to conserve and protect the hymenopterous parasites of the alfalfa caterpillar which attack larvae in the first three instars, Thompson and Steinhaus (1950) have emphasized the importance of timing the application of the nuclear-polyhedrosis virus used in the control of the alfalfa caterpillar.

At times, the activity of the parasites and pathogens appears to be well coordinated. Parasites are known to select hosts free from bacterial (King and Atkinson, 1928), viral (Niklas, 1939), and microsporidian diseases (Masera, 1948). The fungus, *Entomophthora exitialis* Hall and Dunn, is most effective against the spotted alfalfa aphid in the low desert areas of Southern California during the winter months when the parasite, *Praon palitans* Muesebeck, is in diapause, and the fungus is absent during the summer when the parasite is most active (Hall, 1963). Temperature may also be a factor in the integration between pathogens and parasites. During May, in Moorestown, New Jersey, the temperature rarely exceeds 65°F, a situation which is unfavorable for the milky-disease organisms of the Japanese beetle, and the parasite, *Tiphia vernalis* Rohwer, which actively oviposits at this time of the year can therefore survive even in areas of high disease incidence (White, 1943). *Tiphia popilliavora* Rohwer, on the other hand, is active in August when the soil temperature is above 70°F, and at this temperature the disease develops more rapidly and may kill the host before the parasites complete their development.

Incompatibility between pathogens and insect parasites may occur. Fungus epizootics may decimate the host population and upset the existing natural balance between host and parasites (Ullyett and Schonken, 1940). In most cases, however, the parasites and predators migrate rapidly into such areas after the recovery of the host or they may be reintroduced artificially. The effectiveness of the nuclear-polyhedrosis virus has reduced the population of the European spruce sawfly, *Diprion hercyniae,* to such a low level that some parasites have disappeared, but they have been replaced by other introduced parasites which are capable of maintaining themselves at low host population levels (Balch, 1958). In other cases, a pathogen [e.g., *Nosema melolonthae* (Krieg)] may survive an epizootic [e.g., *Beauveria tenella* (Delacroix) Siemaszko on *Melolontha melolontha*] and in spite of the momentary disappearance of the host for several months, may suppress the new host population (Hurpin, 1965).

The biased removal of microsporidian-infected larvae from the biotope by predators decreases the level of infectious inoculum in the habitat and thereby reduces the effectiveness of the microsporidian (Weiser, 1956). In laboratory-reared, self-perpetuating populations of *Anagasta kühniella,* the action of the egg-feeding predatory mite, *Blattisocius tarsalis* (Berlese), precludes the development of bacterial epizootics (Flanders and Hall, 1965).

Different types of pathogens may exist in a single host individual in mutual coexistence or may antagonize or react synergistically with each other (see Vago, 1963). Field observations of insects with multiple infections are not uncommon. The effect of such interactions on microbial control, however, has not been established in most cases. In the armyworm, *Pseudaletia unipuncta,* the granulosis virus acts as a synergist to the nuclear-polyhedrosis virus, and the frequent occurrence of multiple infection in the armyworm during a virus epizootic strongly suggests that the synergistic association plays an important role in the epizootic (Tanada, 1961b).

The nematodes, DD-136 (Dutky, 1959) and *Howardula benigna* (Cobb) (Fronk, 1950), are mutually associated with bacteria which are inoculated by the nematode into the host insect. The bacteria multiply and kill the host and serve as food for the nematodes. These nematodes have been used in microbial control.

There have been some speculations that the presence of one virus in a chronic or latent form may increase the resistance of the insect to a second virus. Such antagonistic associations have been observed in the laboratory (Aruga *et al.,* 1961, 1963a,b; Tanada and Chang, 1964), but whether this phenomenon also occurs in the field has not been established. The effectiveness of a native virus on a native insect population has been discussed in Section VII,A.

In the rhinoceros beetle, *Oryctes* spp., the Heidenreich's and Maya's diseases (apparently virus diseases) inhibit the development of the green muscardine fungus, *M. anisopliae* (Surany, 1960). Antagonism among bacteria and between bacteria and other pathogens also occurs in infected insects (see Tanada, 1963, 1964).

Mixtures of several pathogens have been utilized for insect control. When *B. thuringiensis* and the nuclear-polyhedrosis virus are combined for the control of the alfalfa caterpillar, the insect is killed much more rapidly than with the individual pathogen, but the control is not as complete or as long lasting as that obtained by the application of the virus alone (Thompson, 1958). On the other hand, a similar mixture of bacillus and virus has provided a more efficient means of controlling the great basin tent caterpillar, *Malacosoma fragile,* than with either pathogen by itself (Stelzer, 1965). Mixtures of bacterial species have been applied for the control of *Cacoecia crataegana* Hübner [*Archips crataegana* (Hübner)] with some success (Kudler *et al.,* 1959).

IX. CONCLUSIONS

During the past two decades, considerable progress has been made in the application of microbial pesticides, especially for short-term or temporary control. The increasing knowledge has provided us with a better understanding of the limitations and potentials of microbial control, and the principles of microbial control are being developed and formulated on a sound basis. The effectiveness of microbial control is dependent on a complex of factors, such as the properties of the pathogen and the host, an efficient means of transmission, the physical and biotic environmental factors, and the economic threshold of the host. All of these factors are involved, in one way or another, with both short-term and long-term microbial control. The importance of each factor varies with the pathogen and the insect host. Of major importance for short-term control are the virulence and persistence of the pathogen, the susceptible stages of the host, and the timing and coverage of the pesticidal application.

The principles of long-term control by microbial pesticides are not being discovered and developed as rapidly as were those of short-term control. Most of these principles also apply to the epizootiology of infectious diseases of insects. In long-term control, the significance of the various factors varies much more with the specific pathogen and insect host than in short-term control. For example, in the control of the sawflies which have an arboreal habitat, the virulence, dispersibility, and transmission of the nuclear-polyhedrosis viruses are most significant; in the Japanese beetle, whose larva lives in the soil, the virulence of the milky-disease organism is of secondary importance, and the persistence, propagation, and spread of the pathogen are mainly responsible for long-term control.

We are seriously deficient in our ability to predict the development and occurrence of epizootics in insect populations. With such knowledge, techniques could be devised to initiate and enhance the development of epizootics.

We should anticipate a wider exploitation of the toxins produced by some pathogens. Their isolation, purification, and application, especially in short-term control, should be a valuable addition to insect control. The use of stressors to activate chronic or latent infections already existing in the insect population or in combination with pathogens to increase their effectiveness is another aspect of microbial control with a promising future.

Inasmuch as pathogens are closely associated with parasites and predators in nature, and are generally compatible with most chemical insecticides, they afford extensive opportunities in the integrated control of insect pests. Integration may be applied to short-term or long-term insect control. When the microbial pesticides are used, they generally do not seriously affect the existing parasite–host complex because of their narrow host spectrum, the absence of toxic residues, and the infrequency of secondary pest outbreaks. Moreover, aside from a very few exceptions, most insect pathogens are innocuous to plants and vertebrates.

Acknowledgments

I wish to express grateful appreciation to Mr. Alvin M. Tanabe for his assistance in compiling and preparing the references.

REFERENCES

Adams, J. C., and Hartman, P. A. (1965). *J. Invertebrate Pathol.* **7**, 245.
Afrikian, E. G. (1960). *J. Insect Pathol.* **2**, 299.

Aizawa, K. (1953). *Oyo Dobutsugaku Zasshi* **18,** 143.
Aizawa, K. (1961). *Entomophaga* **6,** 197.
Aizawa, K. (1962a). *J. Insect Pathol.* **4,** 122.
Aizawa, K. (1962b). *J. Insect Pathol.* **4,** 72.
Aizawa, K. (1963). *In* "Insect Pathology, An Advanced Treatise" (E. A. Steinhaus, ed.), Vol. 1, pp. 381–412. Academic Press, New York.
Aizawa, K., and Vago, C. (1959). *Ann. Inst. Pasteur* **96,** 455.
Aizawa, K., Furuta, Y., and Nakamura, K. (1961). *Nippon Sanshigaku Zasshi* **30,** 405.
Aizawa, K., Kawarabata, T., and Sato, F. (1962). *Nippon Sanshigaku Zasshi* **31,** 253.
Anderson, L. D., and Reynolds, H. T. (1960). *J. Econ. Entomol.* **53,** 22.
Andreev, S. V., Martens, B. K., and Molchanova, V. A. (1962). *In* "Radioisotopes and Radiation in Entomology," Proc. Intern. At. Energy Agency, Bombay, 1960, pp. 23–37. Brüder Rosenbaum, Vienna.
Angus, T. A. (1954). *Nature* **173,** 545.
Angus, T. A. (1956a). *Can. Entomologist* **88,** 280.
Angus, T. A. (1956b). *Can. J. Microbiol.* **2,** 122.
Angus, T. A. (1956c). *Can. J. Microbiol.* **2,** 416.
Angus, T. A. (1964a). *Entomophaga, Colloq. Intern. Pathol. Insectes, Paris, 1962* Mem. No. 2, pp. 165–173.
Angus, T. A. (1964b). *J. Insect Pathol.* **6,** 254.
Angus, T. A. (1965). *Bacteriol. Rev.* **29,** 364.
Angus, T. A., and Heimpel, A. M. (1956). *Can. Entomologist* **88,** 138.
Aruga, H. (1963). *In* "Insect Pathology, An Advanced Treatise" (E. A. Steinhaus, ed.), Vol. 1, pp. 499–530. Academic Press, New York.
Aruga, H., Hukuhara, T., Yoshitake, N., and Israngkul Na Ayudhya, A. (1961). *J. Insect Pathol.* **3,** 81.
Aruga, H., Hukuhara, T., Fukuda, S., and Hashimoto, Y. (1963a). *J. Insect Pathol.* **5,** 415.
Aruga, H., Yoshitake, N., and Watanabe, H. (1963b). *J. Insect Pathol.* **5,** 1.
Ashrafi, S. H., Zuberi, R. I., and Hafiz, S. (1965). *J. Invertebrate Pathol.* **7,** 189.
Bailey, L. (1959). *J. Insect Pathol.* **1,** 80.
Bailey, L. (1963). "Infectious Diseases of the Honey-Bee." Land Books, London.
Baird, R. B. (1958). *Proc. 10th Intern. Congr. Entomol., Montreal, 1956* **4,** 689.
Balch, R. E. (1958). *Ann. Rev. Entomol.* **3,** 449.
Balch, R. E., and Bird, F. T. (1944). *Sci. Agr.* **25,** 65.
Baron, S. (1963). *Advan. Virus Res.* **10,** 39.
Beard, R. L. (1944). *J. Econ. Entomol.* **37,** 702.
Beard, R. L. (1945). *Conn. Agr. Expt. Sta., New Haven, Bull.* **491,** pp. 505–583.
Beard, R. L. (1964). *Entomophaga, Colloq. Intern. Pathol. Insectes, Paris, 1962* Mem. No. 2, pp. 47–49.
Beirne, B. P. (1962). *Ann. Rev. Entomol.* **7,** 387.
Bellett, A. J. D. (1965a). *Virology* **26,** 127.
Bellett, A. J. D. (1965b). *Virology* **26,** 132.
Bellett, A. J. D., and Mercer, E. H. (1964). *Virology* **24,** 645.
Benjamin, D. M., Larson, J. D., and Drooz, A. T. (1955). *J. Forestry* **53,** 359.
Bergold, G. H. (1951). *Can. J. Zool.* **29,** 17.
Bergold, G. H. (1958). *In* "Handbuch der Virusforschung" (C. Hallauer and K. F. Meyer, eds.), Vol. 4, pp. 60–142. Springer, Vienna.
Biliotti, E. (1955). *Compt. Rend.* **240,** 1021.

Biliotti, E. (1956a). *Entomophaga* **1,** 45.
Biliotti, E. (1956b). *Entomophaga* **1,** 101.
Billings, F. H., and Glenn, P. A. (1911). *U.S. Dept. Agr. Bur. Entomol., Bull.,* **107.**
Bird, F. T. (1953a). *Can. Entomologist* **85,** 437.
Bird, F. T. (1953b). *Can. J. Zool.* **31,** 300.
Bird, F. T. (1955). *Can. Entomologist* **87,** 124.
Bird, F. T. (1961). *J. Insect Pathol.* **3,** 352.
Bird, F. T. (1964). *Entomophaga, Colloq. Intern. Pathol. Insectes, Paris, 1962* Mem. No. 2, pp. 465–473.
Bird, F. T., and Burk, J. M. (1961). *Can. Entomologist* **93,** 228.
Bird, F. T., and Elgee, D. E. (1957). *Can. Entomologist* **89,** 371.
Blunck, H. (1954). *Z. Angew. Entomol.* **36,** 316.
Borgatti, A. L., and Guyer, G. E. (1963). *J. Insect Pathol.* **5,** 377.
Bras, G., Gordon, C. C., Emmons, C. W., Prendegast, K. M., and Sugar, M. (1965). *Am. J. Trop. Med. Hyg.* **14,** 141.
Breed, R. S., Murray, E. G. D., and Smith, N. R. (1957). "Bergey's Manual of Determinative Bacteriology," 7th Ed. Williams & Wilkins, Baltimore, Maryland.
Briggs, J. D. (1958). *J. Exptl. Zool.* **138,** 155.
Briggs, J. D. (1960). *J. Insect Pathol.* **2,** 418.
Briggs, J. D. (1963). *In* "Insect Pathology, An Advanced Treatise" (E. A. Steinhaus, ed.), Vol. 2, pp. 519–548. Academic Press, New York.
Briggs, J. D. (1964). *In* "The Physiology of Insecta" (M. Rockstein, ed.), Vol. 3, pp. 259–283. Academic Press, New York.
Bucher, G. E. (1957). *Can. J. Microbiol.* **3,** 695.
Bucher, G. E. (1959a). *J. Insect Pathol.* **1,** 331.
Bucher, G. E. (1959b). *J. Insect Pathol.* **1,** 391.
Bucher, G. E. (1960). *J. Insect Pathol.* **2,** 172.
Bucher, G. E. (1963). *In* "Insect Pathology, An Advanced Treatise" (E. A. Steinhaus, ed.), Vol. 2, pp. 117–147. Academic Press, New York.
Bucher, G. E. (1964). *Ann. Entomol. Soc. Quebec* **9,** 30.
Bucher, G. E., and Morse, P. M. (1963). *J. Insect Pathol.* **5,** 289.
Bucher, G. E., and Stephens, J. M. (1957). *Can. J. Microbiol.* **3,** 611.
Bünzli, G. H., and Büttiker, W. W. (1959). *Bull. Entomol. Res.* **50,** 89.
Burgerjon, A. (1957). *Entomophaga* **2,** 129.
Burgerjon, A. (1959). *Entomophaga* **4,** 201.
Burgerjon, A. (1964a). *Entomophaga, Colloq. Intern. Pathol. Insectes, Paris, 1962* Mem. No. 2, pp. 227–237.
Burgerjon, A. (1964b). *Entomophaga, Colloq. Intern. Pathol. Insectes, Paris, 1962* Mem. No. 2, pp. 255–262.
Burgerjon, A., and Biache, G. (1964). *J. Insect Pathol.* **6,** 538.
Burgerjon, A., and de Barjac, H. (1960). *Compt. Rend.* **251,** 911.
Burgerjon, A., and Galichet, P. F. (1965). *J. Invertebrate Pathol.* **7,** 263.
Burgerjon, A., and Grison, P. (1965). *J. Invertebrate Pathol.* **7,** 281.
Burgerjon, A., Grison, P., and Kachkouli, A. (1964). *J. Insect Pathol.* **6,** 381.
Burns, E. C., Wilson, B. H., and Tower, B. A. (1961). *J. Econ. Entomol.* **54,** 913.
Burnside, C. E. (1930). *U.S. Dept. Agr., Tech. Bull.* **149.**
Cameron, J. W. M. (1963). *Ann. Rev. Entomol.* **8,** 265.
Canning, E. U. (1953). *Parasitology* **43,** 287.
Cantwell, G. E., Heimpel, A. M., and Thompson, M. J. (1964a). *J. Insect Pathol.* **6,** 466.

Cantwell, G. E., Knox, D. A., and Michael, A. S. (1964b). *J. Insect Pathol.* **6,** 532.
Chiang, H. C., and Holdaway, F. G. (1960). *J. Econ. Entomol.* **53,** 918.
Clark, E. C. (1956). *Ecology* **37,** 728.
Clark, E. C., and Thompson, C. G. (1954). *J. Econ. Entomol.* **47,** 268.
Clerk, G. C., and Madelin, M. F. (1965). *Brit. Mycol. Soc. Trans.* **48,** 193.
David, W. A. L., and Gardiner, B. O. C. (1960). *J. Insect Pathol.* **2,** 106.
David, W. A. L., and Gardiner, B. O. C. (1965a). *J. Invertebrate Pathol.* **7,** 285.
David, W. A. L., and Gardiner, B. O. C. (1965b). *J. Invertebrate Pathol.* **7,** 347.
Day, M. F. (1965). *J. Invertebrate Pathol.* **7,** 102.
Day, M. F., and Grace, T. D. C. (1959). *Ann. Rev. Entomol.* **4,** 17.
Day, M. F., and Mercer, E. H. (1964). *Australian J. Biol. Sci.* **17,** 892.
de Barjac, H., and Bonnefoi, A. (1962). *Entomophaga* **7,** 5.
Dedonder, R., and Lecadet, M. (1964). *Entomophaga, Colloq. Intern. Pathol. Insectes, Paris, 1962* Mem. No. 2, pp. 197–203.
Doane, C. C. (1960). *J. Insect Pathol.* **2,** 24.
Doane, C. C., and Wallis, R. C. (1964). *J. Insect Pathol.* **6,** 423.
Dowden, P. B., and Girth, H. B. (1953). *J. Econ. Entomol.* **46,** 525.
Dresner, E. (1949). *Contrib. Boyce Thompson Inst.* **15,** 319.
Dresner, E. (1950). *J. N. Y. Entomol. Soc.* **58,** 269.
Duggar, B. M. (1896). *Bull. Illinois State Lab. Nat. Hist.* **4,** 340.
Duncan, J. T. (1926). *Parasitology* **18,** 238.
Dunn, P. H. (1960). *J. Insect Pathol.* **2,** 13.
Dunn, P. H., and Mechalas, B. J. (1963). *J. Insect Pathol.* **5,** 451.
Dutky, S. R. (1959). *Advan. Appl. Microbiol.* **1,** 175.
Dutky, S. R. (1963). *In* "Insect Pathology, An Advanced Treatise" (E. A. Steinhaus, ed.), Vol. 2, pp. 75–115. Academic Press, New York.
Dutky, S. R., and Hough, W. S. (1955). *Proc. Entomol. Soc. Wash.* **57,** 244.
Dutky, S. R., Thompson, J. V., and Cantwell, G. E. (1964). *J. Insect Pathol.* **6,** 417.
Elmore, J. C., and Howland, A. F. (1964). *J. Insect Pathol.* **6,** 430.
Emmons, C. W., and Bridges, C. H. (1961). *Mycologia* **53,** 307.
Eversole, J. W., Lilly, J. H., and Shaw, F. R. (1965). *J. Econ. Entomol.* **58,** 704.
Evlakhova, A. A. (1953). *Dokl. Vses. Akad. Sel'skokhoz. Nauk* **18,** 36.
Evlakhova, A. A. (1954). *In* "Infections and Protozoan Diseases of Insects" (V. I. Poltev, L. V. Alexandrova, A. A. Evlakhova, and M. S. Paveljeva, eds.), Abstr. Repts., Joint Plenary Session Sericult. Apicult., Plant Protection, Vet. Sci. Sect., Vses. Akad. Sel'skokhoz. Nauk, Moscow.
Evlakhova, A. A. (1962). *Proc. 11th Intern. Congr. Entomol., Vienna, 1960* **2,** 861.
Evlakhova, A. A., and Chekhourina, T. A. (1963). *Dokl. Biol. Sci. Sect.* (*English Transl.*) **148,** 199.
Evlakhova, A. A., and Chekhourina, T. A. (1964). *Entomophaga, Colloq. Intern. Pathol. Insectes, Paris, 1962* Mem. No. 2, pp. 137–141.
Falcon, L. A., Leigh, T. F., van den Bosch, R., Black, J. H., and Burton, V. E. (1965). *Calif. Agr.* **19,** 12.
Fassatiova, O. (1964). *Entomophaga, Colloq. Intern. Pathol. Insectes, Paris, 1962* Mem. No. 2, pp. 159–161.
Fast, P. G., and Angus, T. A. (1965). *J. Invertebrate Pathol.* **7,** 29.
Fawcett, H. S. (1944). *Botan. Rev.* **10,** 327.
Fedorintchik, N. S. (1964). *Entomophaga, Colloq. Intern. Pathol. Insectes, Paris, 1962* Mem. No. 2, pp. 51–61.

Feir, D., and Walz, M. A. (1964). *Ann. Entomol. Soc. Am.* **57,** 388.
Fisher, F. E., and Griffiths, J. T., Jr. (1950). *J. Econ. Entomol.* **43,** 712.
Fisher, F. E., Thompson, W. L., and Griffiths, J. T., Jr. (1949). *Florida Entomologist* **32,** 2.
Fisher, F. M., Jr., and Sanborn, R. C. (1962). *J. Parasitol.* **48,** 926.
Fisher, R. A. (1963). *In* "Analytical Methods for Pesticides, Plant Growth Regulators, and Food Additives" (G. Zweig, ed.), Vol. 1, pp. 425–442. Academic Press, New York.
Fisher, R. A. (1965). *Intern. Pest Control* **7,** 8.
Fisher, R. A., and Rosner, L. (1959). *J. Agr. Food Chem.* **7,** 686.
Fitz-James, P. C. (1957). *In* "Spores" (H. O. Halvorson, ed.), Publ. No. 5, pp. 85–92. Am. Inst. Biol. Sci., Washington, D.C.
Flanders, S. E., and Hall, I. M. (1965). *J. Invertebrate Pathol.* **7,** 368.
Fleming, W. E. (1958). *Proc. 10th Intern. Congr. Entomol., Montreal, 1956* **3,** 115.
Franz, J. M. (1956). *Entomophaga* **1,** 107.
Franz, J. M. (1957). *Entomophaga* **2,** 293.
Franz, J. M. (1961a). *Ann. Rev. Entomol.* **6,** 183.
Franz, J. M. (1961b). *In* "Handbuch der Pflanzenkrankheiten" (H. Richter, ed.), 2nd Ed., Vol. 6, pp. 1–302. Parey, Berlin.
Franz, J. M. (1963). *Entomophaga* **8,** 89.
Franz, J. M., and Krieg, A. (1957). *Z. Pflanzenkrankh. Pflanzenschutz* **64,** 1.
Franz, J. M., and Laux, W. (1963). *Entomophaga* **8,** 263.
Franz, J. M., and Laux, W. (1964). *Entomophaga* **9,** 311.
Franz, J. M., and Niklas, O. F. (1954). *Nachrbl. Deut. Pflanzenschutzdienst (Berlin)* **6,** 131.
Franz, J. M., Krieg, A., and Langenbuch, R. (1955). *Z. Pflanzenkrankh. Pflanzenschutz* **62,** 721.
Fronk, W. D. (1950). *J. Econ. Entomol.* **43,** 22.
Gabriel, B. P. (1965). Ph.D. Thesis, Univ. of California, Berkeley, California.
Genung, W. G. (1960). *Florida Entomologist* **43,** 65.
Geršenson, S. M. (1958). *Cesk. Parasitol.* **5,** 105.
Geršenson, S. M. (1959). *Trans. 1st Intern. Conf. Insect Pathol. Biol. Control. Prague, 1958* pp. 197–200.
Geršenson, S. M. (1960). *Probl. Virol. (USSR) (English Transl.)* **5,** 784.
Gilmore, J. E., and Munger, F. (1963). *J. Insect Pathol.* **5,** 141.
Gilmore, J. E., and Munger, F. (1965). *J. Invertebrate Pathol.* **7,** 156.
Gingrich, R. E. (1964). *J. Insect Physiol.* **10,** 179.
Gingrich, R. E. (1965). *J. Econ. Entomol.* **58,** 363.
Grace, T. D. C. (1958). *Science* **128,** 249.
Grace, T. D. C. (1962a). *Nature* **195,** 788.
Grace, T. D. C. (1962b). *Virology* **18,** 33.
Grison, P., and Bourguignon, S. (1965). *J. Invertebrate Pathol.* **7,** 109.
Günther, S. (1959). *Nachrbl. Deut. Pflanzenschutzdienst (Berlin)* **13,** 19.
Gustafsson, M. (1965). *Lantbrukshogskol. Ann.* **31,** 103.
Hadley, C. H. (1948). *U.S. Dept. Agr., Bur. Entomol. Plant Quarantine* **EC–4.**
Hall, I. M. (1954). *Hilgardia* **22,** 535.
Hall, I. M. (1957). *J. Econ. Entomol.* **50,** 551.
Hall, I. M. (1961). *Advan. Pest Control Res.* **4,** 1.
Hall, I. M. (1963). *In* "Insect Pathology, An Advanced Treatise" (E. A. Steinhaus, ed.), Vol. 2, pp. 477–518. Academic Press, New York.

Hall, I. M. (1964). *In* "Biological Control of Insect Pests and Weeds" (P. DeBach, ed.), pp. 610–628. Chapman & Hall, London.
Hall, I. M., and Dunn, P. H. (1958). *J. Econ. Entomol.* **51,** 341.
Hall, I. M., and Dunn, P. H. (1959). *J. Econ. Entomol.* **52,** 28.
Hannay, C. L., and Fitz-James, P. C. (1955). *Can. J. Microbiol.* **1,** 694.
Harvey, J. L. (1960). *Federal Register* **25,** 3207.
Harvey, T. L., and Howell, D. E. (1965). *J. Invertebrate Pathol.* **7,** 92.
Heimpel, A. M. (1955a). *Can. J. Zool.* **33,** 311.
Heimpel, A. M. (1955b). *Can. Dept. Agr. Forest Biol. Div. Bi-monthly Progr. Rept.* **11,** 3.
Heimpel, A. M. (1964). *Entomophaga, Colloq. Intern. Pathol. Insectes, Paris, 1962* Mem. No. 2, pp. 23–33.
Heimpel, A. M. (1965). *Proc. 12th Intern. Congr. Entomol., London, 1964* p. 736.
Heimpel, A. M., and Angus, T. A. (1958). *Can. J. Microbiol.* **4,** 531.
Heimpel, A. M., and Angus, T. A. (1959). *J. Insect Pathol.* **1,** 152.
Heimpel, A. M., and Angus, T. A. (1960). *Bacteriol. Rev.* **24,** 266.
Heimpel, A. M., and Angus, T. A. (1963). *In* "Insect Pathology, An Advanced Treatise" (E. A. Steinhaus, ed.), Vol. 2, pp. 21–73. Academic Press, New York.
Heimpel, A. M., and Harshbarger, J. C. (1965). *Bacteriol. Rev.* **29,** 397.
Herfs, W., and Krieg, A. (1964). *Entomophaga, Colloq. Intern. Pathol. Insectes, Paris, 1962* Mem. No. 2, pp. 263–265.
Hofmaster, R. N. (1961). *J. Econ. Entomol.* **54,** 796.
Huger, A. (1963). *In* "Insect Pathology, An Advanced Treatise" (E. A. Steinhaus, ed.), Vol. 1, pp. 531–575. Academic Press, New York.
Hurpin, B. (1964). *Entomophaga, Colloq. Intern. Pathol. Insectes, Paris, 1962* Mem. No. 2, pp. 41–45.
Hurpin, B. (1965). *J. Invertebrate Pathol.* **7,** 39.
Ignoffo, C. M. (1964). *J. Insect Pathol.* **6,** 318.
Ignoffo, C. M. (1965a). *J. Invertebrate Pathol.* **7,** 217.
Ignoffo, C. M. (1965b). *J. Invertebrate Pathol.* **7,** 209.
Ignoffo, C. M. (1965c). *J. Invertebrate Pathol.* **7,** 315.
Ignoffo, C. M., and Heimpel, A. M. (1965). *J. Invertebrate Pathol.* **7,** 329.
Ignoffo, C. M., Chapman, A. J., and Martin, D. F. (1965). *J. Invertebrate Pathol.* **7,** 227.
Isaacs, A. (1963). *Advan. Virus Res.* **10,** 1.
Isakova, N. P. (1954). *In* "Infections and Protozoan Diseases of Insects" (V. I. Poltev, L. V. Alexandrova, A. A. Evlakhova, and M. S. Paveljeva, eds.), Abstr. Repts., Joint Plenary Session Sericult. Apicult., Plant Protection, Vet. Sci. Sect., Vses. Akad. Sel'skokhoz. Nauk, Moscow.
Isakova, N. P. (1964). *Entomophaga, Colloq. Intern. Pathol. Insectes, Paris, 1962* Mem. No. 2, pp. 175–178.
Ishikawa, Y., and Okino, H. (1965a). *Nippon Sanshigaku Zasshi* **34,** 21.
Ishikawa, Y., and Okino, H. (1965b). *Nippon Sanshigaku Zasshi* **34,** 371.
Issi, I. V., and Maslennikova, V. A. (1964). *Entomol. Rev.* (*USSR*) (*English Transl.*) **43,** 56.
Jafri, R. H. (1965). *J. Invertebrate Pathol.* **7,** 66.
Jaques, R. P. (1961). *J. Insect Pathol.* **3,** 47.
Jaques, R. P. (1964). *J. Insect Pathol.* **6,** 251.
Jaques, R. P., and Fox, C. J. S. (1960). *J. Insect Pathol.* **2,** 17.
Kellen, W. R., and Lewallen, L. L. (1960). *J. Insect Pathol.* **2,** 305.

Kellen, W. R., and Lipa, J. J. (1960). *J. Insect Pathol.* **2,** 1.
Kerner, G. (1959). *Trans. 1st Intern. Conf. Insect Pathol. Biol. Control, Prague, 1958* pp. 169–176.
King, K. M., and Atkinson, N. J. (1928). *Ann. Entomol. Soc. Am.* **21,** 167.
Kishaba, A. N., Shankland, D. L., Curtis, R. W., and Wilson, M. C. (1962). *J. Econ. Entomol.* **55,** 211.
Kodaira, Y. (1962). *Agr. Biol. Chem. (Tokyo)* **26,** 36.
Koidsumi, K. (1957). *J. Insect Physiol.* **1,** 40.
Kovačevič, Ž. (1956). *Anz. Schädlingskunde* **29,** 97.
Kovačevič, Ž. (1958). *Anz. Schädlingskunde* **31,** 148.
Kovačevič, Ž. (1959). *Trans. 1st Intern. Conf. Insect Pathol. Biol. Control, Prague, 1958* pp. 115–119.
Kramer, J. P. (1959). *Entomophaga* **4,** 37.
Krieg, A. (1955). *Naturwissenschaften* **22,** 609.
Krieg, A. (1957). *Arch. Ges. Virusforsch.* **7,** 212.
Krieg, A. (1961a). "Grundlagen der Insektenpathologie. Viren- Rickettsien- und Bakterien-Infektionen." Steinkopff. Darmstadt.
Krieg, A. (1961b). *Mitt. Biol. Bundesanstalt Land- Forstwirtsch. Berlin-Dahlem* **103,** 3.
Krieg, A. (1965). *Mitt. Biol. Bundesanstalt Land- Forstwirtsch. Berlin-Dahlem* **115,** 51.
Krieg, A., and Herfs, W. (1963). *Z. Pflanzenkrankh. Pflanzenschutz* **70,** 11.
Krieg, A., and Herfs, W. (1964). *Entomophaga, Colloq. Intern. Pathol. Insectes, Paris, 1962* Mem. No. 2, pp. 193–195.
Krywienczyk, J. (1963). *J. Insect Pathol.* **5,** 309.
Kudler, J., Lysenko, O., and Hochmut, R. (1959). *Trans. 1st Intern. Conf. Insect Pathol. Biol. Control, Prague, 1958* pp. 73–79.
Kushner, D. J., and Harvey, G. T. (1962). *J. Insect Pathol.* **4,** 155.
Kuyama, S., and Tamura, S. (1965). *Agr. Biol. Chem. (Tokyo)* **29,** 168.
Lamanna, C., and Jones, L. (1961). *J. Bacteriol.* **81,** 622.
Langford, G. S., Vincent, R. H., and Cory, E. N. (1942). *J. Econ. Entomol.* **35,** 165.
Lecadet, M., and Martouret, D. (1962). *Compt. Rend.* **254,** 2457.
Lecadet, M., and Martouret, D. (1964). *Entomophaga, Colloq. Intern. Pathol. Insectes, Paris, 1962* Mem. No. 2, pp. 205–212.
Lecadet, M., and Martouret, D. (1965). *J. Invertebrate Pathol.* **7,** 105.
L'Héritier, Ph. (1958). *Advan. Virus Res.* **5,** 195.
L'Héritier, Ph., and Sigot, A. (1946). *Bull. Biol. France Belg.* **80,** 171.
Lysenko, O. (1959). *Trans. 1st Intern. Conf. Insect Pathol. Biol. Control, Prague, 1958* pp. 109–113.
Lysenko, O. (1963a). *In* "Insect Pathology, An Advanced Treatise" (E. A. Steinhaus, ed.), Vol. 2, pp. 1–20. Academic Press, New York.
Lysenko, O. (1963b). *J. Insect Pathol.* **5,** 83.
Lysenko, O. (1963c). *J. Insect Pathol.* **5,** 89.
Lysenko, O. (1963d). *J. Insect Pathol.* **5,** 94.
Lysenko, O. (1965). *Proc. 12th Intern. Congr. Entomol., London, 1964* p. 717.
McConnell, E., and Richards, A. G. (1959). *Can. J. Microbiol.* **5,** 161.
McEwen, F. L., Glass, E. H., Davis, A. C., and Splittstoesser, C. M. (1960). *J. Insect Pathol.* **2,** 152.
Machay, L., and Lovas, B. (1957). *Biol. Kozlemen.* **5,** 7.
MacLeod, D. M. (1963). *In* "Insect Pathology, An Advanced Treatise" (E. A. Stein-

haus, ed.) , Vol. 2, pp. 189–231. Academic Press, New York.
MacLeod, D. M., and Soper, R. S. (1965). *Proc. 12th Intern. Congr. Entomol., London, 1964* p. 724.
Madelin, M. F. (1960). *Endeavour* **19,** 181.
Madelin, M. F. (1966). *Ann. Rev. Entomol.* **11,** 423.
Marcovitch, S. (1958). *J. Tenn. Acad. Sci.* **33,** 348.
Martignoni, M. E. (1957). *Mitt. Schweiz. Anstalt Forst. Versuchswesen* **32,** 371.
Martignoni, M. E. (1962). *Proc. 23rd Biol. Colloq., Insect Physiol., Oregon State Univ.* pp. 89–110.
Martignoni, M. E. (1964a). *J. Insect Pathol.* **6,** 368.
Martignoni, M. E. (1964b). *In* "Biological Control of Insect Pests and Weeds" (P. DeBach, ed.), pp. 579–609. Chapman & Hall, London.
Martignoni, M. E., and Auer, C. (1957). *Mitt. Schweiz. Anstalt Forst. Versuchswesen* **33,** 73.
Martignoni, M. E., and Milstead, J. E. (1962). *J. Insect Pathol.* **4,** 113.
Martignoni, M. E., and Milstead, J. E. (1964). *J. Insect Pathol.* **6,** 517.
Martignoni, M. E., and Scallion, R. J. (1961). *Nature* **190,** 1133.
Martignoni, M. E., and Schmid, P. (1961). *J. Insect Pathol.* **3,** 62.
Martouret, D. (1961). *Mededel. Landbouwhogeschool Opzoekingssta. Staat Gent* **26,** 1116.
Martouret, D., and Euverte, G. (1964). *J. Insect Pathol.* **6,** 198.
Masera, E. (1948). *Actes 7th Congr. Sericult. Intern., Ales, France, 1948* pp. 551–555.
Masera, E. (1954). *Agr. Venezie* **8,** 714.
Massey, C. L. (1964). *J. Insect Pathol.* **6,** 133.
Mechalas, B. J., and Anderson, N. B. (1964). *J. Insect Pathol.* **6,** 218.
Mechalas, B. J., and Dunn, P. H. (1964). *J. Insect Pathol.* **6,** 214.
Menn, J. J. (1960). *J. Insect Pathol.* **2,** 134.
Metalnikov, S., and Metalnikov, S. S., (1935). *Ann. Inst. Pasteur* **55,** 709.
Metalnikov, S., Hergula, B., and Strail, D. M. (1930). *Intern. Corn Borer Invest. Sci. Rept.* **3,** 148.
Monsour, V., and Colmer, A. R. (1951). *Bacteriol. Proc. (Soc. Am. Bacteriologists)* **63,** 57.
Müller-Kögler, E. (1964). *Entomophaga, Colloq. Intern. Pathol. Insectes, Paris, 1962* Mem. No. 2, pp. 111–124.
Müller-Kögler, E. (1965). "Pilzkrankheiten bei Insekten." Parey, Berlin.
Neilson, M. M. (1965). *J. Invertebrate Pathol.* **7,** 306.
Neilson, M. M., and Elgee, D. E. (1960). *J. Insect Pathol.* **2,** 165.
Nickell, L. G. (1959). *Econ. Botany* **13,** 281.
Niklas, O. F. (1939). *Z. Angew. Entomol.* **26,** 63.
Nirula, K. K. (1957). *J. Econ. Entomol.* **50,** 767.
Norris, J. R., and Burges, H. D. (1963). *J. Insect Pathol.* **5,** 460.
Northrup, Z. (1914). *Mich. Agr. Coll. Expt. Sta. Tech. Bull. 18.*
Novák, V., and Samšiňaková, A. (1964). *Entomophaga, Colloq. Intern. Pathol. Insectes, Paris, 1962* Mem. No. 2, pp. 133–135.
Olthof, Th. H. A., and Estey, R. H. (1963). *Nature* **197,** 514.
Ossowski, L. L. J. (1957). *Ann. Appl. Biol.* **45,** 81.
Ossowski, L. L. J. (1960). *J. Insect Pathol.* **2,** 35.
Paillot, A. (1925). *Rev. Gen. Sci. Bull. Soc. Philomath.* **36,** 206.
Patel, N. G., and Cutkomp, L. K. (1961). *J. Econ. Entomol.* **54,** 773

Pepper, R. E., and Costilow, R. N. (1964). *J. Bacteriol.* **87,** 303.
Phillips, C. M., Bucher, G. E., and Stephens, J. M. (1953). *Can. Entomologist* **85,** 8.
Pimentel, D., and Shapiro, M. (1962). *J. Insect Pathol.* **4,** 77.
Pospelov, V. P. (1938). *In* "Summary of the Scientific Research Work of the Institute of Plant Protection for the Year 1936," Pt. III, pp. 64–67. Lenin Acad. Agr. Sci., Leningrad. (1939). *Rev. Appl. Entomol.* **A27,** 306.
Pospelov, V. P. (1940). *In* "The Beet Weevil and Its Control" (N. M. Kulagin and G. K. Pyatnitzkiĭ, eds.), pp. 45–46. Vses. Akad. Sel'skokhoz. Nauk, Moscow. (1942). *Rev. Appl. Entomol.* **A30,** 66.
Pramer, D. (1965). *Bacteriol. Rev.* **29,** 382.
Pridham, T. G., St. Julian, G., Jr., Adams, G. L., Hall, H. H., and Jackson, R. W. 1964). *J. Insect Pathol.* **6,** 204.
Pyatnitzkiĭ, G. K. (1942). *In* "The Beet Weevil and Its Control" (N. M. Kulagin and G. K. Pyatnitzkiĭ, eds.), pp. 25–37. Vses. Akad. Sel'skokhoz. Nauk, Moscow. (1942). *Rev. Appl. Entomol.* **A30,** 64.
Raun, E. S. (1963). *Iowa State J. Sci.* **38,** 141.
Rhodes, R. A. (1965). *Bacteriol. Rev.* **29,** 373.
Rivers, C. F. (1959). *Trans. 1st Intern. Conf. Insect Pathol. Biol. Control, Prague, 1958* pp. 205–210.
Roberts, D. W. (1964). Ph.D. Thesis, Univ. of California, Berkeley, California.
Rockwood, L. P. (1950). *J. Econ. Entomol.* **43,** 704.
Roegner-Aust, S. (1949). *Z. Angew. Entomol.* **31,** 1.
Rosický, B. (1951). *Vestn. Cesk. Zool. Spolecnosti* **15,** 219.
Sager, S. M. (1960). *J. Insect Pathol.* **2,** 307.
Schaefer, E. E. (1936). *Union S. Africa Dept. Agr. Sci. Bull.* **160.**
Schaerffenberg, B. (1957a). *Anz. Schädlingskunde* **30,** 69.
Schaerffenberg, B. (1957b). *Z. Angew. Entomol.* **41,** 395.
Schaerffenberg, B. (1964). *J. Insect Pathol.* **6,** 8.
Schmidt, L. (1956). *Glasnik Sumske Pokuse* **12,** 105.
Shands, W. A., Simpson, G. W., and Hall, I. M. (1963). *Maine Agr. Expt. Sta. Bull.* **T6.**
Shvetsova, O. I. (1950). *Mikrobiologiya* **19,** 532.
Shvetsova, O. I., and Ts'ai, Hsü-yü. (1962). *Entomol. Obozrenie* **41,** 486.
Sidor, Ĉ. (1959). *Ann. Appl. Biol.* **47,** 109.
Sidor, Ć. (1965). *Sumarskog Lista Broj* **9–10,** 381.
Smirnoff, W. A. (1959). *Can. Entomologist* **91,** 246.
Smirnoff, W. A. (1960). *Can. Entomologist,* **92,** 957.
Smirnoff, W. A. (1961). *J. Insect Pathol.* **3,** 29.
Smirnoff, W. A. (1963a). *J. Insect Pathol.* **5,** 104.
Smirnoff, W. A. (1963b). *J. Insect Pathol.* **5,** 242.
Smirnoff, W. A. (1963c). *J. Insect Pathol.* **5,** 389.
Smirnoff, W. A. (1964). *Entomophaga, Colloq. Intern. Pathol. Insectes, Paris, 1962* Mem. No. 2, pp. 249–254.
Smirnoff, W. A. (1965a). *J. Invertebrate Pathol.* **7,** 71.
Smirnoff, W. A. (1965b). *J. Invertebrate Pathol.* **7,** 387.
Smirnoff, W. A., and Hutchison, P. M. (1965). *J. Invertebrate Pathol.* **7,** 273.
Smirnoff, W. A., and MacLeod, C. F. (1961). *J. Insect Pathol.* **3,** 266.
Smirnoff, W. A., and MacLeod, C. F. (1964). *J. Insect Pathol.* **6,** 537.
Smith, K. M. (1959). *In* "The Viruses" (F. M. Burnett and W. M. Stanley, eds.), Vol. 3, pp. 369–392. Academic Press, New York.

Smith, K. M. (1963). *In* "Insect Pathology, An Advanced Treatise" (E. A. Steinhaus, ed.), Vol. 1, pp. 457–497. Academic Press, New York.
Smith, O. J., Hughes, K. M., Dunn, P. H., and Hall, I. M. (1956). *Can. Entomologist* **88,** 507.
Stairs, G. R. (1964a). *J. Insect Pathol.* **6,** 373.
Stairs, G. R. (1964b). *Virology* **24,** 520.
Stairs, G. R. (1965a). *J. Invertebrate Pathol.* **7,** 427.
Stairs, G. R. (1965b). *Can. J. Microbiol.* **11,** 509.
Steinhaus, E. A. (1948). *J. Econ. Entomol.* **41,** 859.
Steinhaus, E. A. (1949). "Principles of Insect Pathology." McGraw-Hill, New York.
Steinhaus, E. A. (1954a). *Science* **120,** 186.
Steinhaus, E. A. (1954b). *Hilgardia* **23,** 197.
Steinhaus, E. A. (1956a). *Hilgardia* **26,** 107.
Steinhaus, E. A. (1956b). *J. Agr. Food Chem.* **4,** 676.
Steinhaus, E. A. (1957a). *Ann. Rev. Microbiol.* **11,** 165.
Steinhaus, E. A. (1957b). *J. Econ. Entomol.* **50,** 715.
Steinhaus, E. A. (1958a). *Proc. 10th Intern. Congr. Entomol., Montreal, 1956* **4,** 725.
Steinhaus, E. A. (1958b). *Ecology* **39,** 503.
Steinhaus, E. A. (1959a). *Hilgardia* **28,** 351.
Steinhaus, E. A. (1959b). *J. Econ. Entomol.* **52,** 506.
Steinhaus, E. A. (1960a). *J. Insect Pathol.* **2,** 225.
Steinhaus, E. A. (1960b). *Bacteriol. Rev.* **24,** 365.
Steinhaus, E. A., and Dineen, J. P. (1959). *J. Insect Pathol.* **1,** 171.
Steinhaus, E. A., and Leutenegger, R. (1963). *J. Insect Pathol.* **5,** 266.
Steinhaus, E. A., and Thompson, C. G. (1949). *J. Econ. Entomol.* **42,** 301.
Steinkraus, K. H., and Provvidenti, M. L. (1958). *J. Bacteriol.* **75,** 38.
Steinkraus, K. H., and Tashiro, H. (1955). *Science* **121,** 873.
Stelzer, M. J. (1965). *J. Invertebrate Pathol.* **7,** 122.
Stephens, J. M. (1952). *Can. J. Zool.* **30,** 30.
Stephens, J. M. (1957a). *Can. Entomologist* **89,** 94.
Stephens, J. M. (1957b). *Can. J. Microbiol.* **3,** 995.
Stephens, J. M. (1959). *Can. J. Microbiol.* **5,** 203.
Stephens, J. M. (1963a). *In* "Insect Pathology, An Advanced Treatise" (E. A. Steinhaus, ed.), Vol. 1, pp. 273–297. Academic Press, New York.
Stephens, J. M. (1963b). *J. Insect Pathol.* **5,** 61.
Stevenson, J. P. (1959). *J. Insect Pathol.* **1,** 232.
Sulzdalskaya, M. V. (1954). *In* "Infections and Protozoan Diseases of Insects" (V. I. Poltev, L. V. Alexandrova, A. A. Evlakhova, and M. S. Paveljeva, eds.), Abstr. Repts., Joint Plenary Session Sericult. Apicult., Plant Protection, Vet. Sci. Sect. Vses. Akad. Sel'skokhoz. Nauk, Moscow.
Surany, P. (1960). *South Pacific Comm. Tech. Paper* No. 128.
Sussman, A. S. (1951). *Mycologia* **43,** 338.
Sussman, A. S. (1952). *Mycologia* **44,** 493.
Sylvester, C. J., and Costilow, R. N. (1964). *J. Bacteriol.* **87,** 114.
Takahashi, N., Suzuki, A., Miyamoto, S., Mori, R., and Tamura, S. (1963). *Agr. Biol. Chem.* (*Tokyo*) **27,** 583.
Talalaev, E. V. (1958). *Entomol. Rev.* (*USSR*) (*English Transl.*) **37,** 557.
Tamashiro, M. (1960). *J. Insect Pathol.* **2,** 209.
Tamura, S., Takahashi, N., Miyamoto, S., Mori, R., Suzuki, S., and Nagatsu, J. (1963). *Agr. Biol. Chem.* (*Tokyo*) **27,** 576.

Tamura, S., Kuyama, S., Kodaira, Y., and Higashikawa, S. (1964). *Agr. Biol. Chem. (Tokyo)* **28,** 137.
Tanada, Y. (1953). *Proc. Hawaiian Entomol. Soc.* **15,** 235.
Tanada, Y. (1956a). *J. Econ. Entomol.* **49,** 52.
Tanada, Y. (1956b). *J. Econ. Entomol.* **49,** 320.
Tanada, Y. (1959a). *Ann. Rev. Entomol.* **4,** 277.
Tanada, Y. (1959b). *J. Insect Pathol.* **1,** 215.
Tanada, Y. (1961a). *J. Agr. Vet. Chem.* **2,** 114, 157.
Tanada, Y. (1961b). *J. Insect Pathol.* **3,** 310.
Tanada, Y. (1963). *In* "Insect Pathology, An Advanced Treatise" (E. A. Steinhaus, ed.), Vol. 2, pp. 423–475. Academic Press, New York.
Tanada, Y. (1964). *In* "Biological Control of Insect Pests and Weeds" (P. DeBach, ed.), pp. 548–578. Chapman & Hall, London.
Tanada, Y., and Chang, G. Y. (1960). *J. Insect Pathol.* **2,** 201.
Tanada, Y., and Chang, G. Y. (1964). *J. Insect Pathol.* **6,** 500.
Tanada, Y., and Reiner, C. E. (1960). *J. Insect Pathol.* **2,** 230.
Tanada, Y., and Reiner, C. E. (1962). *J. Insect Pathol.* **4,** 139.
Tanada, Y., and Tanabe, A. M. (1965). *J. Invertebrate Pathol.* **7,** 184.
Tanada, Y., Tanabe, A. M., and Reiner, C. E. (1964). *J. Insect Pathol.* **6,** 439.
Tashiro, H. (1957). *J. Econ. Entomol,* **50,** 350.
Tashiro, H., and White, R. T. (1954). *J. Econ. Entomol.* **47,** 1087.
Telenga, N. A. (1959). *Trans. 1st Intern. Conf. Insect Pathol. Biol. Control, Prague,* 1958 pp. 155–168.
Théodoridès, J. (1952). *Vie Milieu* **3,** 288.
Thompson, C. G. (1958). *Proc. 10th Intern. Congr. Entomol., Montreal, 1956* **4,** 693.
Thompson, C. G. (1959). *J. Insect Pathol.* **1,** 189.
Thompson, C. G., and Steinhaus, E. A. (1950). *Hilgardia* **19,** 411.
Thomson, H. M. (1958). *Can. J. Zool.* **36,** 309.
Toumanoff, C. (1931). *Ann. Parasitol. Humaine Comparee* **9,** 462.
Toumanoff, C. (1959). *Ann. Inst. Pasteur* **96,** 108.
Toumanoff, C. (1960). *Ann. Inst. Pasteur* **98,** 367.
Toumanoff, C., and Lapied, M. (1954). *Ann. Inst. Pasteur* **87,** 370.
Toumanoff, C., and Vago, C. (1952). *Compt. Rend.* **235,** 1715.
Trager, W. (1935). *J. Exptl. Med.* **61,** 501.
Ullyett, G. C., and Schonken, D. B. (1940). *Union S. Africa Dept. Agr. Forestry Sci. Bull.* **218.**
Vago, C. (1951). *Rev. Can. Biol.* **10,** 299.
Vago, C. (1952). *Congr. Intern. Patol. Compar., 6th, Madrid, 1952* **1,** 121.
Vago, C. (1953). *Intern. Congr. Microbiol., 6th, Rome, 1953* **5,** 556.
Vago, C. (1956). *Entomophaga* **1,** 82.
Vago, C. (1963). *In* "Insect Pathology, An Advanced Treatise" (E. A. Steinhaus, ed.), Vol. 1, pp. 339–379. Academic Press, New York.
Vago, C., and Kurstak, E. (1965). *J. Microbiol. Serol.* **31,** 282.
van den Bosch, R., and Stern, V. M. (1962). *Ann. Rev. Entomol.* **7,** 367.
Vaňková, J. (1957). *Folia Biol. (Prague)* **3,** 175.
Vaňková, J. (1964). *Entomophaga, Colloq. Intern. Pathol. Insectes, Paris, 1962* Mem. No. 2, pp. 271–291.
Vaughn, J. L., and Faulkner, P. (1963). *Virology* **20,** 484.
Veber, J. (1964). *Entomophaga, Colloq. Intern. Pathol. Insectes, Paris, 1962* Mem. No. 2, pp. 403–405.

Wada, Y. (1957). *Japan. J. Ecol.* **7,** 90.
Wagner, R. R. (1961). *Bacteriol. Rev.* **25,** 100.
Wallis, R. C. (1957). *J. Econ. Entomol.* **50,** 580.
Weiser, J. (1956). *Z. Pflanzenkrankh. Pflanzenschutz* **63,** 625.
Weiser, J. (1957). *Z. Angew. Entomol.* **40,** 509.
Weiser, J. (1961). *Monograph. Angew. Entomol.* **17.**
Weiser, J. (1963). *In* "Insect Pathology, An Advanced Treatise" (E. A. Steinhaus, ed.), Vol. 2, pp. 291–334. Academic Press, New York.
Weiser, J. (1965). *Proc. 12th Intern. Congr. Entomol., London, 1964* p. 726.
Weiser, J., and Lysenko, O. (1956). *Cesk. Mikrobiol.* **1,** 216.
Weiser, J., and Veber, J. (1957). *Z. Angew. Entomol.* **40,** 55.
Welch, H. E. (1963). *In* "Insect Pathology, An Advanced Treatise" (E. A. Steinhaus, ed.), Vol. 2, pp. 363–392. Academic Press, New York.
Welch, H. E. (1965). *Ann. Rev. Entomol.* **10,** 275.
Welch, H. E., and Briand, L. J. (1961). *Can. Entomologist* **93,** 759.
Wellington, W. G. (1962). *J. Insect Pathol.* **4,** 285.
White, R. T. (1940). *J. Econ. Entomol.* **33,** 303.
White, R. T. (1943). *J. N. Y. Entomol. Soc.* **51,** 213.
White, R. T., and Dutky, S. R. (1940). *J. Econ. Entomol.* **33,** 306.
Wolfenbarger, D. A. (1964). *J. Econ. Entomol.* **57,** 732.
Wolfenbarger, D. A. (1965). *J. Invertebrate Pathol.* **7,** 33.
Xeros, N. (1954). *Nature* **174,** 562.
York, G. T. (1958). *Iowa State Coll. J. Sci.* **33,** 123.

3 ELECTROMAGNETIC ENERGY

Stuart O. Nelson

ELECTRICAL CONDITIONING, PRESERVATION, AND PROTECTION INVESTIGATIONS
FARM ELECTRIFICATION RESEARCH BRANCH
AGRICULTURAL ENGINEERING RESEARCH DIVISION
AGRICULTURAL RESEARCH SERVICE
UNITED STATES DEPARTMENT OF AGRICULTURE AND
UNIVERSITY OF NEBRASKA, LINCOLN, NEBRASKA

I. THE ELECTROMAGNETIC SPECTRUM

Various forms of electromagnetic energy have long been recognized to offer possible means for insect control. Recent emphasis on the development of nonchemical control methods has given impetus to more complete exploration of the electromagnetic spectrum in an attempt to discover principles which might be useful in developing new pest control methods.

The electromagnetic spectrum is made up of several kinds of radiation, often thought of as separate types of energy (Fig. 1). All of these radiations—radio-frequency, infrared, visible, ultraviolet, x-ray, and γ-ray—are electromagnetic in nature and differ from one another only in wave-

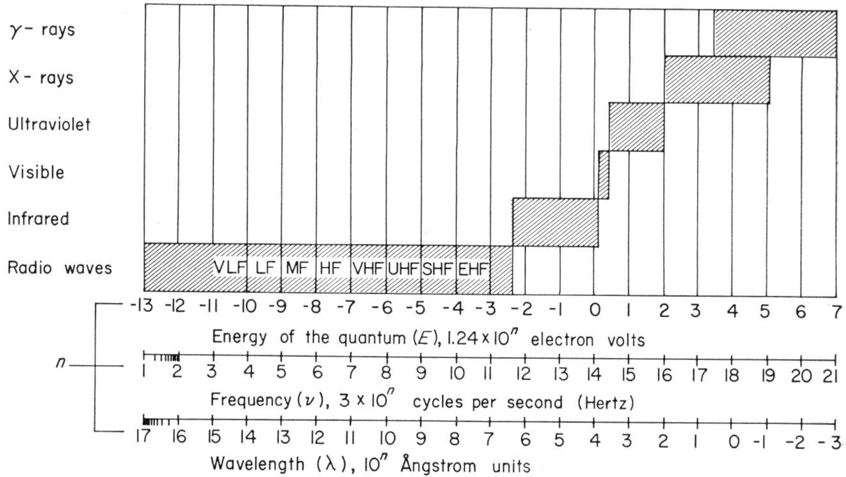

FIG. 1. Electromagnetic spectrum showing approximate ranges of the radiation.

length and related characteristics. All travel through free space as electromagnetic waves with the velocity of light. For any particular wavelength there is an associated frequency of vibration of the electric and magnetic fields defined by the relationship:

$$\nu = c/\lambda \tag{1}$$

where ν is the frequency in cycles per second or hertz (Hz), c is a fundamental physical constant, the velocity of light in vacuum (2.9979×10^8 meters per second), and λ is the wavelength in the same length unit used for the expression of c. There is also an energy associated with each quantum of radiation,

$$E = h\nu \tag{2}$$

where h is another fundamental physical constant, Planck's constant (6.625×10^{-27} erg-second).

Scales are shown in Fig. 1 for energy of the quantum, frequency, and wavelength. The electron volt, a more convenient energy unit than the erg in dealing with high energy radiation, is shown on the energy scale. An electron volt (eV) is the amount of energy acquired by an electron when accelerated by a difference of potential of 1 volt (V). One eV $= 1.602 \times 10^{-12}$ erg. Although angstrom units are shown as the unit for the wavelength scale, other more convenient units are used for particular portions of the spectrum. Meters and centimeters (cm) are used in the radio-frequency range, whereas microns are commonly used in the infrared region, and millimicrons (nanometers) are used as well as angstrom units in the visible and ultraviolet regions. Energies of the quanta are more commonly used than wavelength to describe the ionizing radiations, i.e., high energy ultraviolet, x-rays, and γ-rays.

The scales in Fig. 1 for energy, frequency, and wavelength are logarithmic, and the integers indicated along the scales are exponents of 10. For example, in the radio-frequency region, the very-high-frequency (VHF) range is divided from the ultra-high-frequency (UHF) range at 3×10^8 Hz or 300 megahertz (MHz). The wavelength at this frequency is 10^{10} Å or 1 m, and the energy of the quantum is 1.24×10^{-6} eV. The visible portion of the spectrum shown in Fig. 1 spans the region from somewhat below 4×10^3 to a little below 8×10^3 or about 4000–8000 Å.

The effects of electromagnetic energy on living matter have been studied in detail for certain phenomena such as photosynthesis and the absorption of ionizing radiation, but much still remains unknown concerning the interaction of all types of electromagnetic radiation with biological material. Generally, however, the longer wavelength radiations produce heating effects and the shorter wavelength radiations produce chemical effects, including ionization of the atoms of the absorbing media.

The principles and possibilities of insect control using electromagnetic energy will now be considered for each portion of the spectrum, taking these in the order of increasing quantum energy.

II. RADIO FREQUENCIES

A. General Theoretical Aspects

Beginning at the low energy end of the electromagnetic spectrum, radiation of energy does not become efficient or feasible, because of wavelength and antenna-size relationships, until we reach the radio frequencies. Extremely low frequency radio waves exist in nature, but unless they can be generated easily they cannot be considered for pest control purposes.

The radio-frequency (RF) portion of the spectrum has been extensively explored in the development of radio communications, radar, navigation aids, etc. Therefore sources of RF energy, usually electron-tube oscillators, are available in a wide variety of types and capabilities. RF energy has been employed for many years in the medical field for therapeutic heating of body tissues in the practice of diathermy. Since absorption of RF energy by biological material results in heating of the tissues, its use for pest control has been frequently considered.

Based on RF heating only, control of insects by radiation of RF energy into any sizeable space is impractical (Baker *et al.,* 1956). Calculation of the power required for a special case revealed that tremendous amounts of energy are needed for a range of only a few feet, with power requirements increasing as the square of the distance from the radiating antenna. Concentration of RF energy into a narrow beam reduces the energy requirement, but it would still be substantial.

RF energy can be applied to materials without actually radiating the energy. Materials can be exposed to RF electric fields between electrodes connected to RF power oscillators and will absorb energy directly from the field. For a homogeneous dielectric material, the power dissipated in the dielectric is given by the following relationship:

$$P = K_1 f E^2 \epsilon'' \tag{3}$$

where $K_1 = 0.556$ when P is power in watts/cm^3 (W/cm^3), f is frequency in MHz, E is the electric field intensity in kilovolts/cm (kV/cm), and ϵ'' is the dielectric loss factor of the material.* If P is expressed in W/in.^{3} and E in kV/in., $K_1 = 1.41$. The amount of dielectric heating which occurs during exposure of a material to an RF field therefore depends not only upon factors which can be controlled, such as frequency and field intensity, but also on the dielectric properties of the material, over which we can exercise little or no control. The dielectric properties of most biological materials are frequency-dependent and vary with moisture content and temperature.

The amount of heating that occurs when a dielectric material absorbs RF energy depends upon the power dissipated or rate of energy absorption, and the specific heat and density of the material. If no vaporization of moisture or other changes in state occur and heat losses from the dielectric are negligible, the heating rate in degrees per second may be expressed as follows:

$$dT/dt = K_2 P / c\rho \tag{4}$$

* ϵ'' corresponds to the relative dielectric loss factor of the rationalized mks system of units. The loss factor is the imaginary component of the complex permittivity, $\epsilon = \epsilon' - j\epsilon''$, where ϵ' is the real permittivity or dielectric constant.

where $K_2 = 0.239$ when T is temperature in degrees C, P is in W/cm^3, c is the specific heat, and ρ is the density in gm/cm^3. If P is expressed in W/in.^{3}, T in degrees F, and ρ in lb/ft^3, $K_2 = 1.64$.

B. Theory and Experiments in RF Insect Control

1. *Background*

Several reviews have been published on studies of RF electric fields for insect control (Ark and Parry, 1940; Proctor and Goldblith, 1951; Frings, 1952; Thomas, 1952; Thomas and White, 1959a,b; Peredel'skii, 1956; Nelson and Whitney, 1960; van den Bruel *et al.,* 1960a,b; Nelson, 1962; Watters, 1962). Important principles involving application of this type of energy have also been treated in the literature (Frings, 1952; Thomas, 1952; Thomas and White, 1959a,b; Nelson and Whitney, 1960; Whitney *et al.,* 1961).

Major interest in applying RF energy to insect control has involved treatment of grain, foodstuffs, and wood. The basis for this interest stems from the nature of absorption of RF energy by materials in a high-frequency electric field [Eq. (3)]. For certain combinations of hosts and insects, their respective dielectric properties are favorable for differential or selective absorption of energy, and the insects can be killed without damaging the host material (Thomas, 1952; Nelson and Whitney, 1960; Whitney *et al.,* 1961). If the thermal tolerance of the host is sufficiently lower than that of the infesting organism, heating by conventional means may be a possible control measure. RF heating, however, can be advantageous if the pest organism can be selectively heated, because less energy is required to accomplish control. In addition, RF dielectric heating is much more uniform and faster than conventional heating.

2. *Selective RF Heating*

That differential or selective heating may occur can be seen using a simplified analysis of the situation for stored-grain insects (Nelson and Whitney, 1960; Whitney *et al.,* 1961). The expected heating rate for any substance exposed to a high-frequency electric field is given by Eq. (4), subject to the indicated restrictions. Substituting Eq. (3) into Eq. (4) we have:

$$dT/dt = KfE^2\epsilon''/c\rho \qquad (5)$$

where K is the product of K_1 and K_2 for appropriate units desired. When a quantity of infested grain is exposed to a high-frequency electric field, the insects and the grain kernels will be subjected to the same frequency, so f in Eq. (5) will be the same for both materials. The proportionality constant K will also be the same for both insects and grain. All other factors determining the heating rate may be different for the insects and the grain. The dielectric loss factor, ϵ'', has been measured for some insects (Nelson and Whitney, 1960) and for a variety of grains (Nelson, 1965). Values for the density and specific heat of wheat are available in the literature, and reasonably good estimates can be made for the density and specific heat of the insects. The value of E for the grain can be calculated from known or measured quantities for any particular application. For example, if the grain is exposed in polystyrene boxes with parallel top and bottom between parallel, flat-plate electrodes in contact with the top and bottom of the box, the field intensity in the grain is given as follows:

$$E = V/(d_1 + d_2\epsilon_1/\epsilon_2) \qquad (6)$$

where V is the RF potential difference between the electrodes, d_1 is the depth of the grain measured perpendicular to the plane of the electrodes, d_2 is the combined thickness of the lid and bottom of the polystyrene box, and ϵ_1 and ϵ_2 are, respectively, the permitivities or complex dielectric constants of the grain and polystyrene.

The value of E for the insects is more difficult to obtain, for it depends not only upon the geometry of the situation, but also upon the dielectric constants of both the grain and the insects. We can estimate the value of E for the insects by considering the case of a sphere embedded in an infinite medium (Thomas, 1952; Nelson and Whitney, 1960). For this mathematical model, the field intensity in the sphere is

$$E_1 = E_2[1 - (\epsilon_1 - \epsilon_2)/(2\epsilon_2 + \epsilon_1)] \qquad (7)$$

where the subscript 1 refers to the sphere and the subscript 2 refers to the medium. Separate measurements at 40 MHz and about 75°F yielded permittivities of 6.6 for adult rice weevils and 4.2 for wheat with 13% moisture (Nelson and Whitney, 1960). Assigning the value for the insects to the sphere and the value for the wheat to the medium in Eq. (7) would indicate that the field intensity for the insects is about 0.84 times that for the wheat.* Values obtained for the dielectric loss factor, ϵ'', at 40 MHz for rice weevils and wheat were 2.2 and 0.5, respectively. The bulk density of the wheat and that of the rice weevil are about the same. The specific heat of the wheat is about 0.4. Estimating the specific heat of the rice weevils to be 0.7 and using all of these figures in conjunction with

* Since the measurements of permittivity and loss factor were made on insects in bulk, the values obtained apply in effect to the insect and its share of the surrounding space. Where "field intensity in the insect" is referred to in this discussion, it must be kept in mind that this intensity is a sort of average for the insect and its associated space envelope.

Eq. (5), we can estimate that the heating rate for rice weevils treated in wheat is about 1.8 times the heating rate for the wheat.

Analysis of experiments conducted to test this theory of differential heating showed that the differential heating factor for rice weevils in wheat varied from 2.3 to about 1.2 through a range of exposures which resulted in insect mortality ranging from 20 to 80% (Nelson and Whitney, 1960). The experimental results support the prediction that the insects should heat at a faster rate than the grain, and indicate that the differential heating factor decreases with increasing temperature in the range studied.

Theoretical analyses of factors influencing selective or differential heating of a sphere embedded in an infinite medium show that such phenomena can be expected only when the dimensions of the absorbing particles in a different medium are at least of the order of 1 mm (Schwan and Piersol, 1954) or 0.4 mm (Thomas, 1952). For a comprehensive theoretical treatment of differential heating and frequency-dependent characteristics of the differential heating factor, the reader is referred to the work of Thomas (1951, 1952).

Obviously, the electrical characteristics of the host medium determine to a high degree the success with which RF electric fields may be employed for controlling infesting insects. In general, the RF conductivity of the host material, which is proportional to frequency, and the dielectric loss factor must be lower than those of the insect for any differential heating to be possible (Thomas, 1952). In experiments at frequencies between 1 and 27 MHz with fruits and vegetables, Frings (1952) concluded that there was little possibility of controlling insects in these materials without also heating the plant material. Webber *et al.* (1946), working at a frequency of 11 MHz, found that stored-product insects in packaged flour and dry cereals could be controlled, but they experienced difficulty with arcing and charring of the packaging material and contents in some instances. Control of insects infesting wood as well as insects and mites infesting grain has been achieved by a number of investigators.

3. *Wood-Infesting Insects*

Andreev and Balkashin (1935) reported control of borers in wood using RF treatments of 3–5 minutes. Thomas and White (1959a), working with RF fields of 76 and 37 MHz, showed that powder-post beetles, *Lyctus brunneus* (Stephens), in oak blocks could be controlled with exposures on the order of 1 minute, but found that control could not be achieved with wood temperatures any lower than those required in normal kiln treatment. Conduction of heat from the insects into the wood tends to reduce differential temperatures when exposures of this length are used.

Shorter exposures can be used with more intense fields, but the lethal temperature required increases as the length of exposure is decreased. While RF heating therefore is not competitive with the conventional heating techniques generally used in controlling wood-infesting insects, there may be applications which could be profitably handled using RF treatment. Such applications, suggested by Thomas and White (1959a,b) and Thomas (1960), include treatment of dimensioned stock while moving through factories, particularly where RF equipment is already in use for curing glued joints, and for treatment of paneling *in situ,* where one side of the paneling is inaccessible. These studies showed that an electrode arrangement could be employed on one side only to utilize stray-field heating for effective insect control to a depth of about 1 in. Experiments using microwave equipment operating at a frequency of 2425 MHz (wavelength, 12.4 cm) have also shown promise for wood-borer control according to van den Bruel *et al.* (1960b). At such frequencies, the RF energy can be radiated and directed into the wooden material from one side. Successful treatment of infested woodwork in a house has been reported using this method (Bollaerts *et al.,* 1961).

4. *Mites*

Early studies by Russian investigators showed that grain and flour mites were controlled by exposures to RF electric fields (reviewed by Ark and Parry, 1940). Exposures effective for controlling bean weevils also killed mites when treated in seed (Vishniakova, 1934; Evreinov, 1935). Exposures at frequencies of 7.5 and 27 MHz were found effective in killing several species of mites in wheat (Andreev and Balkashin, 1935).

5. *Stored-Grain Insects*

a. General. Stored-grain insect control using RF electric fields has received considerable attention. The reader is referred to the reviews already cited in Section II,B,1 for accounts of these studies. Established principles are summarized here, however, as they relate to stored-grain insect control possibilities using RF electric fields.

b. Developmental Stages. For most insect species, the adult is more susceptible to control by RF exposure than are the immature stages (Mouromtseff, 1933; Vishniakova, 1934; Kuznetzova, 1937; Frings, 1952; Baker *et al.,* 1956; Nelson and Whitney, 1960; Whitney *et al.,* 1961; Nelson *et al.,* 1966). This phenomenon was also noted by Headlee and Burdette (1929) in pioneering research on effects of high-frequency fields on insects. Frings (1952) noted that older adults were more susceptible to RF fields than younger adults. Indications that young rice weevil, *Sitophilus oryzae* (L.), adults are more resistant than older adults have also been noted in stored-grain insect experiments (Nelson *et al.*, 1966).

Complete mortality of adult rice weevils in wheat can be obtained using RF treatments of a few seconds which raise the grain temperature to about 100°F (38°C). Since normally temperatures of this order alone are not lethal to the insects even when held for several hours, differential heating with the insects selectively absorbing energy at a faster rate is a plausible explanation. For immature rice weevils, temperatures of approximately 140°F (60°C) in RF-treated grain are required for complete mortality. For sublethal exposures, pupal mortality of this species appears to be higher than mortality of the larval and egg stages. Complete mortality of all immature stages requires about the same exposure. Mild exposures of wheat infested with eggs of the rice weevil, granary weevil [*Sitophilus granarius* (L.)], and the lesser grain borer [*Rhyzopertha dominica* (F.)] appear to stimulate subsequent emergence of adults from the treated wheat (Nelson *et al.*, 1966). Differences in susceptibility between adult and larval stages of the confused flour beetle, *Tribolium confusum* Jacquelin duVal, are not marked when treated in wheat shorts (Whitney *et al.*, 1961). Larvae of the cadelle, *Tenebroides mauritanicus* (L.), were found more susceptible than the adult for exposures producing less than 100% mortality of either stage, but both stages required the same exposure for complete mortality (Nelson *et al.*, 1966).

In attempting to explain differences in susceptibility between the adult and immature stages, several factors can be examined. Headlee and Burdette (1929) attributed differences in susceptibility to differences in the specialization of the nerve tissue. Frings (1952) suggested that high current concentrations in the legs of adults might explain the greater susceptibility of the adult insects to high-frequency electric fields (Section II,B,5,d). Since the immature forms are smaller than the adults, the degree of differential heating may be less or even nonexistent if the theoretical minimum size considerations are applicable (Section II,B,2). For internal forms, partial shielding effects of the kernel may be a factor (Section II,B,5,f).

c. Species. Species vary in their response to RF treatment. Some species are more susceptible than others, and as already indicated, developmental stages of the same species respond differently. Based on adult mortality data, comparison of several stored-grain insect species treated in wheat at a frequency of 39 MHz and at a field intensity of 3 kV/in. show the following rank in order of increasing resistance: rice and granary weevils, saw-toothed grain beetle, *Oryzaephilus surinamensis* (L.), confused and red flour beetles, *Tribolium castaneum* (Herbst), dermestids, *Trogoderma parabile* Beal, cadelle, and lesser grain borer. Another interspecies difference concerns the degree of delayed mortality following treatment. A substantial increase in mortality is noted with rice and granary weevils between mortality counts taken 1 day and 1 week after treatment (see Fig. 3), whereas practically no change in mortality occurs thereafter due to the treatment. The lesser grain borer also exhibits substantial delayed mortality, but such mortality increases more during the second week after treatment than during the first week. The confused flour beetle and red flour beetle exhibit less delayed mortality, and most of this occurs between 1 and 3 weeks after treatment (Whitney *et al.,* 1961; Nelson *et al.,* 1966).

d. Injury. Frings (1952) noted and studied in some detail the knockdown of insects when exposed to high-frequency electric fields and concluded that rapid heating in the legs could explain the knockdown effect. Injuries noted by Whitney *et al.* (1961), in stored-grain insects following RF treatment in wheat, lend support to this theory, for injuries were noted in the appendages, particularly the joints of the legs. Injury of the histoblasts has also been observed recently in RF-treated larvae of the yellow mealworm, *Tenebrio molitor* L. (Kadoum, 1963). Larvae which had received sublethal exposures produced adults with badly deformed and missing legs. Similar results were noted by Vande Berg (1965) in wax moths, *Galleria mellonella* (L.), developing from RF-treated larvae.

e. Reproduction. Experiments with rice weevils and confused flour beetles showed that adults which survived RF exposures insufficient for complete mortality were capable of reproducing (Whitney *et al.,* 1961). The more severe treatments, however, greatly reduced the number of progeny produced by a given number of surviving adults. Studies with lesser grain borers showed that the reproduction rate was lowered when adults exposed to RF treatment suffered greater than about 50% mortality (Nelson *et al.,* 1966).

f. Host Medium. Properties of the host medium influence the effectiveness of RF electric fields in controlling stored-grain insects. Confused flour beetles in wheat shorts require treatments producing much higher temperatures in the host medium for control than when the same insects are treated in wheat (Whitney *et al.,* 1961). RF treatment producing grain temperatures under 120°F (49°C) in wheat provided 100% mortality for adult flour beetles, whereas some insects survived treatments resulting in 140°F (60°C) temperatures in wheat shorts. This difference has been explained using measured values of the dielectric properties of the insects and host media in Eqs.(5) and (7). Resulting calculations indicate that no favorable differential heating can be expected for the flour beetles in wheat shorts (Nelson and Whitney, 1960; Whitney *et al.,* 1961).

Because the field intensity to which the insects are subjected depends upon geometrical or spatial factors as well as the dielectric properties of the insects and host material, it appears that the particle size and shape of granular host material might influence lethal exposure levels. Experiments with granary weevils in wheat and corn of the same moisture content revealed that the insects survived better in wheat than in corn at mortality levels below about 80% (Nelson and Kantack, 1966). For higher mortalities there was little difference. Rice weevil adults exposed to RF fields in different sizes of glass beads survived better in smaller-sized glass particles (Nelson *et al.,* 1966).

Early investigations by Pyenson (1933) revealed that certain materials surrounding insects tend to shield them when exposed to high-frequency electric fields. While tissues of grain kernels are not very good electrical conductors, it appears that insect forms inside kernels of grain might be partially shielded because of geometrical factors and differences in dielectric properties which influence the field intensity distribution. Experiments with both rice weevil adults and lesser grain borer adults treated inside and outside wheat kernels indicated that insects treated outside the kernels suffered somewhat higher mortalities than insects of the same age which were exposed while inside the kernels (Nelson *et al.,* 1966). Therefore, it seems that partial shielding by the kernel might be an additional factor to consider in explaining the longer exposures required for control of immature internal forms of some stored-grain insects.

Mortalities of rice weevil adults treated in wheat of 11.4 and 12.8% moisture were reported to be indistinguishable (Whitney *et al.,* 1961), but more recent work indicates that, within the range from 12 to 15%, treatment may be slightly more effective as the moisture content of the wheat increases (Nelson and Kantack, 1966). Moisture content of the host medium after RF treatment has a definite effect on insect survival. Mortalities of insects transferred to higher-moisture wheat following exposure to RF electric fields were lower than those of insects transferred to wheat of lower moisture contents which were still high enough to maintain normal insect activity (Nelson *et al.,* 1966). Since insects treated at lethal exposures lose moisture amounting to a few percent of their weight, it seems reasonable that they should survive better in media of higher moisture content, for they should then be able to regain the lost moisture more rapidly.

g. Frequency. Most studies on stored-grain insect control with RF electric fields have been conducted in the frequency range from 1 to 50 MHz. Often the choice of frequency has been dictated by the operating frequency of commercially available power oscillators designed for other dielectric heating applications. Based on electrical theory, however, and also from experimental results, the frequency for insect control studies using unmodulated fields is not critical as long as the frequency exceeds a certain "transition frequency." Thomas (1952) indicates that this "transition frequency" for most biological materials lies in the range of 1–200 MHz.

Early implications that certain frequencies were more effective for heating insect tissues than plant tissues were based on faulty assumptions, as explained by Thomas (1952). Also the idea that certain selective frequencies exist for certain insects has not been borne out by theory or experiment.

Studies with confused flour beetles in flour and granary weevils in wheat which were exposed to microwaves of 12.25-cm wavelength (2450 MHz) indicated that no differential heating was obtained. Temperatures in excess of 180°F (82°C) and 170°F (77°C), respectively, were required for control of immature stages. In view of these results, it appears that the lower frequency range where differential heating has been demonstrated, the 1–50-MHz range, has more to offer for insect control possibilities. The microwave region should not be abandoned, however, for there may be other ways of applying the energy which would be more effective. For example, application of high intensity pulses might be explored. There is also a wide relatively unexplored region, as far as insect control is concerned, between these two frequency regions.

Consider again Eq. (3) with respect to power absorption in a mixture of materials such as insects and grain. The influence of frequency on power absorption is not as simple as it appears, because both the field intensity E and the dielectric loss factor ϵ'' are somewhat frequency-dependent. The variation, with frequency, of the dielectric constant and loss factor for bulk wheat and rice weevils is illustrated in Fig. 2 for the 1- to 50-MHz frequency range. Since the field intensity depends upon the relative values of the dielectric constants of the different materials, it may vary with frequency depending upon the changes with frequency in the dielectric constants. For purposes of selectively heating insects in grain, a high insect-to-grain loss-factor ratio and a low insect-to-grain dielectric-constant ratio are desired. Knowledge similar to the information in Fig. 2 over a wider frequency range would be helpful in selecting other frequency ranges for investigation of insect control possibilities.

h. Field Intensity. Effects of frequency and field intensity on insect mortality are difficult to distinguish because both influence the heating rate. The three factors, frequency, field intensity, and heating rate, must therefore all be considered in relation to one another. A number of experiments at 10 and 39 MHz with different combinations of field intensities and heating rates indicate that there are subtle frequency effects which depend upon the species and developmental stages of the insects (Nelson *et al.,* 1966). In comparing 10- and 39-MHz treatments of similar heating rates, the 10-MHz treatment was consistently better for some species and stages, whereas the 39-MHz treatment was consistently better for others. For still others, the two frequencies produced similar mortalities.

High field intensities were found much more efficient than low intensities in killing adult rice weevils in wheat at both 10 and 39 MHz, but with immature stages of the same species, high or low intensities produced the same results (Nelson and Whitney, 1960; Whitney *et al.*,

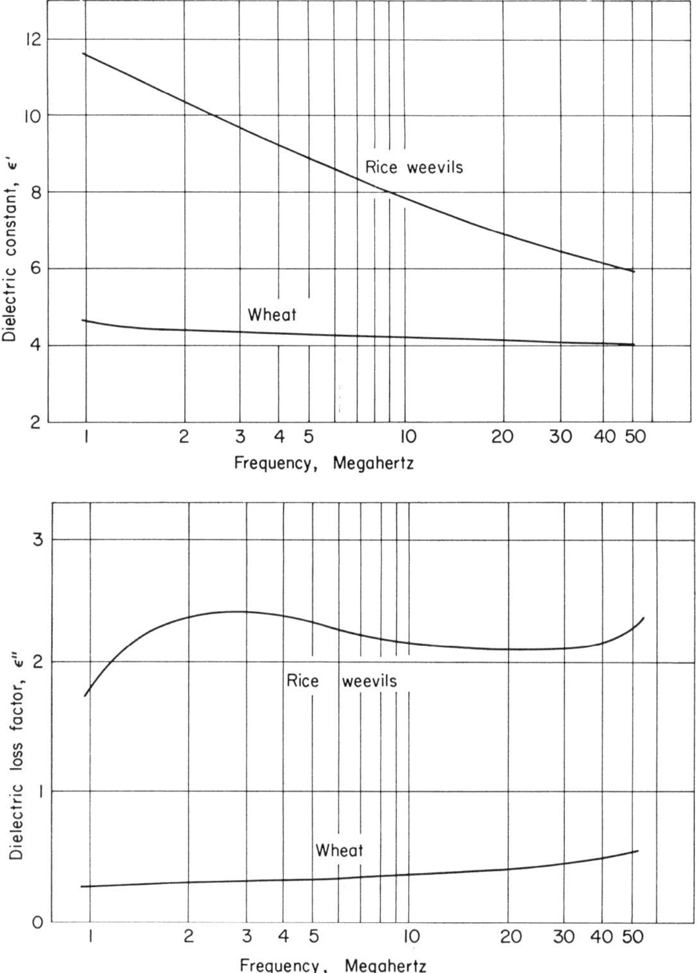

FIG. 2. Frequency dependence of the bulk dielectric constant and loss factor of adult rice weevils and hard red winter wheat (Nelson *et al.*, 1966).

1961). The influence of field intensity for 39-MHz treatment of rice weevil adults in wheat is illustrated in Fig. 3. Temperature of the grain mass immediately after treatment provides a common energy-input basis for comparison of treatments at different field intensities which require different exposure times for comparable energy input. Wide differences in both 1- and 8-day mortalities are evident in Fig. 3 between low and high field intensities. Similar differences duc to field intensity are found for flour beetles and the lesser grain borer adults. Generally, differences attributable to field intensity appear to diminish at intensities greater than about 3 kV/in. For a given frequency, an increase in field intensity increases the heating rate. The more rapid elevation of tempera-

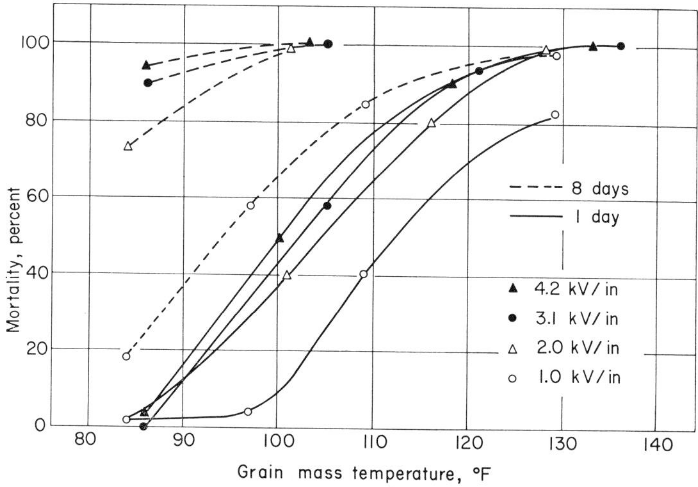

FIG. 3. Mortality of adult rice weevils 1 and 8 days after exposure to radiofrequency electric fields of indicated intensities (Nelson *et al.*, 1966).

ture appears to offer a logical explanation for the greater efficiency of high-intensity treatments which should produce a higher degree of thermal shock in biological organisms. When combined with varying frequency, however, heating rate of the host medium is not necessarily the dominant factor in producing mortality (Nelson *et al.*, 1966).

i. Pulse Modulation. Possible advantages in controlling insects by use of pulse-modulated RF electric fields were suggested by Thomas (1952), though no theoretical basis for any particular optimism was evident. Recently initiated work to check these possibilities has not yet shown any major improvement in efficiency over continuous oscillation for control of adult rice weevils and confused flour beetles in wheat using 39-MHz fields modulated with 5–40-millisecond (ms) pulses at pulse repetition frequencies ranging from 10 to 40 pulses per second (pps) (Nelson *et al.*, 1966). Some improvement in treatment efficiency was noted, however, using 10-ms pulses at a repetition rate of 10 pps and field intensity of 9 kV/in. compared to unmodulated treatment at 3.5 kV/in. for lesser grain borer adults. Much higher field intensities can be used with pulse modulation, even with these relatively long pulse durations. For clean wheat of 13–14% moisture, field intensities for unmodulated RF treatment at a frequency of 40 MHz are limited to about 4 kV/in., or a little higher, because of the tendency for arcing to occur in the grain which chars the kernels. The maximum permissible intensities increase with decreasing grain moisture content. The length of exposure also influences arcing, and moisture escaping from the grain kernels during dielectric heating increases the tendency for arcing. The maximum field intensity which may be employed with pulse modulation appears to increase as the pulse width and pulse repetition frequency are decreased.

Interesting "nonthermal" effects on unicellular organisms have been observed using pulse-modulated RF electric fields (Teixeira-Pinto *et al.*, 1960). Studies of this type have not progressed to the point that it may be determined whether or not any detrimental "nonthermal" effects are possible which might be employed in insect control work. Such possibilities appear worthy of investigation as further information is developed.

C. Mechanism of Lethal Action

Effects other than those resulting from tissue heating have been suspected of producing lethal results. Thomas (1952) concluded that results attributed to "specific effects" other than heating are usually explained by failure to recognize certain fundamental factors, often the existence of differential heating. Complete explanations, however, of some observations are not yet available.

Histological studies on *Blatta* (*Periplaneta*) *orientalis* L. showed that thoracic nerve ganglia were affected by RF exposures (Andreev, 1937). Differences in amino compounds in nerve tissue were noted by Silhacek (1960) between RF-treated and untreated American roaches, *Periplaneta americana* (L.). Symptoms exhibited by treated roaches suggested synaptic blocking as a possible explanation for death following RF exposure. Such effects could be induced by localized absorption of energy and heating in nerve tissue.

Hartzell (1934) examined tissue from ventral nerve ganglia of yellow mealworm larvae which had been exposed to 300-MHz RF electric fields. Microscopic examination "gave the impression that there had been very rapid solution of cells," whereas "coagulation of the cells" was indicated in ganglia of insects exposed to external conventional heating. Electron micrographs of nerve tissue from wax moth larvae were studied by Vande Berg (1965). Some evidence of "vacuolization" was found in nuclei of cells from insects conditioned at 40°F (4°C) prior to RF exposure which did not elevate general body temperatures above room temperature (75°F, 24°C). Neural nuclei of these insects also exhibited an apparent "coagulation," as did those of insects, initially at room temperature, subjected to RF and hot-air oven exposures.

Amino acid analyses conducted on extracts from whole bodies of both RF-treated and untreated yellow mealworm larvae showed that qualitative differences in amino acids were undetectable (Kadoum, 1963). Further investigations with RF-treated larvae of the same species revealed increased weight loss as well as increased oxygen consumption which lasted for a few days before declining to a normal level (Kadoum, 1966). Increases in protein synthesis and amino acid catabolism were also noted, and these phenomena did not appear to be completely explainable on the basis of heating effects alone.

In all of this work, thermal and any "nonthermal" effects on biological material resulting from exposure to RF electric fields are difficult to separate. Efforts to separate effects have been made using pulse modulation and a low duty factor such that the duration of the pulses is a small fraction of the total time of exposure. Some "nonthermal" effects have been reported for unicellular organisms (Teixeira-Pinto *et al.,* 1960), root tip cells (Heller and Teixeira-Pinto, 1959), and gladiolus bulbs (Mickey and Heller, 1964), but, as already pointed out in Section II,B,5,i, nonthermal effects useful for insect control have not yet been observed.

Even with pulse modulation, though average temperatures of organisms may not be elevated appreciably, it would seem that highly absorbing components might still reach higher temperatures during pulses and then cool as heat energy flows to surrounding tissues. Further research is needed to establish the facts and to substantiate any suspected effects not attributable to dielectric heating, particularly as related to the lethal action of RF fields on insects.

D. Practical Aspects

Though considerable experimental work has been conducted showing that RF energy is effective for several insect control applications, no practical-scale applications of significance have been developed. A few pilot experiments employing power oscillators with up to 20-kW output capacity have been tried (Mouromtseff, 1933; Thomas, 1952), and United States patents were issued for the method (Davis, 1934, 1936a,b), but to date the benefits of RF methods have not been deemed sufficient to offset the costs involved for the large-scale installations required for practical use in grain or seed treatment. Dielectric heating has been successfully used in a 120-kW industrial installation for heating bales of feed bags for reuse (*Elec. World,* 1950). The primary objective in this application was the destruction of animal disease organisms.

The economics of RF disinfestation were considered by Thomas (1952), and at that time costs in Britain appeared to be competitive with chemical treatment costs for stored-grain insect control. Costs, of course, are quite dependent upon the size of the installation and the amount of use which the equipment will receive as well as electric energy costs. Cost estimates in the United States for RF treatment of wheat for stored-grain insect control, based on a 200-kW output installation operating 2000 hours/year, indicated a cost of about 3.5 cents per bushel (Nelson and Whitney, 1960; Whitney *et al.,* 1961). Chemical fumigation costs for large elevators were then estimated at 0.5–0.7 cents per bushel, and were therefore much more economical.

It appears that major improvements in efficiency of RF treatment are necessary for the method to become practical for disinfesting grain. The value of nonchemical treatments may increase in the future, and this would improve possibilities for RF control methods. Ionizing radiations also offer good possibilities for practical-scale disinfestation of grain, however, and the economics of these methods can be expected to improve, perhaps more rapidly than those for RF energy, as the nuclear industries develop. Potential use of RF electric fields for insect control in more valuable products than grain, however, may have a brighter future. Ionizing radiations cannot be employed for insect control in seed because of resulting genetic damage produced by exposures of the dosage level required for insect control. RF exposures required for insect control do not damage the germination of wheat when its moisture content is below about 14%. Lowry *et al.* (1954) reported that pink bollworm larvae, *Pectinophora gossypiella* (Saunders), were controlled in cottonseed without damaging seed germination. In fact, RF treatment improves the germination and growth characteristics of several kinds of seed (Nelson and Walker, 1961; Nelson and Wolf, 1964). Development of RF equipment for seed processing might also improve RF insect control possibilities, for dual use of such equipment would make the economics more attractive.

III. INFRARED RADIATION

A. General Theoretical Aspects

Infrared radiation occupies the region of the electromagnetic spectrum lying between the radiofrequencies and the visible portion (Fig. 1). Wavelengths just beyond the red end of the visible spectrum, in the vicinity of 10,000 angstroms (Å) or 1 micron (μ), are frequently termed near-infrared, while the longer wavelength infrared radiation is called far-infrared. The term, "intermediate infrared," is also used to specify the region between near- and far-infrared. Infrared energy is usually associated with heat in that it is radiated by hot bodies and absorption of infrared energy produces heating in the absorbing material. Other electromagnetic radiations also produce heat in absorbing materials. Radiation of infrared energy arises from vibration and rotation phenomena associated with molecules at the surface of the radiating body. The "blackbody radiation" concept is basic to the development of radiation theory. A "blackbody" is a perfect radiator and a perfect absorber of the radiation. Any body at a temperature above absolute zero radiates energy. The quantity of energy radiated is proportional to the fourth power of the absolute temperature (Stefan-Boltzmann law). The radiation and absorption of energy are also dependent upon the characteristics of the surface, which can be described relative to the blackbody surface by a term called the "emissivity" of the surface. For blackbody radiation there is a well-defined wavelength distribution for the radiated energy (Planck's law), and the wavelength of maximum energy radiated is a function of the temperature of the radiator (Wien's law). The wavelength of peak energy becomes shorter as the temperature increases. The total energy radiated, of course, increases in accordance with the Stefan-Boltzmann relation.

Potential uses of infrared radiation for insect control involve two generally different concepts. One utilizes the radiation directly applied to the insects or infested material, whereas the other is based on the insect's suspected ability to sense infrared radiation and man's ingenuity in employing this knowledge in some way to achieve control. Direct applications will be considered first, since these have been chronologically the first possibilities investigated.

B. Direct Application

Tillson (1943) described studies of wheat treatment with infrared energy for insect control and an infrared tunnel designed for grain treatment, which consisted of a conveyor belt with banks of infrared lamps mounted above and below the belt. Temperature measurements at the top and bottom surfaces and in the middle of layers of wheat 0.5- and 0.25-in. deep revealed that a uniform temperature in a 0.25-in. layer could be obtained if the top surface was covered to prevent cooling after treatment. This could be accomplished by carrying the wheat on a conveyor belt into a closed tunnel after exposure to the infrared energy and holding it at 140°F (60°C) for 10 minutes to kill all stored-grain insect life. Such treatments were found to have no effect on germination of the wheat, its baking qualities, or thiamin content. Cost of treatment was indicated to be 0.5 cent per bushel, but the basis of the estimate was not given. Yeomans (1952), describing the same sort of infrared treatment of grain for insect control, concluded that the method was not practical because of high cost and poor penetrating characteristics of infrared radiation.

Frost *et al.* (1944) experimented with infrared treatment of the confused flour beetle, yellow mealworm, and a few other insects. Glass filters were used between the infrared lamps and the specimens to block the visible radiation. They concluded that the insects were killed by absorption of infrared energy which increased the internal body temperature. Little difference was noted between mortality of confused flour beetle adults and larvae receiving the same exposures. Adults of the yellow mealworm turned on their backs when exposed to infrared energy. When restrained so they could not turn over, the adults were killed with much shorter exposures than were required for the larvae. The difference was believed due to higher absorption of infrared energy by the darker adults and longer heat retention because of the more heavily sclerotized body wall of the adults. Flour beetles were protected from the infrared radiation when covered by a 0.25-in. layer of wheat flour.

Infrared energy from lamps has also been studied for effectiveness in killing the pea weevil, *Bruchus pisorum* (L.), in seed peas (Litynski and Wilkojc, 1954). Exposures of 3 minutes to a battery of lamps 25–30 cm from the single-seed layer killed all of the insects without damaging seed germination.

Tilton and Schroeder (1961, 1963) studied the use of infrared radiation from a gas-fired ceramic-panel infrared heater for control of insects infesting rice. The use of a gas-fired infrared source was believed more economical than infrared electric lamps. Immature rice weevils were completely controlled in rice with about 12% moisture by infrared exposures producing a temperature in the rice of 56°C (133°F). Immature lesser grain borers required a grain temperature of 68°C (154°F) for complete control. Intensity of radiation was adjusted by changing the distance between the infrared source and the rice. Length of exposure was also varied. Rapid rates of temperature increase obtained with high intensities appeared more effective; however, later studies indicated that complete mortality might be obtained at lower rice temperatures when lower-intensity, longer exposures were employed. The Angoumois grain moth, *Sitotroga cerealella* (Olivier), was also studied in the later experiments. Rice weevils were the most susceptible, whereas the Angoumois grain moth and the lesser grain borer were more resistant to control by infrared treatment. Results with rice of 14% moisture content indicated that complete mortality might be expected for all three species with rice temperatures in the range from 65° to 70°C (149°–158°F). Mild exposures to infrared radiation were found to increase the emergence of all three species. Agitation of the grain during treatment was suggested as a possible improvement in obtaining uniform treatment.

C. Infrared Sensing

1. *Attraction?*

Beliefs have been expressed that some insects are attracted by infrared radiation. Mosquitoes are attracted to insect traps using tungsten filament incandescent lamps. While such lamps are used mainly for lighting, they emit most of their radiant energy in the near-infrared region. A 500-W quartz infrared lamp, which emits more infrared and a smaller percentage of visible radiation, was found more effective for attracting mosquitoes than a 200-W incandescent lamp in tests comparing traps equipped with the two different lamps (Calderwood and Altman, 1954). Longer-wavelength infrared radiation from a lamp dipped in black paint, however, was found no more effective for attracting mosquitoes than a nonenergized lamp of the same type (Barr *et al.,* 1963). Corn earworm moths, *Heliothis zea* (Boddie), were attracted to an "infrared source" emitting energy between 8 and 13 μ (Callahan, 1965a). The source consisted of a 4-W "mercury arc–argon discharge tube" wrapped with black tape such that no visible radiation was apparent to the human eye. The capture pattern was reported to be correlated with the radiation pattern of the source.

Extensive studies have been conducted on factors influencing attraction of mosquitoes. In field experiments employing heated and clothed dummies, temperature or warmth was found to be the major factor in attracting mosquitoes when the air temperature was below 60°F (15°C) (Brown, 1951). At air temperatures above 60°F, surface moisture on the clothing increased attraction to a greater extent than did increasing the temperature of the attracting bodies. In studies of the attractiveness of human hands to the yellow-fever mosquito, *Aedes aegypti* (L.), skin temperature was found to be an important factor. Warm-skinned hands were more attractive than hands with lower skin temperature (Smart and Brown, 1956).

Studies were conducted by Peterson and Brown (1951) to determine whether or not infrared radiation was a factor in attracting the adult female yellow-fever mosquito to warm objects. Plastic billiard balls, both white and black, were used in the experiments. More mosquitoes were attracted to both black and white balls when their temperatures were in the 100°–110°F (38°–43°C) range than to identical balls which were 20°F (11°C) cooler. When airtight infrared-transmitting (25–40-μ transmission band) windows of thallium bromoiodide were placed between the insect chamber and the heated billiard balls, the response completely disappeared. A large portion of the energy radiated by the balls at these temperatures would not be transmitted through the windows; however, these experiments would tend to show that it was convected heat rather than infrared radiation which made the warm objects attractive to the species studied. The studies by Peterson and Brown (1951) also included experiments designed to test the attractiveness of surfaces of different emissivities to the same species. A cubical container filled with warm water had the faces of the cube treated with different finishes, including polished metal, painted flat black and flat white, black and white enamel, lamp black, varnished, and lacquered. With surface temperatures of 92°F (34°C) and air temperature of 80°F (27°C), no significant differences were found in the number of mosquitoes approaching any of the different treated surfaces, except for the black enameled surface, which attracted 60% more mosquitoes than the white enameled surface and 70% more than the flat black surface. Reflected images of the insects were suggested as a possible explanation for the higher attraction of the black enameled surface. The lack of differences in attractiveness of surfaces with widely differing emissivities provides further evidence that infrared radiation is not involved in the attraction of this species. This conclusion might be questioned, however, since actual infrared emissivities of the different surfaces were not known.

2. *Communication?*

Theories involving infrared radiation have been advanced to explain the remarkable capabilities of males of many moth species in the long-range location of the female. The general bases for such theories were outlined by Duane and Tyler (1950). Temperature measurements obtained from a small thermocouple placed under the thorax of a female cecropia moth, *Hyalophora cecropia* (L.) (*Samia c.*) revealed that wing and leg activity increased the moth's thoracic temperature as much as 11°F (6°C) above the normal resting or ambient temperature. Measurements of the radiation from the moth using an infrared spectrophotometer revealed a radiation pattern closely approximating the blackbody radiation curve for that temperature in the range from 3 to 12 μ. Their data show some irregular deviations in the vicinity of the wavelength of maximum blackbody radiation. If a moth could voluntarily raise her temperature several degrees above that of her surroundings, the radiation from the moth could be detected at considerable distances by a suitable sensitive infrared detector. The possibility that the antennae of male moths might serve as infrared sensors was hinted, since measurements indicated some differences in lengths of antennal hairs, which were of an order of magnitude corresponding to wavelengths in the emission band of the female. Grant (1948) also proposed that insect antennae might serve as infrared detectors, for he observed that the dimensions of the antennal pits were of the proper order of magnitude to act as resonant cavities at wavelengths in the infrared region between 2 and 80 μ.

Laithwaite (1960) proposed a more detailed radiation theory of the assembling of moths. His theory was first inspired by the observation of similarities between certain radar antennae and the structure of the antennae of moths. His observations on the vaporer moth (rusty tussock moth), *Orgyia antiqua* (L.), indicated that, if the moth antennae were functioning as electromagnetic antennae, the spacing of pectinations would indicate an operating wavelength in the infrared region between 20 and 200 μ. Laithwaite described several observations which he did not believe to be satisfactorily explained by olfactory or scent theories, and which appeared to be more logically explained by his radiation theory. Among these was the straight-line approach to the female which males take from long range. The use of both antennae to sense phase differences in radiation would provide the means for a straight-line approach, where it appeared more difficult to account for this on an olfactory basis, since the straight-line approach requires orientation perpendicular to the contour lines of an odor gradient. Observations with the vaporer moth indicated that a nearly direct approach was taken by male moths released 100 yards distant from the virgin female until they came within 10 or 15 yards of the female. The remainder of the approach appeared less direct and was more like that one would expect from a moth following an olfactory gradient. Laithwaite therefore postulated a long-range attraction based on the infrared radiation theory and a short-range attraction founded on olfactory responses.

Another observation which Laithwaite found easier to explain with the radiation theory was the cessation of attraction of the female upon pairing. Scent emitted prior to mating would be expected to continue attracting males following copulation. However, once the female has mated she appears to lose her powers of attraction. In experiments described by Laithwaite, males of the vaporer moth released at a distance of 10 ft. would assemble to an empty box which had earlier held virgin females, to fertilized females, to dead females, to eggs, and to empty female pupal cases. When released 100 yards away, they did not approach these sources of short-range attraction. This evidence further supported his theory that short-range attraction was a scent process, whereas long-range attraction involved emission and detection of infrared radiation, which the female ceased to emit upon pairing. Laithwaite speculated that attraction to light might also be explained through sex attraction involving infrared radiation, which could result from inelastic scattering of visible radiation by water vapor in the air.

Laithwaite's theory and supporting evidence were emphatically and, to a certain extent, convincingly discredited by Kettlewell (1961). In simple experiments with marked moths, some with amputated antennae and some with eyes spotted with black paint, he found that the compound eyes rather than the antennae were involved in attraction of Lepidoptera species to light. After extensive studies, spanning more than 30 years, on the assembling of moths, Kettlewell (1946, 1961) believes that all observations can be satisfactorily explained by the scent diffusion theory. In his opinion the stimulus is airborne, and the male always comes upwind whenever there is any wind at all. Contrary observations are explained by unknown back eddies or outward diffusion in all directions in dead calm.

Kettlewell (1961) noted that there are wide variations in the antennal pectination spacing among species and that there are species of efficient assemblers with no pectinations at all; so the highly developed antennal pectination is not a requisite for long-range assembly. He suggested the possibility of a "repeller scent" as the reason for the cessation of attraction after mating.

Laithwaite (1961) published a reply acknowledging the lack of positive evidence in support of the radiation theory, but further justified the possibilities for its consideration, and pointed out justified criticisms of some of the experimental evidence cited in discrediting the theory.

Callahan (1965a) has postulated further concepts involving the use of infrared radiation by moths. He believes that they locate their mates, as well as scents and host plants, through infrared detection. In addition to the possibility that antennal spines may serve as resonant cavities for infrared detection, he suggests that the compound eyes of nocturnal moths shift slowly from sensitivity in the visible spectrum to longer and longer wavelengths as dark adaptation progresses, and serve as infrared detectors when totally night-adapted (Callahan, 1965c). Thermistor-probe temperature measurements indicated that thoracic temperatures of moths of different noctuid and sphingid species increased from 7° to 20°C (13°–36°F) as the moths vibrated their wings in nonflight activity (Callahan, 1965a). (Temperature differences of this order have been reported by others, including Adams and Heath, 1964.) Infrared bolometer measurements of radiation from the thorax of these same moths indicated peak radiation at wavelengths ranging from 9.10 to 9.59 μ. Callahan's calculations based on these data indicated that infrared radiation from such moths could be detectable at a distance of 1 km, since atmospheric absorption is low in this wavelength region. Callahan (1965b) cited other observations as evidence in support of the theory that far-infrared radiation is involved in insect communication. At present, it appears doubtful that all of these arguments will withstand careful scrutiny. Electrophysiological studies are needed to establish whether or not insects do in fact possess the infrared-sensitive receptors essential for the speculated processes.

Okress (1965) has commented on errors in naive application of electrical waveguide and antenna theory in construction of models of the sensory organs of insects. He proposed more sophisticated models which should be more reliable, but emphasized the difficulties, both experimental and theoretical, which face scientists attempting to verify these ideas.

Miles and Beck (1949) hypothesized that certain insect olfactory receptors are radiation receptors rather than chemoreceptors. In experiments with honeybees, they enclosed honey in gas-tight chambers. One chamber was equipped with an infrared-transmitting window, and another was similarly constructed, but had an infrared-absorbing glass plate behind the infrared-transmitting window. The bees responded better to the chamber in which the window remained transparent to infrared radiation. They attempted to explain the insect's ability to sense the presence of honey vapor by increased loss of infrared energy by radiation from the antennae because of its absorption by the honey vapor molecules. There would appear to be no basis for this argument in radiation theory, for absorption of the radiated energy should in no way influence the radiant heat loss from the emitting body. These observations, however, if valid, would lend support to the infrared theory of olfaction, since infrared radiation would be emitted by the honey vapors and could be detected by the bees if they possessed infrared receptors. Similar, more carefully controlled experiments by Johnston (1953) did not confirm the results of Miles and Beck. In fact, Johnston, who used both the honeybee and the black blowfly, *Phormia regina* (Meigen), concluded that the infrared loss theory of olfaction must be rejected.

Sex-attractant chemicals (pheromones) of several insect species were studied by Wright (1963) to see whether certain low-frequency molecular vibrations might be predominant in any particular group. The structure of known sex lures and attractants was studied using different models to test the hypothesis that such molecular vibrations and rotations provide a physical basis for odor. He found little uniformity in attractant chemicals, and the evidence obtained did not confirm the theory. Wright believed, however, that the vibration theory of odor was "consistent with what is known of the substances, which are demonstrably perceptible to various species by what appears to be an olfactory process."

D. Practical Potential

Direct application of infrared radiation to control insects by heating infested products has already been shown possible (Section III,B). Apparently the potential advantages of such methods have not yet been deemed important enough to justify the cost. Some improvements in efficiency of infrared treatment should be possible as further information is developed on properties of materials which influence absorption of infrared energy. Sources providing the desired wavelength distributions might then be provided for particular applications.

It is still too early to evaluate the possibilities of infrared radiation for insect control in all of its ramifications. Experimental and theoretical approaches to the questions of insect response to infrared energy should provide the information needed to assess control possibilities, but this will likely require a number of years. Infrared technology has developed rapidly in recent years because of its significance as a passive detection method and emphasis on military applications. Instrumentation has been and is being developed which will facilitate the experimental measurements and tests which must be performed.

IV. VISIBLE AND ULTRAVIOLET RADIATION

A. General Theoretical Aspects

Just above the infrared portion of the electromagnetic spectrum, relative to quantum energy, lies that region to which the human eye is sensitive, the visible region. Beyond this lies the ultraviolet region. Both ultraviolet and visible radiation originate when the electrons of excited atoms undergo transitions from one orbital energy level to a lower energy level; the difference in energy is emitted from the atom as a quantum of radiation. Differences in the orbital electron energy levels of the different atoms account for the wide variation in the energy of the quanta and consequent wavelengths of the radiation. Visible radiation, or light, spans the wavelength range from approximately 3800 to about 7700 Å. Colors of the spectrum are associated with wavelength, as indicated in Fig. 4. The quality of any particular light depends upon the quantity of energy of the different wavelengths represented, or the spectral energy distribution. With regard to insect response and control studies, a number of terms useful in describing light are important.

With respect to illumination of a surface, the intensity, without regard to light quality, but evaluated in terms of visual effects, is often measured in foot-candles. The quantity of light incident upon a surface or emitted by a source is often expressed in lumens and has the units of energy per unit time. An illumination of 1 foot-candle is equivalent to incident energy of 1 lumen per square foot. In regard to a light source, a point source of 1 candela will provide at 1-ft distance an illumination of 1 foot-candle, and will radiate into space 4π lumens. Since in practice light sources are not point sources, but have an appreciable area, the brightness of the source is often important. The brightness of a source can be expressed in lumens per square foot or foot-lamberts. The lumen is also employed as the unit for luminous flux in the metric system, where the illumination is expressed in lumens per square centimeter and source brightness in the same units or in lamberts.

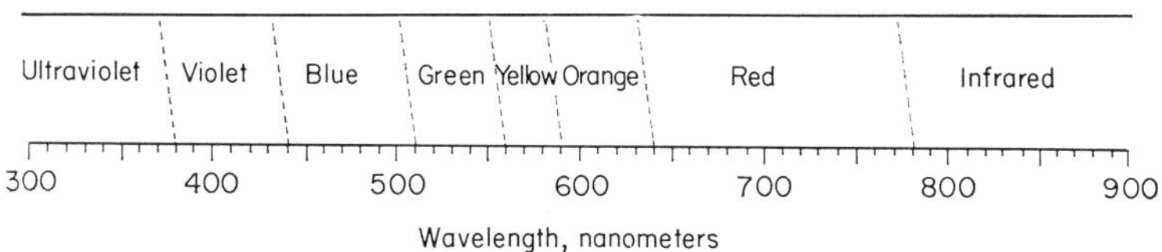

FIG. 4. Colors associated with wavelength in the visible spectrum.

In many studies, absolute measurements of light energy are needed. The irradiance, or incident energy per unit area per unit time, can be expressed in W/cm^2. The radiant emittance from a source is expressed in the same units. Spectral irradiance and spectral radiant emittance are terms defined similarly, except that they are taken per unit wavelength interval.

B. Insect Response

The compound eyes and ocelli of insects are sensitive to much shorter wavelengths than the eyes of man. "Vision" for insects extends well into the near ultraviolet region to below 3000 Å [300 mμ or nanometers (nm)]. The responsive region for insects has been found to range between about 253 and 700 nm, with some variation among species (Dethier, 1953, 1963). In general, the ultraviolet and blue-green regions of the spectrum are the most effective in attracting insects.

Insects respond to visible and ultraviolet (UV) radiation in many ways. A number of classifications for such responses have been set up (Fraenkel and Gunn, 1961; Jander, 1963; O'Brien and Wolfe, 1964). The principal response, and the one of main interest currently in considering control possibilities is that of phototaxis, the movement toward or away from light. Other effects of light on insects, for example, photoperiod and diapause, are also of interest for possible future control methods.

There is a vast amount of scientific literature dealing with insect response to visible and UV radiation. No attempt is made here to review this material in any detail, but some of the pertinent studies and reviews are cited.

Early studies dealt with the influence of light on insect behavior. Many were aimed at determining whether insects possess any form of color vision or the ability to differentiate between light of different wavelengths. Studies were also conducted to determine the range of wavelengths to which insects are sensitive. Reviews of these studies have been published by Bertholf (1931a,b), Weiss (1943, 1946), Wulff (1956), Dethier (1953, 1963), Goldsmith (1961, 1964), Hollingsworth (1961), Burkhardt (1964) and Mazokhin-Porshnyakov (1964). Later studies using electrophysiological techniques to obtain electroretinograms (ERG's) from insects exposed to visible and UV stimuli added a great deal to the information now available on spectral response of insects. These are included in the reviews by Wulff (1956), Dethier (1953, 1963), Goldsmith (1961, 1964), Hollingsworth (1961), Burkhardt (1964), and Mazokhin-Porshnyakov (1964).

Experiments conducted in the visible spectrum by Bertholf (1931a) with the honeybee, *Apis mellifera* L., established that this insect was sensitive to wavelengths as long as 677 nm in the red region, but the peak efficiency for phototaxis was obtained at 553 nm. His results also indicated that bees could distinguish differences in brightness of two illuminated frosted-glass windows when the brightness of one was reduced to 70% of that of the other. The human eye, in tests conducted with the same apparatus, was capable of distinguishing differences in brightness when that of one window was reduced to 90% of the other at levels employed. Similar experiments in the ultraviolet region (Bertholf, 1931b) showed that bees responded to radiation down to 297 nm, but not at 280 nm. A UV-transmitting filter was employed between the monochromatic UV source and the window of the test chamber to exclude visible light, and the intensity of white light at a nearby window of the insect test chamber was adjusted to provide equal attraction of the bees as they approached the two windows from the opposite end of the test chamber. The intensity of white light, adjusted by changing the distance of the lamp from the window, required to balance the response to UV of different wavelengths provided an index of "relative stimulative efficiency." Corrections were applied for differences in UV output at various monochromator settings. The highest stimulative efficiency for the insects was obtained at 365 nm, and when combined with data obtained in the visible region, the peak at 365 nm was much higher than that at 553 nm in the visible region. Studies with *Drosophila* indicated a similar peak response at 365 nm and a lesser peak (by 5.5 times) at 487 nm (Bertholf, 1932).

Weiss *et al.* (1941a,b, 1942, 1943, 1944) tested the response of over 50 different species to wavelengths between 360 and 740 nm. Using incandescent lamps and filters, they attempted to obtain uniform intensities at different wavelengths by calculating the required distance between lamp and filter from data on the spectral analysis of the lamp and transmission characteristics of the filter. When insects were tested at a distance of 1 ft from the filter (illuminated window) they showed a peak response in the 470- to 528-nm range. When tested at a distance of 6 ft from the filter, the peak response occurred in the 365 region with a secondary peak for some insects shown in the 470- to 528-nm region. Weiss (1944, 1946) concluded from his own work, and a review of that of others, that generally the peak stimulation for insects occurs at about 365 nm with a lesser secondary peak at about 492 nm.

The difference in response for insects tested at 1 and 6 ft indicated to Weiss et al. that the wavelength of maximum sensitivity may be a function of intensity. Such a dependence was suggested by Fingerman and Brown (1952) based on experimental group motor response of *Drosophila melanogaster* Meigen. Their data indicated a decrease in wavelength for maximum response with decreasing intensity. While their indicated relative intensity levels for the response comparisons were 1, 0.1, and 0.001, they did not maintain equal intensities at each wavelength for each of these four levels. Their conclusions were criticized by Goldsmith (1961), partly for this reason. Using electrophysiological methods, Goldsmith (1960) found no change in the spectral sensitivity of the compound eye of the worker honeybee through a similar range of intensities above the threshold of retinal action potential. According to Goldsmith (1961), the suspected dependence on intensity of wavelength for maximum sensitivity has not been confirmed and there is still confusion concerning intensity–wavelength relationships for maximum sensitivity.

Other aspects of insect vision which have been studied include the determination of flicker-fusion frequencies and apparent ability of insects to detect differences in the plane of polarized light (Wulff, 1956; Dethier 1953, 1963). Ruck (1958) studied the ERG's of the honeybee, dragonfly *Pachydiplax longipennis* Burmeister, a dipteran fly, *Phormia regina* (Meigen), all diurnal insects with characteristically high flicker-fusion frequencies (190–350/second), and the cockroach, *Periplaneta americana* (L.), a nocturnal insect with a low flicker-fusion frequency (45–60/second). Ruck concluded that there were no generally apparent relationships between flicker-fusion frequency, photic sensitivity, waveform of the ERG, or rate of dark adaptation of the compound eyes or ocelli.

Electrophysiological studies of Goldsmith and Ruck (1958) showed that the dorsal ocelli of worker honeybees appear to have two types of photoreceptors, one maximally sensitive at 490 nm and the other at 335–340 nm. Dorsal ocelli of cockroaches, *Periplaneta americana* (L.) and *Blaberus craniifer* Burmeister, appeared to have only one type of photoreceptor, with peak sensitivity at about 500 nm. The compound eye of the worker honeybee appears to have two receptor systems. When dark-adapted, maximum sensitivity due to a green receptor system appears at about 535 nm with a second smaller peak sometimes showing up at about 345 nm (Goldsmith, 1960, 1961). When adapted to yellow light, however, the ultraviolet receptor system with maximum sensitivity at about 345 nm becomes predominant. As stated earlier, no shift between these two peaks was observed due to changes in light intensity. Since the dark-adapted compound eye of the bee has only a minor peak in the near ultraviolet, Goldsmith (1960) performed a behavioral experiment to determine whether response to near ultraviolet might be evoked through reception of UV by the dorsal ocelli. Ocelli of some bees were covered with an opaque paint and tested in comparison with normal bees for attraction to two windows of a test chamber. One window was illuminated with radiation of 365 nm and the other with light of 546-nm wavelength. Energy transmitted through the windows into the test chamber was adjusted to the same level for both wavelengths. Normal bees showed a strong preference for 365 nm. Bees with painted ocelli took longer to respond, but showed the same high preference for 365 nm indicating that the ocelli are not required for UV phototactic response.

The spectral response of seven stored-grain insect species was tested by Stermer (1959). Insects were trapped below two ground-quartz windows at opposite ends of a large test chamber. Each window was illuminated at a low intensity with monochromatic light of different wavelengths. Peak response for the almond moth, *Cadra cautella* (Walker) (*Ephestìa c.*) Angoumois grain moth, *Sitotroga cerealella* (Olivier), lesser grain borer, *Rhyzopertha dominica* (F.), and red flour beetle, *Tribolium castaneum* (Herbst), was obtained at 500 nm with a secondary peak between 334 and 365 nm. For the Indian-meal moth, *Plodia interpunctella* (Hübner), the major peak was between 334 and 365 nm, and a secondary peak was obtained at 500 nm. The rice weevil, *Sitophilus oryzae* (L.), responded equally well at all wavelengths between 334 and 546 nm. Response of all species was low at 600 nm and still lower at 280 nm. Number of insects responding increased as intensity of the source was increased through seven times initial intensity. No shift in spectral sensitivity was observed with change in intensity except for the almond moth, whose peak response appeared to shift from 546 to 365 nm as intensity increased.

Hollingsworth (1961) studied the response of the pink bollworm moth, *Pectinophora gossypiella* (Saunders), to narrow wavelength bands between 280 and 625 nm using a long test chamber with ground-quartz diffusing plates illuminated by monochromatic sources at each end. In some of these studies, the relative response at each wavelength was obtained in comparison to a 365-nm source. Similar results were obtained in other tests where the insects were exposed to all-different pairings of the various wavelength bands. At the low intensities employed, the pink bollworm moths exhibited peak response at about 515 nm, with a lesser but definite secondary peak at about 365 nm. Very little response was obtained at wavelengths longer than 600 nm and shorter than 300 nm. Using wavelengths of 515, 365, and 405 nm (the wavelength of lowest response between the two peaks), pink bollworm moth response tests were conducted in a larger chamber using intensities of 20, 40, and 80 times that employed in the other tests. With a twentyfold increase in intensity, relative response at the three wavelengths remained the same. At the two higher intensities, however, the maximum response shifted from 515 to 365 nm.

The shifts in wavelength of peak response noted by Hollingsworth (1961) and Stermer (1959) using the same test chamber were in the opposite direction of those attributed to the change in intensity in studies by Weiss (1943). The importance of intensity in wavelength response studies has been noted by many investigators, but the true influence of intensity is complicated by a number of factors and is not presently well understood (Hartsock, 1961; Hollingsworth, 1965).

Group motor response of the boll weevil, *Anthonomus grandis* Boheman, was also studied by Hollingsworth *et al.* (1964a,b). Tests conducted in a Y-shaped chamber, with a low intensity 365-nm stimulus (16×10^{-12} W/cm^2) at the end of one arm and narrow wavelength bands in the range from 315 to 665 nm used successively at the end of the other arm, revealed a fairly broad response peak in the 490–540-nm range. The response curve indicated a possible minor peak at 365 nm. Experiments with seven different intensity levels of 490- and 365-nm radiation up to 496×10^{-12} W/cm^2 showed no shift in wavelength of maximum response, though differences in response at the two wavelengths were no longer significantly different at the higher intensity levels. The threshold level for response of the boll weevil to 365-nm radiation was found to be about 16×10^{-12} W/cm^2, and from this level to 16×10^{-10} W/cm^2, response was linear with the log of intensity (Hollingsworth *et al.*, 1964a).

Electrophysiological studies by Mordue *et al.* (1965) also showed a broad response peak for the boll weevil centered at about 518 nm and agreeing very closely with the behavioristic response curves obtained by Hollingsworth *et al.* (1964a). In addition, ERG responses indicated another peak in the ultraviolet below 382 nm, which was the short-wavelength limit of the equipment used. No difference in response of males and females was observed. The radiant emittance of the source for the ERG studies, $0.2–2.5 \times 10^{-6}$ W/cm^2, was $10^4–10^5$ times greater than for sources used in the group motor-response studies (Hollingsworth *et al.*, 1964a). ERG studies on the cotton bollworm moth, *Heliothis zea* (Boddie), and the tobacco budworm moth, *Heliothis virescens* (F.), also revealed peak responses in the region of 515 nm (Hollingsworth, 1965).

Less conclusive behavioral studies on the spectral response of tobacco hornworm moths, *Manduca sexta* (Johannson) (*Protoparce s.*) were described by Earp and Stanley (1965). Caged insects in a dark chamber were observed through an infrared telescope as they were exposed to narrow wavelength bands in the region from 310 to 578 nm. Response seemed best at the shorter wavelengths below 340 nm.

C. Attraction

Thousands of insect species, particularly nocturnal fliers, are attracted to light. For a partial listing of photopositive species, see Hollingsworth *et al.* (1963). A great many experiments are reported in the literature concerning tests of light sources for attracting insects. Only a few of these can be described here. Some of the various light sources and types of electric lamps are described briefly by Hienton *et al.* (1958) and by Hienton (1961). Early studies in the laboratory and in the field revealed that brilliance, size of luminous area, and color were important factors influencing the attractiveness of lamps for the codling moth, *Carpocapsa pomonella* (L.) Herms and Ellsworth, 1934; Marshall and Hienton, (1935, 1938). Large amounts of ultraviolet in the spectral output of lamps were found to add to their attractiveness. Further studies showed that light in the violet-blue spectral region was the most attractive for the European corn borer, *Ostrinia nubilalis* (Hübner) (*Pyrausta n.*), (Ficht and Hienton, 1941). These studies also indicated that ultraviolet radiation below 320 nm did not add to the effectiveness of lamps for attracting the European corn borer. Increasing the size of the luminous area increased the number of insects attracted, though not in direct proportion. Attraction of moths was found to be almost directly proportional to the intensity of light visible to the human eye. Headlee (1937) experimented with different colored lamps in mosquito traps and found a mercury–argon vapor lamp most efficient in attracting mosquitoes.

Taylor and Deay (1950) evaluated different wavelengths of radiation for effectiveness in attracting European corn borer moths by selecting sources and glass filters to provide outputs peaking at 254, 365, 436, and 525 nm. The 365-nm sources were found to be the most effective lure for capturing corn borer moths in the field. Other sources with a wide range of peak outputs were compared with 365-nm sources in the laboratory, but none with equivalent wattage was found more effective. The 15-W blacklight (15-W BL) fluorescent lamp was the 365-nm source used in these studies. It was identified as the 360 BL lamp. Spectral energy emission of this lamp was presented by Hienton *et al.* (1958, p. 253).

Glick and Hollingsworth (1955) compared in the laboratory the attractiveness of a large number of different types of lamps, each with different spectral outputs, for attracting the pink bollworm moth. Lamps with principal radiant output in the near ultraviolet region attracted the most moths. Field tests confirmed their conclusions, and the 15-W BL fluorescent lamp was found to be the most efficient lamp for attracting the pink bollworm moth and several other species of cotton insects (Glick and Hollingsworth, 1954, 1955). Continued field studies revealed that the 2-W argon glow lamp with peak output at 365 nm, but with less visible radiation than the BL fluorescent lamp, was rather selective in capturing pink bollworm moths. Traps equipped with argon lamps captured fewer total insects, but pink bollworm moths constituted a much larger percentage of the total catch, and a higher percentage of these were female moths (Glick and Hollingsworth, 1956). Other workers have verified the high efficiency of the BL fluorescent lamps in attracting many nocturnal species (Frost, 1958; Belton and Kempster, 1963), though exceptions are observed for some species (Frost, 1954).

The spectral energy distributions of several commercial lamps used in insect attraction studies have been analyzed and compared by Deay *et al.* (1965). The curves for three of these lamps are reproduced in Fig. 5, where comparisons are illustrated for a 75-W tungsten filament incandescent lamp and two different 15-W BL fluorescent lamps. The 15-W BL lamp supplied by the General Electric Company for several years prior to 1961 utilized a phosphor designated as a "conventional type." In that year a change to the Philips phosphor was made in the manufacture of their 15-W BL lamps. The lamps with the Philips phosphor emit more visible radiation, but no substantial differences have been noted between the two types of BL lamps in attraction of insects.

D. Insect Light Traps

Many types of traps have been devised to capture insects attracted to sources of light. Historical development and some of the earlier traps

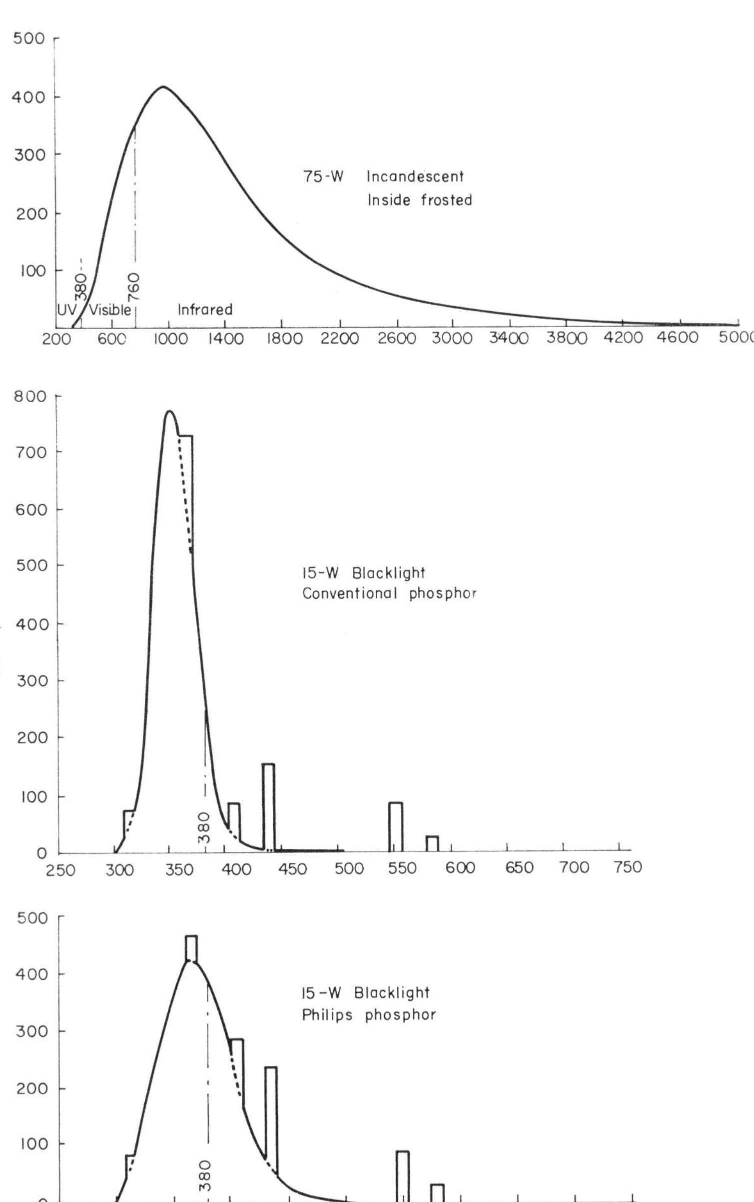

FIG. 5. Spectral energy distribution for three different lamps used in attracting insects (from Deay *et al.*, 1965).

were described by Frost (1952). Hienton (1961) described some of the more popular and successful light traps, and pertinent design characteristics and types of modern traps for survey applications have been presented by Hollingsworth *et al.* (1963). Developments in electric insect-trap design have also been discussed by Stanley (1965). Insect light-trap designs are currently under study and development as interest in control possibilities expands. Some of the future possibilities in trap development have been discussed by Hollingsworth (1965).

Generally, the traps consist of an attractant lamp and a collection device. Frequently, the 15-W BL fluorescent lamp is mounted vertically above a funnel which guides the insects into a collection can in survey traps, where the insects are killed by a suitable volatile agent. Prompt killing of the insects in survey traps is essential to preserve the catch, because beetles captured in the traps soon tear up other species, making identification difficult. Devices are usually provided for rainfall removal in order to keep the catch dry. For control applications, several means of killing the insects have been tried. Electrocuting grids have been used, and, when properly designed, are effective but costly. Cages in which the insects perish naturally, or a reservoir of water with a surface layer of oil or diesel fuel, are simple methods employed in some applications.

Sheet metal baffles are usually employed adjacent to the vertical lamps above the funnel to stop the stronger flyers, though many species are observed to dive into the funnel as they approach the trap. Baffles may serve as places for some of the smaller species to alight and remain, thus reducing the catch for those species. Use of a fan between the funnel and collection container to draw air from the funnel into the collection container is effective in increasing the catch for some species (Herms, 1932; Glick and Hollingsworth, 1956; Deay *et al.,* 1965). Proper placement and operation of the fan tend to reduce escape from the collection container. An air relief vent must be provided below the fan for effective airflow.

Numerous studies have been made to determine importance of different factors influencing effectiveness of light traps for capturing various species. Some of the general conclusions concerning light-trap design for practical application are summarized by Hartsock *et al.* (1965). Few things are as well established as the wavelength region for maximum attractiveness. Many studies have shown increases in trap catch with an increase in the number of lamps, but the increase is generally proportionately smaller for each lamp added to a trap (Glick and Hollingsworth, 1955; Deay *et al.,* 1965). Larger insect catches are generally achieved in traps providing greater exposure of the light source to the surrounding space. Straight-tube fluorescent lamps attract more insects per watt of input energy than do circular tubes, and they require less maintenance attention also (Hartsock *et al.,* 1965). Height of the trap above the ground or the crop influences the catch (Ficht *et al.,* 1940; Glick and Hollingsworth, 1956; Deay *et al.,* 1965). Naturally many environmental factors, such as weather and location with respect to vegetation, influence light-trap catches. The influence of temperature and wind on activity has been studied for some species, and these effects are evident in light-trap collections (Tashiro, 1961).

1. *Survey and Detection Applications*

Light traps have come into widespread use in recent years as an entomological survey device, and as such have been extremely helpful in overall insect control programs. They are used extensively for detection and quarantine work. Tashiro and Tuttle (1959) reported that light traps equipped with BL fluorescent lamps captured 70 times more European chafer beetles, *Amphimallon majalis* (Razoumowsky), than the best available chemical-bait traps, that the light traps were effective over a much longer period, and that they captured beetles on nights when none were seen in flight by observers. Light traps designed especially for European chafer work have been used extensively by the Plant Pest Control Division, ARS, USDA for surveys in connection with quarantine programs.

Light traps, designed especially for pink bollworm survey, have been employed on a large scale to detect possible spread of this species in cotton-growing areas of California (Berry *et al.,* 1959). Argon lamps, earlier found especially useful for this species, were employed as the attractant ultraviolet source. On many occasions, insects unknown to a particular region have been taken in light-trap collections (Glick, 1961). In addition to detection, the study of population changes and migration are other important applications. In this way light traps have been helpful in predicting potential infestations (Glick, 1961) and the need for control measures (Pfrimmer, 1961). Studies over a period of several years by Parencia *et al.* (1962) indicated that light-trap data on the cotton bollworm, tobacco budworm, cotton leafworm, *Alabama argillacea* (Hübner), and the cabbage looper could be useful, in combination with knowledge of factors influencing the population of these cotton insects, in warning cotton growers of impending injurious infestations.

Light traps also show promise in predicting the need for control measures for fruit insects. Madsen and Sanborn (1962) reported that blacklight traps were helpful in studying flight patterns of the codling moth, navel orangeworm, *Paramyelois transitella* (Walker), peach twig borer, *Anarsia lineatella* Zeller, and other fruit and nut insects. Light-trap collections provided earlier notice of codling moth activity than bait pans, and females captured by the traps were reproductively young. Similar observations have been reported from Australia for *Cydia pomonella* L. (Geier, 1960) and from France for *Laspeyresia pomonella* L. (Coutin and Anquez, 1962). The traps are therefore helpful in timing spray applications for control. Use of light traps as an aid in warning of codling moth infestations was also reported by Ehrenhardt (1961). Further optimism was expressed by Sanborn (1962) for the use of light traps for earlier prediction and better timing of spray applications in peach twig-borer control.

Information on insect light-trap collections at various locations throughout the United States is currently tabulated and published weekly for the benefit of entomologists concerned with insect control problems.* The usefulness of insect light traps will continue to expand as further information is obtained relating catches to field populations. Trap use for detection, quarantine, and survey work will also find increased application.

2. *Control Applications*

There has been interest for many years in the possible development of control methods using light. Effectiveness and economy of chemical controls, however, have made active development and testing of light-trap control possibilities less interesting until very recently, when all types of nonchemical control have been examined more carefully.

Early experiments in California using artificial light in an effort to interfere with egg-laying activity of the codling moth showed some possible reduction in damage to apples on trees which were illuminated for a few hours each evening (Herms, 1929). These studies were extended to tests in which the moths were captured in light traps. A reduction in infested apples was observed in trees near light traps equipped with a blue light source, but the results were not encouraging enough to recommend traps for control purposes (Herms and Ellsworth, 1934). A number of experiments on the attraction of codling moths in orchards in Indiana were described by Marshall and Hienton (1935, 1938). These studies concentrated on finding the most effective light source and type of trap. Parrot and Collins (1935) and Collins and Machado (1937), using light traps in orchards in New York, found consistently less codling moth damage to fruit in plots where light traps were used than in similar plots without traps, but cover sprays of insecticides were still required. Hamilton and Steiner (1939) obtained substantial reductions in codling moth damage to apples from trees equipped with light traps during two successive years in southern Indiana.

* Cooperative Economic Insect Report, issued by Plant Pest Control Division, Agricultural Research Service, United States Department of Agriculture.

Use of electric light traps for European corn borer control was studied by Ficht and Hienton (1939) and Ficht *et al.* (1940). Various trap-distribution patterns were employed in field plots of a few acres. Corn borer infestations were greatly reduced in the plots when traps were used, but complete control was not achieved. Large numbers of gravid females were captured by the light traps. The possibility of protecting sweet corn from severe corn borer losses was found, but light traps were not generally recommended for corn borer control on the basis of these results. Taylor and Deay (1950) also found light traps equipped with BL fluorescent lamps and electrocuting grids effective in reducing corn borer infestations near traps installed in corn fields.

Attempts to control certain cotton, corn, and vegetable insects in Texas with 142 blacklight traps distributed over 3000 acres were unsuccessful (Noble *et al.* 1956). Infestations of the cotton bollworm, pink bollworm, and cabbage looper were unaffected by the traps used in this experiment. In another experiment, a single trap in a screened 0.05-acre plot of cotton failed to prevent the buildup of pink bollworm infestation (Glick and Hollingsworth, 1956). Cotton insect control possibilities with light traps has experienced a new surge of interest in the past 2 years. Studies on a 40-acre plot indicated that cotton bollworm populations can be reduced by light traps (*Stanford Res. Inst. J.*, 1965). Reports on large-scale tests conducted in Texas during 1965 should also soon be available.

Experiments on controlling vegetable insects in Indiana showed that light traps equipped with BL lamps were effective in protecting cucumbers from striped and spotted cucumber beetles, *Acalymma vittata* (F.) and *Diabrotica undecimpunctata howardi* Barber (Deay *et al.*, 1959, 1963). Cucumber yields from 60- by 60-ft garden plots sprayed with dieldrin and similar plots protected only by a single trap with five 15-W BL lamps were comparable. Use of light traps and partial insecticide treatment resulted in even better yields. Tomato infestations by the tobacco hornworm, *Manduca sexta* (Johannson) (*Protoparce s.*) and tomato hornworm, *Manduca quinquemaculata* (Haworth) (*Protoparce q.*), were also substantially reduced in these studies (Deay, 1961).

Insect control with light traps has been more successful with tobacco and tomato hornworms in tobacco than in any other application. Observations by Stanley and Dominick (1958) during a 3-year study in Virginia indicated only a slight suppression of the hornworm population, but results reported by Deay (1961) on studies in southern Indiana were more promising. Subsequently, light traps were recommended for hornworm control in Indiana (Purdue Univ., 1961).

Tests in North Carolina in 1961 on recapture of marked hornworm moths showed that these insects move extensively over relatively great distances and indicated that a large area would be desirable for any control study (Stanley *et al.,* 1964). In 1962, an area 12 miles in diameter was selected for a control experiment and traps with one 15-W BL lamp were installed at an average density of three traps per square mile. Reductions in hornworm population ranged from 55% for the female tobacco hornworm to 89% for the male tomato hornworm (Stanley *et al.,* 1964). The population control estimates were supported by egg and larval counts inside and outside the test area. Data during the second season's operation in 1963 showed an overall hornworm population reduction of about 80% compared to fields outside the test area (U.S. Dept. Agr., 1963). Reductions in hornworm population at the center of the area, compared to fields 20 miles outside the area, averaged about 85% in 1964 (Stanley and Taylor, 1965). Late season stalk-cutting practices contributed to the control achieved in 1963 and 1964. The encouraging results of this large-scale light-trap control experiment have initiated other large-scale experiments for control of tobacco, cotton, and vegetable insects on the part of growers and farm groups. Scientists are cooperating in some of these experiments, and much more information on the effectiveness of these methods for control should be available in a few years.

Studies using caged, virgin female hornworm moths in the immediate vicinity of blacklight traps showed that the combined attraction of the sex lure and the light substantially increased the numbers of male moths captured (Hoffman *et al.,* 1966). Presence of two virgin female moths tripled the average catch of males, while ten virgin females increased the catch twelvefold. Experiments with the cabbage looper in California during 1965 showed similar results (Henneberry and Howland, 1966). Combined use of caged virgin females and blacklight traps increased catches to a much greater extent than either type of attractant could account for individually. The effectiveness of light in combination with the pheromone in capturing cabbage looper moths is encouraging for control possibilities.

Methods are also under study for applying chemosterilants to insects captured in light traps. Following capture and treatment, the insects are permitted to escape and rejoin the natural population (Henneberry *et al.,* 1965).

Interesting possibilities for control of the cockchafer, *Lachnosterna bidentata* Burmeister, in Malayan rubber plantations were reported by Rao (1964). The grubs of this species damage cover crops and young rubber trees by feeding on the roots. Emerging adults fly to neighboring forests to feed and mate and return to reinfest the rubber plantations. Experiments designed to capture the beetles during the flight season using blacklight traps showed considerable promise as a control method.

E. Photoperiod

Effects of light on insects in terms of the photoperiod, diapause, and polymorphism are well documented (de Wilde, 1962; O'Brien and Wolfe, 1964; Adkisson, 1965). Many insects enter diapause, a dormant state characterized by complete lack of feeding, increased resistance to low temperature, and a low rate of respiration, to enable survival until the next breeding season. Induction of diapause is principally controlled by the photoperiod, i.e., the relative duration of light and dark periods of the daily cycle. As the day length (light period) diminishes in the late summer or early autumn, some insects enter diapause. Termination of diapause in the spring or early summer can be controlled by either photoperiodic phenomena or temperature, depending upon the species.

The organism thus possesses not only the ability to detect light, but also some sort of "biological clock." The capacity of insects, as well as other animal and plant organisms, to measure time in some way has been the object of numerous studies and observations (Pittendrigh, 1961; Pittendrigh and Minis, 1964). Laboratory studies with the pink bollworm have shown that diapause is induced by light periods of 13 hours or less and prevented by light periods of 13.25 hours or longer (Adkisson, 1964). The nature of the "clock" mechanism triggering insect diapause is largely unknown, though ingenious experiments by Williams and Adkisson (1964) have shown that the brain is involved in the reception and implementation of photoperiodic signals in pupae of the oak silkworm, *Antheraea pernyi* (Guèr.). In spectral studies with the same species, the effective wavelengths for termination of diapause were found to range from 398 to 508 nm in the violet, blue, blue-green region of the spectrum (Williams *et al.,* 1965; Adkisson, 1965). Yellow and red light produced no detectable effects at all.

Beck and Hanec (1960), studying induction of diapause in the European corn borer, found that incidence of diapause increased as temperature and day length decreased. This species, like the pink bollworm, overwinters as a diapausing larva, pupates in the spring, with the moths emerging and laying eggs early in the summer. The young larvae which emerge from the eggs (first generation) feed on the corn plant. In central corn-growing areas of the United States the larvae become full-grown in midsummer, pupate, emerge as moths, and lay eggs resulting in a second generation of borers, which develop and enter diapause in the last larval stage. In the northern regions, where the growing season and day lengths are shorter, diapausing larvae are predominantly from the first generation. The difference is explained by photoperiod action in initiating diapause.

Some evidence indicates that diapausing insects measure the length of the dark period rather than the length of the light period. In work with the pink bollworm, Adkisson (1964, 1965) found in interrupting the dark period with "light breaks" of 1-hour duration at various times, that diapause was prevented only when there was an uninterrupted dark period of from 8 to 11 hours. It therefore appears that the insect uses two light signals, dusk and dawn, in determining whether or not to diapause. Studies by Barker *et al.* (1964) revealed that interrupting the dark period for the imported cabbage worm larvae, *Peris rapae* (L.), with daily photoflashes prevented diapause in pupae which would normally diapause.

These findings suggest possible development of insect control methods using light to prevent normal diapause. If light applications could be provided in the field which would promote emergence of moths too late in the season to produce another generation of overwintering or diapausing larvae or pupae, much of the population would perish during the late fall and early winter.

Other Effects

Effects of light on insect growth and development other than diapause-related phenomena have been reviewed by Ball (1965b). In addition to preferential responses by insects to various wavelengths and intensities, it has been shown that daily exposures of a few minutes to low-intensity longer wavelengths of the visible spectrum can have detrimental effects on insect growth and development (Ball, 1958). These effects were characterized by reduction in life span and weight for German cockroaches, *Blattella germanica* (L.), and by increased maturation periods, decreased adult weights, and increased mortality of the milkweed bug, *Oncopeltus fasciatus* (Dallas). The mechanism of growth inhibition by red and far-red light is not known. Ball (1958) suggested that the action may result from reception of the light by dorsal neurosecretory cells in the brain, since the insects have a relatively thin cuticle. Reception of light by brain tissue in oak silkworm pupae is the explanation offered by Williams and Adkisson in the work already mentioned on photoperiod. Additional evidence of the action of light on nerve tissue was presented by Ball (1965a) in electrophysiological responses obtained from the dark-adapted terminal abdominal ganglia of *Periplaneta americana* (L.).

Another effect of light on insects was recently reported by Riordan (1964), who found that adults of a small chalcid, *Dahlbominus fuscipennis* Zetterstedt, and the mosquito, *Aedes aegypti* (L.), were sterilized by exposures to a high-intensity discharge from a photoflash unit. A single flash sterilized 82–87% of chalcids exposed within 3 mm of the flash tube. Two flashes either killed or sterilized all exposed chalcids.

All mosquitoes were sterilized by a single flash when exposed within 7 mm of the flash tube. Effective wavelengths of the radiation from the flash tube which produce these results are still in question.

G. Practical Potential

There appear to be many possibilities for the eventual development of new insect control methods using visible and ultraviolet radiation. At the present time the most practical application is that of attracting certain economic species and destroying them. Use of light traps with effective attractant sources when applied on a large enough scale may well become practical in protecting certain crops. Experiments designed to control insects over sizeable areas should provide some of the answers concerning practical application of these methods during the next few years. Costs of traps and installation when amortized over their useful lives are very low compared to costs of chemical pesticides currently required for protection of many crops. Research on the development of better radiant attractants and possible combined use with pheromones, chemosterilants, and insecticides may provide even more promising possibilities.

Use of light to control diapause offers intriguing prospects for future control methods. On the basis of laboratory results already available, for example, with the pink bollworm and the European corn borer, it would appear that use of light in the field to extend the day length may be effective in preventing diapause and thereby in reducing the population of susceptible species. Perhaps the mere interruption of dark periods with brief exposures to light may be effective if properly carried out. Many questions remain unanswered relating to the mechanisms of action of light on insect life and development. Ultimate exploitation of control possibilities must await the findings of future research.

V. IONIZING RADIATIONS

A. General

Beyond the ultraviolet portion of the electromagnetic spectrum, in the direction of increasing quantum energy, lies the x-ray region. x-Rays are emitted when orbital electrons in inner shells of atoms near the nucleus change from a higher energy level to a lower level. The energies of x-ray photons are sufficient to produce ionization in biological tissue as well as in gases and other materials. Since x-rays are easily produced in the laboratory by bombarding a target of tungsten or other suitable material in an evacuated tube with a beam of electrons, much research on the effects of ionizing radiation on biological materials has been conducted using x-rays. So-called "soft" x-rays of longer wavelength are less penetrating than the shorter wavelength "hard" x-rays. The hardness of x-rays and corresponding energy of the quanta are dependent upon the energy of the electrons in the x-ray tube.

γ-Rays have the same effect on biological tissue as x-rays; however, γ-rays can be more energetic and therefore more penetrating. γ-Rays are emitted from the nuclei of radioactive materials and arise from nuclear disintegration and other reactions. There is an overlapping in the spectrum of x-rays and γ-rays (Fig. 1). The nature of the two in this region is believed to be the same except for their origin.

Because of the penetrating nature of x-rays, differing absorption characteristics of different materials, and ease of production and detection with sensitized films, x-rays have been widely used, not only in medical diagnostic and therapeutic work, but also for industrial inspection of metals, such as detection of flaws in castings. x-Ray techniques have been developed for detection of hidden insect infestation in grain (Milner *et al.,* 1950, 1952). By using soft x-rays which provide greater contrast, good detail concerning insect forms inside kernels of grain can be obtained. Thus development of internal forms can be studied to obtain information available in no other way.

Application of ionizing radiation to insect control problems appears promising in two ways generally. Such radiations can be applied in doses lethal to the population, or in doses which sterilize the irradiated insects. Radiation-induced sterilization is the subject of the following chapter, and so it will not be treated further here except to note that it is a powerful tool as already demonstrated in the widely acclaimed large-scale eradication of the screwworm using the "sterile-male" method.

In any practical application of ionizing radiation for insect control, the cost of the source of radiation is a major factor. Since production of x-rays is very inefficient, x-radiation has not been considered for any application involving irradiation of large quantities of any infested product. γ-Radiation, which is available in abundance from certain radioisotopes, is much more promising for such applications. β-Rays, which are energetic electrons emitted from nuclei of disintegrating atoms, or electrons artificially accelerated to very high energies, are another type of ionizing radiation which is considered for such applications. Strictly speaking, electrons are particles and therefore not electromagnetic energy in the true sense. Accelerators are available which provide beams of high energy electrons. Accelerated electrons are not as penetrating as γ-rays; however, the radiation from such accelerators can be turned off and on as needed.

There is a good deal of literature on effects of ionizing radiations on insects, for certain insect species are convenient and useful laboratory animals for many such basic studies. No attempt will be made to review here, even generally, the literature relating to these types of studies. Recent reviews have been prepared by Grosch (1962), O'Brien and Wolfe (1964), and Cornwell (1964a). In fact, the literature on application of ionizing radiation for insect control studies is too voluminous to review effectively in this treatment. Instead, certain previous reviews will be cited and findings of certain studies will be summarized to provide a general accumulation of pertinent factors relating to direct insect control with ionizing radiation.

Some clarification may be helpful in regard to units employed to indicate measured dosage of ionizing radiation applied in different studies. One finds in the literature doses given in r (roentgens), rep (roentgens-equivalent-physical), and rads. The roentgen unit is defined in terms of the ionization produced in air by x-rays or γ-rays, specifically, the quantity of electromagnetic radiation which produces one electrostatic unit of ions, of either sign, per cubic centimeter in pure air under standard conditions. It is thus defined in terms of energy absorption per unit volume in air rather than incident energy. The effects of sources of ionizing particles such as electrons can also be evaluated in terms of roentgens. The rep was established to provide a unit of energy absorption for media other than air, and was originally defined as 83 ergs per gram of biological tissue, since the roentgen corresponds to about 83 ergs per gram of air. Energy absorption in different tissues, however, varies with the nature of the tissue and the energy of the photons. The definition of the rep was later revised to 93 ergs per gram of soft tissue, corresponding more nearly to the energy absorbed from photons of moderate energy in soft biological tissue. In order to avoid the confusion resulting from the use of the rep, the rad was finally adopted as the quantity of ionizing radiation which results in the absorption of 100 ergs per gram of irradiated material at the point of interest. Since all three units are of the same order of magnitude, and the doses required to achieve specified biological effects usually vary somewhat with a number of other variables, required doses are often not well-enough defined that the difference between the different radiation dosage units is very important. It is well to understand the difference, however, in cases where close comparisons may be made. The care with which dosimetry was handled in any particular study is another factor to be considered.

B. Stored-Grain Insects

Control of an insect population by direct irradiation appears to have good potential for practical use in control of stored-product insects. Since, economically, grain is probably the most important stored product with which we have serious insect problems, the use of electron and γ-radiation for disinfesting grain has received much attention.

Studies of ionizing radiation for insect control in grain and other products have been reviewed by Cornwell (1959a,b, 1964a, 1966), Cornwell and Bull (1960), Nelson (1962, 1966), O'Brien and Wolfe (1964), Brownell (1964), and Nelson and Seubert (1966). Results of earlier studies are summarized in a publication by the United States Army Quartermaster Corps (1957).

The lethal and sterilizing effects of x-rays, γ-rays, and accelerated electrons are about equivalent in their action on insects (Proctor *et al.*, 1954; Baker *et al.*, 1953a,b). In general, the mortality of insects exposed to ionizing radiation increases with increasing dosage, and complete mortality occurs in a shorter time as the dose is increased. For complete mortality of adults of most stored-grain insect species within a few days, doses of radiation ranging between about 200 and 500 kilorads (krads, 10^3 rads) are required. Complete mortality for many stored-grain insect species occurs a week or two following exposure to doses of the order of 100 krads. Doses ranging from about 10 to 20 krads have been found completely effective for sterilization and cause eventual death of the irradiated population within a few to several weeks.

Hassett and Jenkins (1952) exposed eight different insect species to γ-radiation from Co^{60} using a series of dosages ranging from 16.1 to 322 kiloroentgens (kr). Included were adults and larvae of the black carpet beetle, *Attagenus piceus* (Olivier), and the larder beetle, *Dermestes lardarius* L., and adults of the cigarette beetle, *Lasioderma serricorne* (F.), rice weevil, *Sitophilus oryzae* (L.), lesser grain borer, *Rhyzopertha dominica* (F.), confused flour beetle, *Tribolium confusum* Jacquelin duVal, and powder-post beetle (southern lyctus beetle), *Lyctus planicollis* LeConte, all important stored-product pests, and for comparison the vinegar fly, *Drosophila melanogaster* Meigen. Differences in survival were noted among species; the powder-post beetle, carpet beetle, dermestid larvae, and lesser grain borer were more resistant than the other species. The general conclusion indicated that a dose of 65 kr would constitute a completely lethal treatment and that control could be achieved by stopping reproduction with doses in the 1.6- to 3.2-kr range.

Proctor *et al.* (1954) also found the lesser grain borer somewhat more resistant to γ-radiation than the cigarette beetle and the saw-toothed grain beetle, *Oryzaephilus surinamensis* (L.); the confused flour beetle survival curve fell in an intermediate range. Doses required for complete mortality 1 day after treatment ranged from about 200 to 300 krep for adults of these four species, but a 30-krep exposure was effective for reproductive sterilization. Using accelerated electrons of 2 MeV (10^6 eV) energy, Baker *et al.* (1953a,b) found that doses of 500 and 250 krep provided complete immediate mortality, respectively, for adult confused flour beetles and granary weevils, *Sitophilus granarius* (L.). Treatment at 10 krep prevented eggs of the two species from hatching and completely sterilized the adults. A study in the Soviet Union by Peredel'skii *et al.* (1957) indicated that 5 kr of x-radiation was almost sufficient for sterilization of granary weevils in grain, and it was estimated that 10 kr might be adequate to sterilize all grain and flour insects.

Cornwell *et al.* (1957) reported adult survival data and emergence data for immature stages from tests involving 17 different species of cereal-infesting insects which were exposed to different doses of γ-radiation from Co^{60}. Energies of γ-rays from Co^{60} are 1.2 and 1.3 MeV. Survival curves for adults of eight species irradiated at 50 krep showed that the lesser grain borer and red flour beetle, *Tribolium castaneum* (Herbst), were the most resistant, and the rice weevil and flat grain beetle, *Cryptolestes pusillus* (Schönherr), were the most susceptible of the species compared. Adults of all species were sterilized. Additional tests showed that 6 krep was sterilizing for some species, but not completely effective for others, including the granary weevil and both confused and red flour beetles. Treatment of immature stages at 20 krep reduced emergence of adults from irradiated cultures, inhibited development of young larvae, and resulted in complete sterility among adults which did emerge. Nicholas and Wiant (1959) studied effects of 1-MeV electrons on different species and developmental stages of several grain-infesting species. Results verified the increasing resistance to radiation as metamorphosis progresses from egg to adult. Retardation of larval development due to irradiation was also noted. Three moth species included were found more resistant than various beetle species. Dennis *et al.* (1962), reporting work conducted with a similar accelerator in 1954, found that a dose of 10 krep prevented rice weevil eggs from developing and provided a high degree of adult sterilization. Other studies by Dennis (1961) using γ-radiation from Co^{60}, showed that a few confused flour beetle adults survived 100-kr exposures 2 months after irradiation. The same exposure produced complete mortality after 2 months for larvae of the same species and for adult rice weevils, granary weevils, saw-toothed grain beetles, lesser grain borers, and larvae of the Indian-meal moth, *Plodia interpunctella* (Hübner), and the almond moth, *Cadra cautella* (Walker) (*Ephestia c.*).

The most extensive series of investigations to date on effects of ionizing radiation on insects, with a view toward practical utilization, have been conducted at the Wantage Research Laboratory of the United Kingdom Atomic Energy Authority under the direction of Dr. P. B. Cornwell. While a number of publications on the studies have appeared in the journals, for example, Cornwell (1958, 1959a, 1961), Cornwell and Bull (1960), and a chapter by Cornwell (1964b) included some of the data, much of the information has been available only in a series of Atomic Energy Research Establishment (AERE) reports. This material is to be summarized in a new book now in press (Cornwell, 1966). Detailed studies with different developmental stages showed that a γ-radiation dose of 16 krads would, through sterilization, provide effective control of the rice weevil and granary weevil (Cornwell and Morris, 1959a), confused flour beetle (Banham, 1962), red flour beetle (Crook, 1962), saw-toothed grain beetle, though slight fertility remained, (Jefferies, 1962), and the lesser grain borer and cigarette beetle (Pendlebury *et al.*, 1962b). The 16-krad dose was deemed unlikely to provide control of the tropical warehouse moth (almond moth) and the Indian-meal moth (Pendlebury *et al.,* 1962b). A comparison of γ-irradiation susceptibility of 5 laboratory strains and 30 wild strains of granary weevil showed that, though some of the wild strains appeared somewhat more resistant, the 16-krad dose would still be expected to provide control of this species in commercial channels (Cornwell and Morris, 1959b).

Other studies revealed that high temperatures preceding irradiation sensitized granary weevils to ionizing radiation, as indicated by mortality, while low temperature pretreatment gave some protection (Pendlebury *et al.,* 1962a). During irradiation, effects of temperature were reversed. Temperature was not found to modify susceptibility to sterilization however. Densely crowded cultures of the granary weevil, which resulted in increased metabolic temperatures during larval development, increased the lethal effect of radiation, but did not influence sterilization susceptibility (Martin *et al.,* 1962). A decrease in the efficiency of radiation was observed for granary weevils when the dose was fractionated, i.e., when the dose was given in separate applications with intervening periods of time (Jefferies and Cornwell, 1958; Jefferies and Banham, 1961). Loss of effectiveness due to fractionation of dose varied with number of fractions, interval time, temperature, and development stage of the insects. A comparison of granary weevil irradiation with γ-rays from Co^{60} and 2–3-MeV electrons revealed that no differences could be detected in mortality characteristics of insects exposed to the two different sources in the pupal stage (Bull *et al.,* 1961). Mature adults were found more susceptible to γ-radiation, however, both in mortality sustained and degree of sterilization. Equivalent rates in the application of equal doses, however, were not available with the equipment employed. These studies also indicated little promise concerning possible use of granary weevils as biological dosimeters for engineering studies on radiation disinfestation of grain.

Another phase of the studies conducted at Wantage involved appraisal of sterile-male release methods for control of the flour-mill moth, *Ephestia kuhniella* Zeller. Pertinent biological factors were determined and methods were devised for rearing the large numbers of moths which would be needed (Bull and Wond, 1962), but biological and ecological studies in an infested mill indicated that the species was not suitable for control by sterile-male release (Crook *et al.,* 1960).

Some concern has been expressed that a margin of safety be employed in any practical application of ionizing radiation for sterilization of insects to insure that none capable of reproducing survive, because of possible development of radiation-resistant strains through combined effects of radiation-induced variants and natural selection (Cornwell, 1959a, 1964a). Development of such resistance has not been observed in the laboratory. Studies on control of granary weevil populations with substerilizing doses of 1–14.5 krads of γ-radiation, however, have shown that such doses suppress the populations to very low levels, and, further, that the reproductive potential of nonsterilized insects introduced into the population is depressed due to the presence of sterile males from the irradiated population (Cornwell *et al.,* 1962). Partial protection against reinfestation is thus provided by surviving sterile males when lower doses than those required for complete control (16 krads) are employed.

C. Other Insects

While stored-grain insects have occupied the efforts of most researchers interested in control of stored-product insects with ionizing radiation, other possibilities have also been explored. A 10-krep exposure of adult bean weevil, *Acanthoscelides obtectus* (Say), in navy beans to 2-MeV electrons produced complete mortality, but sterility effects were not reported (Baker *et al.,* 1954). Work on the black carpet beetle, larder beetle, cigarette beetle, and powder-post beetle was mentioned in Section V,B. Effects of Co^{60} γ-rays on a powder-post beetle, *Lyctus brunneus* Steph., the furniture beetle, *Anobium punctatum* (De Geer), and the death-watch beetle, *Xestobium rufovillosum* (De Geer), were studied by Bletchly and Fisher (1957). Irradiation of all three species at doses of 8 kr resulted in completely sterile eggs. Development of 1- to 4-day-old eggs of *Anobium* and *Xestobium* was prevented by 4 kr, but radiation resistance of the eggs increased rapidly as they developed, and doses in excess of 32 kr were required to kill mature eggs of either species. Further studies showed that 10 kr was an effective dose in preventing completion of development and for sterilization (Bletchly, 1961).

γ-Radiation from Co^{60} has been used in tests on Mexican fruit fly, *Anastrepha ludens* (Loew), eggs and larvae in grapefruit (Brownell and

Yudelovitch, 1962). Doses of 5 krads were effective in breaking the life cycle of the insect. The same dose stopped development of eggs and first-instar larvae completely, and though larvae which were mature at time of irradiation pupated, no adult flies emerged from the fruit. Proposals were presented for design of a mobile railway γ-irradiator for treating infested fruit. Mobile irradiators have also been proposed for soil treatment to kill soil insects and nematodes with γ-radiation. It has been suggested that doses of 100 krads would be sufficient to inhibit weed growth as well as sterilize insects and nematodes (Bowen and Smith, 1959).

D. Effects on Products

For any ionizing radiation treatment to be useful, it must accomplish the desired effects on the insects without materially damaging the host product which one is trying to preserve. Fortunately, this is possible in cases of major interest, though it is possible for the radiation sensitivity of the host material to exceed that of the infesting organism, and here radiation would have little to offer for pest control.

Ionizing radiation cannot be recommended for insect control in grain intended for use as seed. Baker *et al.* (1953a,b, 1954) reported that wheat exposed at 100-krep doses of 2-MeV electrons germinated but failed to emerge from the soil, and that seed treated at 10 krep gave rise to seedlings definitely retarded in growth. Similar results were observed for navy beans. More detailed studies of botanical effects of 1-MeV electrons on germination and growth of wheat exposed to doses between 10 and 200 krep showed that doses of 40 krep or more drastically reduced emergence from soil and that even the lowest dosage had detrimental effects on early growth (Soderholm and Walker, 1955).

A number of studies have shown that irradiation of grain at dosage levels required for insect control are not harmful to the milling and baking qualities of cereal grains used for these purposes. Milner and Yen (1956) found that γ-ray doses of 125 krep were not damaging to the bread-making quality of flour milled from irradiated wheat, but doses of 250 krep and more were detrimental. Baker *et al.* (1953a) reported that satisfactory loaves of bread were obtained from both wheat and flour irradiated with 500-krep doses of 2-MeV electrons, and that slight flavor changes were detectable but were not objectionable. γ-Irradiation at doses ranging from 125 to 500 krads induced slight changes in biscuits and scones baked from irradiated wheat, but the changes were of little consequence (Cornwell, 1959b). Cropsey *et al.* (1962) found the threshold for flavor and odor changes in flour milled from electron-irradiated wheat at about 170 krep for 0.6-MeV electrons and at 46 krep for 1-MeV electrons regardless of wheat moisture content. Kraybill (1959) reported that wheat irradiated at 250–500 krads produces a satisfactory loaf of bread with only slight objectionable flavor, and the off-flavor threshold for irradiation of flour lies between 20 and 50 krads. No effect on stability or nutritive value of oats was found at 1 Mrad (10^6 rads), but slight flavor changes were found. A number of spices were satisfactorily irradiated at 150 krads except for cinnamon which had a bitter flavor at this dose but was all right at 75 krads. Brownell and Yudelovitch (1962), in the report already cited, found that no flavor or other undesirable changes were found in taste-panel tests of grapefruit irradiated at doses up to 30 krep.

Effects of ionizing radiation on food products have been studied extensively in connection with the United States Army Quartermaster Corps program on radiation preservation of foods (1957), the United States Atomic Energy Commission program concerned with food pasteurization by radiation, and a food-irradiation research program in the United Kingdom (Ley, 1964). Serious off-flavors can be produced in many products by the several-Mrad doses required for bacterial sterilization, but for the much lower dose levels required for insect sterilization, no flavor changes are noted in most products.

The question of wholesomeness of irradiated foods was, according to Kraybill (1960), confused by publication in the press of misleading information about results of some experimental animal-feeding tests. There is no reliable evidence of any toxicity in irradiated foods, and deleterious effects reportedly due to ingestion of such foods were largely explained on the basis of well-known nutritional deficiencies. Ionizing radiation has little effect on the macronutrients, but fat-soluble vitamins are quite easily destroyed by radiation (Read, 1959). Water-soluble vitamins are destroyed no more by ionizing radiation than by normal cooking procedures. Results of many long-term animal-feeding tests indicate that no significant changes are produced in wholesomeness of foods by irradiation of levels needed for preservation (Schweigert, 1959; Brownell, 1964). Since doses many times smaller than those of interest in food sterilization are required for radiation control of stored-product insects, there seems to be very little question that radiation treatment would impair in any way the quality of such products for consumptive use. Action of the United States Food and Drug Administration in 1963 in approving wheat irradiation for absorbed doses of 20–50 krads with γ-ray sources providing energies of not greater than 2.2 MeV attests to this belief in the safety of using such irradiated material for food purposes. Similar approval for electron irradiation is expected to be forthcoming.

E. Practical Aspects

The problems and techniques involved in applying ionizing radiation on a practical scale for radiation disinfestation of grain and other products and the potential for its development have been treated in great depth, particularly by two of the pioneering leaders in this work, P. B. Cornwell of the United Kingdom and L. E. Brownell of the United States. The interested reader is referred to a number of these excellent treatments rather than subjected to an inadequate summary or lengthy recasting of the pertinent material here. The potential of ionizing irradiation of grain for insect control has been considered in detail by Cornwell and Bull (1960), Cornwell (1961, 1966), and Brownell (1964), and was discussed from a global point of view (Cornwell, 1964a,b, 1965; Brownell, 1964). Design of γ-irradiation facilities have been considered for a number of possible future applications (Brownell *et al.,* 1955; Cornwell and Bull, 1960; Brownell and Yudelovitch, 1962; Horne and Brownell, 1962; Brownell, 1964). Estimated costs of radiation disinfestation and factors influencing these costs have been discussed (Cornwell, 1959a, 1964a; Cornwell and Bull, 1960; Horne and Brownell, 1962; Brownell, 1964). Although costs for radiation treatment are generally believed higher than for conventional chemical fumigation, costs for radiation treatment could become competitive in certain situations. Relative merits of γ-radiation sources and electron accelerators are yet to be evaluated for large-scale practical application, but each may have its place and both types of sources can be made available with the capacity to process grain at the high rates which grain terminals customarily employ in handling grain.

While progress in approaching practical application of ionizing radiation for grain disinfestation has been much slower than anticipated by those working in the field, concrete evidence of interest in the process continues. The United States Atomic Energy Commission has made available to the Department of Agriculture a γ-irradiation plant designed for pilot-scale studies on stored-product insect control. These will be conducted at the USDA Stored-Products Research and Development Laboratory at Savannah, Georgia. A description of the facility and the planned research program was recently outlined by Laudani *et al.* (1965). The Co^{60} bulk-grain irradiator is designed for a capacity range from 1.25 to 5 tons/hour, with a dose of 25 krads at a flow rate of 3.4 tons/hour.

An undertaking which promises to provide an even better test of the potential for large-scale disinfestation was recently approved by the United Nations Special Fund Board, upon the recommendation of the Joint FAO/IAEA Division of Atomic Energy in Agriculture. A 20–70-ton/hour Co^{60} irradiation plant is to be installed in a grain terminal at the port city of Iskerderun, Turkey (Goresline, 1965; Cornwell, 1965). For a number of reasons, this appears to be a well-chosen location for the evaluation of the irradiation process. The khapra beetle, *Trogoderma granarium* Everts, is the principal grain pest, but many other species are present as well. Iskerderun is the collecting center for durum and soft wheat from all of southeast Turkey, and the terminals are equipped with modern grain-handling equipment. Completion of the installation is scheduled for the summer of 1966. A general industrial appraisal and evaluation of the operation is planned, including engineering, grain handling, entomological, and economic analyses.

Within a few years, results from the studies made possible by these two new and interesting advances in the development of radiation processing for insect control should help materially in pointing the way toward utilization of this type of energy for pest control if it can be demonstrated to be economically sound.

ACKNOWLEDGMENTS

The author gratefully acknowledges the helpful suggestions of the following colleagues who reviewed the manuscript for this chapter: H. J. Ball, J. G. Hartsock, T. E. Hienton, J. P. Hollingsworth, and J. M. Stanley.

REFERENCES

Adams, P. A., and Heath, J. E. (1964). *Nature* **201,** 20.

Adkisson, P. L. (1964). *Am. Naturalist* **98,** 357.

Adkisson, P. L. (1965). *Proc. Conf. Electromagnetic Radiation Agr.* p. 30. Illum. Eng. Soc., Am. Soc. Agr. Engrs.

Andreev, S. V. (1937). *Rev. Appl. Entomol.* **A25,** 154.

Andreev, S. V., and Balkashin, B. (1935). *Zashchita Rast.* (**I**), 121.

Ark, P. A., and Parry, W. (1940). *Quart. Rev. Biol.* **15,** 172.

Baker, V. H., Taboada, O., and Wiant, D. E. (1953a). *Mich. State Univ. Agr. Expt. Sta. Quart. Bull.* **36,** 94.

Baker, V. H., Taboada, O., and Wiant, D. E. (1953b). *Agr. Eng.* **34,** 755.

Baker, V. H., Taboada, O., and Wiant, D. E. (1954). *Agr. Eng.* **35,** 407.

Baker, V. H., Wiant, D. E., and Taboada, O. (1956). *J. Econ. Entomol.* **49,** 33.

Ball, H. J. (1958). *J. Econ. Entomol.* **51,** 573.

Ball, H. J. (1965a). *J. Insect Physiol.* **11,** 1311.

Ball, H. J. (1965b). *Proc. Conf. Eelectromagnetic Radiation Agr.* p. 36. Illum. Eng. Soc., Am. Soc. Agr. Engrs.

Banham, E. J. (1962). *U.K. At. Energy Authority Res. Group Rept.* **AERE–R3888.**

Barker, R. J., Cohen, C. F., and Mayer, A. (1964). *Science* **145,** 1195.

Barr, A. R., Smith, T. A., Boreham, M. M., and White, K. E. (1963). *J. Econ. Entomol.* **56,** 123.

Beck, S. D., and Hanec, W. (1960). *J. Insect Physiol.* **4,** 304.

Belton, P., and Kempster, R. H. (1963). *Can. Entomologist* **95,** 832.

Berry, N. O., Blanc, F. L., and Klopfer, S. M. (1959). *Calif. Dept. Agr. Bull.* **48,** 211.

Bertholf, L. M. (1931a). *J. Agr. Res.* **42,** 379.

Bertholf, L. M. (1931b). *J. Agr. Res.* **43,** 703.

Bertholf, L. M. (1932). *Z. Vergleich. Physiol.* **18,** 32.

Bletchly, J. D. (1961). *Ann. Appl. Biol.* **49,** 362.

Bletchly, J. D., and Fisher, R. C. (1957). *Nature* **179,** 670.

Bollaerts, D., Quoilin, J., and van den Bruel, W. E. (1961). *Compt. Rend. 13e Symp. Intern. Phytopharm. Phytiat.,* p. 1435.

Bowen, H. J. M., and Smith, S. R. (1959). *Nature* **183,** 907.

Brown, A. W. A. (1951). *Bull. Entomol. Res.* **42,** 575.
Brownell, L. E. (1964). *U.S. Dept. Agr., Northern Util. Res. Develop. Div., Rept. 2nd Natl. Conf. Wheat Util. Res.*
Brownell, L. E., and Yudelovitch, M. (1962). *Intern. Symp. Radioisotopes Radiation Entomol Pt. I. Use Radioisotopes Tracers. Sect. I. Ecol. Gen. Biol.* p. 193. Intern. At. Energy Agency, Vienna.
Brownell, L. E., Nehemias, J. V., and Bulmer, J. J. (1955). Publ. 1943, 7–40–P under Proj. 1943–7, U.S. At. Energy Comm., Eng. Res. Inst., Univ. of Michigan, Ann Arbor, Michigan.
Bull, J. O. and Wond, T. J. (1962). *U.K. At. Energy Authority Res. Group Rept.* **AERE–R3895.**
Bull, J. O., Wond, T., and Cornwell, P. B. (1961). *U.K. At. Energy Authority Res. Group Rept.* **AERE–R3890.**
Burkhardt, D. (1964). *Advan. Insect Physiol.* **2,** 131.
Calderwood, D. L., and Altman, L. B. (1954). Unpublished observations.
Callahan, P. S. (1965a). *Nature* **206,** 1172.
Callahan, P. S. (1965b). *Ann. Entomol. Soc. Am.* **58,** 727.
Callahan, P. S. (1965c). *Ann. Entomol. Soc. Am.* **58,** 746.
Collins, D. L., and Machado, W. (1937). *J. Econ. Entomol.* **30,** 422.
Cornwell, P. B. (1958). *New Scientist* **4,** No. 79.
Cornwell, P. B. (1959a). *Intern. J. Appl. Radiation Isotopes* **6,** 188.
Cornwell, P. B. (1959b). *J. Sci. Food Agr.* **10,** 409.
Cornwell, P. B. (1961). *Food Irradiation* **1,** A9.
Cornwell, P. B. (1964a). *Proc. Intern. Food Ind. Congr.*, Session 1, Paper 5.
Cornwell, P. B. (1964b). *In* "Massive Radiation Techniques" (S. Jefferson, ed.), p. 141, Wiley, New York. Newnes, London.
Cornwell, P. B. (1965). *Food Irradiation* **6,** A2.
Cornwell, P. B. (1966). "The Entomology of Radiation Disinfestation of Grain—a collection of original research papers." Macmillan (Pergamon), New York.
Cornwell, P. B., and Bull, J. O. (1960). *J. Sci. Food Agr.* **11,** 754.
Cornwell, P. B., and Morris, J. A. (1959a). *U.K. At. Energy Authority Res. Group Rept.* **AERE–R3065.**
Cornwell, P. B., and Morris, J. A. (1959b). *U.K. At. Energy Authority Res. Group Rept.* **AERE–R3163.**
Cornwell, P. B., Crook, L. J., and Bull, J. O. (1957). *Nature* **179,** 670.
Cornwell, P. B., Martin, V. J., Burson, D. M., Bull, J. O., and Pendlebury, J. B. (1962). *U.K. At. Energy Authority Res. Group Rept.* **AERE–R3892.**
Coutin, R., and Anquez, P. (1962). *Advan. Hort. Sci. Appl. Proc. 15th Intern. Hort. Congr., Nice, 1958* **2,** 280.
Crook, L. J. (1962). *U.K. At. Energy Authority Res. Group Rept.* **AERE–R3889.**
Crook, L. J., Bull, J. O., and Cornwell, P. B. (1960). *U.K. At. Energy Authority Res. Group Rept.* **AERE–R3297.**
Cropsey, M. G., Leach, C. M., Ching, T. M., Sather, L. A., and Wiant, D. E. (1962). *Ann. Appl. Biol.* **50,** 487.
Davis, J. H. (1934). U.S. Patent No. 1,972,050.
Davis, J. H. (1936a). U.S. Patent No. 2,040,600.
Davis, J. H. (1936b). U.S. Patent No. 2,064,522.
Deay, H. O. (1961). *U.S. Dept. Agr.* **ARS 20–10.**
Deay, H. O., Taylor, J. G., and Johnson, E. A. (1959). *Proc. N. Central Branch Entomol. Soc. Am.* **14,** 21.

Deay, H. O., Hartsock, J. G., and Barrett, J. R., Jr. (1963). *Proc. N. Central Branch Entomol. Soc. Am.* **18,** 37.
Deay, H. O., Barrett, J. R., Jr., and Hartsock, J. G. (1965). *Proc. N. Central Branch Entomol. Soc. Am.* **20,** 109.
Dennis, N. M. (1961). *J. Econ. Entomol.* **54,** 211.
Dennis, N. M., Soderholm, L. H., and Walkden, H. H. (1962). *U.S. Dept. Agr.* **AMS 531.**
Dethier, V. G. (1953). *In* "Insect Physiology" (K. D. Roeder, ed.), p. 488. Wiley, New York.
Dethier, V. G. (1963). "The Physiology of Insect Senses." Wiley, New York.
de Wilde, J. (1962). *Ann. Rev. Entomol.* **7,** 1.
Duane, J. P., and Tyler, J. E. (1950). *Interchem. Rev.* Spring-Summer p. 25.
Earp, U. F., and Stanley, J. M. (1965). *Proc. Conf. Electromagnetic Radiation Agr.* p. 25. Illum. Eng. Soc., Am. Soc. Agr. Engrs.
Ehrenhardt, H. (1961). *Rev. Appl. Entomol.* **49,** 518.
Elec. World. (1950). **134,** 120.
Evreinov, M. G. (1935). *Elek. Selsk. Khoz.* **I,** 20.
Ficht, G. A., and Hienton, T. E. (1939). *J. Econ. Entomol.* **32,** 520.
Ficht, G. A., and Hienton, T. E. (1941). *J. Econ. Entomol.* **34,** 599.
Ficht, G. A., Hienton, T. E., and Fore, J. M. (1940). *Agr. Eng.* **21,** 87.
Fingerman, M., and Brown, F. A., Jr. (1952). *Science* **116,** 171.
Fraenkel, G. S., and Gunn, D. L. (1961). "The Orientation of Animals." Dover, New York.
Frings, H. (1952). *J. Econ. Entomol.* **45,** 396.
Frost, S. W. (1952). *Penn. State Coll. Agr. Bull.* **550.**
Frost, S. W. (1954). *J. Econ. Entomol.* **47,** 275.
Frost, S. W. (1958). *Proc. 10th Intern. Congr. Entomol., Montreal, 1956* **2,** 583.
Frost, S. W., Dills, L. E., and Nicholas, J. E. (1944). *J. Econ. Entomol.* **37,** 287.
Geier, P. W. (1960). *Nature* **185,** 709.
Glick, P. A. (1961). *U.S. Dept. Agr.* **ARS 20–10.**
Glick, P. A., and Hollingsworth, J. P. (1954). *J. Econ. Entomol.* **47,** 81.
Glick, P. A., and Hollingsworth, J. P. (1955). *J. Econ. Entomol.* **48,** 173.
Glick, P. A., and Hollingsworth, J. P. (1956). *J. Econ. Entomol.* **49,** 158.
Goldsmith, T. H. (1960). *J. Gen. Physiol.* **43,** 775.
Goldsmith, T. H. (1961). *In* "Light and Life" (W. D. McElroy and B. Glass, eds.), p. 771. Johns Hopkins Press, Baltimore, Maryland.
Goldsmith, T. H. (1964). *In* "The Physiology of Insecta" (M. Rockstein, ed.), Vol. I, p. 397, Academic Press, New York.
Goldsmith, T. H., and Ruck, P. R. (1958). *J. Gen. Physiol.* **41,** 1171.
Goresline, H. E. (1965). *Food Irradiation* **6,** A10.
Grant, G. R. M. (1948). *Proc. Roy. Soc. Queensland* **60,** 89.
Grosch, D. S. (1962). *Ann. Rev. Entomol.* **7,** 81.
Hamilton, D. W., and Steiner, L. F. (1939). *J. Econ. Entomol.* **32,** 867.
Hartsock, J. G. (1961). *U.S. Dept. Agr.* **ARS 20–10.**
Hartsock, J. G., Deay, H. O., and Barrett, J. R., Jr. (1965). Paper presented at *Ann. Meeting Entomol. Soc. Am., New Orleans, Louisiana.*
Hartzell, A. (1934). *Contrib. Boyce Thompson Inst.* **6,** 211.
Hassett, C. C., and Jenkins, D. W. (1952). *Nucleonics* **10,** 42.
Headlee, T. J. (1937). *J. Econ. Entomol.* **30,** 309.
Headlee, T. J., and Burdette, R. C. (1929). *J. N.Y. Entomol. Soc.* **37,** 59.

Heller, J. H., and Teixeira-Pinto, A. A. (1959). *Nature* **183,** 905.
Henneberry, T. J., and Howland, A. F. (1966). *J. Econ. Entomol.* **59,** 623.
Henneberry, T. J., Howland, A. F., and Wolf, W. W. (1965). *Proc. Conf. Electromagnetic Radiation Agr.* p. 34. Illum. Eng. Soc., Am. Soc. Agr. Engrs.
Herms, W. B. (1929). *J. Econ. Entomol.* **22,** 78.
Herms, W. B. (1932). *Agr. Eng.* **13,** 292.
Herms, W. B., and Ellsworth, J. K. (1934). *J. Econ. Entomol.* **27,** 1055.
Hienton, T. E. (1961). *U.S. Dept. Agr.* **ARS 20–10.**
Hienton, T. E., Wiant, D. E., and Brown, O. A. (1958). "Electricity in Agricultural Engineering." Wiley, New York. Chapman & Hall, London.
Hoffman, J. D., Lawson, F. R., and Peace, B. (1966). *J. Econ. Entomol.* **59,** 809.
Hollingsworth, J. P. (1961). *U.S. Dept. Agr.* **ARS 20–10.**
Hollingsworth, J. P. (1965). *Proc. Conf. Electromagnetic Radiation Agr.* p. 28. Illum. Eng. Soc., Am. Soc. Agr. Engrs.
Hollingsworth, J. P., Hartsock, J. G., and Stanley, J. M. (1963). *U.S. Dept. Agr.* **ARS 42–3–1.**
Hollingsworth, J. P., Wright, R. L., and Lindquist, D. A. (1964a). *Agr. Eng.* 45, 314.
Hollingsworth, J. P., Wright, R. L., and Lindquist, D. A. (1964b). *J. Econ. Entomol.* **57,** 38.
Horne, T., and Brownell, L. E. (1962). *In* "Radioisotopes and Radiation in Entomology," Proc. Symp., Bombay, 1960. Intern. At. Energy Agency, Vienna.
Jander, R. (1963). *Ann. Rev. Entomol.* **8,** 95.
Jefferies, D. J. (1962). *U.K. At. Energy Authority Res. Group Rept.* **AERE-R3891.**
Jefferies, D. J., and Banham, E. J. (1961). *U.K. At. Energy Authority Res. Group Rept.* **AERE–R3503.**
Jefferies, D. J., and Cornwell, P. B. (1958). *Nature* **182,** 402.
Johnston, J. W., Jr. (1953). *Physiol. Zool.* **26,** 266.
Kadoum, A. M. A. A. (1963). M.S. Thesis, Univ. of Nebraska, Lincoln, Nebraska.
Kadoum, A. M. A. A. (1966). Ph.D. Thesis, Univ. of Nebraska, Lincoln, Nebraska.
Kettlewell, H. B. D. (1946). *Entomologist* **79,** 8.
Kettlewell, H. B. D. (1961). *Entomologist* **94,** 59.
Kraybill, H. F. (1959). *Intern. J. Appl. Radiation Isotopes* **6,** 187.
Kraybill, H. F. (1960). *Nucleonics* **18,** 112.
Kuznetzova, E. A. (1937). *Rev. Appl. Entomol.* **A25,** 154.
Laithwaite, E. R. (1960). *Entomologist* **93,** 133.
Laithwaite, E. R. (1961). *Entomologist* **94,** 95.
Laudani, H., Tilton, E. W., and Brower, J. H. (1965). *Food Irradiation* **6,** A6.
Ley, F. J. (1964). *In* "Massive Radiation Techniques" (S. Jefferson, ed.), Wiley, New York. Newnes, London.
Litynski, M., and Wilkojc, A. (1954). *Roczniki Nauk Roliniczych* **A69,** 625. (1955). *Rev. Appl. Entomol.* **A43,** 389.
Lowry, W. L., Chapman, A. J., Wratten, F. T., and Hollingsworth, J. P. (1954). *J. Econ. Entomol.* **47,** 1022.
Madsen, H. F., and Sanborn, R. R. (1962). *Calif. Agr.* Feb. p. 12.
Marshall, G. E., and Hienton, T. E. (1935). *Agr. Eng.* **16,** 365.
Marshall, G. E., and Hienton, T. E. (1938). *J. Econ. Entomol.* **31,** 360.
Martin, V. J., Burson, D. M., Bull, J. O., and Cornwell, P. B. (1962). *U.K. At. Energy Authority Res. Group Rept.* **AERE–R3893.**
Mazokhin-Porshnyakov, G. A. (1964). *Entomol. Rev.* (*USSR*) (*Engl. Transl.*) **43,** 257.

Mickey, G. H., and Heller, J. H. (1964). *Trans. ASAE* **7,** 398.
Miles, W. R., and Beck, L. H. (1949). *Proc. Natl. Acad. Sci. U.S.* **35,** 292.
Milner, M., and Yen, Y. (1956). *Food Technol.* **10,** 528.
Milner, M., Lee, M. R., and Katz, R. (1950). *J. Econ. Entomol.* **43,** 933.
Milner, M., Lee, M. R., and Katz, R. (1952). *Food Technol.* **6,** 44.
Mordue, D. L., Sittler, O. D., and Hollingsworth, J. P. (1965). *Am. Soc. Agr. Engrs., St. Joseph, Michigan.* Paper No. 65–302.
Mouromtseff, I. E. (1933). *Elec. World* **102,** 667.
Nelson, S. O. (1962). *Trans. ASAE* **5,** 20.
Nelson, S. O. (1965). *Trans. ASAE* **8,** 38.
Nelson, S. O. (1966). *Trans. ASAE* **9,** 398.
Nelson, S. O., and Kantack, B. H. (1966). *J. Econ. Entomol.* **59,** 588.
Nelson, S. O., and Seubert, J. L. (1966). *Sci. Aspects Pest Control.* Publ. 1402 Natl. Acad. Sci., Natl. Res. Council, p. 136.
Nelson, S. O., and Walker, E. R. (1961). *Agr. Eng.* **42,** 688.
Nelson, S. O., and Whitney, W. K. (1960). *Trans. ASAE* **3,** 133.
Nelson, S. O., and Wolf, W. W. (1964). *Trans ASAE* **7,** 116.
Nelson, S. O., Stetson, L. E., and Rhine, J. J. (1966). *Trans. ASAE* **9,** 809.
Nicholas, R. C., and Wiant, D. E. (1959). *Food Technol.* **13,** 58.
Noble, L. W., Glick, P. A., and Eitel, W. J. (1956). *U.S. Dept. Agr.* **ARS** 33–28.
O'Brien, R. D., and Wolfe, L. S. (1964). "Radiation, Radioactivity, and Insects." Academic Press, New York.
Okress, E. C. (1965). *Appl. Opt.* **4,** 1350.
Parencia, C. R., Jr., Cowan, C. B., Jr., and Davis, J. W. (1962). *J. Econ. Entomol.* **55,** 692.
Parrot, P. J., and Collins, D. L. (1935). *J. Econ. Entomol.* **28,** 99.
Pendlebury, J. B., Jefferies, D. J., Banham, E. J., and Bull, J. O. (1962b). *U.K. At. Energy Authority Res. Group Rept.* **AERE–R4003.**
Pendlebury, J. G., Banham, E. J., Cooper, B. E., and Bland, C. M. (1962a). *U.K. At. Energy Authority Res. Group Rept.* **AERE–R3641.**
Peredel'skii, A. A. (1956). *Usp. Sovrem. Biol.* **41,** 228.
Peredel'skii, A. A., Rumiantsev, P. D., Bibergalv, A. V., Rodionova, L. Z., and Pertsovskii, E. S. (1957). *Biofizika* **2,** 209.
Peterson, D. G., and Brown, A. W. A. (1951). *Bull. Entomol. Res.* **42,** 535.
Pfrimmer, T. R. (1961). *U.S. Dept. Agr.* **ARS 20–10.**
Pittendrigh, C. S. (1961). *Harvey Lectures Ser.* **56,** 93.
Pittendrigh, C. S., and Minis, D. H. (1964). *Am. Naturalist* **98,** 261.
Proctor, B. E., and Goldblith, S. A. (1951). *Advan. Food Res.* **3,** 120.
Proctor, B. E., Lockhart, E., Goldblith, S. A., Grundy, A. V., Tripp, G. E., Karel, M., and Bragle, R. C. (1954). *Food Technol.* **8,** 536.
Purdue Univ. (1961). "Directions for the Use of Light Traps to Control Hornworms on Tobacco." Mimeo. Lafayette, Indiana.
Pyenson, L. (1933). *J. N.Y. Entomol. Soc.* **41,** 241.
Rao, B. S. (1964). *J. Rubber Res. Inst. Malaya* **18,** 243.
Read, M. S. (1959). *Proc. Intern. Conf. Preserv. Foods Ionizating Radiations, Cambridge, Mass.*
Riordan, D. F. (1964). *Nature* 204, 1332.
Ruck, P. R. (1958). *J. Insect Physiol.* **2,** 261.
Sanborn, R. (1962). *Western Fruit Grower* **16,** 22.
Schroeder, H. W., and Tilton, E. W. (1961). *U.S. Dept. Agr.* **AMS–445.**

Schwan, H. P., and Piersol, G. M. (1954). *Am. J. Phys. Med.* **33,** 371.
Schweigert, B. S. (1959). Paper presented at *Inter-Am. Symp. Peaceful Appl. Nucl. Energy 2nd, Buenos Aires.*
Silhacek, D. L. (1960). M.S. Thesis, Univ. of Nebraska, Lincoln, Nebraska.
Smart, M. R., and Brown, A. W. A. (1956). *Bull. Entomol. Res.* **47,** 89.
Soderholm, L. H., and Walker, E. R. (1955). *Botan. Gaz.* **116,** 281.
SRI (*Stanford Res. Inst.*) *J.* (1965). **4,** 10.
Stanley, J. M. (1965). *Am. Soc. Agr. Engrs., St. Joseph, Michigan.* Paper No. 65–303.
Stanley, J M., and Dominick, C. B. (1958). *J. Econ. Entomol.* **51,** 78.
Stanley, J. M., and Taylor, E. A. (1965). *Proc. Conf. Electromagnetic Radiation Agr.,* p. 39. Illum. Eng. Soc., Am. Soc. Agr. Engrs.
Stanley, J. M., Lawson, F. R., and Gentry, C. R. (1964). *Trans. ASAE* **7,** 125.
Stermer, R. A. (1959). *J. Econ. Entomol.* **52,** 888.
Tashiro, H. (1961). *U.S. Dept. Agr.* **ARS 20–10.**
Tashiro, H., and Tuttle, E. L. (1959). *J. Econ. Entomol.* **52,** 744.
Taylor, J. G., and Deay, H. O. (1950). *Agr. Eng.* **31,** 503.
Teixeira-Pinto, A. A., Nejelski, L. L., Jr., Cutler, J. L., and Heller, J. H. (1960). *Exptl. Cell Res.* **20,** 548.
Thomas, A. M. (1951). *Tech. Rept.* W/T20. Brit. Elec. Allied Ind. Res. Assoc., Surrey, England.
Thomas, A. M. (1952). *Tech. Rept.* W/T23. Brit. Elec. Allied Ind. Res. Assoc., Surrey, England.
Thomas, A. M. (1960). *Elec. Times* **138,** 121.
Thomas, A. M., and White, M. G. (1959a). *Tech. Rept.* W/T37. Brit. Elec. Allied Ind. Res. Assoc., Surrey, England.
Thomas, A. M., and White, M. G. (1959b). *Wood* (*London*) **24,** 407.
Tillson, E. D. (1943). *Grain* (*Chicago*) p. 3.
Tilton, E. W., and Schroeder, H. W. (1961). *Rice J.* **64,** 23.
Tilton, E. W., and Schroeder, H. W. (1963). *J. Econ. Entomol.* **56,** 727.
U.S. Army Quartermaster Corps. (1957). "Radiation Preservation of Food." U.S. Govt. Printing Office, Washington, D.C.
U.S. Dept. Agr. (1963). "Use of Light Traps for Control of Hornworms on Tobacco." AE and ENT, ARS. Mimeo, November 22.
Vande Berg, J. S. (1965). M.S. Thesis, Univ. of Nebraska, Lincoln, Nebraska.
van den Bruel, W. E., Bollaerts, D., Pietermaat, F., and Van Dijck, W. (1960a). *Parasitica* (*Gembloux*) **16,** 29.
van den Bruel, W. E., Pietermaat, F., Bollaerts, D., and Stefens, P. (1960b). *Mededel. Landbouwhogeschool Opzoekingsstat. Staat Gent* **25,** 1377.
Vishniakova, M. S. (1934). *Elek. Selsk. Khoz.* **5,** 26.
Watters, F. L. (1962). *Proc. Entomol. Soc. Ontario* **92,** 26.
Webber, H. H., Wagner, R. P., and Pearson, A. G. (1946). *J. Econ. Entomol.* **39,** 487.
Weiss, H. B. (1943). *J. Econ. Entomol.* **36,** 1.
Weiss, H. B. (1944). *J. N.Y. Entomol. Soc.* **52,** 267.
Weiss, H. B. (1946). *J. N.Y. Entomol. Soc.* **54,** 17.
Weiss, H. B., Soraci, F. A., and McCoy, E. E., Jr. (1941a). *J. N.Y. Entomol. Soc.* **49,** 1.
Weiss, H. B., Soraci, F. A., and McCoy, E. E., Jr. (1941b). *J. N.Y. Entomol. Soc.* **49,** 149.

Weiss, H. B., Soraci, F. A., and McCoy, E. E., Jr. (1942). *J. N.Y. Entomol. Soc.* **50,** 1.

Weiss, H. B., Soraci, F. A., and McCoy, E. E., Jr. (1943). *J. N.Y. Entomol. Soc.* **51,** 117.

Weiss, H. B., McCoy, E. E., Jr., and Boyd, W. M. (1944). *J. N.Y. Entomol. Soc.* **52,** 27.

Whitney, W. K., Nelson, S. O., and Walkden, H. H. (1961). *U.S. Dept. Agr.* **AMS 445.**

Williams, C. M., and Adkisson, P. L. (1964). *Biol. Bull.* **127,** 511.

Williams, C. M., Adkisson, P. L., and Walcott, C. (1965). *Biol. Bull.* **128,** 497.

Wright, R. H. (1963). *Nature* **198,** 455.

Wulff, V. J. (1956). *Physiol. Rev.* **36,** 145.

Yeomans, A. H. (1952). *In* "Insects, the Yearbook of Agriculture 1952." U.S. Govt. Printing Office, Washington, D.C.

4 RADIATION-INDUCED STERILIZATION

Leo E. LaChance, C. H. Schmidt, and R. C. Bushland

METABOLISM AND RADIATION RESEARCH LABORATORY
ENTOMOLOGY RESEARCH DIVISION
AGRICULTURAL RESEARCH SERVICE
UNITED STATES DEPARTMENT OF AGRICULTURE
STATE UNIVERSITY STATION
FARGO, NORTH DAKOTA

I. INTRODUCTION

The use of sterile insects to control and eradicate insect populations is one of the revolutionary departures in modern entomology. The origin of the idea and the development of the techniques are intimately related to research on the screwworm fly, *Cochliomyia hominivorax* (Coquerel) (see Section VII). However, for broad application, the theoretical ideas should be separated from the applied aspects. For this reason, the present chapter deals first with the theoretical basis for the sterile-male technique and then with the genetic and cytological basis for insect sterilization. After successful applications of the method are described, some difficulties, restrictions, and failures are discussed.

THE THEORY

II. DEVELOPMENT OF THE STERILE-MALE THEORY

The first paper (Knipling, 1955) explaining in detail the theory of insect eradication by the release of sterilized males was not published until 1955, but the idea began to take form in an isolated laboratory in Texas in 1937. At that time, Dr. E. F. Knipling observed that the female screwworm flies appeared to mate only once, and he suggested that if some way could be found to sterilize males without impairing their mating activity, it might be feasible to use these sterilized males to eradicate the isolated population of screwworms in the Southeast (see Section VII). Although the idea was not published for many years, Knipling worked on the mathematical models for the sterile-insect techniques using a number of different approaches (Knipling, 1955, 1959, 1964).

The sterile-insect techniques proposed by Knipling involve two distinct procedures. The first procedure, already mentioned, involves the rearing, sterilization, and release of the sterile insects to mix with and compete for mates with those of the natural population. The second procedure, also proposed by Knipling (1959), involves the sterilization of a portion of the natural population. Although both procedures invoke the sterility principle, they differ greatly in the manner in which the populations are affected. The requirements for applying the procedures in practical insect control are also vastly different. Thus, they should be regarded as two entirely different procedures for managing insect and other pest populations.

The first sterility theory in its simplest form asserts that the introduction of fully competitive sterile organisms into a natural population will reduce the reproductive potential of the natural population in proportion to the ratio of sterile to fertile insects present in the population after insects are released. If the ratio is 1 : 1, and the released sterile insects are fully competitive, the reproductive capacity of the natural population will be reduced by 50%; if the ratio is 9 : 1, the reproductive capacity of the natural population will be reduced by 90%, etc.

The second sterility theory does not involve the release of sterile insects. However, in its simplest form it asserts that the sterilization of a given proportion of both sexes of the natural population will result in two separate effects on the reproductive potential of the population. The sterilized insects cannot reproduce. Thus, from the standpoint of reproduction, this effect would be the same as killing the same proportion of the natural population. However, an added effect is achieved because the sterilized population can in turn nullify the reproductive capacity of a proportionate percentage of the remaining unsterilized individuals in the population.

If 90% of a natural population is sterilized (without adversely affecting mating competitiveness), those sterilized cannot produce progeny. In addition, these sterile individuals can in turn reduce the reproductive capacity of the remaining 10% by 90%, leaving 1% to reproduce. The combined effect will be to nullify the reproductive capacity of the total population by 99%. In contrast, the normal procedure of killing 90% will leave 10% to reproduce. Also, in contrast with the sterile-insect-release system, if enough sterile insects are released so that the total population consists of 90% sterile and 10% fertile insects, the initial effect will be the same as that achieved by destroying 90% of the population.

To contrast the difference between the sterile-insect-release method and the conventional method of killing insects, it is pointed out that the continued use of the same treatment for killing insects in subsequent generations will have the same percentage effect regardless of population density, whereas the release of a constant number of sterile insects will cause a higher and higher percentage of control as the natural population density declines and the ratio of sterile to fertile insects increases. The relative effects of these various procedures for controlling reproduction in insect populations will be considered in more detail with the help of hypothetical models.

A. Characteristic Trends of an Insect Population Subjected to the Release of Sterile Insects

For the first theoretical model involving release procedure (Table I), it is assumed that the natural population exists in an isolated area containing a stable population of 2 million insects with a 1 : 1 ratio of

TABLE I

THEORETICAL POPULATION DECLINE IN EACH SUBSEQUENT GENERATION WHEN A CONSTANT NUMBER OF STERILE MALES ARE RELEASED AMONG A NATURAL POPULATION OF 1 MILLION FEMALES AND 1 MILLION MALES[a]

Generation	Number of virgin females in the area	Number of sterile males released each generation	Ratio of sterile to fertile males competing for each virgin female	Percentage of females mated to sterile males	Theoretical population of fertile females each subsequent generation
F_1	1,000,000	2,000,000	2 : 1	66.7	333,333
F_2	333,333	2,000,000	6 : 1	85.7	47,619
F_3	47,619	2,000,000	42 : 1	97.7	1,107
F_4	1,107	2,000,000	1,807 : 1	99.95	Less than 1

[a] From Knipling (1955).

males to females in equilibrium with the environment and with the biotic potential canceled out by environmental resistance. Each generation, 2 million sterile males would be released in this area to compete equally for mates. With equal competition, two-thirds of the native females would make sterile matings, and the reproductive potential of the mixed population would be reduced to that extent. In the next generation there would be only 333,333 normal insects. Then, if the rate of male release was maintained constant at 2 million, the reproducing insects in the second generation would be outnumbered 6 : 1, and only one-seventh could reproduce. This would give a second generation population of 47,619, which would be outnumbered 42 : 1; with equal competition 97.7% of these matings would be sterile. The further reduced progeny (1107) would then be subjected to the overwhelming ratio of 1807 : 1. No fertile matings would be expected because by the law of chance less than one of the normal insects would be expected to mate with another normal insect.

A stable insect population that exactly replaces itself each generation is an artificial situation. A population changes in numbers each generation, and the interaction of all the complex factors constituting the environmental resistance determines whether it increases or decreases. It is more realistic to assume that under favorable conditions insects will increase in number until intraspecific competition becomes a limiting factor. Knipling (1964) estimated that for many pest species a fivefold increase per generation might be expected in the absence of cultural or chemical control practices.

With a fivefold rate of increase, the introduction into a normal population of sterile insects at a rate of 2 : 1 would not bring about control; it would merely reduce the rate of increase. Thus Knipling, in 1964, presented a model showing the population trend after an overflooding of 9 : 1 when the rate of normal increase was fivefold each generation (see Table II).

A total of 45 million sterile insects (in uniformly sustained releases) were required to bring about eradication in five generations. Only a 50% reduction occurred in the F_1 generation, but the reduction increased to 74% between the F_1 and F_2 generations. The rate of decline increases progressively as the ratio of sterile to fertile insects increases.

Against many pest species, 90% control can be achieved with applications of insecticides. However, in contrast to the effects produced by the release of sterile insects, the percentage mortality after treatment is independent of the number of insects in the treated area, but completely dependent on the rate of treatment. Therefore, 90% control of insects can be attained, whether there are 1 million or only 100 insects in the unit area. Thus, in many circumstances it is more economical to control the initial population with chemicals and to repeat treatments against subsequent generations as long as the population can be reduced more cheaply by insecticides than by sterile males. When the population is sufficiently low so that a 9 : 1 overflooding with sterile insects is less expensive than 90% control by chemicals, release of sterile insects can begin and be continued with progressively greater effect until eradication is attained. The effect of such an integrated program is shown in another model taken from Knipling (1964) (see Table III). In this theoretical

TABLE II

TREND OF AN INSECT POPULATION SUBJECTED TO CONTROL BY THE SUSTAINED RELEASE OF COMPETITIVE STERILE INSECTS WHEN 90% OF THE TOTAL POPULATION IN THE FIRST GENERATION CONSISTS OF STERILE INSECTS THAT HAVE BEEN RELEASED

	Insects per unit area			
Generation	Number of fertile insects	Number of sterile insects	Ratio of sterile to fertile insects	Number of insects reproducing
Parent	1,000,000	9,000,000	9 : 1	100,000
F_1	500,000	9,000,000	18 : 1	26,316
F_2	131,580	9,000,000	68 : 1	1,907
F_3	9,535	9,000,000	944 : 1	10
F_4	50	9,000,000	180,000 : 1	0

TABLE III

RELATIVE TRENDS OF INSECT POPULATIONS SUBJECTED TO REPEATED APPLICATIONS OF INSECTICIDES ALONE, COMPARED WITH A PROGRAM COMBINING APPLICATIONS OF INSECTICIDES FOR THREE GENERATIONS WITH SUBSEQUENT RELEASE OF STERILE INSECTS

	Numbers insects per unit area		
Generation	Normal trend (increase rate, 5×)	Control by insecticides at 90% level	Insecticide treatments followed by sterile-insect releases
Parent	1,000,000	1,000,000	1,000,000
F_1	5,000,000	500,000	500,000
F_2	25,000,000	250,000	250,000
F_3	125,000,000	125,000	125,000 : 1,125,000 sterile
F_4	125,000,000	62,500	62,500 : 1,125,000 sterile
F_5	125,000,000	31,250	16,450 : 1,125,000 sterile
F_6	125,000,000	15,625	1,190 : 1,125,000 sterile
F_7	125,000,000	7,812	0

treatment, eradication would be achieved by three applications of insecticides followed by the release of 5 million sterile insects compared with the 45 million sterile insects required for the model shown in Table II.

In actual practice it is not essential to maintain the uniformly high release rate shown in Table II. If the number of insects released is reduced from 9 to 1 million beginning with the F_3 generation, eradication would be achieved by the F_5 generation at an expenditure of only 3 million insects over three generations instead of 18 million over two. This principle of reducing the release rate of sterile insects when the natural population is at a low level is being applied in the current screwworm-eradication program in the Southwest (see Section VII).

B. Characteristic Trends of an Insect Population Subjected to Sterilization Treatments

It seems important to show the trend of an insect population when the natural population is sterilized and to show how this trend differs from the trends of populations subjected to control by the use of conventional killing agents or by the release of a constant population of sterile insects. These differences are shown in Table IV.

The first column shows the assumed normal trend of an uncontrolled population (also shown in Table III). The second column shows the trend of the population subjected to 90% control by a killing agent each generation (also shown in Table III). The third column shows the trend when 90% of the insects of the natural population are sterilized

TABLE IV

Characteristic Trends of Hypothetical Insect Populations Subjected to Various Types of Control[a]

	Number insects per unit area			
Generation	No control (increase rate 5×)	Control by killing agent at 90% level	Control by sterility agent at 90% level	Control by release of sterile insects, 9 : 1 overflooding ratio for initial generation
Parent	1,000,000	1,000,000	1,000,000	1,000,000
F_1	5,000,000	500,000	50,000	500,000
F_2	25,000,000	250,000	2,500	131,580
F_3	125,000,000[b]	125,000	125	9,535
F_4	125,000,000	62,500	6	50
F_5	125,000,000	31,250	0	0

[a] After Knipling (1964).
[b] Assumed maximum density.

each generation. The fourth shows the trend when the initial release rate of sterile insects results in an overflooding ratio of 9 : 1 (also shown in Table II). We will now compare the effects of the three methods of insect population control. Sterilization of 90% of the natural population each generation results in a much greater reduction in the population each generation than is achieved by the killing agent. However, the two methods have one feature in common. The effect remains constant each generation, irrespective of the population density. The difference in the two procedures is that the killing method remains constant at the 90% level, whereas the sterility method when applied to the natural population remains constant at the 99% level because of the extra effect on reproduction achieved by the sterile insects. In contrast with the two methods discussed, the sterile-insect-release system (Table IV, column 4) becomes progressively more effective as the natural population declines.

When the relative effects of the sterile-insect-release method (column 4) are compared, the release of enough sterile insects to reduce reproduction by 90% is shown to be less effective initially than the sterilization of the natural population at the 90% level (column 3.) The reason has already been discussed. When the natural population is sterilized, a double effect is produced. However, the sterile-insect-release procedure becomes progressively more effective, and after a few generations the rate of decline exceeds the rate of decline achieved by sterilizing the natural population.

Since the two sterilization procedures affect population trends in different ways, there should be circumstances where the integration of the two procedures would have merit, just as was proposed for integrating the use of an insecticide and the release of sterile insects. For example, by noting column 3 (sterilization of the natural population), it might become more economical to employ sterile-insect releases in lieu of another chemosterilant treatment when the natural population has been reduced to the theoretical level of 50,000 following the initial treatment. To reach the same theoretical level by the use of insecticides, it would be necessary to apply insecticide treatments for five or six generations.

III. COMPONENTS OF STERILITY

A major obstacle to any discussion of radiation-induced sterility in insects is a lack of agreement on what sterility is. Unfortunately, it means different things to different people, perhaps because there are various causative factors. By definition, sterility is the inability to produce offspring; a sterile individual cannot contribute a genome to the next generation. However, one school of thought disagrees with this definition because it does not include individuals who *do* produce progeny which are in turn incapable of reproduction. Admittedly there is a point here. It is possible for progeny to inherit a factor or condition from the parents which render the progeny incapable of reproduction; the point is whether the original parent or the barren offspring (or both) should be termed "sterile." For the present discussion, we shall define sterility as the inability to produce offspring and proceed to examine the various means by which sterility can be induced.

Sterility may be caused by (1) infecundity in females, (2) aspermia or sperm inactivation in males, (3) inability to mate, or (4) dominant lethal mutations in the reproductive cells of either the male or the female. All these conditions lead to sterility and can be produced in insects by ionizing radiation. Obviously, not all are equally suitable for the sterile-male technique of insect control.

The only kind of sterility used with any success thus far is sterility based on dominant lethal mutations in the sperm of the male and infecundity of the simultaneously released females. Consequently, we shall devote considerable space to sterility based on dominant lethals, but we cannot exclude from discussion the other types of sterility. Some are potentially useful for practical purposes, and several types of sterility tend to occur in any radiation study with insects.

When both sexes are exposed to radiation, sterility of the two sexes may be caused by different factors. In any one insect, sterility may be due to a combination of two or more factors. For example, a treated female can initially produce eggs with dominant lethals and later on become infecund. A treated male very often transmits sperm with dominant lethals in the initial matings and later becomes aspermic.

IV. PHYSIOLOGICAL AND CYTOGENETIC BASIS FOR VARIOUS KINDS OF STERILITY

A. Infecundity

A loss of fecundity, which is equivalent to a depression in egg production, has been repeatedly observed after treatment of female insects with ionizing radiation (Annan, 1955; Grosch, 1962; LaChance and Leverich, 1962), with ultraviolet radiation (Steen, 1934), and with ingested radioisotopes (Grosch *et al.,* 1956; Grosch, 1959).

The production of eggs in insects is largely dependent on the differentiation of oocytes from oogonia and the proper function of nutritive cells (trophocytes or nurse cells). Severe damage to the oogonia can result in permanent infecundity, and the nutritive cells can also be damaged by irradiation (Grosch *et al.,* 1956; Erdman, 1961; LaChance and Bruns, 1963). At certain times during egg maturation, the nutritive cells are extremely sensitive to irradiation; at other times they are remarkably resistant. In fact, after these cells become fully differentiated and attain a high degree of polyploidy, even large doses of radiation will not affect the growth and production of eggs that appear normal (Grosch and Sullivan, 1954), though the eggs may contain dominant lethal mutations. In mosquitoes, the time elapsing between intake of a blood meal and the radiation treatment largely determines whether the irradiated females will produce eggs with dominant lethal mutations (no hatch) or no eggs. Terzian and Stahler (1958) observed that when female *Aedes aegypti* (L.) were irradiated 4 hours after a blood meal only 10 kr were required to produce infecundity, whereas irradiation 42 hours after a blood meal required 100 kr to produce the same effect.

B. Inability to Mate

Several instances are known in which radiation treatments were diagnosed as producing sterility and later it was found that the treated insects were unable to mate. Although it is debatable whether a treatment that debilitates an insect to the point where mating is unlikely or impossible really can be called sterilization, the result is nevertheless the same. In studies with irradiated males of *Rhodnius prolixus* Stål, Baldwin and Shaver (1963) found that the sterility of some crosses was really the result of infecundity on the part of the females. Although the females had not been irradiated, some treatments of the males interfered with pairing and coitus, which was expressed as infecundity in females. Any treatment that affects the mating activity (or some other aspect of reproductive physiology) of the males may lead to infecundity in untreated females if transfer of sperm or proper length of copulation are prerequisites for egg laying. Lowered fecundity of untreated females mated to irradiated males is often observed in the Lepidoptera, where the large doses of radiation required to sterilize males are suspected of affecting some aspect of mating behavior (see Sections V and IX).

C. Sperm Inactivation

Sperm inactivation can be due to loss of fertilizing capacity or loss of motility of the sperm. Although it is commonly thought that a significant amount of sperm inactivation resulting from radiation treatments would cause the sterile-male technique to fail, this is not necessarily true. When the females are polygamous, sterile males with inactivated sperm are useless because future matings with fertile males would completely negate the effect of previous sterile matings. However, when the females mate only once, mating with a sterile male containing inactivated sperm *might* have the same effect as mating with a sterile male that transmits active sperm with dominant lethal mutations. The adequacy of sperm inactivation as the basis for sterility depends largely on whether sperm transmission is required to insure that the female will refrain from further mating. Recent studies on the housefly (*Musca domestica* L.) (Riemann *et al.*, 1967) indicate that sperm transfer is not necessary to initiate the monogamous response in females; therefore, sterility based on sperm inactivation is potentially useful. For the present, however, we can examine the limited information about radiation-induced sperm inactivation in insects.

Reliable data comparing the doses of radiation that induce dominant lethal mutations with those that result in sperm inactivation are mainly available for species that reproduce parthenogenetically. In these species, the two effects can be separated easily and accurately. Most investigators of sperm inactivation in parthenogenetic insects agree that inactivation of sperm by irradiation does not occur until complete dominant lethality is attained (Stancati, 1932; Maxwell, 1938; Whiting, 1938a,b; Heidenthal, 1945; Clark *et al.*, 1957; Lee, 1958; Henneberry, 1964). However, the phenomenon may not become apparent immediately after treatment. For example, Clark *et al.* (1957) found that when males of *Habrobracon juglandis* (Ashmead) were irradiated with α-particles, 6700 rep induced dominant lethals in the sperm, but 187,000 rep produced no evidence of sperm inactivation for 3 days, a small amount after 7 days, and complete inactivation after 10 days. In fact, the dose required to inactivate sperm also stunned the adults and caused early death.

In species which require fertilization for egg development, sperm activity is difficult to evaluate without using special cytological techniques. Generally, it is difficult to determine whether an irradiated male transmits sperm with dominant lethal mutations, inactivated sperm, or any sperm at all. All three possibilities result in nonhatching eggs.

In males of *A. aegypti*, 10,000 r induced dominant lethals in all sperm, and sperm motility was not affected (Terzian and Stahler, 1958). Even at doses of 30,000 r, no loss of sperm motility occurred, and no evidence of deterioration appeared for at least 20 days after irradiation. In fact, even at 50,000 r the sperm remained normal in appearance and motile as long as the mosquitoes were able to survive the radiation dose. Death of the males followed soon after the sperm defects appeared.

The fertilizing capacity of mouse and rat sperm also are unaffected by large doses of radiation (see Russell, 1954 for references). Valenta *et al.* (1963) studied the effect of radiation doses ranging from 10 to 150,000 r on the male sexual cells of some domestic animals (bull, ram, rabbit, dog) and of man. They found that vitality and motility of sperm were affected only by doses of 1000–10,000 r—relatively large doses for mammals.

In earlier radiation studies, adult male *Drosophila melanogaster* Meigen were almost sterilized by doses of 5000–10,000 r, but there was no evidence of sperm inactivation because the sperm fertilized the eggs and participated in the zygote formation (Muller and Settles, 1927; Demerec and Kaufmann, 1941). However, more recent studies with *D. melanogaster* (Yanders, 1959; Lefevre and Jonsson, 1962) suggest that some sperm can be inactivated by fairly low doses of radiation, and Yanders (1964) concluded that some aspect of sperm behavior, possibly motility, was affected by x-rays at doses in the range commonly used in genetic and sterility studies.

Other investigations also indicate that comparatively moderate radiation treatments have deleterious effects on sperm activity. Jaynes and Godwin (1957) found that when males of the white-pine weevil, *Pissodes strobi* (Peck), were irradiated with 10 and 20 kr and mated to untreated females, the females produced no viable offspring. However, when the untreated females had previously mated with untreated males before mating with irradiated males, they produced more viable offspring at the higher doses. Apparently, the more heavily irradiated sperm were less competitive.

Although sperm with dominant lethal mutations can apparently fertilize eggs and usually compete with untreated sperm, it is not certain that all species of insects can be sterilized by radiation without inactivating some of the sperm. Furthermore, for field application, the minimum dose of radiation necessary to induce lethal mutations in insect sperm should be used. Clearly, we are not dealing with an agent with which, if a small dose is desirable, a larger dose would be better.

D. Dominant Lethal Mutations

Dominant lethal mutations can be induced in germ cells of insects by electromagnetic radiations (γ-rays and x-rays) as well as by particulate radiations such as neutrons (Baker and von Halle, 1954) and α-particles (Clark *et al.,* 1957). These mutations in insect sperm have been used successfully in several eradication programs. It seems appropriate, therefore, to discuss the way irradiation acts to produce lethal mutations, the cytogenetic basis, and the method by which the production of offspring is prevented.

For insects, Muller coined the term "dominant lethal mutation" in the same classic paper in which he reported the discovery of the mutagenic effects of radiation in *Drosophila* (Muller, 1927). A dominant lethal mutation is a nuclear change that can effect the death of the zygote, even though it is introduced by only one of the germ cells that unite at fertilization. This definition of dominant lethal mutation will be used throughout this chapter. Basically, a dominant lethal mutation does not hinder the maturation of the affected cell into a gamete or the participation of the gamete in the formation of a zygote, but prevents the zygote from developing to maturity. Lethal mutations are not lethal to the treated cell; they are lethal to its descendant—the zygote it forms. What is the nature of dominant lethal mutations? Practically all investigators agree that radiation-induced dominant lethal mutations arise as a result of chromosome breaks in the treated cell (Muller, 1940; Pontecorvo, 1941, 1942; Lea and Catcheside, 1945; Glass, 1955; Lee, 1958; Abrahamson and Herskowitz, 1957; LaChance, 1967). However, we cannot yet completely exclude the possibility that a point mutation may have a dominant lethal effect. It is equally difficult to prove that such a point mutation is not the result of a very minor and undetectable chromosome break or rearrangement.

Several types of chromosome changes induced by radiation are lethal. For example, a simple chromosome break that interrupts the continuity of the chromosome would be lethal if the break did not rejoin before cell division. Chromosome fragments are lost at cell division because they lack a centromere which ordinarily insures their inclusion in a daughter nucleus. In addition, certain chromosome breaks that rejoin, but not in the original fashion, also have a dominant lethal effect. These asymmetrical exchanges result in acentric fragments *and* dicentric chromosomes, which in turn yield fragments and genetic imbalance as a result of breakage–fusion–bridge cycles during cleavage divisions in the zygote. (For background information on bridge formation and structural rearrangements, see Muller, 1954.) Chromosome bridge formation is usually a sufficient condition for lethality because, during early cleavage, unbalanced chromosome sets are formed in which many genetic loci occur in a single dose or not at all.

Studies on diverse insect species have shown that radiation-induced dominant lethal mutations are characterized by the presence of chromosome bridges and fragments between dividing nuclei in the egg (Sonnenblick, 1940; Sonnenblick and Henshaw, 1941; Whiting, 1945a,b; Fahmy and Fahmy, 1954, 1955; von Borstel, 1955).

In a recent study (LaChance and Riemann, 1964), adult males or females of the screwworm fly were treated with γ-radiation at doses that induced high rates of dominant lethals. The developing embryos were fixed and studied for evidence of chromosome damage. When males were treated, many chromosome aberrations were found in the first two cleavage divisions. Often embryonic development had ceased at this point. When females were irradiated and mated to untreated males, many chromosome aberrations were found during the first two meiotic divisions (which occur after the egg is laid and before cleavage begins), as well as during cleavage divisions. It was obvious that the female pronucleus would be deficient for large amounts of genetic material.

At what time must dominant mutations kill the affected carrier? A lethal mutation can be expressed at any stage of the insect's life cycle, but death usually occurs before hatching because the insect must surmount a number of developmental crises during the embryonic period. Cytogenetic studies indicated that death generally occurs before blastoderm formation, usually during the early divisions (von Borstel, 1955, 1961; Atwood *et al.,* 1956; Whiting *et al.*, 1958; von Borstel and Rekemeyer, 1959; LaChance and Riemann, 1964), but sometimes at much later stages of development (Tazima, 1960). Posthatching deaths represent a very small fraction of the total number of lethal mutations expressed (Demerec and Fano, 1944; Catcheside and Lea, 1945; Whiting *et al.,* 1958; Hadorn, 1961; Grosch, 1962; LaChance and Riemann, 1964). Some lethal mutations expressed after the egg stage were observed in *H. juglandis* by Atwood *et al.* (1956) and also in the boll weevil, *Anthonomus grandis* Boheman, by Lindquist *et al.* (1964).

The reasons why embryos die as a result of chromosome breaks have been discussed at length elsewhere (LaChance, 1967). Chromosome breakage in gametes is expected to lead to chromosome imbalance in the zygote and often to breakage–fusion–bridge cycles. Embryonic death is associated with a depression in the mitotic rate in the developing embryo, with complete cessation of mitosis often occurring in the second or third cleavage division. This depression in the rate of mitosis may be related to the presence of chromosome bridges. Death of the embryo is often accompanied by polyploid cleavage nuclei (Demerec and Fano, 1944), which indicates that DNA synthesis may persist for some time after mitotic division has ceased.

E. Aspermia

Aspermic males do not have a supply of mature sperm to transmit to females during mating. Aspermia is observed in male insects after exposure to sufficient radiation to inhibit the spermatogenic cycle. The spermatogenic cycle can be interrupted at any stage of the insect's life cycle with different results. For example, irradiation of a male insect where the testes contain only spermatogonial cells (usually in the very early larval or nymphal stages) could destroy these cells; the resulting adults would then be devoid of mature sperm. However, irradiation of male insects whose testes contain a variety of reproductive cell stages produces quite different results.

V. TYPE OF STERILITY INDUCED IN VARIOUS MALE INSECTS

The type of sterility produced by the irradiation of male insects is determined largely by the types of cells present in the testes when the males are exposed to ionizing radiation. For example, at certain life stages the testes of some insects contain an assortment of reproductive cells (spermatogonia, spermatocytes, spermatids, and sperm); other insects have testes which contain mostly one kind of germ cell at any given time.

Sado (1961) in a thorough cytological study of the silkworm *Bombyx mori* (L.) found that at a given stage the contents of the testes were mostly limited to one or two types of germ cells, as shown in the following tabulation:

Life stage of insect	Most abundant cell stage in testes
Hatching to first instar	Primary spermatogonia
Second to third instar	Secondary spermatogonia
Third to fourth instar	Spermatocytes in early meiotic prophase
Fourth to fifth instar	Spermatocytes in early and late meiotic prophase
Fifth instar to spinning	Same as above with a few early spermatids
Pupae	Spermatids and fully formed mature spermatozoa
Emerged adults	Only fully formed spermatozoa

After he identified the kind of germ cell that was being exposed by the irradiation of males at different stages in the life cycle, Sado (1961) found that irradiation between the first and third instar resulted in a marked degree of cell death because the testes contained large numbers of secondary spermatogonia which are extremely sensitive to radiation. Presumably, at this stage a large dose would kill all the cells, and the adult males would be aspermic. However, small doses did not reduce fertility because the destruction of some cells was compensated by regeneration of undamaged spermatogonia (primary). These regenerated spermatogonia developed into functional sperm in 20 days. Thus, irradiation of silkworm larvae with small doses before the larvae had reached the third instar did not produce sterility, even through many germ cells had been killed. The early fifth instar, when most germ cells are in the late spermatocyte stage, was the most radiosensitive stage. Even 500 r produced a marked reduction in fecundity, i.e., reduction in the average number of eggs laid by females mated to irradiated males. He concluded that irradiation produced abnormal late meiotic stages, especially late prophase, which resulted in a reduced number of sperm and the formation of nonfunctional sperm. Evidently, when these males were mated to untreated females the mating did not elicit the same response from the females because they laid reduced numbers of eggs (behaved like virgins). Irradiation of silkworm pupae and adults is, of course, expected to produce sterility by the induction of dominant lethal mutations in the spermatids and sperm.

In their studies on the European corn borer, *Ostrinia nubilalis* (Hübner), Chaudhury and Raun (1966) found that the testes in various larval and pupal stages contained reproductive cells that were predominantly all in the same stage. Therefore, irradiation of this species should yield equally interesting results. Unfortunately, few studies on other Lepidoptera combine cytological studies with radiation sterilization experiments.

Many insects, even before they reach the adult stage, contain both pre- and postmeiotic germ cells in their testes. Thus at irradiation the testes contain a variety of cell types. Consequently, the adult males may exhibit several kinds of sterility in sequence. In his extensive cytological studies of irradiated male screwworm flies, Riemann (1967) found gross chromosome damage in the dividing meiotic cells of the testes 2–3 hours after irradiation with 6200 r (sterilizing dose). After 24 hours, all germ cells in the anterior portion of the testis were degenerating—these included all gonial cells and young primary spermatocytes. However, even at this sterilizing dose, secondary spermatocytes and those primary spermatocytes that had completed their growth phase were not killed though they had suffered gross chromosomal aberrations. The majority of the spermatids formed from these cells failed to develop into normal-appearing sperm. He concluded that when a dose of 6200 r was given to pupae, the adult male's entire supply of transferable sperm was derived from cells that were spermatids or immature sperm at the time of treatment. In fact, the premeiotic cells in the testes were killed by far smaller doses than were required to induce lethal mutations in the postmeiotic stages. The young primary spermatocytes were very radiosensitive; they were mostly killed by a dose of only 100 r. Most secondary spermatogonia were killed by 500 r. Total destruction of primary spermatogonia occurred in a range of doses from 1500 to 3000 r.

Welshons and Russell (1957) also found that doses of radiation which sterilized adult males of *D. melanogaster* by inducing dominant lethal mutations in postmeiotic stages caused the death of all cells in the premeiotic stages at the time of treatment.

Riemann and Flint (1967) showed that when the testes were removed from male boll weevils 10 days after they received a sterilizing dose (inducing dominant lethals in the sperm), no spermatogonia could be found in any testes that had received over 6000 r (less than the sterilizing dose), and none were found in some testes that had received only 2000 r. In the boll weevil, substerilizing doses of radiation produced gaps in the stages of spermiogenesis, an indication of an interruption in normal cell development for some time after irradiation, probably as a result of death among the spermatogonial cells. Even if all the cells in the germarium (spermatogonia) were not destroyed, several days would probably elapse before any new mature sperm would be formed by the males. However, the males usually contained bundles of seemingly mature sperm which represented "old" sperm, i.e., those that were in a postspermatogonial stage at the time of irradiation.

VI. TYPE OF STERILITY DESIRED FOR INSECT CONTROL PROGRAMS

What kind of sterility is most desirable in terms of producing sterile insects for use in control programs? It is impossible to list the ideal type of sterility for all species; each insect must be evaluated independently. In order to identify the desired kind of sterility, a great deal must be known concerning the reproductive physiology of the species.

For species that are polygamous, or those that require sperm to remain monogamous, the sterile males must transmit sperm containing dominant lethal mutations that are competitive with the sperm contributed by normal males. The sterile males should contain a sufficient store of mature sperm or spermatids at the time of irradiation to last through the number of matings normally experienced by the males of this species in the field. If after a given number of matings the male eventually runs out of sperm and becomes aspermic, his efficiency in the program becomes zero because he cannot contribute competitive sperm. If the male does not become aspermic, the sperm contributed in later matings should certainly not arise from cells that were premeiotic at the time of treatment. This point is of considerable practical importance since transfer of sperm derived from surviving gonial cells usually results in mostly viable offspring (fertility recovered). It is a fairly simple task to determine whether a given dose of radiation permits survival of any spermatogonial cells. It is also easy to ascertain how many matings an irradiated male can undertake and still transmit sperm.

Although the types of sterility described can also be used with species that are monogamous, other possibilities could be effectively utilized.

Males that contribute immotile sperm or no sperm at all would be equally effective, providing the act of copulation is sufficient to prevent the female from entering into further matings (see Section IV,C). Of course, in monogamous species it is difficult to imagine that any one male participates in numerous matings when the females accept only one mate and the sexes are present in equal numbers.

One requirement common to all types of induced sterility is that the general vigor, longevity, sexual competitiveness, and mating behavior of the males should not be severely impaired by the radiation treatment.

THE APPLICATIONS

VII. SCREWWORM ERADICATION PROGRAMS

The sterile-male technique of insect control is intimately associated with research on the biology and control of the screwworm. The screwworm is not only the major example of the practical application of the method over a wide geographic area, but it is the first insect used in radiation research with eradication as an objective. Therefore, an account of the developments in the screwworm eradication program seems appropriate.

Screwworms have been a pest of livestock in the southwestern United States since the livestock industry began in this area, but accurate biological research dates from the study of Cushing and Patton (1933) in which it was shown that the species was distinct from the secondary screwworm, *Cochliomyia macellaria* (F.), with which it had been confused. Previously, entomologists had believed that the screwworm was a facultative parasite with larvae capable of growing in wounds, but most commonly found breeding in carcasses.

When the screwworm was proved to be an obligatory parasite which, in nature, must complete at least the first half of its larval growth in the wound of a living animal, research was greatly stimulated, particularly in the southern field laboratories of the United States Department of Agriculture.

Coincident with the discovery of the obligatory parasitism of the insect was an extension of its range from the southwestern to the southeastern United States. Screwworms first appeared in Georgia in 1933, presumably as the result of importation of infested cattle from the Southwest. The infestation spread into peninsular Florida, where the winter climate was sufficiently mild to permit winter survival.

Figure 1 shows the distribution of screwworms in the United States in 1935–1957. Screwworms are tropical and subtropical insects native to the Western Hemisphere. The southwestern infestation is the northern fringe of a population that extends throughout Latin America to the temperate zone of South America. In the United States, year-round infestations were restricted in the coldest years to peninsular Florida and the most southern parts of states bordering on Mexico, but almost every summer screwworms extended their range of normal migration to the limits outlined on the map. Some years, the shipping of infested livestock caused localized infestations much farther north, and serious damage

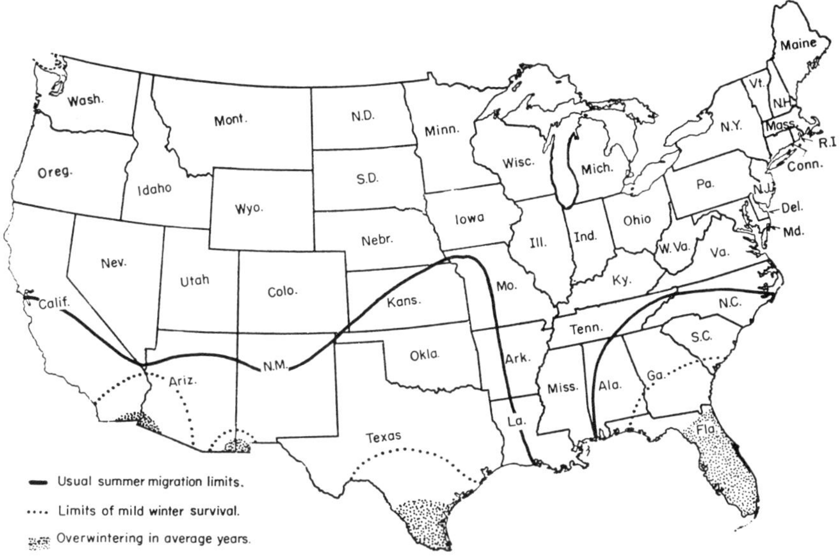

FIG. 4.1. Screwworm distribution in the United States.

from the screwworm was occasionally reported from New Jersey, Illinois, Iowa, and South Dakota.

During 1935–1937, the United States Department of Agriculture, in cooperation with agricultural agencies of the southeastern states, conducted an intensive campaign to eliminate the newly introduced pest. Chemicals were applied to wounds, and good animal husbandry was encouraged. These efforts greatly reduced the incidence of wound infestation, and the insect population was brought to a low level. However, enough survived in wild hosts and neglected livestock to prevent eradication.

During this period, Knipling was working in the Southeast studying the field biology of the screwworm. In 1937 he was reassigned to the

Menard, Texas laboratory of the United States Department of Agriculture where Melvin and Bushland (1940) had established a thriving laboratory colony of screwworms. The larvae were grown under incubator conditions on a nutritional medium composed of ground meat, blood, and water, with formaldehyde as a preservative. During routine laboratory maintenance of the colony, it was noted that the insects began mating at the age of 2 days, but that the flies were seldom observed in copulation after 4 days though males continued to pursue females. Knipling speculated that the females mated only once and suggested to his associates that if the males could be sterilized without impairing their mating activity, sterilized males might be used to eradicate the isolated population in the Southeast. There, as noted, the population was restricted to peninsular Florida by winter cold, and it could be brought to very low numbers by good animal management and chemical control.

When the Livestock Insects Laboratory of the Entomology Research Division was established in Kerrville, Texas in 1946, Knipling directed that research on the mating habits of screwworms should be intensified and that tests with a variety of organic chemicals be initiated in a search for a male sterilant. Since none of the compounds tested were effective, the results were not published.

Next, A. W. Lindquist of the Corvallis, Oregon laboratory called attention to the report by Muller (1950) in which it was stated that ionizing radiation could cause male sterility by inducing dominant lethal mutations in the sperm. Knipling wrote to Muller describing the screwworm problem and inquiring about the possibility of producing competitive sterile males by irradiation. When the reply was encouraging, tests were begun with x-rays (Bushland and Hopkins, 1951) and γ-radiation (Bushland and Hopkins, 1953).

x-Rays and γ-rays proved equally effective, and either adults or pupae could be sterilized. The most efficient method was to irradiate pupae 2 days before adult emergence. A dose of 2500 r sterilized males, and 5000 r sterilized females. Longevity of adults was somewhat reduced by sterilization, but pronounced lethal effects were not noted with radiation doses below 20,000 r. Females treated with 5000 r produced a few infertile eggs, but no eggs developed after irradiation at 7500 r. The 7500-r dose was adopted as standard for subsequent field tests and eradication programs. In laboratory experiments, the radiation-induced sterility was permanent, and the sterilized males were competitive with normal males in cage-mating experiments.

The laboratory procedures were so successful that field studies were undertaken in Florida. Screwworms were brought to the vanishing point by the release of 100 sterile males per square mile per week on Sanibel

Island near Fort Myers, Florida, but eradication could not be proved because the test area was not sufficiently isolated to prevent immigration of a few fertilized females from nearby untreated areas. The clinching experiment was performed on the Island of Curacao, where Baumhover *et al.* (1955) achieved eradication on that isolated 170-square-mile island.

Three years of field work and pilot testing then followed in Florida to develop the efficient mass-rearing techniques and the field procedures required for the production and distribution of as many as 50 million flies per week. This was the estimated number required to treat the entire overwintering area at the rate of 400 sterile males (plus an equal number of sterile females) per square mile per week. By 1957 the work had progressed sufficiently so that the Animal Health Division (then called the Animal Disease Eradication Division) of the United States Department of Agriculture could undertake a cooperative eradication program with livestock officials of the southeastern states (Bushland, 1960a,b).

In the fall of 1957, when funds became available for the southeastern program, a contract was let for a rearing facility to produce 50 million flies weekly at Sebring, Florida. In the interim, before the rearing facility was complete, the Entomology Research Division pilot plant at Orlando, which had the capacity to produce 2 million sterile flies weekly, was to be used for methods development, training, and preliminary releases in selected locations. However, the unusually severe winter freezes in December, 1957 and January, 1958 apparently killed all the screwworms north of Orlando, Florida and greatly reduced the overwintering population in the southern half of the peninsula.

Production was therefore started immediately in the Orlando pilot facility. By January, releases of sterile flies at the rate of 1 million per week were initiated. Then a crash program at Orlando increased the weekly output until it was 14 million flies per week by the time rearing was transferred to the completed facility at Sebring in July. Because the Orlando production was insufficient to blanket the entire overwintering area, the flies were released in an experimental barrier across the peninsula north of Orlando. As warmer weather came in February and March, a few screwworms were found in northern Florida and southern Georgia, probably as a result of transportation of infested cattle. The immediate vicinity of the local infestations was "hot-spotted" by special releases of sterile flies. These emergency procedures prevented a buildup of screwworms and limited the infested territory largely to the area south of Orlando. By 1959, 1 year ahead of the most optimistic expectations, the program was successfully completed.

In 1960, after Florida had been free of screwworms for more than a year, western Tennessee became infested, apparently as a result of shipment of cattle from the Midwest. Next, screwworms appeared in Alabama, either as a result of normal migration from Tennessee or because of the movement of animals. Only small numbers of sterile flies were available from a standby colony maintained at the Kerrville, Texas laboratory. However, localized treatment with the flies was sufficient to contain the population until it was eliminated in the fall.

In the summer of 1961, screwworms again appeared near Pensacola, Florida, apparently introduced by shipment of infested cattle in spite of the inspection and insecticide spray stations at major highways crossing the Mississippi River. Cattle were sprayed with insecticides, and sterile flies were released locally. The incipient outbreak was eliminated.

The spectacular success of the barrier of sterile flies used in Florida during the spring of 1958 and the effective localized use of limited numbers of sterile insects to contain and eliminate subsequent reinfestations in Alabama and Florida encouraged workers and substantiated the belief that screwworm eradication might be extended to include the southwestern United States.

The Southwest actually had a rather similar situation. In peninsular Florida the original plans had been limited to the eradication of the population isolated during the winter months in an area of only 50,000 square miles which could be treated as a unit. In the southwestern United States, most winter survival occurred in the lower Rio Grande Valley of Texas. Screwworms were not known to survive in New Mexico during the coldest winters, but in average years winter infestations did occur in a few square miles of southwestern New Mexico and an adjacent area in Arizona. A somewhat larger overwintering area was located in southwestern Arizona and southern California. (In the arid Southwest there is negligible west-to-east migration because of unfavorable climate outside the irrigated valleys). The area south of San Antonio, Texas where screwworms might be expected to overwinter included about 50,000 square miles, an eradication zone no bigger than was involved in Florida.

It seemed likely that treatment at the rate of 800 flies per square mile per week in southern Texas could bring about eradication in 1 or 2 years. Thereafter, it was possible that the flies could be diverted to a barrier zone 100 miles wide, extending on either side of the Rio Grande River from Brownsville, Texas to Albuquerque, New Mexico, to protect eastern New Mexico, all of Texas, and the Midwest from migrating flies. Then a weekly distribution over this barrier of 100 sterile males per square mile per week would leave a waiting population of sterile males sufficient to overwhelm the progeny of any migrating fertilized females coming from the other side of the Rio Grande River. Alert livestock producers would be asked to protect their livestock from screwworm infestation and to report infestations for special "hot spot" releases such as had been successful in the Southeast.

The entire plan, recognized as a large-scale experiment, was estimated to cost 5 million dollars annually. However, livestock producers believed that the annual losses due to screwworms in Texas and the lower Midwest were about 100 million dollars. The cost–benefit ratio appeared to justify an experiment that seemed to have such a good chance of success. Livestock producers in Texas and the adjoining states organized the Southwest Animal Health Research Foundation. Three million dollars in freewill donations were collected from producers to match federal funds until state appropriations could become available. In February, 1962, Congress made funds available to start the program. In February, 1964, the second anniversary of the program, officials of the United States Department of Agriculture, Texas, and New Mexico jointly declared that screwworms had been eradicated from the test area, though a few local infestations were detected north of the Rio Grande at seasons of the year when the weather favored migration. Treatment of emergency "hot spots" was quickly effective, and from the fall of 1963 to the summer of 1965 no self-sustaining population of screwworms appeared to exist in any locality north of the Rio Grande River.

Much of the information about the progress of the eradication program came through the voluntary cooperation of livestock growers working with vocational agriculture teachers and county agents of the Texas Agricultural Extension Service. Animal inspectors of the United States Department of Agriculture and the Texas Animal Health Commission, county agents, and vocational agriculture teachers distributed small bottles of preservative in mailing kits addressed to the Screwworm Eradication Headquarters at Mission, Texas. Whenever livestock owners found infested cattle (myiasis), they treated the wound with a larvicide and forwarded samples of the larvae to the Insect Identification Center where taxonomists of the Animal Health Division identified the larvae.

The records of myiasis for the Texas eradication program compiled from the samples from livestock owners are presented in Table V. The data were obtained by the Animal Health Division and the Texas Agricultural Extension Service and reported in weekly progress reports by D. Deterling, J. H. Roethe, and W. Newton. Values shown under the heading "nonscrewworms" are the numbers of samples of other species submitted for identification. Larvae that commonly develop in carcasses are sometimes found in wounds of living animals, where they usually feed on necrotic tissue. When this tissue is consumed, the larvae may burrow into living muscle and are then frequently confused with screwworms. Thus any appraisal of progress in screwworm control required that samples of all wound-infesting larvae be evaluated by specialists. Among those found were *C. macellaria* (in warm weather), *Phormia regina* (Meigen) (in colder months), occasionally the genera *Phoenicia, Sarcophaga,* and *Cynomyia,* and other species, even the housefly. Since none of these maggots are affected by the release of sterile screwworms, the number of nonscrewworm samples should not vary greatly from year to year with uniform reporting. The values therefore provide an indication of the general efficiency of the reporting system.

TABLE V

Records of Myiasis from Texas, 1962–1965

	Samples of screwworms				Samples of nonscrewworms			
	1965	1964	1963	1962	1965	1964	1963	1962
January–March	4	7	217	451	546	739	544	199
April–June	201	80	1,247	17,241	1,967	2,509	3,172	1,283
July–September	117	12	824	19,322	609	600	996	349
October–December	144	124	2,628	12,470	915	1,409	1,853	601
Total	466	223	4,916	49,484	4,037	5,257	6,565	2,432
Correction factor	1.63	1.25	1.00	2.70	—	—	—	—
Corrected total	760	279	4,916	133,607	—	—	—	—
Percent control	99.94	99.98	99.63	90.00	—	—	—	—

Only a few collection kits had been distributed to livestock growers when the program started in February, 1962, but by the end of that year, an active information and education program implemented chiefly by the efforts of county agricultural extension agents and vocational agriculture teachers had put sample vials in the hands of the majority of livestock producers. A total of 2432 nonscrewworm samples were submitted in 1962. In 1963, a record number of 6565 were identified, which was interpreted as a peak of producer participation in the program. Therefore the data from 1963 were used to establish correction factors in computing percentage control.

Since extensive rancher reporting of infestations started with the beginning of the program in 1962, the degree of control effected that year is only an average of grower opinions. The consensus of those reports was that infestations of screwworms had decreased 90% in 1962 over the average of preceding years. In 1962, livestock growers submitted 49,484 samples of screwworms and 2432 samples of nonscrewworms; in 1963 when the sampling program was firmly established, 6565 cases of nonscrewworms were reported, a ratio that gave a correction factor of 2.70. Thus if reporting had been as good in 1962 as it was in 1963, about 133,607 screwworm samples would have been submitted in 1962. If the rough average of 90% control for the first year is correct, then the infestations in domestic animals treated by ranchers in preceding years in Texas must have averaged about 1,300,000 annually. During 1964, only 223 samples of screwworms were submitted; 5257 samples of other maggots were returned. The ratio of 6565 to 5257 gives a correction factor of 1.25. Then if reporting had been as thorough in 1964 as it was in 1963, 279 samples of screwworms would have been reported.

The relatively few screwworms reported during 1964 were found in widely scattered areas of Texas, but most of the cases occurred within 300 miles of areas of known screwworm activity in northern Mexico. The infestations were therefore attributed to fertile flies migrating from Mexico. However, such migration exceeded the previous estimate of maximum flight of fertilized flies (about 100 miles).

A special test was therefore made with dye-marked sterile screwworms released at one point on the Colorado River in Texas. Hightower *et al.* (1965) reported capturing these marked flies in their most distant trap, which was 180 miles from the release point. Since laboratory observations indicated that sterilized flies are not quite as vigorous as normal insects and screwworms reared on artificial media are not as robust as those grown in wounds on living animals, the trapping experiment completely supported the theory that all the 1964 cases in Texas originated in flies migrating from Mexico.

During 1965, a total of 466 samples of screwworms were submitted, all within the flight range of northern Mexico. However, in two separate localities the infestations were concentrated. Field surveys showed that these infestations had probably bred in both localities a few generations before they were eliminated. Both were north of the barrier zone where routine releases of sterilized flies were made, so lack of rancher inspection probably permitted the local buildups.

Although the number of screwworms in Texas increased in 1965, the area involved was smaller. During 1964, screwworms were found in 65 Texas counties; in 1965 only 53 counties were involved. This reduction in infested area was accomplished by an extension of the fly releases into northern Mexico.

As the result of the new knowledge about migration, sterile-fly releases were made as far as 300 miles south of the Rio Grande River. However, production of sterile flies has been limited to about 80 million flies per week during the cold months and to 110 million flies per week in warm weather because of budgetary considerations. The barrier zone thus cannot be of uniform width or uniformly populated. Rates of release vary from 200 to 800 flies per square mile per week, and the barrier varies in width from 50 to 300 miles, depending on geography, weather, and screwworm breeding in nature.

Field observations published by Barrett (1937) and Lindquist (1955) indicated that the normal migration of screwworms was 35 miles per week during the most favorable seasons of the year, and studies in outdoor cages suggested that the maximum longevity of flies in nature during periods of warm weather was about 3 weeks. It was thus surprising to find flies migrating 300 miles.

Monro (1963a), in commenting on Knipling's mathematical models, suggested that control resulting from the release of sterile insects might be augmented by a "flushing" effect. If the native population already filled the most favorable ecological niches, the addition of sterile insects should crowd the natives into a less suitable environment where natural mortality would be higher. This effect could be an important added factor of extermination if the entire area inhabited by an insect species was saturated with sterile released insects. In fact, such releases of screwworms did appear to increase the rate of migration out of the saturated area. Another aspect of the flushing effect might be the pursuit by sterile males of inseminated females. Baumhover (1965), in a report on sexual aggression in screwworms, found that in cages this pursuit is so vigorous the females died of exhaustion. Other less obvious crowding effects may contribute to increased migration: Lindquist (1955) theorized that native populations of screwworms seldom exceed 100 pairs of flies per square mile, even during outbreak conditions; Hightower (1963) found that, during summer drought conditions, both sterile and native screwworms rested at night only on the terminal twigs of trees overhanging water courses. Also, before screwworm females are ready to oviposit, they visit animals to feed on wound fluids. In the absence of sterile-insect releases, an observer might see only a few females during a whole day's observation of attractive wounds; in areas receiving from 200 to 800 sterile flies per square mile per week, several females can usually be observed feeding together. Perhaps crowding in a feeding or resting situation may contribute to population pressure.

In 1965, Congress appropriated funds to extend screwworm eradication to Arizona and California, and agencies of those states made matching funds available. Screwworm eradication is therefore under way in the last remaining infested portions of the United States. Releases of sterile flies are being made in those areas and in adjacent regions of northern Mexico. These eradication procedures have been so consistently successful that the screwworm will be eliminated from the southwestern United

States. However, as long as Mexico remains infested there will be a continuing problem of combating migrating flies and their progeny.

Arrangements are now being developed between officials of Mexico and the United States for an extension of the release area to the Isthmus of Tehuantepec, which is only about 150 miles wide at the narrowest point. The cost of maintaining a 1200-mile-long barrier zone from Texas to California is about 5 million dollars per year. With an annual benefit-to-cost ratio of 20 : 1, the expense is economically justified, but it will be much less expensive over the long term to eliminate flies over most of Mexico and maintain a narrow barrier in the southern part of that country. If appropriate international agreements could be reached, the next step in screwworm eradication would be the elimination of screwworms from Central America with a barrier at the Isthmus of Panama.

VIII. FIELD TRIALS FOR CONTROL

A. Diptera

1. *Fruit Flies*

The demonstration of the successful use of the sterile-male technique against the screwworm on the island of Curacao and in the southeastern and southwestern programs stimulated research on the use of this technique against several species of fruit flies. Fruit flies are of great economic importance and are relatively easy to rear in large numbers in the laboratory. Thus, they were good candidates.

a. The Mediterranean Fruit Fly. The first field test was conducted against the Mediterranean fruit fly, *Ceratitis capitata* (Wiedemann), on the island of Hawaii in 1959–1960. A total of 18.7 million laboratory-reared Mediterranean fruit flies irradiated with 10,000 r in the late pupal stage were released at weekly intervals in an isolated mountainous area (altitude 3900–6000 ft) during a 13-month period beginning in June, 1959 (Steiner *et al.,* 1962). No effect was noted in 1959, but during the last 4 months of the test (May through August, 1960), the wild popluation was reduced by an average of 90%. A greater reduction could not be obtained because the test area was not sufficiently isolated to keep flies from migrating in from other infested areas of the island. During the last 2.5 months of the test, based on studies made with marked irradiated flies, the average ratio of sterile to wild flies was only 3.2 : 1. No detectable reduction in the population of the Mediterranean fruit fly was noted in the other areas in Hawaii during the summer of 1960. Also, once the releases were discontinued, the infestation in the test area was back to its usual level after 2 generations (Steiner, 1965). The possibility of suppressing the natural population by overflooding the whole area with sterile Mediterranean fruit flies was thus demonstrated.

Another field trial with the Mediterranean fruit fly was initiated in Puntarenas, Costa Rica in July, 1963. The test area was a small peninsula of about 1 square mile which sustained a high population of wild flies. The sterile flies used in the test were laboratory reared and the pupae were irradiated at 10,000–12,500 r. No deleterious effects on adult longevity and mating were noted (Katiyar and Valerio, 1963), but males were less than one-third as competitive as normal males. During 1963, sterile flies were released at rates ranging from 0.5 to 2.2 million per month. In 1964, the ground releases were increased to 6.5 million per month. By May, 1964 there was a moderate reduction in the percentage of infested fruit, and a fivefold reduction in the number of larvae per fruit (from 3.13 to 0.6), but the effect was not clear cut because two insecticide applications combined with protein hydrolyzate (which acts as an attractant) were made—one in February and one in March, 1964—to keep the fly population at a reasonable level (Katiyar, 1965). Releases were continued in 1965, and the infestation was further reduced to 0.08 larvae per fruit in the test area compared with 4.72 in the untreated control area (W. E. Stone, personal communication). These results were sufficiently encouraging so that a United Nations Special Fund Project was begun in 1965 in an effort to eradicate the Mediterranean fruit fly from Costa Rica, Nicaragua, El Salvador, and Panama.

Other feasibility studies are underway in Israel to determine whether the Mediterranean fruit fly can be eradicated from citrus growing areas. In 1963, over 3 million flies were irradiated in the pupal stage and tagged with either P^{32} or Calco Blue, but recapture of released flies was extremely low. In 1964, improved rearing techniques produced a more vigorous fly, and recaptures of 10% were obtained (Nadel, 1965).

b. The Oriental Fruit Fly. The first field trial with the oriental fruit fly, *Dacus dorsalis* Hendel, was made on the 33-square mile island of Rota, one of the Mariana Islands located 37 miles northeast of Guam. The flies were reared and irradiated in the pupal stage with 10,000–12,000 r in the Honolulu Fruit Fly Laboratory of the United States Department of Agriculture. About 3 million pupae a week were shipped by commercial jet to Guam, held there at 27°C, and released by plane over Rota.

Between September, 1960 and August 1, 1962, a total of 531 million fruit fly pupae that produced 403 million flies was shipped and released. Fortunately, a genetic strain of flies with white markings could be used in conjunction with traps baited with methyleugenol to evaluate the seasonal fly abundance and the effectiveness of the releases (Steiner *et al.*,

1962). The fruit infestations caused by the oriental fruit fly could not be depressed; it was impossible to flood the island with sufficient numbers of sterile flies; 16 million flies a week would have been needed to obtain a 10 : 1 ratio when the native population was at its lowest level (Steiner, 1965).

A second field experiment was tried on Guam during the summer of 1963 to take advantage of two typhoons that had destroyed most of the tree fruit hosts of the oriental fruit fly. Again, the white-marked strain was used. After an initial release of 1.2 million sterile flies at the north end of the island where some fruit fly hosts had survived the typhoon, only two wild flies were recaptured. Thus, the natural population was at a very low ebb, and an attempt at eradication was begun. A total of 17 million flies was released, and the oriental fruit fly was eradicated. Two wild flies were caught in March, 1964, but they were probably brought in before the island could be quarantined from further infestations. This eradication of the oriental fruit fly on Guam is highly important. As Steiner (1965) pointed out, it "demonstrated the unusual opportunity for eradication that is available in the sterile-release method when a species is under severe stress because of unfavorable environmental conditions."

An attempt to eradicate the oriental fruit fly from the islands of Saipan, Tinian, and Aguijan in the western Pacific was begun in February, 1964. Over 90 million flies from Hawaii were released in the 90-square mile test area during a 5-month period (Steiner, 1965). However, the longevity of the sterile flies was greatly reduced, partly because of the long air trip from Hawaii and partly because of the handling of the pupae and flies on the ground. This latter difficulty was especially marked on Saipan where ground releases were made by native personnel, an indication that entomologists are needed to supervise ground releases. In spite of these difficulties, eradication was nearly achieved on Tinian. In September, 1964, the releases were discontinued (L. D. Christenson, personal communication).

c. The Melon Fly. Eradication of the melon fly, *Dacus cucurbitae* Coquillett, was achieved on the island of Rota in 1963. During 1960–1961, fly populations were normally higher from January to September. Therefore, the melon fly population was reduced nearly 75% before the first release in September, 1962 by spraying the border vegetation in the host areas (concentrated in fewer than 25 farms) with a protein hydrolyzate bait containing malathion (Steiner *et al.*, 1965).

The pupae were reared and irradiated at a mean dose of 10,000–10,500 r in a cobalt-60 source in Hawaii. From September 24, 1962 to July 4, 1963, over 305 million irradiated pupae were shipped weekly by commercial jet to Guam; there 190 million adults were distributed by ground releases, and another 90 million were released from the air on Rota. In some releases, the sterile flies were tagged with dyes; thus the number recaptured in traps baited with cuelure indicated the ratio of sterile to normal males as well as their density on the island. The last infested fruit was sampled in December, 1962, less than 3 months after the beginning of the releases. A few wild females were still in the area by February 1963 since 38 nonfertile eggs were found.

In this eradication, test fly losses were high, both on the ground where predators congregated about the release sites and in the air where the wind often carried the flies away from the island. A reduction of 50% in longevity and increased wing damage to the sterile flies were also noted (Steiner *et al.,* 1965). A few wild melon flies were caught on Rota in 1964, apparently blown in from Guam. Therefore, an additional 6 million sterile flies were released to prevent reestablishment (Steiner, 1965). Obviously, vigilance is necessary to keep a clean area from reinfestation.

d. The Mexican Fruit Fly. Several small field tests were made with the Mexican fruit fly, *Anastrepha ludens* Loew, in Mexico at Santa Rosa and San Carlos in 1960–1962, based on the findings of Rhode *et al.* (1961) that adults of the Mexican fruit fly were sterilized when 12-day-old pupae were exposed to 5000 r of gamma radiation. The flies were reared and irradiated at 7000 r in Mexico City. In 1960, 1,172,000 sterile flies were released at Santa Rosa in a semi-isolated hacienda containing about 1 square mile of host fruit, including grapefruit. The number of larvae per pound of fruit decreased. Slightly better results were obtained the following year when 838,000 sterile flies were released, and similar results were obtained in the San Carlos area (Christenson, 1963). These field tests were only moderately successful because the native fly population was very high, the test areas were not sufficiently isolated, and the number of sterile flies was not high enough to overwhelm the natural population. However, they strongly suggested that the sterile-male method would be completely effective as long as a high ratio of sterile to wild flies could be maintained.

e. The Queensland Fruit Fly. The Queensland fruit fly, *Dacus tryoni* (Froggatt), is found in eastern Australia where it is an important pest of fruit crops. The possibilities of using the sterile-male technique on the fringe areas of the distribution are promising because the populations are small, fluctuate a great deal, and are limited to semi-isolated areas. The first field test took place in two small towns in New South Wales—Manilla (1962–1963) and Warren (1963–1964)—(Monro, 1965). The 8-day-old pupae were irradiated with a dose of 5000 rad (Monro, 1963b). Control was not achieved in the Manilla area because the projected fly production could not be maintained, and the releases were discontinued. However, the limited program caused about a 50% reduction in egg hatch from loquats collected in the spring of 1963 compared with those collected from the control areas.

The second field trial in the spring of 1963 at Warren covered an area of less than 2 square miles. Releases of about 120,000 flies per week continued until March, 1964. At the peak of the season, less than 5% of the fruit was infested compared with 75% in the check town. As Monro pointed out, "Eradication is difficult to demonstrate in such towns . . . because they are in contact with outside populations which seem to move into them along the water courses." There is no doubt that, with isolated populations, eradication could be achieved (Monro, 1965).

3. *The Sheep Blowfly*

The sheep blowfly, *Lucilia sericata* Meigen, has a low population density in Great Britain, usually less than 1900 per square mile. In laboratory experiments, Donnelly (1960) showed that both adult males and females could be sterilized when 6- to 8-day-old pupae were irradiated with 4000 rep. However, adult longevity was reduced by 40% compared with the controls.

A field test was made in 1956 on Holy Island off the Northumberland coast, 60 miles south of Edinburgh. The natural population on the 2 square-mile island was estimated to be about 4000. Nearly 175,000 pupae irradiated at 6000 rep at Harwell were allowed to emerge at three release points on the island from July through September, 1956. Releases were made at weekly intervals, but adult emergence was poor, especially during the last few weeks. The following year rearing procedures were improved, and the releases were resumed. Also, the sterilizing dose was raised to 7000 rep in the mistaken belief that the previous dose had been insufficient. Overall emergence was better the second year, 55% vs. 49%, and the poor emergence did not recur (MacLeod and Donnelly, 1961).

Consistently poor weather both summers prevented a good assessment of the progress of the releases. However, tests in 1958 indicated that no decrease in the fly population had occurred. Donnelly (1965) discussed some possible causes of the failure in these tests. Losses in the sterile population may have been caused by predation at the release sites, low emergence rates (especially during the 1956 test), and vulnerability of the irradiated insects to adverse weather conditions. The released flies also were not as competitive as the normal flies since they could barely achieve two copulations compared with an average of 6–12 for normal flies.

3. *The Housefly*

Several field tests have been made with chemosterilized flies, but only two have been reported with flies sterilized with γ-rays.

The first field trial was attempted in a small rural area (about 500 acres) in the Italian province of Latium. Male houseflies irradiated in the pupal stage with 2000 r were released over a 4-month period beginning in March, 1961. The population of female flies was greatly reduced during May and June, but it increased rapidly once the releases were discontinued (Rivosecchi, 1962). The incomplete success of the field trial was attributed to at least two major factors: (1) the test area was not sufficiently isolated and (2) the 2000-r dose was too low [Rivosecchi (1962) reported that fertile eggs were obtained even at ratios of 30 : 1 : 1]. A dose of about 2850 r is necessary to induce lethal mutations in all sperm of the housefly (Schmidt *et al.,* 1964).

The second test was begun in 1965 on the island of Grand Turk in the Bahama Islands as part of an integrated program for the control of houseflies. This program included the application of insecticides and chemosterilants as well as the release of flies sterilized by γ-rays. Pupae were reared and irradiated with 6000 r at the Gainesville, Florida laboratory and shipped by air to Grand Turk, where the adults were released at a rate of 500,000–1,000,000 per week (G. C. LaBrecque, personal communication). Sterility as high as 95% was noted in the natural population after the releases. Unfortunately, however, the releases were discontinued in May, 1965 because scheduled air service was terminated. This test was a qualified success since it demonstrated that some degree of control could be achieved. However, incomplete knowledge of housefly behavior in the field was a distinct handicap.

4. *Mosquitoes*

The effect of radiation on over a half-dozen species of mosquitoes has been studied. However, small field tests have been attempted with only three species. None can be regarded as highly successful, but a great deal of useful information resulted, and the lessons learned can be applied with advantage to other species of mosquitoes and to other insects.

a. The House Mosquito. Ramakrishnan *et al.* (1962) showed in a laboratory experiment that male pupae of *Culex pipiens fatigans* Wiedemann irradiated at 7700 r produced sterile adults. When these males were placed in cages with normal males and females at 2 : 1 : 1 ratio, 38–40% of the egg rafts were nonviable.

A small field trial was therefore made in a semi-isolated Indian village of Uttar Pradesh to assess the effect of releasing sterilized *C. p. fatigans* on the wild population (Krishnamurthy *et al.,* 1962). Only

24,000 sterile adults were released over 6 weeks. Despite this limited release, a 6% reduction in the viability of egg rafts sampled from the experimental area was noted, compared with the control area. Better evaluation was hampered by the lack of a practical method of estimating the natural adult population. Some problems encountered in this field experiment were: difficulty in rearing, handling, and irradiating large numbers of mosquito pupae; insufficient ecological data to properly evaluate the test; and poor public relations with the villagers. In spite of these complications, Krishnamurthy *et al.* (1962) were optimistic. They concluded that the sterile-male technique offered possibilities for mosquito control when and if these problems were solved.

b. The Common Malaria Mosquito. In some preliminary studies on the possible use of the sterile-male technique with *Anopheles quadrimaculatus* Say, Davis *et al.* (1959) showed that doses of 8800–12,900 r applied in the pupal or adult stage produced sterility in both sexes.

In the fall of 1959, a field test was begun on two small islands of about 2 square miles in the southern part of Lake Okeechobee in Florida. Mosquitoes colonized in the laboratory were sterilized in the pupal stage with 12,000 r of gamma radiation. The adult females were removed and the males were transferred to cages 24–48 hours after emergence, transported 180 miles by car, and released on the island. A total of 328,900 males, averaging 3700 per square mile per week, was released during the 11-month test. Only a small amount of sterility occurred during the fall of 1959, when the natural population was in a seasonal decline, and the releases had no effect when the natural population increased during the summer of 1960 (Weidhaas *et al.,* 1962). In September, 1960, the test was discontinued because both islands were flooded in the aftermath of a hurricane.

A second test was made near Lake Panasoffkee, Florida. A total of 104,700 sterile males was released from September 23 to December 2, 1960, at two sites in the breeding area. No decrease in mosquito population or any induced sterility in the wild females were conclusively demonstrated (Weidhaas *et al.,* 1962). The number of sterilized males was insufficient to overwhelm the natural population, and the males may have been inadequate and unable to mate with the wild females. The factors that may influence the ability of sterilized male *A. quadrimaculatus* to inseminate wild females were discussed by Dame and Schmidt (1962).

An additional 2-year study was made by Dame *et al.* (1964) in the same area with both irradiated and chemosterilized male mosquitoes. The behavior of the colonized males was significantly different from that of wild males and made them incapable of incorporating themselves sufficiently into the wild population; they were sexually vigorous and competitive with wild males for colony females and sexually compatible with wild females near the release site, but they were unable to disperse and find the wild females. First-generation male progeny of wild population sterilized before release were able to find and mate with wild females.

c. *The Yellow-Fever Mosquito.* In preliminary tests, McCray *et al.* (1961) showed that pupae of *A. aegypti* exposed to 10,500–17,500 r in a cobalt-60 γ-ray point source were completely sterilized. (With this type of source, the pupae are placed between two small concentric rings around the point source; in the usual γ-irradiator, the sample is in the center and the cobalt-60 strips are on the periphery.) The cage studies were sufficiently encouraging to warrant a field test in two areas in the western suburbs of Pensacola, Florida in 1960. The test was continued in 1961. A laboratory strain was used in the 1960 release; a Pensacola-derived strain was used in 1961. The pupae were passed through a pupae separator, and the male pupae were irradiated and shipped by air to Pensacola where they were distributed twice a week (in 1960) and once a week (in 1961) to the release sites. There the pupae were placed in old automobile tires half-filled with water. In 1961, the irradiation procedure was changed because of the large variation in doses (9400–18,750 r) as well as the decrease in emergence and longevity that occurred at doses above 12,000 r. By changing the geometry of the pupal holders in relation to the point source, the limits of the dose range were narrowed to 8800–9500 r, and pupal survival was increased to 94% (Fay *et al.,* 1963).

From July to November, 1960, a total of 3.9 million pupae was distributed; from April to October, 1961, the number was increased to 6.7 million. The results were inconclusive because populations decreased in some test areas and in some check areas (Morlan *et al.,* 1962). In the first test area, the releases reduced but failed to eradicate natural populations, probably because of reinfestation. It was concluded that before the sterile-male technique can be adapted for mosquito control, additional investigations of mosquito biology are required, especially with regard to male dispersal under field conditions.

B. Lepidoptera

1. *The Codling Moth*

The codling moth, *Carpocapsa pomonella* (L.), is one of the most important pests in the apple and pear growing areas of the world. Investigations were begun in Summerland, British Columbia, in 1956 to determine whether the sterile-male technique could be applied to this insect. Proverbs and Newton (1962a) reported that exposure of mature pupae (within one day of emergence) to 40,000 rads of γ-radiation induced dominant lethality in about 99% of the sperm without affecting emergence, longevity, or mating behavior. These results were confirmed by Hathaway (1966). Three field-cage orchard experiments were therefore initiated during 1961–1962.

Large screened cages were placed over dwarfed apple trees, and irradiated male moths exposed to 40,000 rads as pupae were placed with normal moths, both males and females, in the cages in the following ratios: 10 : 1 : 1 and 20 : 1 : 1. The reproductive potential based on larval progeny was reduced 85 and 97%, respectively. However, when irradiated females were present, the results were less favorable; the reduction was only 57 and 96% (Proverbs and Newton, 1962b). Hathaway (1966) also obtained a greater reduction in the F_1 generation in his field-cage tests when irradiated females were not present. Apparently, irradiated females have a small but detrimental effect in small-cage tests.

Results were sufficiently promising, so that a small semi-isolated orchard was selected for field evaluation. The first release occurred in 1962. Rearing difficulties were encountered during the first year, and this resulted in insufficient overflooding (8 : 1) of the native population. In 1963, a 20 : 1 ratio was achieved, and less than 0.5% of the fruit was injured compared with 7% in 1962. Almost complete eradication was achieved in 1964, with fruit injury down to 0.05% in spite of reinfestation from nearby orchards. In 1964, another test was begun in an isolated apple orchard in the Okanagen Valley in British Columbia. Injury to the fruit dropped from 60 to 2% after release of both sterile male and female moths (Proverbs, 1965). These tests indicated that more rapid rearing and irradiating techniques must be developed before large field tests can be undertaken.

A cobalt-60 irradiation facility is now being installed at the United States Department of Agriculture Laboratory in Yakima, Washington, and an improved tray rearing method has been developed to produce the necessary number of moths for large field tests. As many as 100 larvae per square foot of tray space can be reared on an artificial medium based on fortified wheat germ diet (J. F. Howell, personal communication).

2. *The Navel Orangeworm*

The navel orangeworm, *Paramyelois transitella* (Walker), is an important pest of walnuts and almonds in California. Husseiny (1963) reported that complete sterility was obtained in both sexes when 8-day-old pupae were exposed to 50,000 rads of γ-radiation. At this dose, mating, egg laying, and longevity did not seem to be affected, and sterilized males were competitive in cage tests. The addition of sterile females only to normal males and females also reduced the number of viable eggs just as effectively as the introduction of sterile males alone. However, when both were present, the reduction was four times greater (Husseiny and

Madsen, 1964), an interesting finding since, with the codling moth (Proverbs and Newton, 1962b), the presence of irradiated females was detrimental.

A small field test was initiated in a 5-acre walnut grove after the development of a simple, economical method of mass rearing. Both sterile males and females were released. Most females mated but once under field conditions; only 8% mated twice. The results are encouraging, and further small field tests are underway.

C. Coleoptera

1. *The Field Cockchafer* (*White Grubs*)

The field cockchafer, *Melolontha vulgaris* (F.), is a serious pest of root vegetables in Switzerland; damage is caused by the white grubs. Horber (1963a) realized that the biology of this insect was such that it would be a good candidate for the sterile-male technique. Flight period occurs only during a few weeks of every third year, and the male precedes the female; also, the appearance of the first beetles can be predicted accurately. Laboratory studies indicated that the adults were sterilized at a relatively low dose of 3000 r of x-rays with no untoward side effects.

Two small field tests (74 acres) were conducted in a farming area in northwestern Switzerland—the first in 1959 and the second in 1962. Adult males collected in light traps were irradiated at a dose of 3300 r with x-rays, tagged with a white dye, and released at the appropriate time. In the 1959 test, 3109 tagged, irradiated, adult males were released; in 1962, the number was increased to 8594. The white grub population was reduced 80% in the 1959 test area compared with the control areas. Complete eradication was obtained in 1962 (Horber, 1963a,b). These field trials are significant for the following reasons: (a) they represent the first successful test with a Coleopteran pest; (b) releases did not depend on artificial rearing; (c) only males irradiated in the adult stage were released; and (d) the sterile-male technique was successfully used in an area not completely isolated.

THE RESEARCH NEEDS

IX. COMPARATIVE RADIOSENSITIVITY OF VARIOUS INSECT SPECIES

In Sections VI and VIII, we mentioned the sterilizing doses required for a few insect species. Now we will consider in more detail (1) the range of doses required to sterilize different insect species, (2) some possible reasons for the difference in doses required, and (3) the implications of these factors for the sterile-male technique of insect control. Table VI is a summary of the published research concerning differential radiosensitivity of insect species, but it is not a complete survey of all radiation work performed on insects.

In assembling Table VI, we became acutely aware of the many pitfalls in accepting the data without reservation. For example, there are different kinds of sterility; but in assembling data for over 60 species, we were not sure that we were always comparing the same kind of sterility. The sterility for most of the species reported in Table VI is almost certainly the result of dominant lethal mutations in the sperm. However, irradiation may sometimes have produced males that either did not mate or did not transmit sperm. It is even possible that the sperm transmitted were immotile. A few of the studies published in foreign journals were available to us only in abstracts, which made it difficult to estimate the kind of sterility and the accuracy of the dosimetry. Wherever possible, we have used the amount of radiation that induced lethal mutations in 98–99% of the sperm as the sterilizing dose. This figure is often considerably lower than the sterilizing dose quoted by the authors which was based on 100% sterility. Thus Table VI will be used as a central point of reference in our discussion of the differential radiosensitivity of insects, with the understanding that differences in stage irradiated, dosimetry, or type of sterility studied could make our comparisons somewhat hazardous.

Lee (1958) stated that at equal doses of radiation there was no significant difference in the average number of dominant lethals induced in the sperm of the honeybee and *Drosophila*. Heidenthal (1945) made a similar comparison between lethal mutations induced in the sperm of *H. juglandis* and *D. melanogaster*. The large number of species listed in Table VI permits comparisons between insects of the same order and between orders, and reveals several interesting points. Most Dipteran species can be sterilized with doses under 10 kr. However, within this order a threefold difference in sterilizing doses was noted. The Hymenoptera seem to require about 6–10 kr to induce sterility. Also most Coleopteran species investigated are sterilized by 4–10 kr, with the possible exception of the Khapra beetle and the Chinese rice weevil. Differences in dosimetry may account for the larger doses required in these studies. The Lepidoptera, however, require very large doses of radiation to produce sterility.

What are some reasons for the vast differences in dose required to sterilize the various species of insects? Of course most insects differ in size, physiology, chromosome number, chromosome size, and in many other ways. However, some consistent differences exist between radiosensitive and radioresistant species. For example, the resistant Lepidoptera

TABLE VI

COMPARATIVE RADIOSENSITIVITY OF INSECT SPECIES. ESTIMATED DOSES OF γ-RAYS OR X-RAYS REQUIRED TO "STERILIZE" MALES

Scientific name	Common name	Stage treated	Sterilizing dose (r)[a]	Reference
DIPTERA				
Musca domestica L.	Housefly	4-Day-old pupae	2,850	Schmidt *et al.* (1964)
Musca domestica L.	Housefly	Pupae 2–3 days before emergence	3,000	Sacca (1961)
Glossina morsitans Westwood	Tsetse fly	Pupae	6,000	Potts (1958)
Haematobia irritans (L.)	Horn fly	Pupae	5,000	Lewis and Eddy (1964)
Stomoxys calcitrans (L.)	Stable fly	Pupae	5,000	R. L. Harris (unpublished)
Hippelates pusio Loew	Eye gnat	Adult males 1–1.5 days old	4,500–5,000	Flint (1965)
Hypoderma lineatum (de Villers)	Cattle grub	Pupae	2,500–5,000	Drummond (1963)
Cochliomyia hominivorax (Coquerel)	Screwworm fly	5-Day-old pupae	2,500	Bushland and Hopkins (1953)
Cochliomyia hominivorax (Coquerel)	Screwworm fly	1-Day-old adult	5,000–7,000	LaChance and Crystal (1965)
Lucilia sericata Meigen	Sheep blowfly	Pupae	3,000–4,500 rep	Donnelly (1960)
Drosophila melanogaster Meigen	Fruit fly	Adult males	11,000–12,000	Catcheside and Lea (1945)
Anastrepha ludens (Loew)	Mexican fruit fly	12-Day-old pupae	3,000	Rhode *et al.* (1961)
Dacus oleae (Gmelin)	Olive fruit fly	Pupae	12,000	Melis and Baccetti (1960)
Dacus oleae (Gmelin)	Olive fruit fly	Adults	15,000–18,000 rads	Thomou (1963)
D. dorsalis Hendel	Oriental fruit fly	Pupae	10,000	Steiner *et al.* (1962)
D. cucurbitae Coquillett	Melon fly	Pupae	9,500–10,000	Steiner *et al.* (1965)
D. tryoni (Froggatt)	Queensland fruit fly	8-Day-old pupae	4,000 rads	Waterhouse (1962); Monro (1963b)
Ceratitis capitata (Wiedemann)	Mediterranean fruit fly	Pupae 3–4 days before emergence	10,000	Steiner and Christenson (1956)

Anopheles quadrimaculatus Say	Common malaria mosquito	Pupae or adults	8,865–12,900	Davis *et al.* (1959)
A. maculipennis atroparvus von Thiel	—	Adults	6,000	Frizzi and Jolly (1961)
Culex fatigans Wiedemann	House mosquito	Pupae	7,700	Ramakrishnan *et al.* (1962)
C. tarsalis Coquillett	—	Pupae	15,000	Darrow and Eddy (unpublished)
Aedes aegypti (L.)	Yellow fever mosquito	24-Hour-old pupae	8,000–10,000	Weidhass and Schmidt (1963); Fay *et al.* (1963)
Psila rosae (F.)	Carrot rust fly	15-Day-old pupae	2,800	McClanahan (1965)
HYMENOPTERA				
Apis mellifera L.	Honeybee	Adults	7,000–11,000	Lee (1958)
Habrobracon juglandis (Ashmead) (=*Bracon hebetor* Say)	—	Adults	8,000–10,000	Whiting (1938b); Heidenthal (1945)
Mellitobia sp.	—	Adults	10,000	Kerschner (1946)
Mormoniella vitripennis (Walker)	—	Adults 1–2 days old	6,000	Mortimer and von Borstel (1963)
COLEOPTERA				
Anthonomus grandis Boheman	Boll weevil	Adults	7,000–10,000	Davich and Lindquist (1962)
Sitophilus sasakii Takahashi	Rice weevil	Adults	10,000	Laviolette and Nardon (1963)
S. granarius (L.)	Grain weevil	Adults	6,000 rads	Cornwell and Bull (1960)
S. oryzae (L.)	Rice weevil	7-Day-old adults	7,500–10,000	Hoover *et al.* (1963)
Pissodes strobi (Peck)	White-pine weevil	Adults	5,000	Jaynes and Godwin (1957)
Callosobruchus chinensis L.	Chinese rice weevil	Adults	~42,000	Quraishi and Metin (1963)
Tribolium confusum Jacquelin duVal	Confused flour beetle	Old pupae	4,000	Erdman (1962)
		Adults	4,000	
T. castaneum (Herbst)	Red flour beetle	Old pupae	5,000	Erdman (1962)
		Adults	5,000	

TABLE VI—(*Continued*)

Scientific name	Common name	Stage treated	Sterilizing dose (r)[a]	Reference
Epilachna varivestis Mulsant	Mexican bean beetle	Pupae	4,000	Henneberry *et al.* (1964)
		Adults	8,000	
Trogoderma granarium Everts	Khapra beetle	Pupae	15,000	Kansu (1962)
Lasioderma serricorne (F.)	Cigarette beetle	Adults	15,000	Harvey (1963)
Attagenus piceus (Olivier)	Black carpet beetle	Adults	15,000	Harvey (1963)
Ips confusus (LeConte)	California five-spined ips	Adults	10,000	Stark (1963)
Melolontha vulgaris F.	White grubs cockchafers	Adults	3,000	Horber (1963b)
Rhyzopertha dominica (F.)	Lesser grain borer	Adults	16,000 rads	Pendelbury *et al.* (1962)
LEPIDOPTERA				
Pectinophora gossypiella (Saunders)	Pink bollworm	Pupae	30,000	Ouye *et al.* (1964)
Ostrinia nubilalis (Hübner)	European corn borer	1-Day-old males	32,000	Walker and Brindley (1963)
Heliothis virescens (F.)	Tobacco budworm	Late pupae	~45,000	Flint (unpublished)
Carpocapsa pomonella (L.)	Codling moth	Pupae	40,000 rads	Proverbs (1964)
Anagasta kuehniella (Zeller)	Mediterranean flour moth	Pupae	45,000 rads	Bull (1963)
Paramyelois transitella (Walker)	Navel orangeworm	Mature pupae	50,000 rads	Husseiny (1963)
Diatraea saccharalis (F.)	Sugarcane borer	Adult males	25,000–30,000	D. Walker (unpublished)
Platynota stultana (Walsingham)	Omnivorous leaf roller moth	24-Hour-old males	32,000	S. W. Jacklin *et al.* (unpublished)
Plodia interpunctella (Hübner)	Indian-meal moth	?	50,000	Harvey (1963)
Porthetria dispar (L.)	Gypsy moth	9-11 Day-old pupae	20,000	Godwin *et al.* (unpublished)
Bombyx mori (L.)	Silkworm	?	20,000	Astaurov (1937)
Cadra cautella (Walker)	Almond moth	Adults	16,000 rads	Pendelbury *et al.* (1962)

Gnorimoschema operculella (Zeller)	Potato tuberworm	Adults	12,000–24,000	Elbadry (1965)
[*Phthorimaea operculella* (Zeller)]		8-Day-old pupae	12,000–24,000	
Laphygma exigua (Hübner)	Cotton moth	Pupae	10,000–11,000	Rasulov (1963)
Chloridea obsoleta F.	Bollworm (European species)	Pupae	8,000	Andreev *et al.* (1962)
[*Heleothis armigera* (Hübner) (Old World)]				
Thaumetopoea pithyocampa Schiff	Pine processionary moth	15-Day-old pupae	4,000	Baccetti and Zocchi (1962)
ORTHOPTERA				
Blatella germanica (L.)	German cockroach	Old nymphs	3,200	Ross and Cochran (1963)
HEMIPTERA				
Rhodnius prolixus Stål	—	Adults	20,000	Baldwin and Shaver (1963)
HOMOPTERA				
Planococcus citri (Risso)	Citrus mealybug	Adults	~16,000	Brown and Nelson-Rees (1961)
OTHER ARTHROPODS				
Tetranychus telarius (L.)	Carmine spider mite	1-Day-old males	32,000	Henneberry (1964)
Amblyomma americanum (L.)	Lone star tick	Adults	2,500	R. O. Drummond *et al.* (unpublished)

[a] Unless otherwise noted.

are characterized by: (1) high chromosome number (Virkki, 1963), (2) small chromosomes, (3) holokinetic chromosomes (diffuse centromeres) (Suomalainen, 1953), and (4) heterogametic females. The radiosensitive Diptera, in contrast, have (1) low chromosome numbers, (2) larger chromosomes, (3) localized centromeres, and (4) heterogametic males. Are any of these factors associated with differences in radiosensitivity? It is impossible to say at this time—we can only speculate.

Differences in chromosome number alone do not seem to account for the variation in radiosensitivity. Although the chromosome number is unknown for many species reported in Table VI, the sugarcane borer with a haploid number of 17, and the European corn borer, with a haploid number of 31 (Guthrie *et al.,* 1965), can both be sterilized by about 30 kr. Adult housefly males with six pairs of chromosomes are sterilized with less irradiation than is required for the adult male screwworm flies which also have six pairs of chromosomes. Most mosquito species that have only three chromosome pairs require larger doses than either the housefly or screwworm fly. However, differences in chromosome number may be responsible for divergent results in radiation studies of closely related species. For example, Erdman (1963) found that the frequency of dominant lethal mutations induced by 2 kr were about twice as high in *T. confusum* as in *T. castaneum*. These two species differ by only one pair of chromosomes. Additional comparisons are futile until the interphase nuclear volumes or nuclear DNA content of some of these species is known. This approach proved most illuminating in the study of plant radiosensitivity (Sparrow *et al.,* 1963).

If chromosome breakage and loss of chromosome parts are the basis of dominant lethal mutations, then insect species with holokinetic chromosomes (diffuse centromeres) might be very resistant to the induction of dominant lethal mutations because all fragments would be retained in the daughter nuclei. Chromosome exchanges would be expected to produce lethal mutations. In fact, both the Lepidoptera and Hemiptera (Hughes-Schrader and Ris, 1941) have chromosomes with diffuse centromeres, and species of these orders are indeed relatively radioresistant.

Tazima (1960) showed that most dominant lethal mutations induced in the silkworm cause death of the zygote late in embryonic development, unlike dominant lethal mutations in other species, which cause death during the early stages of embryonic development. This reversal seems to set the Lepidoptera apart from the other insect orders. The difference in time of death of the zygotes may be merely a reflection of difference in chromosome structure, but it may also be intimately related to the radioresistance of the Lepidoptera.

Differences in radiosensitivity between insect species is of more than just academic interest. Obviously any insect suffers more somatic damage at 35 kr than at 3 kr; therefore, insects that require large doses of radiation to produce sterility may be more subject to serious somatic damage. However, escape from somatic damage is not guaranteed, even if the insect can be sterilized with low doses of radiation.

X. CONSIDERATIONS AFFECTING THE SUCCESS OF FUTURE FIELD TRIALS

Some factors that should be considered before initiating a large field trial were discussed by Knipling (1955). They will be discussed in the following sections.

A. A Practical Method of Inducing Sterility

A sufficient number of laboratory studies must be made to determine the optimum dose and the optimum stage of development for sterilization so that the mating behavior, competitiveness, and longevity of the test insects will not be seriously affected. Some of these points were discussed in Sections V, VI, and IX.

A misconception exists to the effect that only a dose which induces 100% dominant lethal mutations in the sperm should be used. When the percentage of dominant lethal mutations is plotted versus the dose, the slope of the curve flattens out at the larger doses, so that a great increase in dose is required to obtain only a small increase in dominant lethality. This larger dose is often accompanied by an increase in the harmful effects on emergence, competitiveness, and longevity. A level of radiation which produces 98% dominant lethal mutations in the sperm (and total infecundity in the females if both are released) is acceptable, especially since there is usually no danger of recovery of fertility in the male (see Section VI). With the sheep blowfly (Section VIII), a minimum dose of 4000 r could have been used. For the housefly, doses in the neighborhood of 3000 r rather than 6000 r should have been tried.

B. Knowledge of the Components of Sterility

The age at which the pupae or adults should be subjected to radiation must be determined with a fair degree of accuracy because of sensitivity to radiation changes (see Section VI). For example, laboratory tests in Southern Rhodesia were greatly handicapped because the age of the tsetse fly pupae that were collected was not known (G. J. W. Dean and

S. Wortham, unpublished). Irradiation of younger pupae decreased emergence and survival of the male flies, which emphasized the urgent need for the satisfactory rearing method still lacking for this species.

The possible harmful side effects which may be induced by radiation should be kept in mind; Riemann and Flint (1967) discovered that the gut epithelium of the boll weevil was very sensitive to irradiation, and that mortality was actually associated with cell death in the gut. Also, the type of sterility should be known, whether it be dominant lethal mutations, aspermia, or other types of sterility, for it may explain some results obtained in small cage tests. It is not sufficient to know that a female has deposited sterile eggs; the investigator should also know whether mating and a transfer of sperm have taken place. Mating competition tests with different ratios of sterile males to normal males are important and necessary, and the insects must be checked to determine whether mating has actually occurred. In some of the longer-lived species, the number of matings the sterile male can undergo and still transfer sperm must be known. This was one of the problems encountered with the sheep blowfly (see Section VIII).

C. Rearing Methods

For a large field trial, an economical method must be available for rearing large numbers of insects. For smaller tests, this is not mandatory, as Horber (1963b) clearly demonstrated with the white grub. However, lack of mass-rearing procedures should not deter laboratory studies on other phases of the program.

A great many problems are encountered in the transition from laboratory colonies to mass rearing, whether artificial or natural diets are used. An example is the screwworm eradication program. The mechanical details necessary for mass rearing should be keyed to the habits of the insect. Thus, development of procedures involves many studies of the nutritional requirements for production of large numbers of vigorous insects at the lowest possible cost. To reduce insect handling and keep personnel costs to a minimum, the facilities must be highly mechanized. Insect handling should also be minimized, not only during rearing and sterilization, but also before releases because excessive handling may be detrimental to the insect, depending on the species involved. With *C. p. fatigans,* Krishnamurthy *et al.* (1962) reported that handling resulted in high mortality.

The possibility of expansion of rearing facilities should be kept in mind; facilities that are too limited to permit adequate and rapid expansion—when and if the need arises—can seriously jeopardize the field test. This was the case with one fruit fly program in the Pacific.

Adequate quantitative information should be available on the density of the natural population, and the time of occurrence of the low level of the population cycle is especially important. Thus we return again to the biology and ecology of the insect which is being considered a candidate for the sterile-male technique.

D. Quantitative Information on Natural Populations

Usually, little information is available on the biology of insects, particularly in the field. This is true even of the ubiquitous housefly. The behavior (reproductive potential, competitiveness, flight range, and longevity) of the sterile insect should be known, and should also be compared with the behavior of the natural population of the test area. Certainly some information can be obtained in the laboratory and from small outdoor cage tests, but tests are desirable because insect behavior may vary in the two situations. Such differences in behavior were encountered in the studies with the common malaria mosquitoes (see Section VIII).

Population densities are usually based on information derived from trapping studies, in which suitable attractants are combined with the release and recapture of tagged insects. Progress on the successful use of the sterile-male technique with fruit flies was especially rapid because of the availability of specific attractants such as methyleugenol, medlure, cuelure, and siglure. Much research is needed on attractants for other insect species. Also, a prime requisite in insect tagging is the use of a non-injurious, long-lasting tag that can be easily recognized. The ultimate, of course, is a genetically marked insect which is not found to any great degree in the natural population of the test area and which can be easily reared. The fortuitious discovery of a white thorax mutant strain of the oriental fruit fly was a tremendous help in the eradication program on Guam (Section VIII); when the marked strain was used in conjunction with the attractant traps, it provided a constant monitoring of the sterile population and permitted rapid calculation of the ratios of sterile to normal flies in the field.

The size of the natural population must be estimated carefully to determine the number of sterile insects required for a successful program. But the reproductive potential of the species must also be assessed as accurately as possible. With some insect species, the natural population can increase at fantastic rates under very favorable conditions, and can ruin a release experiment, as was the case with the oriental fruit fly on Rota (Section VIII).

A widespread notion still exists that the sterile-male technique is applicable only to monogamous species. This is not true because sterility based on the production of dominant lethal mutations works just as well on species that mate several times (see Section VI). The successes obtained with the fruit flies (Section VIII) certainly attest to that fact.

E. Other Considerations

Practical methods must be available for reducing the natural population to levels that are manageable with sterile males. Knipling pointed out that the major advantage of the sterile-male technique is that it becomes more and more effective as the natural population decreases. Insecticide applications can be used alone or with baits to reduce the natural populations to workable levels before releases of sterile insects. Unfavorable weather conditions, such as cold spells and typhoons, can also be used to advantage.

Once a population is controlled or eradicated, great care must be taken to prevent reinfestations from outside sources: continue releases for a period of time; set up barrier zones; initiate quarantine measures; and keep rearing facilities on a standby basis for possible emergencies.

Before a large release program is initiated, the cost of the program, rearing, releasing, and evaluating should be justified; that is, the cost should be no more than the losses caused by the pests plus the cost of current control measures. The expense of keeping the area free from reinfestations once eradication has been achieved must also be considered.

Not all insects of economic importance are candidates for the sterile-male technique; sometimes other methods of inducing sterility in the native population (i.e., chemosterilants) may be more advantageous. Knipling (1964) pointed out that the "sterile insects to be released must not cause undue losses to crops or livestock, or create hazards for man that outweigh the benefits of achieving or maintaining population control." For example, in the field test with mosquitoes (Section VIII), females were not released. However, such problems do not arise if the males are harmless or methods are available for separating the sexes.

XI. CONCLUSIONS

Despite the many studies of sterilization induced by radiation as a method for insect control, and the successful demonstration of the method on at least five insect species in the last decade, our knowledge of the technique is still in its infancy. The possibilities for a broad application must await further study, for we are merely on the threshhold of realizing the full potential of this approach to insect control. The frustrations and disappointments will be frequent, the difficulties numerous, but the rewards great.

REFERENCES

Abrahamson, S., and Herskowitz, I. H. (1957). *Genetics* **42,** 405.
Andreev, S. V., Samoilova, Z. I., Martens, B. K., and Ivanskii, N. L. (1962). *Zashch. Rast.* **9,** 25.
Annan, M. E. (1955). *J. Heredity* **46,** 177.
Astaurov, B. L. (1937). *Biol. Zh.* **6,** 3.
Atwood, K. C., von Borstel, R. C., and Whiting, A. R. (1956). *Genetics* **41,** 804.
Baccetti, B., and Zocchi, R. (1962). *Redia* **47,** 161 (Ital., English Sum.).
Baker, W. K., and von Halle, E. (1954). *Science* **119,** 46.
Baldwin, W. F., and Shaver, E. L. (1963). *Can. J. Zool.* **41,** 637.
Barrett, W. L., Jr. (1937). *J. Econ. Entomol.* **30,** 873.
Baumhover, A. H. (1965). *J. Econ. Entomol.* **58,** 544.
Baumhover, A. H., Graham, A. J., Bitter, B. A., Hopkins, D. E., New, W. D., Dudley, F. H., and Bushland, R. C. (1955). *J. Econ. Entomol.* **48,** 462.
Brown, S. W., and Nelson-Rees, W. A. (1961). *Genetics* **46,** 983.
Bull, J. O. (1963). *Intern. At. Energy Agency Tech. Rept. Ser.* **21,** 47.
Bushland, R. C. (1960a). *Advan. Pest Control Res.* **3,** 1.
Bushland, R. C. (1960b). *Advan. Vet. Sci.* **6,** 1.
Bushland, R. C., and Hopkins, D. E. (1951). *J. Econ. Entomol.* **44,** 725.
Bushland, R. C., and Hopkins, D. E. (1953). *J. Econ. Entomol.* **46,** 648.
Catcheside, D. G., and Lea, D. E. (1945). *J. Genet.* **47,** 1.
Christenson, L. D. (1963). *Intern. At. Energy Agency Tech. Rept. Ser.* **21,** 31.
Clark, A. M., Rubin, M. A., and Fluke, D. (1957). *Radiation Res.* **7,** 461.
Cornwell, P. B., and Bull, J. O. (1960). *J. Sci. Food. Agr.* **11,** 754.
Cushing, E. C., and Patton, W. S. (1933). *Ann. Trop. Med. Parasitol.* **27,** 539.
Dame, D. A., and Schmidt, C. H. (1962). *Proc. Ann. Meeting New Jersey Mosquito Exterm. Assoc.* **49,** 165.
Dame, D. A., Woodard, D. B., Ford, H. R., and Weidhaas, D. E. (1964). *Mosquito News* **24,** 6.
Davich, T. B., and Lindquist, D. A. (1962). *J. Econ. Entomol.* **55,** 164.
Davis, A. N., Gahan, J. B., Weidhaas, D. E., and Smith, C. N. (1959). *J. Econ. Entomol.* **52,** 868.
Demerec, M., and Fano, U. (1944). *Genetics* **29,** 348.
Demerec, M., and Kaufmann, B. P. (1941). *Am. Naturalist* **75,** 366.
Donnelly, J. (1960). *Entomol. Exptl. Appl.* **3,** 48.
Donnelly, J. (1965). *Proc. 12th Intern. Congr. Entomol., London, 1964* pp. 253–254.
Drummond, R. O., (1963). *Intern. J. Radiation Biol.* **7,** 491.
Elbadry, E. (1965). *Ann. Entomol. Soc. Am.* **58,** 206.
Erdman, H. E. (1961). *Intern. J. Radiation Biol.* **3,** 183.
Erdman, H. E. (1962). *Nature* **195,** 1218.
Erdman, H. E. (1963). *J. Exptl. Zool.* **153,** 141.
Fahmy, O. G., and Fahmy, M. J. (1954). *J. Genet.* **52,** 603.
Fahmy, O. G., and Fahmy, M. J. (1955). *Genetics* **53,** 181.
Fay, R. W., McCray, E. M., and Kilpatrick, J. W. (1963). *Mosquito News* **23,** 210.
Flint, H. M. (1965). *J. Econ. Entomol.* **58,** 555.
Frizzi, G., and Jolly, M. S. (1961). *Atti Assoc. Genet. Ital.* **6,** 285.
Glass, B. (1955). *Genetics* **40,** 252.
Grosch, D. S. (1959). *Atompraxis* **5,** 290.

Grosch, D. S. (1962). *Ann. Rev. Entomol.* **7,** 81.

Grosch, D. S., and Sullivan, R. L. (1954). *Radiation Res.* **1,** 294.

Grosch, D. S., Sullivan, R. L., and LaChance, L. E. (1956). *Radiation Res.* **5,** 281.

Guthrie, W. D., Dollinger, E. J., and Stetson, J. F. (1965). *Ann. Entomol. Soc. Am.* **58,** 100.

Hadorn, E. (1961). "Developmental Genetics and Lethal Factors," 355 pp. Wiley, New York.

Harvey, J. M. (1963). *U.S. At. Energy Comm., Div. Isotopes Develop., Div. Biol. Med.* **TID–7684,** 55.

Hathaway, D. O. (1966). *J. Econ. Entomol.* **59,** 35.

Heidenthal, G. (1945). *Genetics* **30,** 197.

Henneberry, T. J. (1964). *J. Econ. Entomol.* **57,** 672.

Henneberry, T. J., Smith, F. F., and McGovern, W. L. (1964). *J. Econ. Entomol.* **57,** 813.

Hightower, B. G. (1963). *J. Econ. Entomol.* **56,** 498.

Hightower, B. G, Adams, A. L., and Alley, D. A. (1965). *J. Econ. Entomol.* **58,** 373.

Hoover, D. L., Floyd, E. H., and Richardson, H. D. (1963). *J. Econ. Entomol.* **56,** 584.

Horber, E. (1963a). *Mitt. Schweiz. Landwirtsch.* **11,** 145.

Horber, E. (1963b). *In* "Radiation and Radioisotopes Applied to Insects of Agricultural Importance," Proc. Symp., Athens, 1963, pp. 313–332. Intern. At. Energy Agency, Vienna.

Hughes-Schrader, S., and Ris, H. (1941). *J. Exptl. Zool.* **87,** 429.

Husseiny, M. M. (1963). *Dissertation Abstr.* **24,** 1758.

Husseiny, M. M., and Madsen, H. F. (1964). *Hilgardia* **36,** 113.

Jaynes, H. A., and Godwin, P. A. (1957). *J. Econ. Entomol.* **50,** 393.

Kansu, I. A. (1962). *Z. Angew. Entomol.* **49,** 224.

Katiyar, K. P. (1965). *Intern. At. Energy Tech. Rept. Ser.* **44,** 20.

Katiyar, K. P., and Valerio, S. J. (1963). *In* "The Application of Nuclear Energy to Agriculture," Ann. Rept., 1963, pp. 38–52. Inter-Am. Inst. Agr. Sci., Turrialba, Costa Rica.

Kerschner, J. (1946). *Anat. Record* **96,** 556.

Knipling, E. F. (1955). *J. Econ. Entomol.* **48,** 459.

Knipling, E. F. (1959). *Science* **130,** 902.

Knipling, E. F. (1964). *U.S. Dept. Agr.* **ARS 33–98,** 54 pp.

Krishnamurthy, B. S., Ray, S. N., and Joshi, G. C. (1962). *Indian J. Malariol.* **16,** 365.

LaChance, L. E. (1967). *In* "Genetics of Insect Vectors of Disease," Chapt. 21 (J. Wright and R. Pal, eds.). Elsevier, Amsterdam.

LaChance, L. E., and Bruns, S. B. (1963). *Biol. Bull.* **124,** 65.

LaChance, L. E., and Crystal, M. M. (1965). *Genetics* **51,** 699.

LaChance, L. E., and Leverich, A. P. (1962). *Genetics* **47,** 721.

LaChance, L. E., and Riemann, J. G. (1964). *Mutation Res.* **1,** 318.

Laviolette, P., and Nardon, P. (1963). *Bull. Biol. France Belg.* **97,** 305.

Lea, D. E., and Catcheside, D. G. (1945). *J. Genet.* **47,** 10.

Lee, W. R. (1958). *Genetics* **43,** 480.

Lefevre, G., Jr., and Jonsson, U. B. (1962). *Genetics* **47,** 1719.

Lewis, L. F., and Eddy, G. W. (1964). *J. Econ. Entomol.* **57,** 275.

Lindquist, A. W. (1955). *J. Econ. Entomol.* **48,** 467.

Lindquist, D. A., Gorzycki, L. J., Mayer, M. S., Scales, A. L., and Davich, T. B. (1964). *J. Econ. Entomol.* **57,** 745.
McClanahan, R. J. (1965). *Can. Entomologist* **97,** 1042.
McCray, E. M., Jr., Jensen, J. A., and Schoof, H. F. (1961). *Proc. Ann. Meeting New Jersey Mosquito Exterm. Assoc.* **48,** 110.
MacLeod, J., and Donnelly, J. (1961). *Entomol. Exptl. Appl.* **4,** 101.
Maxwell, J. (1938). *Biol. Bull.* **74,** 253.
Melis, A., and Baccetti, B. (1960). *Redia* **45,** 193.
Melvin, R., and Bushland, R. C. (1940). *J. Econ. Entomol.* **33,** 850.
Monro, J. (1963a). *Science* **140,** 496.
Monro, J. (1963b). *Intern. At. Energy Tech. Rept. Ser.* **21,** 33.
Monro, J. (1965). *Intern. At. Energy Tech. Rept. Ser.* **44,** 22.
Morlan, H. B., McCray, E. M., Jr., and Kilpatrick, J. W. (1962). *Mosquito News* **22,** 295.
Mortimer, R. K., and von Borstel, R. C. (1963). *Genetics* **48,** 1545.
Muller, H. J. (1927). *Science* **66,** 84.
Muller, H. J. (1940). *J. Genet.* **40,** 1.
Muller, H. J. (1950). *Am. Scientist* **38,** 33, 126.
Muller, H. J. (1954). *In* "Radiation Biology" (A. Hollaender, ed.), Vol. 1, Pt. 1, pp. 351–474. McGraw-Hill, New York.
Muller, H. J., and Settles, F. (1927). *Z. Induktive Abstammungs-Vererbungslehre* **43,** 285.
Nadel, D. J. (1965). *Intern. At. Energy Tech. Rept. Ser.* **44,** 25.
Ouye, M. T., Garcia, R. S., and Martin, D. F. (1964). *J. Econ. Entomol.* **57,** 387.
Pendelbury, J. B., Jeffries, D. J., Banham, E. J., and Bull, J. O. (1962). *At. Energy Res. Estab.* (*Gt. Brit.*) *Rept.* **R 4003,** 23 pp.
Pontecorvo, G. (1941). *J. Genet.* **41,** 195.
Pontecorvo, G. (1942). *J. Genet.* **43,** 295.
Potts, W. H. (1958). *Ann. Trop. Med. Parasitol.* **52,** 484.
Proverbs, M. D. (1964). *Can. Entomologist* **96,** 143.
Proverbs, M. D. (1965). *Western Fruit Grower* **19,** 19.
Proverbs, M. D., and Newton, J. R. (1962a). *Can. Entomologist* **94,** 1162.
Proverbs, M. D., and Newton, J. R. (1962b). *J. Econ. Entomol.* **55,** 934.
Quraishi, M. S., and Metin, M. (1963). *In* "Radiation and Radioisotopes Applied to Insects of Agricultural Importance," Proc. Symp. Athens, 1963, pp. 479–484. Intern. At. Energy Agency, Vienna.
Ramakrishnan, S. P., Krishnamurthy, B. S., and Ray, S. N. (1962). *Indian J. Malariol.* **16,** 357.
Rasulov, F. K. (1963). *Khlopkovodstvo* **7,** 41.
Rhode, R. H., F. Lopez, D., Eguisa, F., and J. Telich, G. (1961). *J. Econ. Entomol.* **54,** 202.
Riemann, J. G. (1967). *Ann. Entomol. Soc. Am.* **60,** 308–320.
Riemann, J. G., and Flint, H. M. (1967). *Ann. Entomol. Soc. Am.* **60,** 298–308.
Riemann, J. G., Moen, D. J., and Thorson, B. J. (1967). *Insect Physiol.* **13,** 407–418.
Rivosecchi, L. (1962). *Riv. Parasitol.* **23,** 71.
Ross, M. H., and Cochran, D. G. (1963). *Ann. Entomol. Soc. Am.* **56,** 256.
Russell, W. L. (1954). *In* "Radiation Biology" (A. Hollaender, ed.), Vol. 1, Pt. 2, pp. 825–859. McGraw-Hill, New York.
Sacca, G. (1961). *Rend. Ist. Super. Sanita* **24,** 5.
Sado, T. (1961). *Proc. Symp. Genet. Effects Radiation, Japanese J. Genetics* **36,** Suppl., 136.

Schmidt, C. H., Dame, D. A., and Weidhaas, D. E. (1964). *J. Econ. Entomol.* **57,** 753.

Sonnenblick, B. P. (1940). *Proc. Natl. Acad. Sci. U.S.* **26,** 373.

Sonnenblick, B. P., and Henshaw, P. S. (1941). *Proc. Soc. Exptl. Biol. Med.* **48,** 74.

Sparrow, A. H., Schairer, L., and Sparrow, R. C. (1963). *Science* **141,** 163.

Stancati, M. F. (1932). *Science* **76,** 197.

Stark, R. W. (1963). *U.S. At. Energy Comm., Div. Tech. Inform., Biol. Med.* No. 1, **TID-4200** (Abstr. KIC139), p. 165.

Steen, E. B. (1934). *Proc. Indiana Acad. Sci.* **43,** 224.

Steiner, L. F. (1965). *Intern. At. Energy Tech. Rept. Ser.* **44,** 28.

Steiner, L. F., and Christenson, L. D. (1956). *Proc. Hawaiian Acad. Sci., 1955–1956* pp. 17–18.

Steiner, L. F., Mitchell, W. C., and Baumhover, A. H. (1962). *Intern. J. Appl. Radiation Isotopes* **13,** 427.

Steiner, L. F., Harris, E. J., Mitchell, W. C., Fujimoto, M. S., and Christenson, L. D. (1965). *J. Econ. Entomol.* **58,** 519.

Suomalainen, E. (1953). *Hereditas* **39,** 88.

Tazima, Y. (1960). *Proc. Symp. Genet. Effects Radiation. Japanese J. Genetics* **36,** Suppl., 50.

Terzian, L. A., and Stahler, N. (1958). *Biol. Bull.* **115,** 536.

Thomou, H. (1963). *In* "Radiation and Radioisotopes Applied to Insects of Agricultural Importance," Proc. Symp., Athens, 1963, pp. 413–424. Intern. At. Energy Agency, Vienna.

Valenta, M., Kolousek, J., and Fulka, J. (1963). *Intern. J. Radiation Biol.* **6,** 81.

Virkki, N. (1963). *J. Agr. Univ. Puerto Rico* **47,** 102.

von Borstel, R. C. (1955). *Genetics* **40,** 107.

von Borstel, R. C. (1961). *In* "Progress of Photobiology," Proc. 3rd Intern. Congr. Photobiol., Copenhagen, 1960 (B. C. Christensen and B. Buchmann, eds.), pp. 243–250. Elsevier, Amsterdam.

von Borstel, R. C., and Rekemeyer, M. L. (1959). *Genetics* **44,** 1053.

Walker, J. R., and Brindley, T. A. (1963). *J. Econ. Entomol.* **56,** 522.

Waterhouse, D. F. (1962). *Intern. J. Appl. Radiation Isotopes* **13,** 435.

Weidhaas, D. E., and Schmidt, C. H. (1963). *Mosquito News* **23,** 32.

Weidhaas, D. E., Schmidt, C. H., and Seabrook, E. L. (1962). *Mosquito News* **22,** 283.

Welshons, W. J., and Russell, W. L. (1957). *Proc. Natl. Acad. Sci. U.S.* **43,** 608,

Whiting, A. R. (1945a). *Am. Naturalist* **79,** 193.

Whiting, A. R. (1945b). *Biol. Bull.* **89,** 61.

Whiting, A. R., Caspari, S., Koukides, M., and Kao, P. (1958). *Radiation Res.* **8,** 195.

Whiting, P. W. (1938a). *Proc. Penn. Acad. Sci.* **12,** 74.

Whiting, P. W. (1938b). *Genetics* **23,** 562.

Yanders, A. F. (1959). *Genetics* **44,** 545.

Yanders, A. F. (1964). *Genetics* **49,** 309.

5 CHEMOSTERILANTS

Wendell W. Kilgore

AGRICULTURAL TOXICOLOGY AND
RESIDUE RESEARCH LABORATORY
UNIVERSITY OF CALIFORNIA, DAVIS, CALIFORNIA

I. INTRODUCTION

For centuries man has used every possible means to control insect pests, progressing from nets, swatters, and fire to potent synthetic insecticides. In the past he has wanted to destroy the pests quickly, but now he sometimes wants to keep them alive so that they can aid in their own destruction. Such self-destruction may be accomplished by sexually sterilizing but not killing the pest species with certain types of chemicals. Chemicals which have this unique property are often referred to as chemosterilants.

Sterilization as a method for insect control is based on the successful eradication of the screwworm fly *Cochliomyia hominivorax* (Coquerel), from the island of Curaçao and from the southeastern part of the United States through the release of adult male flies sterilized by irradiation. The idea of introducing sterile males into a natural population of screwworm flies to obtain control of the species was first proposed in 1937 by Knipling (1960), discussed in detail in Chapter 4. Briefly, Knipling's "sterile-male principle" for insect control consists of rearing and releasing sexually sterile males in numbers greater than exist in a natural population until control or eradication of the natural population is accomplished. If sufficiently large numbers of sterile males are released, they have a decided advantage in competing with the normal males for females in the natural population.

Many chemicals have now been found which are comparable to irradiation in that they deprive insect species of their ability to reproduce. Several of these compounds, when administered orally or by contact to a wide spectrum of insects, produce irreversible sterility without significant adverse effects on mating behavior or length of life. If safe and effective ways of using them can be developed, chemosterilants offer possibilities for achieving insect control not previously considered possible. They offer several distinct advantages over the conventional methods of insect control. According to Knipling (1962), "Insects that are sterilized cannot reproduce and thus such effect is equivalent to killing the insects. In addition, the sterile insects compete with the normal insects to further decrease chances for reproduction. This bonus effect represents the greatest advantage over the conventional way of controlling insect populations. In addition, however, sterile insects in the population are capable of limiting reproduction because of 'time' and 'space' effects, factors which are absent in the killing procedure."

In the broadest sense, insect chemosterilants can be defined as chemicals which deprive insect species of their ability to reproduce. There are several stages at which the reproductive cycle may be disrupted and thus prevent the production of an F_1 generation. The insects may not deposit eggs, the eggs may not hatch, the larvae may not pupate, or the pupal development may not be complete. Any chemical which interrupts the steps in this process and prevents the development of progeny may be classified as a chemosterilant. Compounds which disrupt the reproductive process of female insects are commonly referred to as female chemosterilants, and those which disrupt the male reproductive process are know as male chemosterilants. Chemicals which disrupt both male and female reproduction are called male–female chemosterilants.

II. HISTORICAL DEVELOPMENT

Sterilization with chemicals is actually not a new concept for insect control since it was considered as early as 1937 by Knipling (1960). In recent years, however, chemosterilants have received considerable attention because of the possibility of using them to replace x-ray and γ-ray irradiation for the sterilization process. Many investigators can be credited with pioneering research in the field of insect chemosterilants, but it appears that Goldsmith and co-workers (Goldsmith *et al.,* 1948; Goldsmith and Frank, 1952; Goldsmith, 1955), working with *Drospholia,* were the first investigators to report that certain chemicals could cause the retardation of ovarian development in insects, resulting in the sterilization of the female sex. Almost simultaneously, Mitlin and his associates (Mitlin *et al.,* 1954, 1957; Konecky and Mitlin, 1955; Mitlin, 1956) reported on the effects of chemicals on the ovarian development of houseflies. Ascher (1957a,b, 1958) working independently, also reported during this period on ovarian inhibitors in both houseflies and mosquitoes.

Recognizing the feasibility of using chemicals rather than irradiation for sterilizing insects, the Entomology Research Division of the Agricultural Research Service, United States Department of Agriculture established a screening program in 1958 (LaBrecque *et al.,* 1960) to evaluate chemicals for sterilant activity in houseflies. Compounds showing some activity in the housefly were also tested on two species of mosquitoes.

As a result of this screening program, several chemicals with sterilant activity were found. Amethopterin (I), Methotrexate; *N*-{*p*-([(2,4-diamino-6-pteridinyl)methyl]methylamino)benzoyl}glutamic acid, and 5-fluorouracil (II) were among the first chemicals tested which showed some promise.

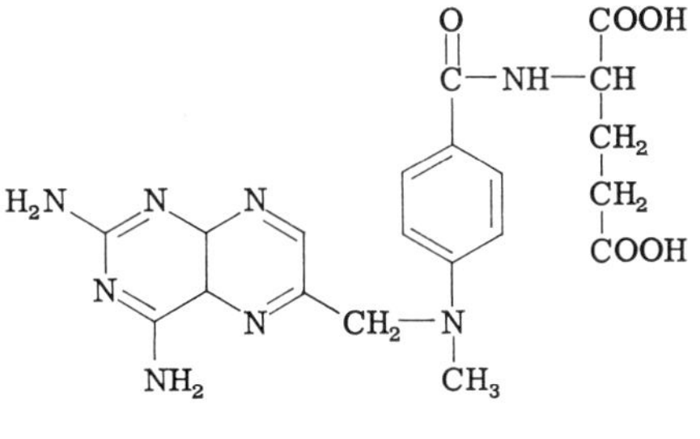

Amethopterin
(I)

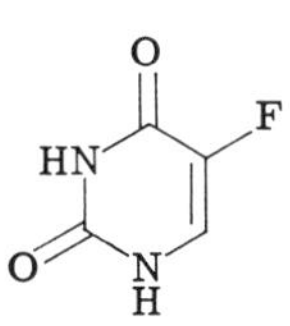

5-Fluorouracil
(II)

Both chemicals have been used extensively in man as therapeutic agents for the treatment of certain types of tumors, and it was not too surprising to learn that they also affect other rapidly proliferating tissues, such as the gonads of insects. Amethopterin sterilizes houseflies without interrupting oviposition at dosages as low as 0.005 gm per 50 gm of diet. The eggs which are deposited simply do not hatch. At higher dosages, oviposition is interrupted and no eggs are deposited (LaBrecque *et al.,* 1960). Further tests have shown that amethopterin sterilizes female flies but does not affect the males. Likewise, Kilgore and Painter (1966) have reported that the antimetabolite 5-fluorouracil sterilizes only females.

In later studies LaBrecque (1961) found a group of chemicals which seemed to have exceptional promise in sterilizing both male and female houseflies, mosquitoes, and stable flies. The compounds were all alkylating agents and each contained several aziridinyl (ethylenimine) groups. The common and chemical names of these chemosterilants are apholate (III), 2,2,4,4,6,6-hexakis(1-aziridinyl)-2,2,4,4,6,6-hexahydro-1,3,5,2,4,6-triazatriphosphorine; aphomide (IV), *N,N'*-ethylenebis[*p,p*-bis(1-aziridinyl-*N*-methylphosphinic amide]; and aphoxide, also called TEPA (V), tris(1-aziridinyl)phosphine oxide.

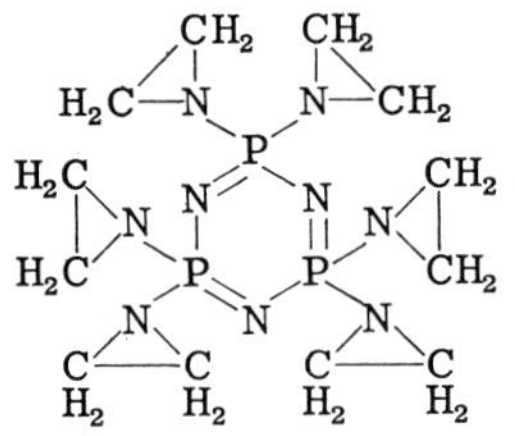

Apholate
(III)

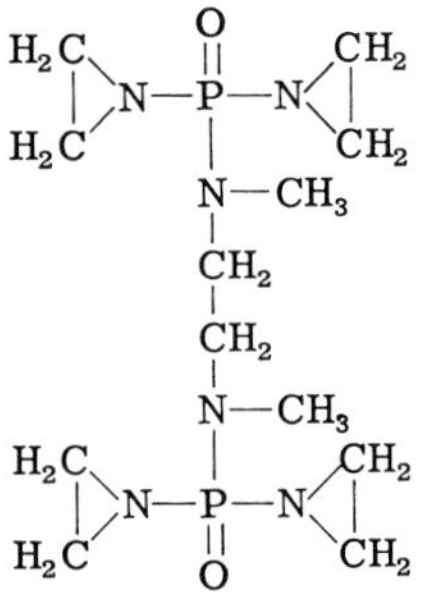

Aphomide
(IV)

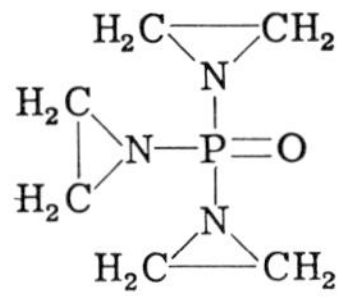

Aphoxide (TEPA)
(V)

When these compounds were fed to adult houseflies at concentrations of 0.5–1.00% in the food, they caused sterility in both sexes. The sterilizing effect caused by apholate was most apparent when the chemical was fed to adult flies immediately upon emergence. When the female flies were given treated food and mated with normal males, they deposited only a few eggs; these were not viable. Also, when normal females were mated with treated males, almost the normal number of eggs was deposited, but none hatched. In small-cage experiments, treated flies competed satisfactorily with normal flies, which indicated that apholate did not interfere with their mating habits. In large-room experiments in which normal flies were compared to treated flies, it was found that flies given a choice of treated or untreated diets eventually produced 13–121 pupae, whereas flies given an untreated diet produced about 40,000 pupae.

From the results of these early experiments, a whole new area of investigation has been opened. Testing programs are now underway in laboratories throughout the world, and compounds having sterilant activity in insects are being reported with increasing frequency.

III. TESTING PROCEDURES

Potential chemosterilants may be administered to insects by using any one of a number of procedures. The technique most widely used, however, is one in which the candidate chemical is added to the diets of adult insects. The treated diets may be given to the insects for short or long periods, depending on the circumstances and the information desired. But once the treated food has been consumed, the insects should be observed periodically for possible reproductive abnormalities. Usually oviposition is inhibited or eggs which are deposited are not viable. Such observations are then related to the amount of chemical in the food and sterilant activity. Other techniques which have been employed with varying degrees of success are larval feeding, topical application, and microinjection.

The housefly has been used almost exclusively as the primary test insect for large-scale screening programs, although other insects can also be used. Undoubtedly, ease of rearing and handling has been a factor in the selection of the housefly as the test insect by most laboratories. In the secondary testing of a chemical, several other insect species have been used, especially the screwworm fly and several species of mosquitoes.

A. Adult Feeding

LaBrecque *et al.* (1960) have tested several hundred compounds for sterilant activity on adult houseflies by using a continuous-feeding technique. In their screening program, the fly food (6 parts sugar, 6 parts powdered nonfat dry milk, 1 part powdered egg) was mixed with acetone containing the chemical in suspension or solution. The amount of test chemical added to the food ranged from less than 0.1 to 1.0%. After drying (24 hours) the food was then pulverized and placed in emergence cages containing 100 pupae. Colonies given untreated food were maintained as controls. The cages were examined periodically to determine the number of emerged flies and to observe any toxic effect caused by the treated food.

Oviposition medium was placed in each cage just preceeding the oviposition period and was examined daily for deposition of eggs. The viability of any eggs deposited was then determined by inspecting the medium for larvae 2 days after oviposition. Larvae which emerged were reared to adults and observed for abnormalities.

Other treated-food techniques for houseflies have also been used with considerable success. Kilgore and Painter (1962) found that recovery occurs when houseflies are fed a diet containing 5-fluorouracil for 36–48 hours. Consequently, sterility is partial and only temporary. As a result, Painter and Kilgore (1964a) classified a number of chemicals as either temporary or permanent sterilants. In their experiments, 200 recently emerged flies were used for each test. The flies were given water *ad lib* and 1.0 gm of food (6 parts finely powdered sucrose, 6 parts dried skim milk, 1 part powdered egg) treated with the candidate chemical. Tests were run with food containing 0.01, 0.1, and 1.0% of the chemicals. After the treated dry diets were consumed, the flies were given a normal milk diet (100 ml of canned milk, 400 ml of water, and 0.3 ml of 37% formaldehyde). If the compound showed any promise of effectiveness, intermediate levels were then used to determine the lowest level necessary to produce sterility.

Eggs were collected twice daily from the surface of the oviposition medium and placed in petri dishes containing 1.5% solidified agar. In most instances, 100 eggs were examined and placed in groups of 10 on the surface of the agar. The petri dishes were then incubated at 37°C for at least 3 days, although most of the eggs which hatched did so within the first 6–8 hours after plating. The larvae were collected several times each day and the number of eggs which hatched was recorded for determination of the percentage hatch.

Murvosh *et al.* (1964), also using houseflies, have successfully used granulated sugar rather than a dry milk diet as the carrier for chemicals being tested for sterilant activity. These workers presented the insects with treated diets and water only for 3 days. After this 3-day period, a paper cup containing untreated fly food (dried milk diet) was placed in each cage. The treated sugar was not removed. Oviposition medium (CSMA, Chemical Specialties Manufacturing Association) in a paper cup was placed in each cage on the seventh or eighth day. Eggs were then collected and a sample of 100 eggs from each cage was counted onto a small piece of damp black cloth placed on moist larval medium in a rearing container. The eggs were then examined for hatch 2 days later. When the eggs hatched, the larvae crawled from the cloth into the rearing medium. Pupal development was determined 5 days later by using a flotation procedure.

Variations of these methods for administering chemicals orally to houseflies, as well as to other insects, have now been used by a number of investigators. To determine whether a compound is a male or female sterilant, or both, the two sexes are separated immediately upon emergence and each group is fed the treated food for 24–48 hours. Once the diets are consumed the males and females are placed in the same cage and the experiment is continued as described above.

Crystal (1963), working with the screwworm fly, has used both honey and sugar syrup as carriers for the test chemicals. Also, Burden and Smittle (1963) have successfully tested a number of compounds for sterilant activity in cockroaches by using the treated-food technique. The chemicals were added to the food of the roaches in much the same way as has been described for houseflies. However, second-nymphal-instar cockroaches were used rather than the adults. Weidhaas *et al.* (1961) have reported that mosquitoes can be sterilized by feeding the adult insects treated-honey diets.

B. Topical Applications

1. *Treated Surfaces—Residual Application*

Many investigators have found that chemosterilants can also be effective when applied externally to adult insects. Weidhaas (1962) reported that both male and female mosquitoes, *Anopheles quadrimaculatus,* could be sterilized by exposing them for 4 hours to deposits of TEPA (V) (10 mg/ft^2) on glass surfaces. This type of exposure was equally effective against 24-hour-old virgin females and males, 4-day-old males, and females collected in nature which were artifically inseminated in the laboratory. Some of the females collected in nature had already laid from one to four batches of eggs when they were collected for treatment. The females collected in the field were exposed to surfaces of masonite board treated with TEPA. Deposits on the masonite were prepared by dissolving TEPA in ethanol and applying a sufficient amount of solution with a small paint brush to give a film of 500 mg/ft^2. For the glass surfaces, the TEPA was dissolved in absolute methanol and 1 ml of solution was pipeted into a 3-in. glass petri dish. The petri dish was rotated until the methanol evaporated leaving a deposit of 10 mg/ft^2.

In stable flies, *Stomoxys calcitrans* (L.), apholate (III) induces sterility when adults are exposed to residual films of the compound in glass jars (Harris, 1962). An exposure period of 48 hours was required when the flies were placed in a half-pint jar containing a 10-mg film of apholate. However, a 100-mg film per jar was effective with an exposure period of 1 hour. Treated jars, stored indoors, were effective up to 24 weeks. The inside surface of the jar was coated by placing 1 ml of carbon tetrachloride containing the desired amount of apholate in the jar and turning it until the solvent evaporated. A control was prepared by evaporating carbon tetrachloride alone in the jar.

Meifert *et al.* (1963) have investigated the effectiveness of residual applications of apholate (III), TEPA (V), 5-fluoroorotic acid, and metepa (VII) in producing sterility in houseflies. They found that TEPA and metepa at 250 mg/ft^2 on glass surfaces sterilized both male and female flies during an exposure period of 2–4 hours. The treated glass surfaces were effective up to 30 days. Neither apholate nor 5-fluoroorotic acid was effective when applied at dosages ranging from 10 to 250 mg/ft^2 with exposure periods of 2, 3, or 4 hours. In all of these experiments, the chemosterilants were dissolved in methanol and applied to the surfaces of pint mason jars. For even distribution, 2.5 ml of the chemosterilant solution was placed in the jar, and the jar was then rotated so that the solution covered its entire inner surface. After evaporation of the solvent (24 hours later) five males or females less than 2 hours old were placed in each of the treated jars. The jars were covered with organdy treated with a repellent to prevent the flies from resting on the cloth surface. After the desired exposure period, the flies were mated with five virgin flies of the opposite sex.

2. *Dusting, Dipping, and Spraying*

Chamberlain (1962) has demonstrated that adult screwworm flies can be sterilized by dusting them with apholate (III). In his experiments, anesthetized flies were shaken for 10 seconds in 2-oz vials containing either a small quantity of undiluted apholate or formulations prepared with pyrophyllite. Prior to returning the insects to the holding containers, excess dust was removed from the flies by dropping them on paper towels. Both sexes were completely sterilized by the undiluted apholate dust and by a 50% formulation. Even at a 10% concentration, a high degree of sterility resulted when both sexes were dusted. There were indications, however, that the undiluted dusts caused some mortality above the normal expectations.

Dusting experiments were also successful when prepupae were used. Dusts were applied by allowing the prepupae of a definite age to crawl in undiluted apholate or in dilutions prepared with pyrophyllite. Complete control of reproduction was obtained when the prepupae were treated with undiluted apholate, regardless of whether only one or both sexes were from treated prepupae. At lower concentrations (25 and 50%) almost 100% sterilization resulted when both sexes were from treated prepupae. However, at these dust concentrations females were apparently more affected than the males.

Many dipping experiments have been conducted with both larvae and pupae. Chamberlain (1962) dipped prepupae of the screwworm fly in various solutions of apholate dissolved in a variety of carrier solvents (benzene, water, 50% acetone–50% water), with some containing emulsifiers. The dipping period varied from 10 seconds to 30 minutes. Younger prepupae were more susceptible to reproduction control than older prepupae. When pupae were dipped, control of reproduction was not consistent. In most of the tests with pupae, treatment with apholate caused a 40–60% reduction in reproduction. This effect was caused mainly by the reduction in the number of eggs which were deposited rather than to an increase in the number of sterile eggs. However, the sterility of the eggs was noticeably different in some experiments. Concentrations of 1 and 2% apholate in an emulsifiable solution resulted in no more than 88% control of reproduction, even when the pupae were exposed for as long as 30 minutes.

Gouck *et al.* (1963b) have used the dipping technique for testing a number of chemicals on housefly larvae as well as screwworm larvae. The results were not encouraging since the chemicals which were tested had essentially no effect on the adult insects when the larvae were dipped into the solutions. In these experiments, full-grown housefly larvae were dipped in a 1% emulsion, suspension, or solution of the candidate chemical in water. The larvae were placed in the emulsion for 30 seconds and then removed and placed in rearing media. In the case of the screwworm fly, full-grown larvae were placed in a screen-wire dipper and submerged for about 5 seconds in a 1% solution of the test chemical in acetone or other solvent. They were then drained and placed briefly on paper towels to remove excess liquid. After this treatment the pupae were allowed to continue their life cycle through the adult stage. Perhaps the compounds would have been more effective had higher concentrations of the chemicals been used or if the exposure period had been longer. It is worth noting that adult male or female Mexican bean beetles, *Epilachna varivestis* Mulsant, dipped in an aqueous solution of 0.5% apholate, were completely sterilized (Henneberry *et al.*, 1964). They were also sterilized following confinement for 48 hours on foliage sprayed with the same concentration.

Spraying as a means of testing chemicals for sterilant activity has been limited to only a few studies. This may be due, in part, to the difficulty of preparing adequate spray formulations. However, Cressman (1963) has successfully demonstrated that citrus red mites, *Panonychus citri* McGregor, can be sterilized by spraying them with 0.03–0.1% formulations of apholate, TEPA, and aphomide. The fecundity of the treated females and the viability of the eggs and larvae were reduced following the spray treatment.

Crystal (1965a) has developed an aerosol-generating apparatus for treating screwworm flies *en masse* with chemosterilants. He found that sexual sterility was induced after 6 minutes when both sexes were treated with thio-TEPA or tretamine.

IV. SPECIES AFFECTED

A. Spectrum

At the present time, testing procedures (Section III) involving the various species of insects vary from laboratory to laboratory since standardized methods have not been adopted. Therefore, it is difficult to evaluate critically data reported by investigators now working in this field. However, during the last decade it has become apparent that chemosterilants can interrupt the reproductive cycle of a large number of insect species. Both insect and mite species examined for possible sterilization with chemicals are listed in Table I. It is quite clear from this impressive list that chemosterilization is not restricted to any particular group or order, and that both biting and sucking insects may be affected.

Chemicals which have chemosterilant activity when added to the diets of adult insects may or may not be effective when applied topically. In addition, some chemicals may sterilize only adults while others may be sterilants only during the larval stage of development. The method of application and the period of development at which an insect is sterilized are extremely important from a practical standpoint, but at the present time the demonstration of possible chemosterilization at any stage of development or by any method of application in a large variety of species is probably more important. There is no doubt, as illustrated in Table I, that many arthropods have been and can be sterilized with chemicals, and that the number will be greatly expanded as more species are tested.

B. Resistance

The development of resistance by insects to conventional insecticides has been a serious problem in the past and undoubtedly it will continue to be one in the future. At present, many investigators are also concerned with the possibility of the development of resistance to chemosterilants. This fear is well justified since many of the more promising compounds are alkylating agents and therefore potential mutagens. Information now available indicates that some chemosterilants do cause the induction of lethal mutations (Section VI,A). Thus, it is quite possible that certain undesirable side effects, including resistance, could develop through continued exposure of insects to substerilizing dosages of these compounds. Unfortunately, adequate long-term experiments which will help to elucidate this problem are not yet available. In one report, however, Hazard *et al.* (1964) found that *Aedes aegypti* (L.) developed some resistance to apholate after 11 generations. Since the resistant strain was only four to five times less sensitive to apholate than the normal strain, these investigators felt that the maximum degree of resistance which developed was not a limiting factor.

V. CHEMICALS AFFECTING REPRODUCTION

A. CHEMICAL TYPES

Chemicals tested for chemosterilant activity vary widely in structure, but for the convenience of the reader they are grouped here into three major categories: alkylating agents, antimetabolites, and miscellaneous chemicals. A partial list of the chemicals which have been tested is given in Table II.

1. *Alkylating Agents*

Alkylating agents are compounds which are highly reactive and combine readily with a variety of chemical and biochemical substances. With respect to biological systems, the most important feature of these chemicals is that of alkylation. Alkylation is simply the replacement of a hydrogen atom by an alkyl group or a substituted alkyl group, such as aminoalkyl, hydroxyalkyl, and thioalkyl. The alkyl moiety may be only a methyl (CH_3) group, or it may be a more complicated group. The most important factor here is that a hydrogen atom must be replaced with a carbon atom linkage for an alkylation to have taken place. Mechanistically, an alkylating agent behaves as a carbonium ion reacting as an electrophile.

Examples of reactive groups in compounds that can serve as alkylating groups are (the arrow indicates the site of bond cleavage to produce the alkylating agent):

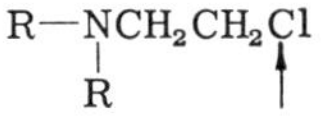

R—N(R)CH$_2$CH$_2$—Cl (arrow at the C—Cl bond)

Chloroethylamines
(nitrogen mustards)

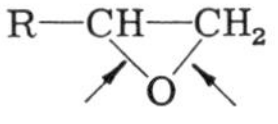

R—CH—CH$_2$ (epoxide ring through O; arrows at the C—O bonds)

Epoxides

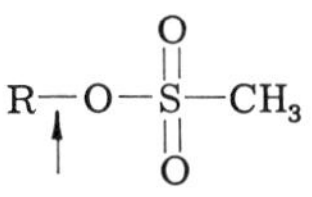

R—O—S(=O)(=O)—CH$_3$ (arrow at the R—O bond)

Mesyloxy esters

R—N(CH$_2$)(CH$_2$) (aziridine ring; arrows at the N—C bonds)

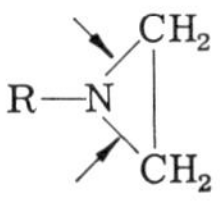

Aziridines
(ethylenimines)

TABLE I

INSECTS AND MITES AFFECTED BY REPRODUCTION INHIBITORS

Order	Common name	Scientific name	Reference
Orthoptera	Cockroach		
	German cockroach	*Blattella germanica* (Linnaeus)	Burden and Smittle (1963); Kenaga (1965); Parish and Arthur (1965b);
Hemiptera	True bug		
	Milkweed bug	*Oncopeltus fasciatus* (Dallas)	Simkover (1964)
Homoptera	Aphid		
	Pea aphid	*Acyrthosiphon pisum* (Harris)	Bhalla and Robinson (1966)
Lepidoptera	Moths		
	Cabbage looper	*Trichoplusia ni* (Hübner)	Henneberry and Kishaba (1966); Howland *et al.* (1965, 1966)
	Fall armyworm	*Spodoptera frugiperda* (J. E. Smith)	Young and Cox (1965)
	Gypsy moth	*Porthetria dispar* (Linnaeus)	Collier and Downey (1965)
	Pink bollworm	*Pectinophora gossypiella* (Saunders)	Ouye *et al.* (1965)
Coleoptera	Beetles		
	Boll weevil	*Anthonomus grandis* Boheman	Davich *et al.* (1965); Hedin *et al.* (1964); Lindquist *et al.* (1964); Parish and Arthur (1965b); Ridgway *et al.* (1966)
	Confused flour beetle	*Tribolium confusum* Jacquelin duVal	Kenaga (1965)
	Cucumber beetle	*Diabrotica balteata* LeConte	Creighton *et al.* (1966)
	Japanese beetle	*Popillia japonica* Newman	Ladd (1966)
	Mexican bean beetle	*Epilachna varivestis* Mulsant	Henneberry *et al.* (1964)
	Plum curculio	*Conotrachelus nenuphar* (Herbst)	Roach and Buxton (1965)
Diptera	Flies		
	Face fly	*Musca autumnalis* DeGeer	Hair and Adkins (1964)

Fruit fly	*Drosophilia melanogaster* Meigen	Goldsmith and Frank (1952); Goldsmith *et al.* (1948, 1950); Simkover (1964)
Mexican fruit fly	*Anastrepha ludens* (Loew)	Chang and Bořkovec (1966); Benschoter (1966); Shaw and Riviello (1962a, 1965)
Oriental fruit fly	*Dacus dorsalis* Hendel	Bořkovec (1962); Keiser *et al.* (1965)
Housefly	*Musca domestica* Linnaeus	Bořkovec *et al.* (1964); Chang (1965); Chang and Bořkovec (1964); Chang *et al.* (1964); Dame and Schmidt (1964); Fye *et al.* (1965, 1966); Geering *et al.* (1965); Gouck (1964); Gouck and LaBrecque (1964); Gouck *et al.* (1963a,b); Hansens (1965); Hansens and Granett (1965); Kenaga (1965); Kilgore and Painter (1962, 1964, 1966); Konecky and Mitlin (1955); LaBrecque (1961, 1963); LaBrecque and Gouck (1963); LaBrecque *et al.* (1960, 1962a,b, 1963a,b); Meifert *et al.* (1963); Mitlin (1956); Mitlin and Baroody (1958a,b); Morgan and LaBrecque (1962, 1964);

TABLE I—(*Continued*)

Order	Common name	Scientific name	Reference
			Murvosh *et al.* (1964); Painter and Kilgore (1964a,b, 1965); Parish and Arthur (1965a,b); Piquett and Keller (1962); Plapp *et al.* (1962); Ratcliffe and Ristich (1965); Schmidt *et al.* (1964); Simkover (1964)
	Mediterranean fruit fly	*Ceratitis capitata* (Wiedmann)	Keiser *et al.* (1965)
	Melon fly	*Dacus cucurbitae* Coquillett	Keiser *et al.* (1965)
	Screwworm fly	*Cochliomyia hominivorax* (Coquerel)	Chamberlain (1962); Chamberlain and Barrett (1964); Chamberlain and Hamilton (1964); Chamberlain and Hopkins (1960) Crystal (1963, 1964a,b,c, 1965a,b); Gouck *et al.* (1963a)
	Sheep blowfly	*Lucilia sericata* (Meigen)	Yeoman and Warren (1965)
	Stable fly	*Stomoxys calcitrans* (Linnaeus)	Chamberlain and Barrett (1964); Chamberlain and Hamilton (1964); Parish and Arthur (1965b)
	Mosquitoes		
	Common malaria mosquito	*Anopheles quadrimaculatus* Say	Dame and Schmidt (1964); Schmidt *et al.* (1964); Weidhaas (1962); Weidhaas *et al.* (1961)
	Yellow-fever mosquito	*Aedes aegypti* (Linnaeus)	Altman (1963); Dame and Schmidt (1964); Hazard *et al.* (1964);

			Weidhaas (1962); Weidhaas *et al.* (1961)
		Culex tarsalis Coquillett	Plapp *et al.* (1962)
Acarina	Mites		
	Citrus red mite	*Panonychus citri* (McGregor)	Jeppson *et al.* (1966)
	European red mite	*Panonychus ulmi* (Koch)	Harries (1963)
	Pacific mite	*Tetranychus pacificus* (McGregor)	Jeppson *et al.* (1966)
	Two-spotted spider mite	*Tetranychus telarius* (Linnaeus)	Harries (1961, 1963); Smith *et al.* (1965)

TABLE II

COMPOUNDS AFFECTING THE REPRODUCTION IN INSECTS[a]

Compound	References
Alkylating agents: Aziridines	
Aphomide	Cressman (1963); LaBrecque (1961); Shaw and Riviello (1962b); Weidhaas *et al.* (1961)
Apholate	Chamberlain (1962); Cressman (1963); Harris (1962); LaBrecque (1961); Morgan and LaBrecque (1962); Shaw and Riviello (1962b); Weidhaas (1962); Weidhaas *et al.* (1961)
Aziridine, 1,1′-dithiobis-	Bořkovec and Woods (1963); Roehrborn (1962)
Aziridine, 1,1′-dithiobis [2-methyl-	Bořkovec and Woods (1963)
Aziridine, 1,1′-sulfinylbis-	Bořkovec and Woods (1963)
Aziridine, 1,1′-sulfinylbis [2-methyl-	Bořkovec and Woods (1963)
Aziridine, 1,1′-sulfonylbis [2-methyl-	Bořkovec and Woods (1963)
Metepa	LaBrecque (1961); LaBrecque *et al.* (1963a)
Morzid	LaBrecque *et al.* (1960)
Phosphine oxide, bis(2-methyl-1-aziridinyl)phenyl	LaBrecque *et al.* (1963a)
Phosphine sulfide, tris(2-methyl-1-aziridinyl)-	LaBrecque *et al.* (1963a)
TEPA	Cressman (1963); LaBrecque (1961); Morgan and LaBrecque (1962); LaBrecque *et al.* (1962b); Shaw and Riviello (1962b); Weidhaas (1962); Weidhaas *et al.* (1961)
Thio-TEPA	LaBrecque *et al.* (1960); Piquett and Keller (1962)
Tretamine	Fahmy and Fahmy (1954); LaBrecque *et al.* (1960) Piquett and Keller (1962); Roehrborn (1962)
Alkylating agents other than aziridines	
Acridine, 6-chloro-9-({3-[(2-chloroethyl)ethylamino]propyl}amino)-2-methoxy-, dihydrochloride	Carlson and Oster (1961)
Benzoic acid, *o*-({4-[bis(2-chloroethyl)amino]-*o*-tolyl}azo)-	Purdom (1960)
Butyric acid, 4-{*p*[bis(2-chloroethyl)amino]-phenyl}-	Shaw and Riviello (1962a,b)

TABLE II—(*Continued*)

Compound	References
Diethylamine, 2,2′-dichloro-*N*-methyl-, hydrochloride	LaBrecque *et al.* (1960); Loebbecke and von Borstel (1962); Mitlin *et al.* (1957); Mitlin and Baroody (1958a)
Hydroquinone, 2,5-bis{[(2-chloroethyl)amino]-methyl}-	LaBrecque *et al.* (1960)
Methanesulfonic acid, 1,4-dimethyltetra-methylene ester	Roehrborn (1959)
Methanesulfonic acid, ethyl ester	Fahmy and Fahmy (1961)
Methanesulfonic acid, methyl ester	Fahmy and Fahmy (1961)
Myleran	LaBrecque *et al.* (1960)
Antimetabolites: Purines and pyrimidines	
Hydrouracil, 5-methyl-	LaBrecque *et al.* (1960)
Orotic acid	LaBrecque *et al.* (1960)
Purine, 2,6-diamino-	Mitlin and Baroody (1958a)
1*H*-Pyrazolo[3,4-d]pyrimidine, 4-amino-, sulfate	Shaw and Riviello (1962a,b)
4-Pyrimidineacetic acid, 1,2,3,6-tetrahydro-2,6-dioxo-	LaBrecque *et al.* (1960)
Pyrimidine, 2,4-diamino-5-(*p*-chlorophenyl)-6-ethyl-	Mitlin and Baroody (1958a)
Uracil	LaBrecque *et al.* (1960)
Uracil, 5-bromo-	LaBrecque *et al.* (1960)
Uracil, 4,5-diamino-, sulfate	LaBrecque *et al.* (1960)
Uracil, 5-fluoro-	Kilgore and Painter (1962); LaBrecque *et al.* (1960)
Uracil, 5-methyl-	LaBrecque *et al.* (1960)
Uracil, 6-methyl-	LaBrecque *et al.* (1960)
Antimetabolites other than purines and pyrimidines	
β-Alanine, 2-(thienyl)-	Mitlin and Baroody (1958a)
Butyric acid, 2-amino-4-(ethylthio)-, DL-	Mitlin and Baroody (1958a)
Glutamic acid, 4-aminopteroyl-	Goldsmith *et al.* (1948)
Glutamic acid, *N*-(*p*-{[(2,4-diamino-6-pteridinyl)-methyl]amino}benzoyl)-	Goldsmith and Frank (1952); Goldsmith *et al.* (1950); Goldsmith (1952); Mitlin *et al.* (1957); Piquett and Keller (1962)
Glutarimide, 3-[2-(3,5-dimethyl-2-oxocyclo-hexyl)-2-hydroxyethyl]-, acetate	Shaw and Riviello (1962b)
Leucine, L-	LaBrecque *et al.* (1960)
Methionine	LaBrecque *et al.* (1960)
Methotrexate	Goldsmith (1952); LaBrecque *et al.* (1960); Mitlin and Baroody (1958a); Piquett and Keller (1962)

TABLE II—(*Continued*)

Compound	References
3-Pyridinemethanol, 5-hydroxy-4,6-dimethyl-	Mitlin and Baroody (1958a)
Tyrosine	LaBrecque *et al.* (1960)
Miscellaneous	
Acetamide, *N*-methyl-	Mitlin and Baroody (1958a)
Acetamide, (1-naphthoyl)-	LaBrecque *et al.* (1960)
Acetamide, (2-naphthyloxy)-	LaBrecque *et al.* (1960)
Acetic acid, chloro-, 6-pentyl-*m*-tolyl ester	LaBrecque *et al.* (1960)
Acetic acid, 2-phenylhydrazide	LaBrecque *et al.* (1960)
Acetic acid, thymyloxy-	LaBrecque *et al.* (1960)
Acetin, tri-	LaBrecque *et al.* (1960)
Acetoacetanilide	LaBrecque *et al.* (1960)
Acetonitrile, diphenyl-	LaBrecque *et al.* (1960)
Acetophenone	LaBrecque *et al.* (1960)
Acrylic acid, 2-benzyloxyethyl ester	LaBrecque *et al.* (1960)
Adipic acid, bis(3-hydroxybutyl) ester, diacetate	LaBrecque *et al.* (1960)
Aldrin	LaBrecque *et al.* (1960)
Benzeneboronic acid	LaBrecque *et al.* (1960)
Benzimidazole	Mitlin and Baroody (1958a)
p-Benzoquinone	Mitlin and Baroody (1958a)
p-Benzoquinone, 2-methoxy-5-methyl-	Mitlin and Baroody (1958a)
Butyric acid, 2-*o*-tolyloxy-	LaBrecque *et al.* (1960)
Carbamic acid, ethyl ester	Rapoport (1947)
Carbamic acid, methyl ester	Mitlin and Baroody (1958a); Rapoport (1947)
Chlorbenside	Ascher and Hirsch (1961)
Colchicine	Chamberlain and Hopkins (1960); Mitlin *et al.* (1957); Mitlin and Baroody (1958a); Piquett and Keller (1962)
Coumarin	Mitlin and Baroody (1958b)
m-Dioxane, 5-butyl-5-ethyl-2-methyl-2-(*p*-methoxyphenethyl)-	Shaw and Riviello (1962b)
m-Dioxane, 5-butyl-5-ethyl-2-nonyl-	Shaw and Riviello (1962b)
m-Dioxane, 2-(*p*-chlorophenyl)-5-ethyl-4-propyl-	Shaw and Riviello (1962b)
Ethanol, 1,1-bis(*p*-chlorophenyl)-2,2,2-trifluoro-	Ascher (1957a)
Eugenol	LaBrecque *et al.* (1960)
Heliotrine	Clark (1959, 1960)
Hydroquinone, 2,5-dimethyl-	Mukherjee (1961)
Imidodicarboxylic acid, dihydrazide	Mitlin and Baroody (1958a); Piquett and Keller (1962)
Indole-3-acetic acid	LaBrecque *et al.* (1960)
Indole-3-butyric acid	LaBrecque *et al.* (1960)
Lasiocarpine	Clark (1960)
Monocrotaline	Clark (1960)

TABLE II—(*Continued*)

Compound	References
1-Naphthaleneacetic acid	LaBrecque *et al.* (1960)
2-Naphthol, 1-nitroso-	LaBrecque *et al.* (1960)
Nicotinic acid	LaBrecque *et al.* (1960)
Phenacetin	LaBrecque *et al.* (1960)
Piperonyl butoxide	LaBrecque *et al.* (1960); Mitlin and Baroody (1958b)
Potassium arsenite	Mitlin and Baroody (1958b)
Propanol, 1,1-bis(*p*-chlorophenyl)-2,2,3,3,3-pentafluoro-	Ascher (1957a)
Propionic acid, 2-*o*-tolyloxy-	LaBrecque *et al.* (1960)
Pseudourea, 2-(1-naphthylmethyl)-2-thio-hydrochloride	LaBrecque *et al.* (1963a)
Pyrethrin	Tenhet (1947)
Riboflavine	LaBrecque *et al.* (1963a)
Tetradifon	Ascher and Hirsch (1961)
s-Triazine-2-thiol,4,6-diamino-	LaBrecque *et al.* (1963a)
Urea, 1-(1-naphthyl)-2-thio-	LaBrecque *et al.* (1963a)
Urea, 1-phenyl-2-thio-	Mitlin and Baroody (1958b)
Urea, 2-thio-	LaBrecque *et al.* (1963a) Mitlin and Baroody (1958b); Piquett and Keller (1962)

[a] Bořkovec (1964). Reprinted by permission of the copyright owner.

One structural feature common to these reactive groups is carbon bonded to an electronegative element; in some, the electronegative element is in a strained 3-membered ring.

Compounds grouped into this particular category and which show chemosterilant activity are mostly aziridine derivatives. Apholate (III), TEPA (V), thio-TEPA (VI), and metepa (VII) all contain aziridinyl groups and all show considerable promise as insect chemosterilants.

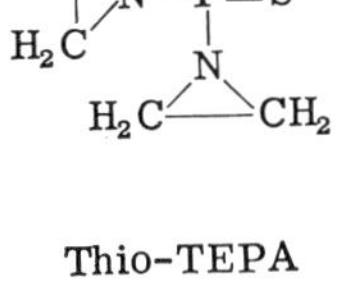

Thio-TEPA

(VI)

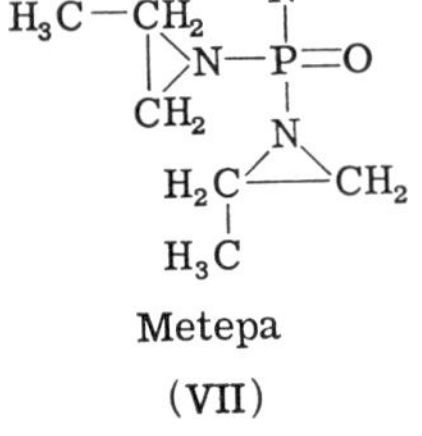

Metepa

(VII)

Unlike some of the antimetabolites, these chemicals induce permanent sterility in most insects and often sterilize both sexes. An undesirable feature of these chemicals, however, is that they are very toxic to higher animals and many can be absorbed through the skin. On the other hand, external absorption is an important feature when considering them for use as chemosterilants. A more detailed discussion of the biological activity and structure–activity relationship of these chemicals can be found in Sections V,B and VI.

2. *Antimetabolites*

An antimetabolite is a chemical which is structurally related to a biologically active substance. If the difference is not too large, the metabolic machinery of a living organism may not be able to distinguish between the two substances and use the antimetabolite as well as the normal metabolite. In bacteria, for example, 5-fluorouracil will replace a large percentage of the normal metabolite uracil in RNA (ribonucleic acid) when the organisms are grown in media containing the antimetabolite. Likewise, amethopterin, a folic acid analog, interferes with the formation of vitamin B_c (folic acid) in higher organisms. In some instances, adverse effects may not be apparent; in others, a metabolic process may be slowed down or even completely inhibited, depending on the antimetabolite and the system involved.

Many antimetabolites are now commercially available, and a large number of them have been tested for sterilant activity in houseflies as well as in other insect species. A few do sterilize insects, but they are not considered as promising as the more potent alkylating agents. Also, there is evidence to indicate that some antimetabolites induce only temporary sterility in insects and that the females are more susceptible than the males, a factor which must be considered when employing the sterile-male principle.

3. *Miscellaneous Groups*

Placed in this category are many chemicals which are structurally unrelated, but which show some promise as chemosterilants. Not included in Table II are two groups of chemicals which are also effective as chemosterilants and worthy of further consideration here. One group consists of the dimethylamine analogs of two aziridinyl compounds, and the other consists of several triphenyltin derivatives.

Hempa (VIII), hexamethylphosphoramide, and hemel (IX), hexamethylmelamine, which are effective male housefly chemosterilants, are dimethylamine analogs of TEPA (V) and tretamine (X), respectively (Chang *et al.,* 1964). However, VIII and IX differ from the aziridinyl compounds in their low mammalian toxicity and in that they are not alkylating agents. A comparison of their structures is given below.

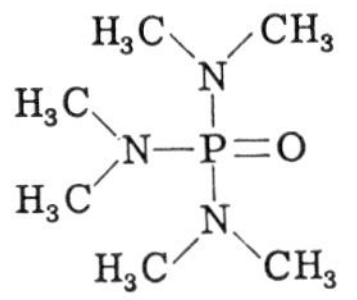

Hempa

(VIII)

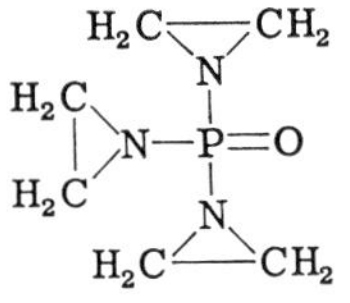

TEPA

(V)

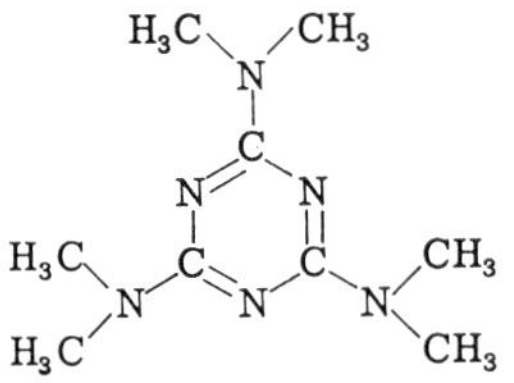

Hemel

(IX)

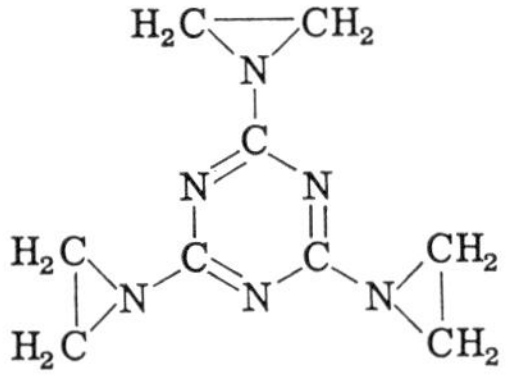

Tretamine

(X)

Numerous triphenyltin derivatives also act as reproduction inhibitors in the housefly and have been investigated extensively by Kenaga (1965). He found that the more active compounds, such as triphenyltin hydroxide, alkyltriphenyltin, and bis(triphenyltin) sulfide have three phenyl groups in common, in addition to a fairly labile fourth group attached to tin. These chemicals will sterilize both males and females, but the females are affected at much lower concentrations than the males. From these observations, it would appear that the triphenyltin moiety might have some promise as a reproduction inhibitor.

B. Structure–Activity Relationships

1. *TEPA–Hempa Analogs*

Chang and Bořkovec (1965) have determined the sterilizing potency of two series of TEPA–hempa analogs administered to male houseflies. The first series tested consisted of bis(1-aziridinyl)alkylaminophosphone oxides (XI, r = Me, Et, Pr, i-Pr, Bu, and octyl).

The second series consisted of TEPA–hempa intermediates (V, VIII, XII, XIII).

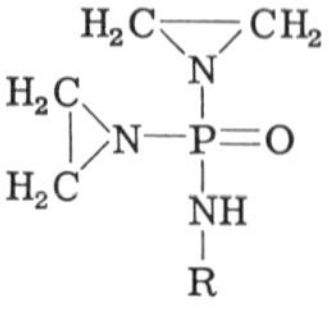

(XI)

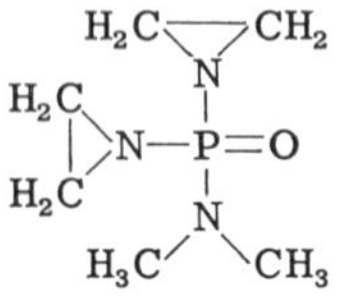

(XII)

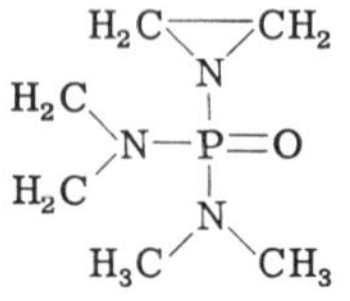

(XIII)

In the first series, the sterilizing activity of the compound decreased with the increasing length of the carbon chain in R, with the exception of the isopropyl compound, which was more effective as a chemosterilant than the ethyl compound. In the second series of compounds, the progressive replacement of aziridinyl groups in (V) with dimethylamino groups led to a regular decrease in the sterilizing activity.

2. *TEPA Analogs*

Four bicyclic aziridinylphosphine oxide analogs of TEPA (3-oxa-6-azabicyclo[3.1.0]hexane, cyclopentenimine, cyclohexenimine, and cycloheptenimine) have been prepared and evaluated as potential insect chemosterilants by Woods and Bořkovec (1965). In addition, these investigators studied the effect of a negative group on the aziridine ring with the phosphoramide of 2-carboethoxyaziridine. A series of alkylaminobisaziridinylphosphine oxides were also prepared and tested.

It was found that the carbon substitution in TEPA lead invariably to a decrease in the housefly sterilizing activity of the compound. The effect appeared to be independent of the electronic characteristics of the substituent. Some of the alkylaminobisaziridinylphosphine oxides, in contrast to the trisaziridinylphosphine oxides, exhibited sterilizing activities as high or higher than TEPA.

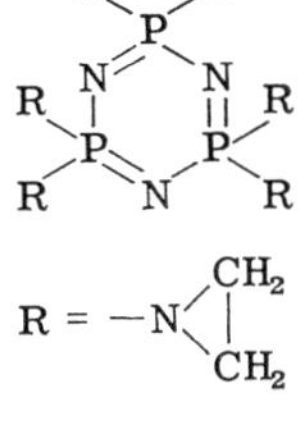

$R = -N(CH_2)_2$ (aziridinyl)

Apholate

(III)

$R = -N(CH_3)_2$

$R = -Cl$

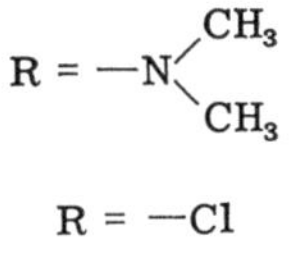

Substituted groups

3. *Apholate Analogs*

Castle and Ristich (1965) and Ristich *et al.* (1965) have evaluated a number of apholate (III) analogs in the housefly for chemosterilant activity. Tests on two complete homologous series based on chlorine or dimethylamino substitutions for the aziridinyl groups were conducted. They found:

1. Two $-N\langle{}^{CH_2}_{CH_2}$ groups were required for sterilant activity.
2. An increase from two to five $-N\langle{}^{CH_2}_{CH_2}$ groups improved sterilant activity.
3. Four $-N\langle{}^{CH_2}_{CH_2}$ groups were required in the chlorine series for effective sterilization.

The difference in activity between the two series appeared to be related to water solubility. The aziridinyl-substituted dimethylamino analogs were all water soluble, whereas the chloro analogs did not dissolve in water until at least four aziridinyl groups were present.

These workers also found that monosubstituents in the apholate molecule, other than chlorine or dimethylamine, did not alter activity. On the other hand, reduced activity was evident when substitutions on the aziridinyl groups were made. Tetrameric apholate containing eight aziridinyl groups, instead of the six in apholate, did not provide any improvement in activity.

C. Metabolism and Mechanism of Action

1. *Antimetabolites*

a. 5-Fluorouracil. Studies made by Kilgore and Painter (1962) have shown that the antimetabolite 5-fluorouracil (II) induces only temporary sterility in adult houseflies when the insects are fed treated diets immediately upon emergence for a period of 36–48 hours. Most of the antimetabolite is excreted as waste material, but a very small yet significant quantity is incorporated into the eggs. By using 5-fluorouracil-2-C^{14} they established that there was a correlation between the amount of C^{14} which accumulated in the eggs and egg viability (Fig. 1).

In later experiments in which treated and untreated males were mated with both treated and untreated females, it was clearly demonstrated that 5-fluorouracil is only a female sterilant (Kilgore and Painter, 1966). Measurements of the radioactive material in the eggs also showed that the male did not contribute to the sterility of the female (Fig. 2).

5-Fluorouracil is inhibitory to many living organisms, and often this

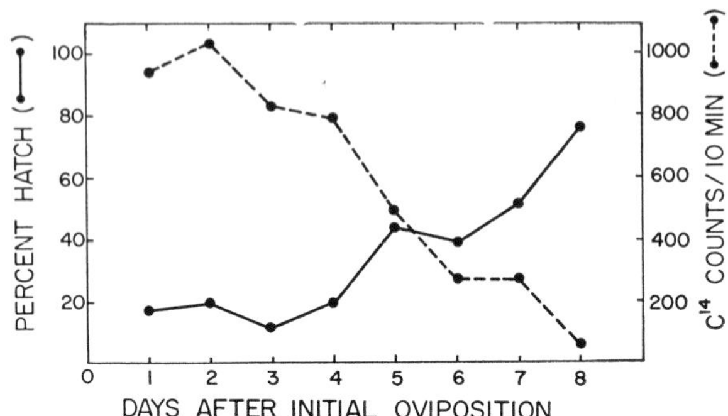

FIG. 1. Correlation between amount of C^{14} present in eggs and egg viability. Eggs were deposited by houseflies fed a diet containing 0.05% 5-fluorouracil-2-C^{14}.

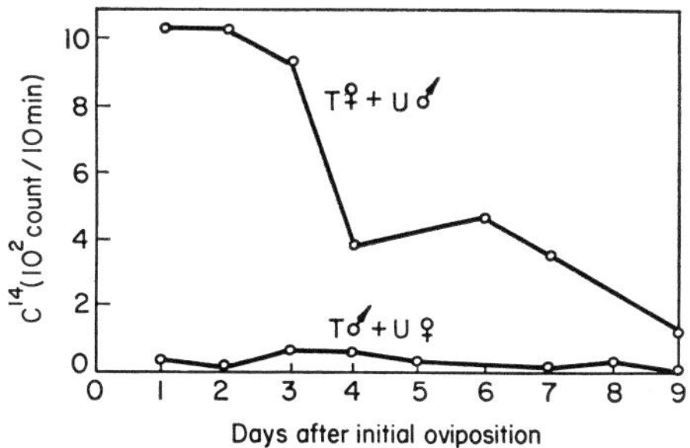

FIG. 2. Radioactivity present in eggs of houseflies fed diets containing 5-fluorouracil-2-C^{14}. T♀ × U♂: treated females mated with untreated males. T♂ × U♀: treated males mated with untreated females. (Kilgore and Painter, 1966; reprinted by permission of the copyright owner.)

toxic effect is caused by the incorporation of the antimetabolite into ribonucleic acid (RNA). Horowitz and Chargaff (1959) found that 5-fluorouracil was incorporated into bacterial RNA to the extent of nearly half the normally present uracil. The analog is incorporated into Ehrlich ascites carcinoma as well as normal tissues (Chaudhuri *et al.,* 1958), and it can replace from 28 to 47% of the normal base in tobacco mosaic virus RNA (Gordon and Staeheln, 1959).

Apparently, 5-fluorouracil is also toxic to housefly eggs as a result of its incorporation into egg RNA (Kilgore and Painter, 1966). An examination of the cellular components of the eggs revealed that approximately 20% of the total amount of C^{14} from the labeled 5-fluorouracil could be extracted with cold perchloric acid, and 80% with hot perchloric acid. Normally, cold perchloric acid extracts the components of the metabolic pool, including precursors of nucleic acid metabolism, while hot perchloric acid extracts and hydrolyzes large molecular components, such as RNA and DNA (deoxyribonucleic acid). Paper chromatograms of the cold perchloric acid extracts showed at least three discrete radioactive spots; two of the spots had R_f values identical to 5-fluorouracil-2-C^{14} and 5-fluorouridylic acid-2-C^{14}. The third spot was not identified but was presumed to be a metabolic intermediate. Paper chromatograms further showed that only one component, identical to 5-fluorouridylic acid-2-C^{14}, was present in the hot perchloric acid extracts.

As an extension of this study, purified egg RNA was prepared and examined for the presence of metabolites derived from 5-fluorouracil. Paper chromatograms of purified RNA (hydrolyzed in alkali) showed that only one radioactive component was present, and this compound had a migration rate identical to 5-fluorouridylic acid in three different solvent systems. Also, it cochromatographed with 5-fluorouridylic acid on a Dowex-2-formate resin column. Acid hydrolysis of the purified RNA yielded only one radioactive component identified as 5-fluorouracil-2-C^{14}, the compound expected if the RNA contained 5-fluorouridylic acid.

It is apparent from these studies that insects, at least houseflies, can incorporate 5-fluorouracil into the RNA of their eggs following the consumption of a diet containing the antimetabolite. Since the amount of antimetabolite incorporated into the eggs is directly related to egg viability, the mode of action of 5-fluorouracil as a chemosterilant is probably due to the replacement of a portion of the uracil in the egg RNA. In support of this hypothesis, it is known that once the antimetabolite is depleted the eggs are again viable.

A proposed scheme of the metabolism of 5-fluorouracil in houseflies is shown in Fig. 3.

b. 5-Fluoroorotic Acid. It has been suggested that, in vertebrates, 5-fluoroorotic acid inhibits the growth of certain tumors by interfering with the normal conversion of orotic acid to orotidylic acid, an intermediate in the *de novo* synthesis of the uracil moiety of RNA and the cytosine moiety of DNA. However, such detailed information is not yet available for insects.

Painter and Kilgore (1965) fed newly emerged houseflies diets containing 0.5% 5-fluoroorotic acid-2-C^{14} for 48 hours to determine if any labeled material would be incorporated into the ovaries and testes of the insects. In Fig. 4 it can be seen that some of the labeled compound, or a metabolic product, was incorporated into the ovaries up to 2 days after the feeding had been discontinued. Following this increase, a relatively rapid loss of radioactivity occurred until a low plateau was reached 8

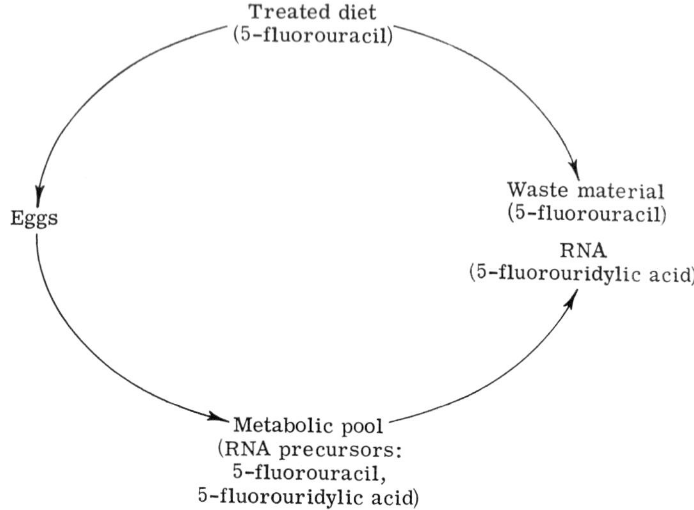

FIG. 3. Metabolism of 5-fluorouracil in houseflies.

days after the end of the feeding period, indicating that a small but significant quantity still remained in the ovaries.

Since the testes did not at any time contain a significant quantity of C^{14}, the measurements were discontinued after 5 days. The incorporation of C^{14} from 5-fluoroorotic acid-2-C^{14} into the ovaries but not the testes, which indicates more effect on the female reproductive system than the male, is in agreement with the report of Crystal (1963) that this antimetabolite is a female sterilant for screwworm flies. Crystal found that 5-fluoroorotic acid reduced fecundity to zero when it was fed only to females for 1 day prior to mating. However, it had little or no effect when fed to males only.

From these data it is difficult to establish the biochemical effects of 5-fluoroorotic acid in insects, other than that it causes them to deposit few, if any, eggs under controlled conditions. It is possible that this antimetabolite has at least two types of toxic effects when used as a chemosterilant. It may interfere with some hormonal mechanism of the females because, at elevated levels of the antimetabolite in the diets, atrophy of the ovaries occurs and no more eggs are deposited. This effect may be unrelated to the deposition of a few viable and nonviable eggs caused by lower levels of the compound. The latter may be associated with the nucleic acid metabolism in the eggs since 5-fluoroorotic acid does interfere in the conversion of orotic acid to orotidylic acid in other organisms. Undoubtedly, further research will help to clarify these points.

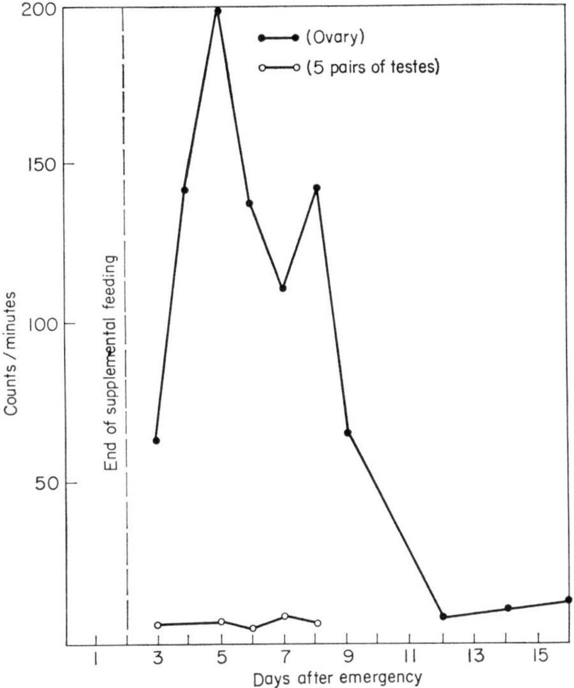

FIG. 4. C^{14} incorporation in the ovaries and testes of house flies fed 0.5% 5-fluoroorotic acid for 48 hours after emergences. (Painter and Kilgore, 1965; reprinted by permission of the copyright owner.)

2. *Alkylating Agents*

a. Metepa. Plapp *et al.* (1962) have studied the metabolism of metepa-P^{32} (VII) in houseflies and in larvae and adults of the mosquito, *Culex tarsalis* Coquillett. In their experiments with houseflies, 2- to 4-day old females of an insecticide-susceptible strain and two organophosphate-resistant strains were used. The flies were treated topically on the tip of the abdomen with metepa-P^{32} dissolved in acetone, or by injecting a water solution of the labeled compound into the thorax. Each fly received a dosage of about 100 μg metepa/gm fly weight. The adult mosquitoes were either fed a 5% sugar–water solution containing 100 ppm of metepa or allowed to feed on male white mice treated intraperitoneally with 100 mg/kg of metepa. Fourth-instar mosquito larvae were exposed to the chemical by placing them in a solution of 100 ppm of metepa for 16 hours.

These investigators found that the chemical was rapidly absorbed in houseflies following topical administration and rapidly detoxified following absorption. About 50% was degraded within 2 hours. Injected flies degraded the chemical at an even more rapid rate; degradation was more than 95% complete in 24 hours. The rates of degradation were similar in the susceptible fly strain and in the organophosphate-resistant strains. They also found that much of the excreted radioactive material appeared to be unmetabolized metepa, although an unknown compound and inorganic phosphate were also detected. Of the material excreted, more than 90% of the total P^{32} recovered in 24 hours was eliminated within 8 hours of treatment.

In the mosquito larvae, the chemical was also degraded rapidly; it was almost completely degraded within 48 hours. However, measurable amounts of radioactivity remained in the insects throughout the pupal and adult stages. The adult mosquitoes also degraded the chemical rapidly but much of the label remained in the insects. From these experiments it appears that metepa is degraded by both mosquitoes and houseflies and that one of the degradation products, possibly phosphate, is reincorporated into the normal metabolites.

In this same study (Plapp *et al.,* 1962) the metabolism of metepa in mice was also investigated. The results showed that the labeled chemical was excreted almost as rapidly in mice as in houseflies. Most of the P^{32} was excreted in the urine and was collected within 12 hours after treatment. Only a small amount was recovered in the feces. In the excretory products, unaltered metepa and one major breakdown product, presumably inorganic phosphate, were detected by paper chromatography.

Dame and Schmidt (1964) have conducted a rather comprehensive study of the uptake and metabolism of metepa-P^{32} in the housefly and in two species of mosquitoes. In their experiments, the common malaria mosquito, *Anopheles quadrimaculatus* Say, the yellow-fever mosquito, *Aedes aegypti* (L.), and the housefly, *Musca domestica* L., were exposed to residual deposits, treated food, or larval media containing the labeled compound.

The labeled compound was rapidly absorbed from glass surfaces by both species of mosquitoes as well as by the houseflies. The houseflies and the malaria mosquitoes absorbed about 7 μg per insect during a 4-hour exposure on surfaces treated with 10 mg metepa-P^{32}/ft^2. The yellow-fever mosquitoes absorbed only 2.5 μg per insect. Houseflies lost 89% of the radioactive material within 24 hours, but a greater percentage of the label remained in adults which had been exposed to the compound during the larval stage or during a 3-day adult feeding period than in adults exposed to residual deposits.

Distribution of the labeled metepa was rapid and general in these insects. The label was found to be proportionately distributed according to relative weights in the various body sections which were investigated. However, in residual exposures, proportionately more radioactivity was found in the legs of the mosquitoes. The distribution did not change over prolonged periods, but in females some P^{32} accumulated in the ovaries. This slight accumulation may have been due to the normal accumulation of phosphate during ovarian development.

An extensive study of the rate of absorption, excretion, and metabolism of metepa-P^{32} has also been made by Chamberlain and Hamilton (1964) in the screwworm fly, *Cochliomyia hominivorax* (Coquerel), and the stable fly, *Stomoxys calcitrans* (L.). These investigators found that, 6 hours after exposure to the sterilant, the screwworm fly had absorbed only half as much labeled material in proportion to its size as the stable fly. To a considerable extent this explained the differences previously noted in the dosages required to sterilize the two species.

The excretion rate of the label by the screwworm fly was twice that of the stable fly, and the metabolism by the stable fly was about twice as fast as for the screwworm fly. It was determined that the principal metabolite was phosphoric acid, and that it appeared in greater quantities in the screwworm fly than in the stable fly. A more-detailed chromatographic analysis of the fecal material from screwworm flies fed metepa-P^{32} revealed that other metabolites were also present. At least four of the unknown metabolites had been present as contaminants in the purified metepa-P^{32} used in the study, but these unknowns all increased in quantity in the fecal material to above the levels found in the original sample. In addition, a new compound appeared for the first time as a distinct spot. Unfortunately, none of these metabolites, or degradation products, has been identified. From their *in vitro* studies, Plapp *et al.* (1962) have suggested that one product may be either the di- or monoacid derivative of metepa. A proposed degradative pathway for metepa and similar alkylating agents is shown in Fig. 5.

b. Thio-TEPA. Parish and Arthur (1965b) have studied the metabolism of thio-TEPA (VI) in four species of insects. The chemical was administered topically at 100 mg/kg to the German cockroach, *Blattella germanica* (L.); housefly, *Musca domestica* (L.); and boll weevil, *Anthonomus grandis* Boheman. With the exception of the boll weevil, maximum absorption had occurred in all insects by 4 hours after the treatment.

The only chloroform-soluble metabolite recovered from the insects was TEPA, the oxygen analog of thio-TEPA. In German cockroaches, houseflies, and stable flies, the amount of thio-TEPA decreased with time after treatment. However, the opposite was true for boll weevils.

Parish and Arthur also found that white rats readily metabolize thio-

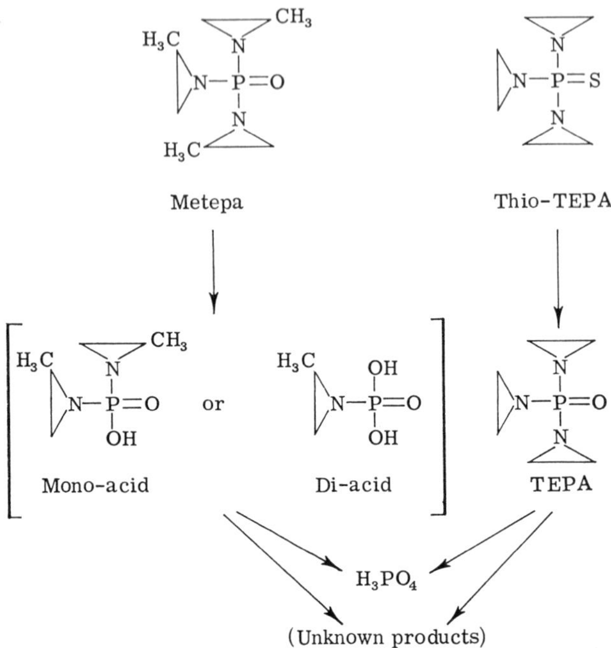

FIG. 5. Proposed degradative pathway of metepa, TEPA, and thio-TEPA in insects.

TEPA. Products were eliminated in both the urine and feces, and TEPA was the predominant compound recovered in the chloroform extracts. Three water-soluble products were also isolated from the urine of the rats, but were not identified. These results confirm earlier work by Craig *et al.* (1958a), who found that rats treated with thio-TEPA excreted TEPA as the main urinary metabolite. Parish and Arthur further reported that the liver and kidney contained higher concentrations of radioactive materials than fat, leg muscle, brain, reproductive organs, or blood.

From these reports, it appears that thio-TEPA is converted to TEPA which is subsequently degraded to inorganic phosphate and several unknown derivatives (Fig. 5).

c. Apholate. It has been reported (Painter and Kilgore, 1964a) that apholate (III) induces permanent sterility in adult houseflies following brief exposures to treated diets (Fig. 6). In addition, eggs deposited by the apholate-treated insects contain very low levels of DNA (Kilgore and Painter, 1964). The amount of DNA found in viable and nonviable eggs (deposited by apholate-treated houseflies) during the normal embryogenesis period is shown in Fig. 7. The absence of DNA synthesis in the de-

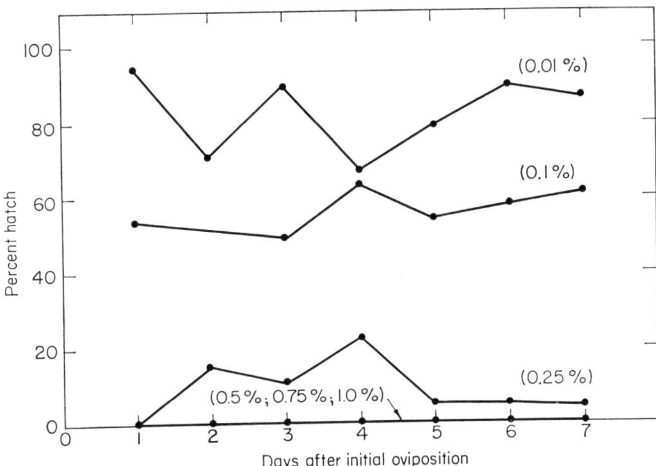

FIG. 6. The percent hatch of eggs obtained from houseflies fed a diet containing 0.01, 0.1, 0.25, 0.5, 0.75, and 1.0% apholate. The flies were given the treated diets for 48 hours immediately following emergence. (Painter and Kilgore, 1964a; reprinted by permission of the copyright owner.)

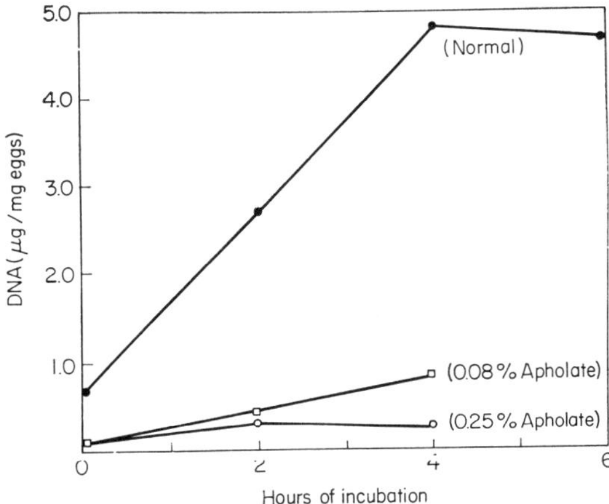

FIG. 7. Synthesis of DNA (acid-insoluble fraction) in normal and nonviable housefly eggs. ●, DNA content of normal eggs deposited by flies given an untreated diet; □, DNA content of nonviable eggs deposited by flies given a diet containing 0.08% apholate; ○, DNA content of nonviable eggs deposited by flies given a diet containing 0.25% apholate. The sampling interval used in these experiments was not short enough to show the characteristic initial lag of DNA synthesis.

veloping eggs caused by apholate is also accompanied by the loss of the ability of the eggs to form lactic acid dehydrogenase. It was also found by these investigators that the activity of the enzyme glucose-6-phosphate dehydrogenase remains unaffected.

Ionizing radiation, mitomycin C, and nitrogen mustard inhibit the synthesis of DNA in many biological systems. However, it is not yet clear whether apholate has a direct or indirect effect on DNA synthesis in the developing eggs. A direct effect could occur by transfer of the chemical from the ovaries to the eggs. In contrast, ovarian damage, hormonal imbalance, or any one of a number of metabolic abnormalities caused by the chemical could have an indirect effect on the systhesis of DNA in the eggs.

Painter and Kilgore (1964b) have also reported that the nucleotide composition of egg RNA from untreated- and apholate-treated houseflies differs markedly. The RNA from eggs of apholate fed houseflies is significantly lower in adenylic acid and somewhat higher in uridylic acid compared to the RNA from normal eggs. Considering these data and the popularity of the concept of alkylation of the guanine moiety in nucleic acids by a number of alkylating agents, it seems possible that apholate could have a direct effect on the eggs. However, until further evidence is obtained a definite conclusion can not be drawn.

VI. BIOLOGICAL EFFECTS

A. Insects

Frequently, elevated dosages of chemosterilants completely inhibit the formation of eggs in treated insects because of irreversible damage caused to the ovaries. Damage to insect ovaries by chemicals was observed as early as 1948 by Goldsmith *et al.,* and later it served as the subject of several additional reports (Goldsmith and Frank, 1952; Goldsmith, 1955; Konecky and Mitlin, 1955; Mitlin, 1956; Mitlin *et al.,* 1957). Mitlin *et al.* (1957) found that both aminopterin and nitrogen mustard when fed to houseflies prevented ovarian development and caused sterility in the female sex. This effect in aminopterin-treated insects was not reversed by feeding either thymidine or folic acid along with the antimetabolite. In a later study, Mitlin and Baroody (1958a) reported that 15 of 26 antitumor compounds caused the inhibition of ovarian growth in houseflies. Painter and Kilgore (1965) have also reported that the antitumor compound 5-fluoroorotic acid prevents ovarian development in houseflies. Photographs of ovaries from 7- to 10-day-old female houseflies, some fed a diet containing 1.0% 5-fluoroorotic acid for 48 hours are shown in Fig. 8. The ovaries from the treated flies compared to nor-

FIG. 8. Photomicrographs of the ovaries of 7 to 10-day-old houseflies: A, from flies fed normal diet throughout adult life; B, from flies fed 1.0% 5-fluoroorotic acid in the diet for 48 hours after emergence. (Painter and Kilgore, 1965; reprinted by permission of the copyright owner.)

mal insects are very small and do not appear to contain developing eggs. On the other hand, according to Painter and Kilgore, ovaries from houseflies fed 0.5% 5-fluoroorotic acid appear normal in size, color, and texture but they vary considerably in egg formation.

Morgan and LaBrecque (1962) have observed that apholate, administered in the food of adult female houseflies at a 1.0% concentration for a period up to 240 hours, inhibits but does not completely eliminate ovarian development. In a similar study with TEPA and metepa, Morgan and LaBrecque (1964) found that ovarian development was inhibited in some flies between the twenty-fourth and forty-eighth hour after the treatment was initiated. In other flies, however, ovarian development was not impaired.

Crystal and LaChance (1963) have reported that concentrations of tretamine as low as 0.5% completely inhibit ovarian growth in screwworm flies when administered to newly emerged females. However, when tretamine was administered to flies 24-hours old, it had little effect on ovarian growth.

Shaw and Riviello (1962a) have investigated the effect of Chloroambucil, 4-{*p*-[bis(2-chloroethyl)amino]phenyl} butyric acid, on the testes and 4-amino-1*H*-pyrazolo[3,4-*d*]pyrimidine sulfate on the ovaries of the Mexican fruit fly, *Anastrepha ludens* (Loew). They found that Chlorambucil almost completely prevented growth of the testes. Likewise, 4-amino-1*H*-pyrazolo(3,4-*d*)pyrimidine sulfate prevented ovarian development.

Often chemosterilants, particularly at low levels, do not inhibit gonadal development in insects, although the insects may be sterilized by the chemicals and deposit nonviable eggs. Many of the chemosterilants contain alkylating groups and cause biological changes closely resembling those produced by irradiation. The production of mutations, chromosome aberrations, and induction of sterility and cancer in animals are all biological effects induced by both alkylating agents and irradiation.

Several studies have now been conducted to analyze the cytogenic effects of chemosterilants having alkylating properties on insect fecundity and fertility. LaChance and Crystal (1963) found that tretamine, 2,4,6-tris(1-aziridinyl)-*s*-triazine, a benzoquinone derivative, 2,5-bis(1-aziridinyl-3,6-bis)2-methoxyethoxy-*p*-benzoquinone, and thio-TEPA, tris(l-aziridinyl)phosphine sulfide, all induce dominant lethal mutations in the oocytes of the screwworm fly. They also found that fewer dominant lethal mutations were induced in prophase oocytes than in oocytes in the metaphase or in the anaphase stage of meiotic division. In a more detailed study, LaChance and Riemann (1964) demonstrated that tretamine induces dominant lethal mutations in both oocytes and sperm of the screwworm fly and that embryonic development generally stops after only a few cleavage divisions. From similar studies, Rai (1964a,b) concluded that in mosquitoes (*Aedes aegypti,* L.) apholate produces aberrations in the somatic chromosomes and induces dominant lethal mutations in the sperm.

B. Mammals

Many insect chemosterilants are well-known neoplastic agents and considerable information concerning their biological effects in mammals is available (Hayes, 1964; Gaines and Kimbrough, 1964; Eriksson, 1965).

Toxicological data pertaining to four of the more promising chemosterilants are given in Table III.

In general, insect chemosterilants having alkylating properties cause delayed effects in mammals, which are accompanied by a selective action against proliferating tissues. Tissues frequently damaged by these chemicals are hematopoietic cells of the bone marrow and lymphoid tissues, the intestinal mucosa, germ cells, embryos, and tumors.

VII. PRACTICAL ASPECTS

That it is theoretically and physically possible to control insect populations by employing the sterile-male principle is now well established with screwworm flies (see Chapter 4). This successful application of the sterile-male principle along with the discovery that chemicals can also sterilize insects has prompted the question "How can chemosterilants be used effectively in the field?". Such a question is understandable since the use of a chemosterilant could provide a potential 99% control rather than the 90% expected from an insecticide (Lindquist, 1961). The development of suitable compounds and acceptable application methods, and a better understanding of insect mating habits are areas of research which must be fully explored before any serious effort can be made to use chemosterilants on a commercial basis. These problems are not insurmountable since the chemical industry has overcome similar though not identical problems in the past. It is quite likely through continued research by Universities, Governmental Agencies, and the chemical industry that relatively safe and specific chemosterilants will be developed for widespread field use.

Assuming that suitable chemosterilants are available, there are several ways in which they can be and have been used to control insect populations in the field. Under certain conditions, some of the relatively toxic compounds now under study could also be used.

A. Contact, Oral-Ingestion Methods

1. *Direct Contact with Sprays and Dusts*

a. Chemosterilants Only. It is conceivable that chemosterilants could be applied to native insect populations as liquid sprays or dusts in much the same manner as conventional insecticides are now applied. This technique would be especially useful for treating ponds, lakes, reservoirs, and open ditches as a procedure for sterilizing water-inhabiting forms of certain insects. Weidhaas (1962) has demonstrated very effectively that mosquito larvae (*Aedes aegypti,* L.) when reared in water containing

TABLE III

TOXICITY OF CHEMOSTERILANTS[a]

	Toxic dosage in rats				Approximate therapeutic dosage in man		
	Single dosage		Repeated daily doses				
Compound	Oral LD_{50} (mg/kg)	Dermal LD_{50} (mg/kg)	mg/kg/day	Effect	mg/kg	Doses	Route
Metepa[b]	136	183	10.0 (×104; oral)	Reduced white blood cells, growth suppression and death in 44–94 days	—	—	—
Metepa[b]	136	183	5.0 (×104; oral)	Sterility, testicular atrophy and 1 of 12 died in 89 days	—	—	—
Metepa[b]	136	183	2.5 (×197; oral)	Subfertility and partial testicular atrophy	—	—	—
Metepa[b]	136	183	1.25 (×197; oral)	None	—	—	—
TEPA	37[b]	87[b]			1.2–2.0	Total	Intra-Muscular[c]
Apholate	98[b]; 90[e]	400–800[b]	15.0 (×98; in food)	Reduced white blood cells, growth suppression, testicular atrophy and 1 of 5 died in 49 days[d]	0.6	3–6	Oral[e]

Apholate	98[b]; 90[e]	400–800[b]	7.5 (×98; in food)	Growth suppression and testicular atrophy[d]	0.6	3–6	Oral[e]
Apholate	98[b]; 90[e]	400–800[b]	3.7 (×77; in food)	Growth suppression and testicular atrophy[d]	0.6	3–6	Oral[e]
Tretamine	ca. 1	—	0.2 (×1; intraperiton)	Temporary sterility[f]	0.12	Total	Intravenous[c]
Tretamine	ca. 1	—	0.2 (×21; intraperiton)	Continuing sterility[g]	0.04–0.08	4	Oral[h]
Tretamine	ca. 1	—	0.05 (×41; intraperiton)	Temporary sterility, and without atrophy[f]	0.3–0.6	Total	Oral[c]

[a] Hayes (1964); reprinted by permission of the copyright owner.
[b] Gaines and Kimbrough (1964).
[c] Sykes *et al.* (1956).
[d] Gaines, T. B., unpublished data.
[e] Olin Mathieson Chemical Co.
[f] Craig *et al.* (1958b).
[g] Jackson and Bock (1955).
[h] Hall (1962).

10 ppm TEPA do develop into sterile adults. Unfortunately, actual field studies with mosquito larvae as well as other water-inhabiting forms of insects are not available.

b. Insecticide–Chemosterilant Combination. Many investigators have advocated the use of chemosterilants in combination with conventional insecticide sprays. Such an approach would certainly be more acceptable to the agriculturists who are concerned with immediate crop damage and used to seeing a rapid decline in insect numbers after spraying with an insecticide. By using a combination spray, a dual purpose could be served. First, the insect population would be reduced to protect the growing crops; and secondly, those few insect pests which remained would be sterilized, providing a further reduction in insect numbers.

2. *Direct Contact with Residues*

Another approach to the sterilization of insects would be to expose them to residual amounts of chemosterilants applied to various types of surfaces. This method has already been successfully tested in the laboratory by a number of workers, but its usefulness in the field has not been explored. It is however worthy of consideration. By using this method, the chemicals could be applied as sprays to places where the insects tend to congregate or breed (i.e., decaying materials, manure piles, etc.) or to areas of heavy infestation. Rather than using a spray to apply the chemicals for residual deposits, pretreated glass plates, masonite sheets, boards, etc., could also be used. Chemicals used in this manner would have to be able to sterilize the insects on contact after only a brief exposure. Some of the chemosterilants now available do sterilize insects when they are exposed to treated sufaces, but the time required to do so ranges from 4 to 48 hours, depending on the insect and dosage involved.

Even though this method seems feasible, it would be far more effective if the insects could be attracted to a selected area, which could be accomplished through the use of attractants and lures. Sex attractants (see Chapter 6) should be particularly useful since they would be more specific than other types of attracting agents.

3. *Oral Ingestion*

Tests have shown that some insects can be sterilized in the field by ingesting food containing a chemosterilant in much the same manner as they are now sterilized in the laboratory. By using this approach, treated food is placed in infested areas for the insects to eat. Presumably the natural food could also be sprayed with the chemosterilant.

LaBrecque *et al.* (1962b, 1963a) and Gouck *et al.* (1963a) in field experiments have found that cornmeal baits are useful as carriers for chemosterilants. In an isolated refuse dump in the Florida Keys, LaBrecque *et al.* (1962b) obtained excellent control of houseflies by using cornmeal baits containing 0.5% aphoxide. In these experiments, applications were made weekly for 9 consecutive weeks, except during the second week. After 4 weeks the housefly population was reduced from 47 per grid to 0. In a later study, LaBrecque *et al.* (1963a) found that the housefly population in a poultry house could be reduced sharply by applying cornmeal baits containing 0.5% metepa to the poultry droppings. Applications were made on a weekly basis for 9 weeks, then semiweekly. Of the three carriers tested (cornmeal, granulated sugar, and vermiculite granules), cornmeal was the most effective.

Gouck *et al.* (1963a) also found that houseflies could be controlled in the field with a chemosterilant incorporated into cornmeal baits. These workers applied the cornmeal baits containing 0.75% apholate to a dump once each week for 7 weeks, and then five times each week for 5 weeks. Following the applications, they found that the fly population decreased from 68 per grid to 5–20 during the first 7 weeks. During the following 5 weeks, the population remained between 0 and 3 per grid.

All of these experiments clearly indicate that treated baits can be used effectively in the field to provide local control of insect populations.

C. Sterile-Male Release Method

Insects could be sterilized in the laboratory with chemosterilants in much the same manner as the screwworm fly was sterilized with irradiation (Bushland, 1960) and then released in large numbers to mate with the existing population. This method might be useful for some species, but it would not be particularly advantageous for insect forms which could cause tremendous crop damage in a short period or be a public nuisance when released in large numbers. An attractive alternative would be to combine the sterile-male release method on a reduced scale with other methods of control. For instances, the sterile-male release procedure could be employed in conjunction with the application of conventional insecticides. The insecticides could be applied first to reduce the number of insect pests in the natural population, and then the sterile insects could be released in far fewer numbers to mate with those escaping the insecticide applications.

VIII. CONCLUSION

Within the last decade the idea of sterilizing insects with chemicals has advanced from theory to actual practice. As a result of research conducted by numerous investigators, a variety of chemicals have been found which can interrupt the reproductive cycles of a large number of insect species. It now appears that chemosterilization as a method for insect control has unlimited possibilities, particularly since the use of a chemosterilant could provide a potential 99% control of an insect species compared to the 90% expected with a conventional insecticide.

REFERENCES

Altman, R. M. (1963). *Am. J. Hyg.* **77,** 221.
Ascher, K. R. S. (1957a). *Science* **125,** 938.
Ascher, K. R. S. (1957b). *Riv. Malariol.* **36,** 209.
Ascher, K. R. S. (1958). *Experientia* **14,** 8.
Ascher, K. R. S., and Hirsch, I. (1961). *Riv. Malariol.* **40,** 139.
Benschoter, C. A. (1966). *J. Econ. Entomol.* **59,** 334.
Bhalla, O. P., and Robinson, A. G. (1966). *J. Econ. Entomol.* **59,** 378.
Bořkovec, A. B. (1962). *Science* **137,** 1034.
Bořkovec, A. B. (1964). *In* "Residue Reviews" (F. A. Gunther, ed.), Vol. 6, pp. 87–103. Springer, Berlin.
Bořkovec, A. B., and Woods, C. W. (1963). *Advan. Chem. Ser.* **41,** 47.
Bořkovec, A. B., Chang, S. C., and Linburg, A. M. (1964). *J. Econ. Entomol.* **57,** 815.
Burden, G. S., and Smittle, B. J. (1963). *Florida Entomologist* **46,** 229.
Bushland, R. C. (1960). *Advan. Vet. Sci.* **6,** 1.
Carlson, E. A., and Oster, I. I. (1961). *Genetics* **46,** 856.
Castle, R. E., and Ristich, S. S. (1965). *Am. Chem. Soc. 150th Meeting,* Paper 55, p. 22A.
Chamberlain, W. F. (1962). *J. Econ. Entomol.* **55,** 240.
Chamberlain, W. F., and Barrett, C. C. (1964). *J. Econ. Entomol.* **57,** 267.
Chamberlain, W. F., and Hamilton, E. W. (1964). *J. Econ. Entomol.* **57,** 800.
Chamberlain, W. F., and Hopkins, D. E. (1960). *J. Econ. Entomol.* **53,** 1133.
Chang, S. C. (1965). *J. Econ. Entomol.* **58,** 669.
Chang, S. C., and Bořkovec, A. B. (1964). *J. Econ. Entomol.* **57,** 488.
Chang, S. C., and Bořkovec, A. B. (1965). *Am. Chem. Soc. 150th Meeting,* Paper 64, p. 26A.
Chang, S. C., and Bořkovec, A. R. (1966). *J. Econ. Entomol.* **59,** 102.
Chang, S. C., Terry, P. H., and Borkovec, A. B. (1964). *Science* **144,** 57.
Chaudhuri, N. K., Montag, B. J., and Heidelberger, C. (1958). *Cancer Res.* **18,** 318.
Clark, A. M. (1959). *Nature* **163,** 731.
Clark, A. M. (1960). *Z. Vererbungslehre* **91,** 74.
Collier, C. W., and Downey, J. E. (1965). *J. Econ. Entomol.* **58,** 649.
Craig, A. W., Fox, B. W., and Jackson, H. (1958a). *Biochem. J.* **69,** 1.
Craig, A. W., Fox, B. W., and Jackson, H. (1958b). *Nature* **181,** 353.
Creighton, C. S., Cuthbert, E. R., Jr., and Reid, W. J., Jr. (1966). *J. Econ. Entomol.* **59,** 163.
Cressman, A. W. (1963). *J. Econ. Entomol.* **56,** 111.
Crystal, M. M. (1963). *J. Econ. Entomol.* **56,** 468.
Crystal, M. M. (1964a). *J. Econ. Entomol.* **57,** 606.
Crystal, M. M. (1964b). *J. Econ. Entomol.* **57,** 726.
Crystal, M. M. (1964c). *Exptl. Parasitol.* **15,** 249.

Crystal, M. M. (1965a). *J. Econ. Entomol.* **58,** 678.

Crystal, M. M. (1965b). *J. Med. Entomol.* **2,** 317.

Crystal, M. M., and LaChance, L. E. (1963). *Biol. Bull.* **125,** 270.

Dame, D. A., and Schmidt, C. H. (1964). *J. Econ. Entomol.* **57,** 77.

Davich, T. B., Keller, J. C., Mitchell, E. B., Huddleston, P., Hill, R., Lindquist, D. A., McKibben, G., and Cross, W. A. (1965). *J. Econ. Entomol.* **58,** 127.

Eriksson, H. (1965). *Arkiv Zool.* **44,** 543.

Fahmy, O. G., and Fahmy, M. J. (1954). *J. Genet.* **52,** 603.

Fahmy, O. G., and Fahmy, M. J. (1961). *J. Genet.* **46,** 1111.

Fye, R. L., Gouck, H. K., and LaBrecque, G. C. (1965). *J. Econ. Entomol.* **58,** 446.

Fye, R. L., LaBrecque, G. C., and Gouck, H. K. (1966). *J. Econ. Entomol.* **59,** 485.

Gaines, T. B., and Kimbrough, R. D. (1964). *Bull. World Health Organ.* **31,** 737.

Geering, Q. A., Brooker, P. J., and Parsons, J. H. (1965). *J. Econ. Entomol.* **58,** 574.

Goldsmith, E. D. (1952). *Proc. 64th Ann. Meeting Am. Assoc. Econ. Entomol., Dec.* p. 41.

Goldsmith, E. D. (1955). *Federation Proc.* **14,** 59.

Goldsmith, E. D., and Frank, I. (1952). *Am. J. Physiol.* **171,** 726.

Goldsmith, E. D., Tobias, E. B., and Harnly, M. H. (1948). *Anat. Record* **101,** 93.

Goldsmith, E. D., Tobias, E. B., and Harnly, M. H. (1950). *Ann. N.Y. Acad. Sci.* **52,** 1342.

Gordon, M. P., and Staeheln, M. (1959). *Biochim. Biophys. Acta* **36,** 351.

Gouck, H. K. (1964). *J. Econ. Entomol.* **57,** 239.

Gouck, H. K., and LaBrecque, G. C. (1964). *J. Econ. Entomol.* **57,** 663.

Gouck, H. K., Meifert, D. W., and Gahan, J. B. (1963a). *J. Econ. Entomol.* **56,** 445.

Gouck, H. K., Crystal, M. M., Bořkovec, A. B., and Meifert, D. W. (1963b). *J. Econ. Entomol.* **56,** 506.

Hair, J. A., and Adkins, T. R., Jr. (1964). *J. Econ. Entomol.* **57,** 586.

Hall, T. C. (1962). *New Engl. J. Med.* **266,** 129, 178, 238, 289.

Hansens, E. J. (1965). *J. Econ. Entomol.* **58,** 944.

Hansens, E. J., and Granett, P. (1965). *J. Econ. Entomol.* **58,** 157.

Harries, F. H. (1961). *J. Econ. Entomol.* **54,** 122.

Harries, F. H. (1963). *J. Econ. Entomol.* **56,** 438.

Harris, R. L. (1962). *J. Econ. Entomol.* **55,** 882.

Hayes, W. J., Jr. (1964). *Bull. World Health Organ.* **31,** 721.

Hazard, E. I., Lofgren, C. S., Woodard, D. B., Ford, H. R., and Glancey, B. M. (1964). *Science* **145,** 500.

Hedin, P. A., Cody, C. P., and Thompson, A. C., Jr. (1964). *J. Econ. Entomol.* **57,** 210.

Henneberry, T. J., and Kishaba, A. N. (1966). *J. Econ. Entomol.* **59,** 156.

Henneberry, T. J., Smith, F. F., and McGovern, W. L. (1964). *J. Econ. Entomol.* **57,** 813.

Horowitz, J., and Chargaff, E. (1959). *Nature* **184,** 1213.

Howland, A. F., Vail, P., and Henneberry, T. J. (1965). *J. Econ. Entomol.* **58,** 635.

Howland, A. F., Vail, P., and Henneberry, T. J. (1966). *J. Econ. Entomol.* **59,** 194.

Jackson, H., and Bock, M. (1955). *Nature* **175,** 1037.

Jeppson, L. R., Jesser, M. J., and Complin, J. O. (1966). *J. Econ. Entomol.* **59,** 15.

Keiser, I., Steiner, L. F., and Kamasaki, H. (1965). *J. Econ. Entomol.* **58,** 682.

Kenaga, E. E. (1965). *J. Econ. Entomol.* **58,** 4.
Kilgore, W. W., and Painter, R. R. (1962). *J. Econ. Entomol.* **55,** 710.
Kilgore, W. W., and Painter, R. R. (1964). *Biochem. J.* **92,** 353.
Kilgore, W. W., and Painter, R. R. (1966). *J. Econ. Entomol.* **59,** 825.
Knipling, E. F. (1960). *Sci. Am.* **203,** 54.
Knipling, E. F. (1962). *J. Econ. Entomol.* **55,** 782.
Konecky, M. S., and Mitlin, N. (1955). *J. Econ. Entomol.* **48,** 219.
LaBrecque, G. C. (1961). *J. Econ. Entomol.* **54,** 684.
LaBrecque, G. C. (1963). *Advan. Chem. Ser.* **41,** 42.
LaBrecque, G. C., and Gouck, H. K. (1963). *J. Econ. Entomol.* **56,** 476.
LaBrecque, G. C., Adcock, P. H., and Smith, C. N. (1960). *J. Econ. Entomol.* **53,** 802.
LaBrecque, G. C., Meifert, D. W., and Smith, C. N. (1962a). *Science* **136,** 388.
LaBrecque, G. C., Smith, C. N., and Meifert, D. W. (1962b). *J. Econ. Entomol.* **55,** 449.
LaBrecque, G. C., Meifert, D. W., and Fye, R. L. (1963a). *J. Econ. Entomol.* **56,** 150.
LaBrecque, G. C., Meifert, D. W., and Gouck, H. K. (1963b). *Florida Entomologist* **46,** 7.
LaChance, L. E., and Crystal, M. M. (1963). *Biol. Bull.* **125,** 280.
LaChance, L. E., and Riemann, J. G. (1964). *Mutation Res.* **1,** 318.
Ladd, T. L., Jr. (1966). *J. Econ. Entomol.* **59,** 422.
Lindquist, A. W. (1961). *J. Wash. Acad. Sci.* **51,** 109.
Lindquist, D. A., Gorzycki, L. J., Mayer, M. S., Scales, A. L., and Davich, T. B. (1964). *J. Econ. Entomol.* **57,** 745.
Loebbecke, E. A., and von Borstel, R. C. (1962). *Genetics* **47,** 853.
Meifert, D. W., Fye, R. L., and LaBrecque, G. C. (1963). *Florida Entomologist* **46,** 161.
Mitlin, N. (1956). *J. Econ. Entomol.* **49,** 683.
Mitlin, N., and Baroody, A. M. (1958a). *J. Econ. Entomol.* **51,** 384.
Mitlin, N., and Baroody, A. M. (1958b). *Cancer Res.* **18,** 708.
Mitlin, N., Konecky, M. S., and Piquett, P. G. (1954). *J. Econ. Entomol.* **47,** 932.
Mitlin, N., Butt, B. A., and Shortino, T. J. (1957). *Physiol. Zool.* **30,** 133.
Morgan, P. B., and LaBrecque, G. C. (1962). *J. Econ. Entomol.* **55,** 626.
Morgan, P. B., and LaBrecque, G. C. (1964). *J. Econ. Entomol.* **57,** 896.
Mukherjee, M. C. (1961). *Sci. Cult.* (*Calcutta*) **27,** 497.
Murvosh, C. M., LaBrecque, G. C., and Smith, C. N. (1964). *J. Econ. Entomol.* **57,** 89.
Ouye, M. T., Graham, H. M., Garcia, R. S., and Martin, D. F. (1965). *J. Econ. Entomol.* **58,** 927.
Painter, R. R., and Kilgore, W. W. (1964a). *J. Econ. Entomol.* **57,** 154.
Painter, R. R., and Kilgore, W. W. (1964b). *Bull. Entomol. Soc. Am.* **10,** 175.
Painter, R. R., and Kilgore, W. W. (1965). *J. Econ. Entomol.* **58,** 888.
Parish, J. C., and Arthur, B. W. (1965a). *J. Econ. Entomol.* **58,** 699.
Parish, J. C., and Arthur, B. W. (1965b). *J. Econ. Entomol.* **58,** 976.
Piquett, P. G., and Keller, J. C. (1962). *J. Econ. Entomol.* **55,** 261.
Plapp, F. W., Jr., Bigley, W. S., Chapman, G. A., and Eddy, G. W. (1962). *J. Econ. Entomol.* **55,** 607.
Purdom, C. E. (1960). *Biochem. Pharmacol.* **5,** 206.

Rai, K. S. (1964a). *Cytologia (Tokyo)* **29,** 346.
Rai, K. S. (1964b). *Biol. Bull.* **127,** 119.
Rapoport, I. A. (1947). *Byul. Eksperim. Biol. i Med.* **23,** 198.
Ratcliffe, R. H., and Ristich, S. S. (1965). *J. Econ. Entomol.* **58,** 1079.
Ridgway, R. L., Gorzycki, L. J., and Lindquist, D. A. (1966). *J. Econ. Entomol.* **59,** 143.
Ristich, S. S., Ratcliffe, R. H., and Perlman, D. (1965). *J. Econ. Entomol.* **58,** 929.
Roach, S. H., and Buxton, J. A. (1965). *J. Econ. Entomol.* **58,** 802.
Roehrborn, G. (1959). *Z. Vererbungslehre* **90,** 457
Roehrborn, G. (1962). *Z. Vererbungslehre* **93,** 1.
Schmidt, C. H., Dame, D. A., and Weidhaas, D. E. (1964). *J. Econ. Entomol.* **57,** 753.
Shaw, J. G., and Riviello, M. S. (1962a). *Science* **137,** 754.
Shaw, J. G., and Riviello, M. S. (1962b). *Ciencia (Mex.)* **22,** 17.
Shaw, J. G., and Riviello, M. S. (1965). *J. Econ. Entomol.* **58,** 26.
Simkover, H. G. (1964). *J. Econ. Entomol.* **57,** 574.
Smith, F. F., Boswell, A. L., and Henneberry, T. J. (1965). *J. Econ. Entomol.* **58,** 98.
Sykes, M. P., Philips, S., and Karnofsky, D. A. (1956). *Med. Clin. N. Am.* **40,** 837.
Tenhet, J. N. (1947). *J. Econ. Entomol.* **40,** 910.
Weidhaas, D. E. (1962). *Nature* **195,** 786.
Weidhaas, D. E., Ford, H. R., Gahan, J. B., and Smith, C. N. (1961). *Proc. Ann. Meeting New Jersey Mosquito Exterm. Assoc.* **48,** 106.
Woods, C. W., and Bořkovec, A. B. (1965). *Am. Chem. Soc. 150th Meeting,* Paper 54, p. 21A.
Yeoman, G. H., and Warren, B. C. (1965). *Vet. Record* **77,** 922.
Young, J. R., and Cox, H. C. (1965). *J. Econ. Entomol.* **58,** 883.

6 PHEROMONES

H. H. Shorey and L. K. Gaston

UNIVERSITY OF CALIFORNIA
RIVERSIDE, CALIFORNIA

I. INTRODUCTION

Some form of communication is essential to the survival and propagation of most insect species. This need for communication is obvious among those groups, such as bees, ants, and termites, that live in social colonies with a division of labor. Less obvious is the fact that most insects, being bisexual, must communicate with each other in order to come together for mating. And still less obvious, but probably equally essential to the survival of the species involved, is the communication that occurs among certain insects causing them to aggregate for mutual defense (Eisner and Kafatos, 1962) or at a suitable breeding site (Section III,B).

The communication systems of insects seem to be particularly vulnerable to our attempts to control those insects by behavioral means. Wright (1964c) has coined the term "metarchon" to represent external stimuli artificially introduced into the environment of an organism for the purpose of modifying its behavior by elicitation of an inappropriate response or inhibition of an appropriate one. The possible approaches to behavioral control include the use of physical as well as chemical stimuli.* Among chemical stimuli are included behavior-inducing chemicals emanating from suitable food sources and oviposition sources in the environment, and those chemicals released by other members of the same species that promote aggregation, mating, or other behavior. This chapter is restricted to a consideration of the possibilities of insect control afforded by manipulation of the chemical communication which causes aggregation among members of a given species.

Biological and chemical characteristics of pheromones will not be discussed in this chapter, except as they relate in an immediate way to the problem of designing behavioral control programs. For a more general treatment of pheromones, the reader is referred to a recent book by Jacobson (1965) and to reviews by Wilson and Bossert (1963) and Karlson and Butenandt (1959).

II. PRINCIPLES OF BEHAVIORAL CONTROL

The behavioral control approach is based on the premise that insect behavior is not reasoned. Instead, the behavior displayed following reception of a particular stimulus is determined by the genetic makeup of the given species. This behavior often appears reasoned because it is so appropriate to the given circumstances. Even though the basic behavior patterns are rigid and instinctive, the timing of the patterns and the manner in which the organism displays them is modified according to the state of a large number of internal (physiological) and external (environmental) factors. So little is presently known about the factors that control and modify insect behavior that the behavior appears plastic and reasoned, even though it is not.

Therefore, although disguised by an appearance of great plasticity, the basic communication behavior that is essential to the survival of a given species is genetically determined, rigid, and instinctive. If the researcher can find a way to block or direct these essential behavior patterns—and if the species does not have a genetic constitution enabling it to substitute another behavior pattern for the blocked ones—then we have a possible means for population control of the species.

* Physical stimuli are divided into two categories: (1) *vibrations* propagated through a medium (a portion of these frequencies are called sound by man); and (2) *electromagnetic radiations* propagated at the speed of light and denoted according to their wavelengths as x-rays, ultraviolet, visible, infrared, micro, or radiowaves. Chemical stimuli are divided into two categories: (1) *odorous* compounds, which travel from the source to the receptor in the vapor state; and (2) *gustatory* compounds, which travel from the source to the receptor in solution.

Often more than one pathway of communication is open to a given species. Thus, in one communication sequence between the sexes of *Drosophila melanogaster* (Meigen) prior to mating, a variety of physical, chemical (odorous and gustatory), and visual stimuli may be employed (Spieth, 1952). The nature of these stimuli is still poorly understood. However, an important phenomenon is recognized: depending upon the level of excitation of the insects, certain of the stimuli can be reduced in intensity or omitted, but the total of the communication sequence still results in copulation between the partners. This occurs because, in the absence of certain stimuli, other of the stimuli operating at a higher level or for a longer period can embody the same final message of a suitable mate. This is the principle of the summation of heterogeneous stimuli first proposed by Seitz (1940).

Because of summation of heterogeneous stimuli, if two or more stimuli are capable of causing a given behavior, and if we only block or control one of the stimuli, the other(s) may be (1) expressed at a higher level (more intensely) and (2) selected for genetically, resulting in increasing resistance of the species to behavioral control. A potential danger in designing a behavioral control program is that only a single communication stimulus might appear to control a particular behavior, but a second, perhaps unexpressed, avenue may be available in the genetic makeup of the species. In many mosquito species, the obvious communication from female to male within the swarm is by wingbeat-produced sound (Wishart and Riordan, 1959). Kliewer *et al.* (1966) have determined that a primary means of communication from females, stimulating mating behavior in males of *Culiseta inornata,* a nonswarming mosquito that mates on the substrate, is by odorous chemicals. They suggested that it is unlikely that chemical communication is limited to one or a few species of mosquitoes. It is more reasonable to assume that some species are quite dependent on this means of communication, while in others such a mechanism would be of less or perhaps no importance. If this were so, the blocking of sonic communication in certain species in which this was the only obvious means of communication between the sexes would not cause a lasting population control, but would instead apply pressure, selecting odorous chemical communication as a principal, rather than minor, avenue.

III. INSECT COMMUNICATION BY PHEROMONES

The term "pheromone" was coined by Karlson and Butenandt (1959) to represent those chemicals that are secreted into the external environment by an animal and that elicit a specific reaction in a receiving individual of the same species. These chemicals have been referred to also as "ectohormones." Wilson (1963) divided pheromones into two general classes depending on their modes of action: (1) one which gives a releaser effect: this entails a more or less immediate and reversible change in the behavior of the recipient; and (2) one which gives a primer effect: this triggers a chain of physiological events in the receiving organism. The latter class, which generally operates through gustatory stimulation by the chemicals and controls such phenomena as caste determination and reproductive control in social Hymenoptera (ants and bees) and Isoptera (termites), will not be considered further here.

Behavior-releasing pheromones are typically odorous and act directly on the central nervous system of the recipient. General types of behavior that may be released include trail following, alarm, sexual activity, aggregation, dispersal, and territoriality (Wilson and Bossert, 1963). The restriction of this chapter to a consideration of aggregation-releasing pheromones does not imply that behavioral control with other classes of pheromones is not possible. However, most speculation to date on behavioral control possibilities with pheromones has centered on the manipulation of that communication which causes insects to aggregate for mating, feeding, or oviposition. Two classes of pheromones that promote aggregation will be considered: (1) sex pheromones, which are released by one sex only and trigger behavior patterns in the other sex that facilitate mating; and (2) general aggregation pheromones, which may be released by only one sex but cause approach responses by individuals of both sexes of the species.

A. Sex Pheromones

Sex pheromones are commonly referred to in the literature as "sex attractants" or "sex lures." These latter terms are misleading in that they imply that the odorous chemicals cause attraction alone. This is a great oversimplification. The responses of the male of the silkworm moth *Bombyx mori* (L.) to the sex pheromone released by the female consist of a fixed chain or sequence of several components that replace one another or are added in a fixed order (Schwinck, 1955a). Immediately above the threshold for stimulation, antennal movements and wing vibration may be released; if the concentration of odorous molecules is not increased, no further behavior patterns may occur. At a higher concentration of pheromone, an oriented approach to the pheromone source is released. This orientation behavior occurs on a level at which, or before which, it would appear possible to obtain behavioral control of the population, and will be considered in greater detail below. If a sufficiently higher level of pheromone stimulation is reached, mating behavior—

including clasper extension and attempts to copulate with nearby objects —is released. This hierarchy of behavior patterns with increasing stimulation by sex pheromones appears to be general. It is evident that male responses to the odorous chemical released by the female are considerably more complex than simple attraction to the odor source.

With few exceptions, when sex pheromones are used by a species to cause orientation and approach of a sexual partner from a distance, the pheromone is released by the female and guides the male. This makes the design of behavioral control programs utilizing the pheromone difficult, or at least indirect. The female is the sex that directly propagates the species, and the laying of fertile eggs by the female must be prevented. With a control program directed against the pheromone receivers, the males, a large portion of males might have to be neutralized before any reduction in female fecundity were noticed. This would depend in large measure on the mating frequency of the males.

Sex pheromones are produced by males of certain insect species; in most cases these serve as short-range mating stimulants after the individuals of opposite sexes have come into close proximity (refs. in Jacobson, 1965). At close range, we feel pessimistic about the chances of obtaining effective behavioral control, since other communication stimuli may substitute for the pheromone (principle of summation of heterogeneous stimuli). An exception to this is the boll weevil, *Anthonomus grandis* Boheman, the males of which release a pheromone that causes the females to approach from some distance (Keller *et al.,* 1964).

1. *Chemistry*

a. Naturally Occurring Compounds. During the past 10 years, microchemical analytical techniques have advanced to a level of sophistication making possible the isolation and identification of insect sex pheromones. The identified sex pheromones are:

Silkworm: *trans*-10-*cis*-12-hexadecadienol (I) (Butenandt *et al.,* 1959, 1961a,b).

$$CH_3(CH_2)_2CH{=}CHCH{=}CH(CH_2)_8CH_2OH$$

(I)

Gypsy moth, *Porthetria dispar* (L): 10-acetoxy-*cis*-7-hexadecenol (II) (Jacobson *et al.,* 1960).

$$CH_3(CH_2)_5\underset{\substack{|\\ O-\underset{\substack{\|\\ O}}{C}CH_3}}{C}HCH_2CH{=}CH(CH_2)_5CH_2OH$$

(II)

Pink bollworm, *Pectinophora gossypiella* (Saunders): 10-propyl-*trans*-5,9-tridecadienyl acetate (III) (Jones *et al.,* 1966).

$$[CH_3(CH_2)_2]_2C{=}CH(CH_2)_2CH{=}CH(CH_2)_3CH_2O\underset{\underset{O}{\|}}{C}CH_3$$

(III)

Cabbage looper, *Trichoplusia ni* (Hübner): *cis*-7-dodecenyl acetate (IV) (Berger, 1966).

$$CH_3(CH_2)_3CH{=}CH(CH_2)_5CH_2O\underset{\underset{O}{\|}}{C}CH_3$$

(IV)

Honeybee queen, *Apis mellifera* L.: 9-keto-*trans*-2-decenoic acid (V) (Gary, 1962).

$$CH_3\underset{\underset{O}{\|}}{C}(CH_2)_5CH{=}CHCOOH$$

(V)

The supposed identification of the female sex pheromone of the American cockroach, *Periplaneta americana* (L.), by Jacobson *et al.* (1963) was proved to be incorrect by Day and Whiting (1964).

With a partial exception of the honeybee queen sex pheromone (V), the isolation, identification, synthesis, and demonstration of biological activity of the five insect sex pheromones explored to date does not support a major thesis of Wright (1964b, 1965a)—that, probably, each pheromone is not a specific chemical. He suggested that the finding of a single compound as the natural lure, as in the gypsy moth, is unusual, and that most insects probably generate an attractive scent that requires the simultaneous presence of several different molecules for activity. Further, he chastizes recent writers who have apparently taken it for granted that female insects can attract their mates by secreting some one, unique, attractive chemical, or just possibly a family of related chemicals, each of which can be attractive all by itself.

Wright (1964b) stated that crude extracts of female insects have often been found to be highly active, but that the activity disappeared when the material was purified in an attempt to isolate the attractive compound. We are not familiar with such loss of activity for sex pheromones, except in the case of the queen honeybee, in which activity has been reduced, *but not lost,* when compound (V) was isolated from other materials contained in queen mandibular gland extracts (Gary, 1962). However, a number of reports exist to the contrary; certain purified pheromones lose activity when recombined with other materials in the crude extract (Section V,B,2).

b. Other Compounds Having Sex Pheromone Activity. Once a sex pheromone is identified, a great deal of interest is centered on the synthesis and testing of biological activity of closely related compounds. This is done for a variety of reasons, including a desire to elucidate the mechanism of olfaction in insects through comparative studies of chemical structure and biological activity, and an attempt to find an inexpensive compound having pheromone activity that can be used in insect control programs. The search for related compounds has proved fruitful in relation to the gypsy moth sex pheromone (II). Jacobson (1960) and Jacobson and Jones (1962) synthesized a potent homolog differing from the natural pheromone only in having two more carbon atoms in the skeleton. This material, prepared from ricinoleyl alcohol, is called "gyplure" (VI):

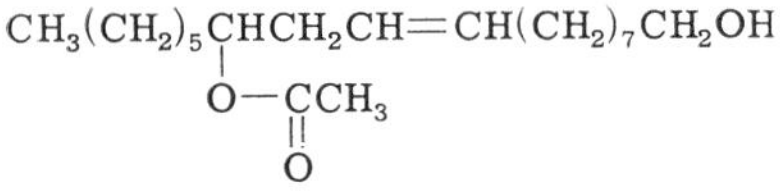

$$\begin{array}{l} CH_3(CH_2)_5\underset{\displaystyle O-\underset{\displaystyle \overset{\|}{O}}{C}CH_3}{\underset{|}{C}}HCH_2CH{=}CH(CH_2)_7CH_2OH \end{array}$$

(VI)

Gyplure is much less expensive to synthesize than the natural pheromone. However, it has not been reported whether the price differential is sufficient to make up for the fact that gyplure is 100-fold less active than the natural pheromone when tested for catches of males in field traps (Jacobson and Jones, 1962). In general, minor modifications of the structure of the silkworm moth or gypsy moth sex pheromones result in complete or almost complete loss of biological activity. These studies are reviewed by Jacobson (1965).

2. *Species Specificity*

It has been commonly assumed that an important characteristic of insect sex pheromones is specificity of biological activity, with males of other species not responding to the pheromone emitted from the female of a given species. This assumption of specificity is usually referred to in reviews as one of the principal characteristics of insect sex pheromones (Jacobson and Beroza, 1964; Wright, 1965a). Such an assumption has appeal because reproductive isolation between closely related species would be most efficiently achieved by isolating mechanisms operating at the sensory and behavioral levels, preventing the two sexes of the related species from approaching each other for mating. Despite these assertions, the literature contains numerous records of a lack of species specificity, and even genus specificity, among insect sex pheromones (Barth, 1937; Görnitz, 1949; Götz, 1951; Kettlewell, 1942, 1943, 1946; Rau and Rau, 1929; Schneider, 1962; Schwinck, 1953, 1955b; Shorey *et al.*, 1965). As early as 1942, Kettlewell (1942) stated: "There is a long list of insects whose specific scents attract a second species, either nearly related or otherwise . . . ." Schneider (1962), using electrophysiological and behavioral assay techniques, found cross-responsiveness of males of seven species of Saturniidae (Lepidoptera) to the pheromones produced by the females, and concluded that chemical differences in pheromones among those species must be slight. Following quantitative gas-chromatographic and behavioral assays of pheromones in four species of Noctuidae (Lepidoptera), Shorey *et al.* (1965) found no behavioral specificity for the female sex pheromones of the cabbage looper versus the alfalfa looper, *Autographa californica* (Speyer), and the bollworm, *Heliothis zea* (Boddie), versus the tobacco budworm, *H. virescens* (F.).

It appears, then, that too little information is available to allow generalizations about sex pheromone specificity. Specificity cannot be assumed when a trap, baited with the female sex pheromone of a given species and placed in the field, catches males of that species alone. An equally logical assumption would be that closely related males that might respond to the pheromone were not present in the same area and at the same time of year as the species in question. Also, differing host specificities, environmental aggregation stimuli, and specific times of day at which mating occurs may serve to maintain species in isolation from one another (Alexander, 1964).

The nonspecificity among sex pheromones of certain related insect species may be advantageous in designing behavioral insect-control programs. Pheromones utilized for control may be directed against a group of pests, with each group comprising those species that respond to each other's pheromones.

3. *Environmental and Physiological Considerations*

Peak breeding periods of certain insect species having one generation per year may be adjusted so that sexual activity occurs during a period of only a few weeks. Other species, having shorter life cycles and overlapping generations, may breed and utilize sex pheromone communication throughout most of the year, depending on the suitability of climatic conditions. The time of day when females release sex pheromones and when the males respond to the pheromones may be very specific for each species (Casida *et al.*, 1963; Götz, 1951; Kettlewell, 1946; Shorey and Gaston, 1965a; Shorey, 1966). Reproducible times of day and times of year when mating (and thus sex pheromone behavior) takes place are important considerations, because behavioral control programs might have to be in effect only during those times. The importance of other environmental and physiological factors that influence sex pheromone behavior should be determined for each species under consideration.

B. Aggregation Pheromones

Much interest has been centered recently on pheromones that are produced by the bark beetles and ambrosia beetles (Coleoptera, Scolytidae), which are the initial invaders of the phloem of suitable host trees and which cause an influx of males and females into the same area. For most species studied, the pheromone has been found to be produced by only one sex:

By females in:
- *Dendroctonus frontalis* Zimm. (Gara *et al.*, 1965)
- *D. ponderosae* Hopkins (Wood and Lanier, unpublished, cited by Wood *et al.*, 1966)
- *D. pseudotsugae* Hopkins (Rudinsky, 1963)
- *Scolytus quadrispinosus* Say (Goeden and Norris, 1964)
- *Trypodendron lineatum* (Oliv.) (Rudinsky and Daterman, 1964)

By males in:
- *Ips confusus* LeC. (Wood, 1962)
- *I. ponderosae* Sw. (Kliefoth *et al.*, 1964)

The producing sex has not been determined in *Ips avulsus* Eichh., *I. grandicollis* Eichh., and *I. calligraphus* Germ. (Vité *et al.*, 1964).

The general behavior, worked out for *I. confusus* (Wood *et al.*, 1966, and refs. therein), is summarized as follows: Certain males initially attack a suitable host tree. While burrowing in the phloem tissue, the male secretes pheromone into the lumen of the hindgut, where it becomes incorporated into the fecal pellets. Pheromone released from the pellets serves to orient flying populations of males and females to the vicinity. The females are stimulated to enter the galleries of pheromone-producing males, while stimulated males construct new galleries when they encounter the host tree. Because the release of pheromone by the males depends on defecation which, in turn, depends on the successful invasion of new host tissue, this phenomenon may be interpreted as an efficient survival mechanism, since it guides the population only toward breeding material which had already proved suitable (Pitman and Vité, 1963).

Although species specificity has been stated to be a characteristic of several bark beetle aggregation pheromones, we do not consider that specificity has been adequately proved. Vité *et al.* (1964) found evidence of varying degrees of lack of species specificity among the pheromones of several *Ips* and *Dendroctonus* species.

IV. INSECT ORIENTATION TO AN ODOR SOURCE

In attempting to design a behavioral control program to prevent or manipulate approaches of insects to a pheromone source, a central and critical question arises. This is, simply, to what does the insect orient when it moves from some distant point, whether meters or kilometers away, in order to arrive at the area of pheromone emission?

Early workers were greatly disturbed concerning the great distances over which certain insects appeared to communicate. Fabre became convinced that some stimulus other than odor must be involved (Teale, 1961). Such confusion and disbelief is reasonable, if one assumes that the only possible mode of orientation to odors is by sensing a concentration gradient of odor molecules and following the gradient from an area of low molecular density through an increasing density to the highest density at the source.

Concentration of molecules diffusing from a point source, in an expanding sphere, follows the inverse cube law. Orientation to increasing molecular concentration in an odor gradient is true orientation to a chemical (chemotaxis) and appears to be a rare phenomenon among insects (Jander, 1963). This method of orientation could not operate over more than a few centimeters, because an odor gradient in nature could not conceivably remain intact over a greater distance. Disruptive and turbulent airflows occur everywhere that a temperature differential exists among objects. In open air, it is air movement turbulence and not diffusion gradients that establishes the distribution of odor molecules (Bossert and Wilson, 1963).

A. Theories of Orientation

Several theories have been advanced to attempt to explain how an insect can orient to a pheromone source if it can not follow a uniform increase in molecular concentration.

1. *Positive Anemotaxis*

The view accepted by most workers is that insects orienting to an odor source follow the air currents bearing the odor until they reach the source. The orientation is directly to the air currents (positive anemotaxis) and not to the chemical. The pheromone serves as a releaser for the anemotactic behavior. In the absence of stimulation, when the insect has "lost" the odor-bearing air stream, anemotactic behavior ceases, and a random, crosswind flight behavior occurs, until the insect once more encounters the odorous air and again behaves anemotactically (Dethier, 1957).

Steiner (1953), after observing the behavior of *Geotrupes stercorarius* L. (Coleoptera, Scarabaeidae), a dung beetle, proposed that the method of orientation to dung from a distance is positive anemotaxis released by the appropriate odor.

Even the orientation of *Drosophila melanogaster* Meigen adults to food odors in "still" air was shown by Steiner (1954) to be actually accomplished by positive anemotaxis. He determined that the lower threshold for orientation by these insects to air currents occurs at an air velocity of 1.7 cm/second.

Classic studies by Schwinck (1954, 1955a, 1958) were conducted in an arena in which males of the silkworm moth were allowed to run free. A living female emitting pheromone was placed on one portion of the arena and air currents were directed from another portion. When stimulated by pheromone, the males ran toward the air source, not toward the female.

These studies are further supported by experimental results with cabbage looper males in wind tunnels having air velocities from about 4.5 to 100 cm/second. The proportion of males moving upwind increased when the air was impregnated with pheromone that had been extracted from female moths (Shorey and Gaston, 1965c).

The literature contains numerous additional observations supporting the view that a principal means of distance orientation of insects to pheromones is through a release of positive anemotaxis (Casida *et al.*, 1963; Holbrook *et al.*, 1960; Kettlewell, 1946, 1961; Wood, 1962; Vité *et al.*, 1964; review by Jacobson, 1965).

a. Mechanism of Anemotaxis. It has been generally acknowledged that day-flying insects can maintain an upwind orientation by visually sensing their direction of sideslip with respect to the substrate below them. A question has been raised as to how positive anemotaxis can operate in nocturnal insects that do not have visual contact with the ground (Wright, 1958). There are at least three possible mechanisms for anemotactic orientation in the dark. First, except under artificially controlled conditions, the environment is not totally dark, and enough light may still be available at night, especially in the sky, to allow visual stabilization of anemotaxis. Second, infrared radiation is sensed and oriented to, at least by the American cockroach (Miles, 1966), a buprestid beetle, *Melanophila acuminata* Degeer (Evans, 1964), and certain night-flying moths (Callahan, 1965d); the sensing of the apparent direction of movement of infrared "images" of the objects over which the flying insect passes may be effective in stabilizing upwind flight. Third, the ability to see the ground may not be necessary for many insects at all (Schwinck, 1958). Visual contact with the ground would be necessary to maintain anemotactic orientation in a smooth airflow, but there is no such thing in nature as a smooth airflow. The direction of movement of certain hair sensilla, out of phase with the direction of movement of the insect body, when the insect is in the usual disrupted, turbulent airflow may communicate wind direction. Orientation to air movement by this third method might help to explain the seemingly erratic, hesitant, zigzag approach of many insects orienting to an odor source.

2. *Filamentous Nature of the Odor Cloud*

Wright (1958) proposed an interesting theory of orientation to an odor source based on the fact that disruptive and shearing air flows create a nonuniform, filamentous odor cloud instead of a uniform distribution of odor molecules. He proposed that insects flying in the right direction, toward the odor source, receive sensory information as a series of pulses caused by passing through alternate high and low molecule density. As the insect penetrates toward the source, the average intervals between pulses will tend to be shorter, and the insect will keep a fixed flight path. With the odor absent or with pulses occurring at increasing intervals, a zigzag flight path would be manifested. This theory has, as yet, received no experimental support.

3. *Infrared Radiation*

Various workers have advanced theories of long-distance orientation of male moths to their potential mates by infrared radiation (Duane and Tyler, 1950; Laithwaite, 1960; Callahan, 1965a–d), usually because of the feeling that it is impossible for an insect to orient through a field of odor particles for any distance, and that the only possible means of communication which could remain effective in intensity over those distances is radiation. These infrared theories have been criticized on practical grounds by Dethier (1957) and Kettlewell (1961).

Callahan (1965d) has shown that male bollworm moths perceive and approach a source of far-infrared radiation in total darkness. Callahan proposes that in nature the males similarly orient to a receptive female. Since the female is warmed when she vibrates her wings, the peak wavelength of infrared radiation from the thorax is shifted upward from the peak wavelength of objects at ambient night temperatures. Callahan (1965a,b) also argues that the sex pheromones in nocturnal moths function only as an "identification substance" at short range. Our impression is that responses by certain insects to infrared radiation have been verified experimentally (Callahan, 1965d; Miles, 1966; Evans, 1964) and are probably involved in certain phases of nocturnal orientation. Since positive anemotaxis has been demonstrated so adequately (Section IV, A,1) to be a general phenomenon, we propose that this is the primary means of orientation to pheromone-emitting mating partners from a distance, and that infrared radiation may function as a means by which the insect can sense the terrain over which it navigates and by which short-range orientation and courtship behavior may be released.

B. Effective Distances for Orientation to Sex Pheromones

The statements in the literature concerning effective distances measured in kilometers for certain moth sex pheromones have been criticized (Dethier, 1957) for a very good reason. When a male moth is marked and released at a great distance from the pheromone source, there is no way to know what proportion of its flight back to the source might have consisted of a random, appetitive flight behavior, unstimulated by pheromone.

At present, there is no way to estimate maximum effective distances except on a theoretical basis. This has been done recently for the sex pheromone of the gypsy moth. After considering the dispersal rate of odor molecules in flowing, turbulent air, Bossert and Wilson (1963) calculated that, in a wind of 100 cm/second, males 4560-m downwind from the source could be activated. Behavioral thresholds for activation of males in winds of 300 and 500 cm/second were calculated to be 2420 and 1820 m, respectively. The active zone is in the shape of a long ellipsoid. These calculations were based on finding by Collins and Potts (1932) that a few males flew at least 2.3 miles to traps baited with 12–15 gypsy moth females. However, as indicated above, such determinations of great effective distances are subject to considerable doubt.

Effective distances for sex pheromones depend most on two biological parameters: (1) the threshold concentration for male stimulation, and (2) the release rate from females.

1. *Thresholds for Male Stimulation*

The lower thresholds of molecular concentration of pheromones needed to cause initial male activation have been established for several insect species by laboratory studies. Higher concentrations may (or may not) be required to cause positive anemotactic responses in the field.

Working with the cabbage looper, Shorey and Gaston (1964) impregnated filter paper with 10^{-5} ♀ equivalent (about 10^{-5} μg) and placed it in an airflow (Gaston and Shorey, 1964) of 5 liters/minute passing over male moths. This concentration was near the threshold for stimulation, with about one-half the males responding by vibrating their wings in 30 seconds, when physiological and environmental conditions were optimized. Release of the pheromone from filter paper occurred at a rate of about 8% per minute during the first several minutes following impregnation (Shorey and Gaston, 1965b). The threshold concentration for stimulation under the above conditions is calculated to be about 2×10^{-7} μg/liter of air.

There is no doubt that the most sensitive, specialized, insect olfactory cells, including those responding to sex pheromones, do work at behavior threshold concentrations like a molecule-counting device. Boeckh *et al.* (1965) calculated that, over a 30-second period, 1200 molecular impacts on antennal receptor cells are required to elicit a behavior response in males of the silkworm.

What appear to be impossibly low threshold concentrations of sex pheromone required for stimulation of silkworm moth males have been reported by Butenandt and Hecker (1961). They placed a glass stirring rod to a depth of 2 cm in a solution of 10^{-12} μg/ml of synthetic silkworm pheromone (I) in petroleum ether. When the rod was held 1 cm in front of male antennae, 50% of the males responded within 1–2 seconds. We calculate that, at most, 0.03 ml of solution remained on the rod; this volume would contain about 76 sex pheromone molecules, a very small proportion of which could volatilize in 2 seconds. The probability that even 1 volatilized molecule would impinge on a receptor site of an antenna of 50% of the males within that time is extremely low. Similar difficulties arise when one attempts to interpret the extremely low threshold concentration of 10^{-12} μg for the gypsy moth (Jacobson and Jones, 1962) and 10^{-14} μg for the American cockroach (Jacobson *et al.*, 1963). However, insufficient experimental data were given by Jacobson and co-workers to allow an evaluation of their results.

2. *Female Release Rates*

The mechanisms of synthesis, transport, and release of sex pheromones in and from their specialized glands are essentially unknown. Steinbrecht (1964) compared the release rate from living female silkworm moth glands with the rate from filter paper impregnated with the pheromone. The activity of the glands was so much greater than the filter paper extract that he concluded that most of the active pheromone must be situated on the gland surface ready for dissipation when the gland is everted. He calculated that more sex pheromone activity is produced by the living female than can be accounted for, even if all the pheromone is on the surface of the gland. Interestingly, Steinbrecht (1964) determined that the average female silkworm moth contains 1.5 μg of pheromone on the day of emergence. This is far greater than the approximately 10^{-2} μg/♀ calculated earlier on the basis of the isolation by Butenandt *et al.* (1959); the difference probably owes to the unavoidable losses entailed in purifying the pheromone.

The above confusing situation concerning the release rate of sex pheromone by the silkworm moth female has also been found to exist for the cabbage looper moth female. Shorey and Gaston (1965b) found two opposing results. Intact cabbage looper females released pheromone causing more male activity than did ether extracts, equivalent to 1 ♀, impregnated on filter paper. This should indicate a loss of at least 8% of the original pheromone content per minute from intact females. However, living females that had been releasing pheromone for as long as 30 minutes showed no reduction in the quantity of pheromone remaining in their abdomen tips. Reported pheromone concentrations in mature cabbage looper females vary from about 1 μg (Shorey and Gaston, 1965b) to about 2 μg (Berger, 1966).

Possible explanations for these inconsistencies suggested by Shorey and Gaston (1965b) are as follows:

1. A pheromone potentiator may be released from living females, but not from extract-impregnated filter papers.

2. Much of the pheromone may be stored as an inactive precursor, which is converted rapidly to active pheromone to replace that which is lost.*

3. In addition to the sex pheromone, a male inhibitor may be present in the ether extracts of female abdomen tips (Section V,B,2).

V. POTENTIAL USES OF PHEROMONES FOR INSECT CONTROL

Pheromones may be used in an insect control program in two possible ways: (1) population density surveys, or (2) direct behavioral control.

Pheromones have great potential usefulness in surveys to determine the presence or abundance of the species in question, so that other control

* This rate of synthesis is well within the known rates of enzyme reactions (Bergmeyer, 1963), which frequently are as high as 100–500 μmoles/minute/mg. Using the estimated pheromone release rate from a cabbage looper moth as 8%/minute and a pheromone concentration of 1 μg per gland, approximately 0.1 μg of pheromone may be released per minute. This corresponds to 4.4×10^{-10} moles/minute. The gland is approximately 800 μ in diameter and 40 μ thick. Assuming a density of 0.9 and an enzyme concentration of 0.01%, the calculated enzyme activity must be 220 μmoles/minute/mg to replenish the volatilized pheromone.

measures can then be exercised. The pheromones may be used to monitor the effectiveness of insect control programs, even when they are not used as a direct agent of control in themselves. These uses of sex pheromones have been reviewed by Jacobson (1965) and will not be considered further here.

Direct usage of pheromones for behavioral control is based on an intimate knowledge of the physiology of the insects and may be divided into two categories: (1) stimulation of behavior and (2) inhibition of behavior. The first category is based on the demonstrated ability of the pheromones to cause anemotactic orientation from a distance. By using the pheromone to cause approaches to a trap or other source selected by man, population control may be obtained. The second category involves permeation of the atmosphere with pheromones so that orientation of insects to the normal pheromone sources in nature is prevented.

The potential success of these methods is a matter of speculation, since to date no effective means of control of insect populations utilizing pheromones has been reported.

A. Causing Orientation

1. *Pheromone Alone as the Orientation Source*

a. Orientation to an Inappropriate Host. The bark beetle aggregation pheromones are typically produced only after the initial beetles have fed on an appropriate host tree, and they serve to orient large segments of the dispersing population to the same tree. This was interpreted earlier as an efficient survival mechanism, since under natural conditions the pheromone guides the population in flight only toward breeding material that proved suitable (Section III,B).

It is the presence of the pheromone, supplemented by visual stimuli from vertical objects, that promotes mass aggregation of the southern pine beetle, *Dendroctonus frontalis* (Gara *et al.,* 1965). In the presence of the pheromone, aggregation and attack by the beetles will occur even on a nonsuitable host. Gara *et al.* (1965) manipulated attacking populations of southern pine beetles by attaching recently infested logs to preselected trees. The flying beetle population concentrated on those trees, attacked them, and were killed because the trees were resistant. Gara *et al.* (1965) also demonstrated that flying southern pine beetles and beetles of *Ips* species could be induced to orient to, land on, and bore into a sweet gum (*Liquidambar styraciflus* L.) tree, an inappropriate host, when an air stream was directed against its trunk from a box containing infested logs.

Properly designed pheromone-emitting traps, incorporating the necessary visual stimuli, probably can be used as efficiently as resistant trees to destroy bark beetle populations.

b. Orientation to a Trap. The problems of developing trapping techniques for insect control will be discussed from the aspect of female sex pheromones, although many of the considerations apply to aggregation pheromones and male-produced sex pheromones also.

For many insect species, particularly in the Lepidoptera, destruction of males at a female sex pheromone-baited trap is the most obvious of the potential means for population control by pheromones, and superficially it appears very reasonable that the method should be successful. The pheromone source could consist of living virgin females, extracts from such females, or a synthetic compound. The principle is based on competition of pheromone from the trap with the pheromone from wild females. Enough males must orient to the trap and be destroyed to prevent most of the wild females from being inseminated. On the basis of theoretical population models, Knipling and McGuire (1966) have proposed that a trapping technique using pheromones could be effective when population levels of the pest species are very low and when a favorable ratio of female equivalents at the trap to wild females in the field can be displayed.

At the trap, males may be destroyed in a variety of ways. The most common technique for destruction used to date has been to coat the trap with a sticky material that ensnares and immobilizes the males on contact. These materials have some disadvantages, in that they may lose their sticky properties after continuous exposure to air or in cold weather, and they may become ineffective when coated with the bodies of trapped males (Jacobson, 1963; Holbrook *et al.,* 1960). Another method of male destruction at the trap is by means of insecticides. Graham and Martin (1963) found that calcium cyanide was not repellent to pink bollworm adults and, therefore, could be used effectively to kill males entering pheromone-baited traps. Other volatile or contact insecticides should be evaluated for similar uses.

Extensive trials using pheromone-baited traps have not demonstrated that this method can give control of populations of the nun moth, *Porthetria monacha* L., the gypsy moth, and the grapevine moth, *Lobesia botrana* (Schiff.) (reviewed by Jacobson, 1965). It is obvious from the failure of such experiments that the pheromone traps did not outcompete wild females. The conclusion to be drawn is that a great deal of basic biological study is necessary before one can intelligently devise a pheromone trapping program to successfully prevent insemination of wild females. Factors to be considered include: (1) an evaluation of sites in nature to which both sexes might orient and at which they might tend to congregate before pheromone communication occurs; (2) the range over which pheromone communication is effective; (3) circadian rhythms of female pheromone release and male pheromone responsiveness; (4) flight ranges of males and of mated females; (5) frequency of mating of males and females; (6) seasonal and geographic distribution of breeding populations of the species; and (7) interaction of the pheromone with other chemical or physical environmental stimuli that also affect orientation behavior (Section V,A,2).

A large source of difficulty in early attempts at this means of population control may have been that the traps were equipped with live females as the pheromone source. Females of a given species typically release pheromone at a characteristic time of day (Section III,A,3). Therefore, caged females would not be expected to release pheromone before wild females, and in their confined state they might not release pheromone until later than many wild females. This difficulty can be overcome with the development of synthetic pheromones, which may be used in two important ways to outcompete wild females. First is a timing advantage, whereby synthetic pheromones can orient responsive males before wild females begin to release pheromone. Second, depending on cost, synthetic pheromones can be used in large amounts, at hundreds or thousands of female equivalents per trap.

c. Orientation to a Sterilization Source. Depending on many variables, including the mating behavior of the insect species considered, sterilization of the males approaching a pheromone source may be a more effective means of population control than destruction of the males (Knipling, 1959, 1960). The males might be caused to contact a chemical sterilant displayed at the pheromone source and then be returned to the field. Subsequent matings of these males with normal females would result in the production of nonviable eggs.

Obviously, males sterilized in this manner must be capable of competing with nonsterile males. In particular, they must be able to display normal behavior patterns at the usual threshold level when subsequently exposed to the sex pheromone from wild females. An increase in the threshold concentration of pheromone necessary to induce orientation and mating behavior might cause a program of this type to be no more effective than if the males had been killed outright. Henneberry *et al.* (1966) determined the pheromone responsiveness of cabbage looper males following their exposure to the chemical sterilant TEPA. Males that had ingested TEPA required 10-fold more female sex pheromone to cause activation than untreated males. Males that had been sterilized by topical sprays of TEPA were as responsive to the pheromone as the controls.

2. *Pheromone Plus Light as the Orientation Source*

A number of researchers into the behavior of nocturnal Lepidoptera have noticed that light is more effective in orienting pheromone-stimulated males than is the pheromone source itself. The attempts of Fabre to study the nocturnal behavior of males of the great peacock moth, *Saturnia pyri* (Schiff.), in the presence of virgin females were frustrated because the males were diverted from their orientation toward the females by the flame of the candle or lamp that he used for illumination. He concluded that ". . . with creatures so madly enamoured of the radiant flame, precise and prolonged experiment becomes unfeasible the moment the observer requires an artificial illuminant" (Teale, 1961). Similarly, Rau and Rau (1929) noted that when males of certain species of Saturniid moths had been aroused to activity by the pheromone from the female, they reacted usually by flying to an incandescent light, even though they had to pass the cage containing females to do so.

Shorey and Gaston (1965c) constructed a flight tunnel in such a manner that males of the cabbage looper, stimulated to orient toward the source of a current of air containing female sex pheromone, moved in a direction opposite from that of males stimulated to orient toward an incandescent light. The light intensity, measured at the center of the tunnel, was roughly one-hundredth that of full moonlight. When the light and pheromone odor were present in the tunnel at the same time, male orientation toward the pheromone source was completely abolished, and most of the males congregated adjacent to the light source.

The biological significance of this dominance of light over pheromone source as an orienting mechanism for pheromone-stimulated moths is obscure. However, an application of the phenomenon has been demonstrated by Henneberry and Howland (1966). They captured more males of the cabbage looper in blacklight traps operated 20 ft from cages containing virgin females than in blacklight traps operated 1–2 miles distant from the females. When the cage of virgin females was placed on the light trap, the number of males captured in the light trap was increased by approximately 20-fold.

Howland (personal communication) speculated that, in a large enough area to prevent migrating moths from nullifying the results, one blacklight trap baited with 96 virgin females per 6 acres should remove enough males from the population that the majority of the females would remain unmated. The expected result would be adequate protection of the crop in the center of that area from caterpillar damage.

Obviously, in very large acreages it would become highly impractical to continue to use virgin females as the pheromone source. Therefore, the recent successful identification of the cabbage looper sex pheromone is very timely. Before the availability of synthetic pheromone, it was necessary to use living cabbage looper females rather than female pheromone-gland extracts because of the superiority of the former in stimulating male activity (Section IV,B,2).

B. Preventing Orientation

1. *Use of Pheromone*

a. Physiological Basis. A general phenomenon that has been noted during observations of male behavior in response to insect sex pheromones is adaptation. Adaptation is indicated if the response to a test stimulus is attenuated by a previous conditioning stimulus. It may occur at the receptor level, at which reactivity of the antennal sense cells is reduced following pheromone stimulation (Boeckh *et al.,* 1965), and it also undoubtedly occurs at higher nervous centers. The final result is that the threshold concentration of pheromone necessary to induce a response is higher for a time following a period of exposure.

It appears probable that it is adaptation that limits male moths to a comparatively short period of time whenever carrying out active searching for the female (Kettlewell, 1961). Similarly, the sexual activity of males of *Culiseta inornata* (Diptera, Culicidae) subsided over several minutes of pheromone exposure (Kliewer *et al.,* 1966). Renewed activity could not be stimulated by pheromone exposure until several hours later. Males of the cabbage looper cannot be used again as reliable bioassay indicators of pheromone concentration until several hours after a first exposure (Ignoffo *et al.,* 1963; Shorey *et al.,* 1964; Shorey and Gaston, 1964).

Since adaptation is a general, accepted phenomenon, how do insects maintain orientation to the odor source over a distance? How does odor stimulation maintain positive anemotaxis if the odor stimulation ceases through adaptation? One probable answer relates to the increasing concentration of odor molecules encountered by an insect as it moves toward the odor source. If the concentration increases at a sufficiently rapid rate to compensate for increasing adaptation, positive orientation to air currents may continue. Another possible means of reducing adaptation proposed by Wilson and Bossert (1963) with respect to ants following trails is the zigzag motion of the follower insect, causing it to move on and off the trail and thus varying the level of stimulation repeatedly.

b. Proposed Method of Control. A number of investigators have suggested recently that, if sufficient synthetic pheromone were spread over large areas, causing the air to be permeated to a sufficiently high level, the additional increment of pheromone contributed by wild females would be imperceptible to males and the males would thus never find and inseminate the females (Babson, 1963; Beroza and Jacobson, 1963; Wright, 1964a–c, 1965a,b). The proposed method has been called the "male confusion" technique. This term may be misleading, implying that males will be stimulated to activity by the pheromone, but will be incapable of orienting to normal females because pheromone is everywhere in the air. "Male inhibition" technique would be a more appropriate term, because, owing to the sensory and central nervous system adaptation discussed above (Section V,B,1,a), male responses to the synthetic and natural pheromone probably would become inhibited all together.

The amount of pheromone required for inhibition must be determined experimentally. Wright (1965a) proposed that a concentration of pheromone 10^5 higher than the threshold required to induce a response would be necessary to saturate the receptor organs. The economic feasibility of such a program will depend on two factors: (1) the cost and (2) the biological activity of the chemical.

c. Experimental Results. Intuitively, it appears very reasonable that because insects adapt so easily to pheromones, the male inhibition technique should be successful. However, the first experiment to demonstrate the feasibility of the technique failed. An attempt to control gypsy moth populations by broadcasting liquid and granular formulations of gyplure (VI) by air over a 400-acre island met with no success (Burgess, 1964). This was interpreted to owe to a masking effect of active *cis*-gyplure by a contamination with inactive *trans*-gyplure and possibly related by-products (Waters and Jacobson, 1965). This explanation does not appear reasonable, however, because such a masking agent should enhance rather than interfere with male inhibition (Section V,B,2). Details of the experiment were not given, and we might assume that since gyplure is 100-fold less active than the natural pheromone in the field (Section III,A,1,b), perhaps not enough material was volatilized into the air per unit of time to cause inhibition of male responsiveness to the natural pheromone from wild females.

2. *Use of Other Chemicals*

The structure of pheromones cannot be varied greatly before the biological activity for a given species is lost. However, in some cases, certain nonpheromone chemicals may be found that act on the antennal sense cells or the central nervous system to cause adaptation to the natural pheromone. This was demonstrated by Boeckh *et al.* (1965) with the honeybee queen pheromone (V) and caproic acid. Both chemicals stimulate the same antennal sense cells, although 10^4 more molecules of caproic acid than pheromone per unit volume of air are required to cause an equal intensity of response. Paired cross-stimulation with caproic acid first and pheromone second, or vice versa, shows the second response to be smaller than the first. Thus, the receptor's adaptation to one of the two substances is also effective for the other.

The following materials have been shown to cause loss of biological activity or "masking" of *cis*-gyplure: 20% or more "crude" *trans*-gyplure, 7% or more of ricinoleyl alcohol, and an inactive substance present in technical-grade methylene chloride (Jacobson, 1963; Waters and Jacobson, 1965). It would be extremely interesting to determine by what mechanism these materials block pheromone responses. The possibility that each chemical destroys the pheromone seems remote, so the only other apparent mechanisms would be adaptation, causing inhibition of responsiveness of gypsy moth males, or repellency, preventing males from entering gyplure-baited traps. If adaptation were the case, the contaminants alone, or even the impure mixtures, should be investigated for possible usefulness in preventing male orientation to normal females. In particular, ricinoleyl alcohol is an inexpensive material, and we wonder why its exciting potential for inhibition of gypsy moth males has not been fully investigated.

Casida *et al.* (1963) found that males of the introduced pine sawfly, *Diprion similis* (Hartig), did not orient to a variety of crude extracts of female sex pheromone. When certain of the extracts were purified by column or gas chromatography, the biological activity was restored. Jacobson and Smalls (1966) determined that an acetone extract of female American cockroaches did not cause male activity; silicic acid chromatography of the extract yielded an active fraction. Masking of activity in crude extracts of the tobacco budworm and cotton bollworm was suggested by Berger *et al.* (1965); however, this interpretation was contested by Shorey and Gaston (1967), who were unable to repeat the results of Berger and co-workers and found crude extracts from these species to possess biological activity.

The possibility of finding compounds (either structurally related or unrelated to natural pheromones) that, by adaptation of pheromone receptors or an effect on orientation mechanisms, prevent male responses to females in nature has great appeal.

VI. SUMMARY

The potentials for behavioral control of insect populations have been developed almost entirely on theoretical grounds. Much of the reason for this is that the tools have only recently become available to allow us to determine experimentally the feasibility of the proposed methods. Of the five known insect sex pheromones, only three are of economic pests, and the identification of two of these was reported in 1966. With recent technological advances in microanalytical chemistry, many more pheromones of the type that cause orientation behavior in receiving insects will undoubtedly be indentified in the next few years.

Others classes of pheromones, including those that stimulate alarm, trail following, territoriality, and dispersal, were not considered in this review. However, certain of these pheromones also may be good candidates for behavioral control studies.

Besides the impetus given by technological advances, behavioral control studies have been stimulated by the desire to develop alternative methods to supplement conventional means of insect control by chemicals. Pheromones have a number of appealing qualities. At the concentrations used, they will be very selective, probably having no biological effect on animals other than small groups of closely related target insects. For this reason, there probably will be no mammalian health hazards or residue problems. It does not appear that insects will develop a resistance to pheromones unless they have an alternate means of communication in their genetic repertoire. If an alternate means is available, it may be expressed at a higher level and selected for genetically, resulting in species resistance to the behavioral control.

Proposed methods of behavioral insect control have been divided into two categories: (1) those causing active orientation of the receiving insects to areas or devices where they can be sterilized or destroyed, and (2) those inhibiting active orientation by the receiving insects. When several different chemicals are known to cause or to inhibit pheromone responses of a given species, the best chemical to be used will be that for which the cost of production divided by the biological activity is a minimum.

REFERENCES

Alexander, R. D. (1964). *Symp. Roy. Entomol. Soc. London* **2,** 78.

Babson, A. L. (1963). *Science* **142,** 447.

Barth, R. (1937). *Zool. Jahrb. Abt. Allgem. Zool. Physiol. Tiere* **58,** 297.

Berger, R. S. (1966). *Ann. Entomol. Soc. Am.* **59,** 767.

Berger, R. S., McGough, J. M., and Martin, D. F. (1965). *J. Econ. Entomol.* **58,** 1023.

Bergmeyer, H. U. (1963). "Methods of Enzymatic Analysis." Academic Press, New York.

Beroza, M., and Jacobson, M. (1963). *World Rev. Pest Control* **2,** 36.

Boeckh, J., Kaissling, K. E., and Schneider, D. (1965). *Cold Spring Harbor Symp. Quant. Biol.* **30,** 263.

Bossert, W. H., and Wilson, E. O. (1963). *J. Theoret. Biol.* **5,** 443.

Burgess, E. D. (1964). *Science* **141,** 526.

Butenandt, A., and Hecker, E. (1961). *Angew. Chem.* **73,** 349.

Butenandt, A., Beckmann, R., Stamm, D., and Hecker, E. (1959). *Z. Naturforsch.* **14b,** 283.

Butenandt, A., Beckmann, R., and Hecker, E. (1961a). *Z. Physiol. Chem.* **324,** 71.

Butenandt, A., Beckmann, R., and Stamm, D. (1961b). *Z. Physiol. Chem.* **324,** 84.

Callahan, P. S. (1965a). *Ann. Entomol. Soc. Am.* **58,** 727.

Callahan, P. S. (1965b). *Ann. Entomol. Soc. Am.* **58,** 746.

Callahan, P. S. (1965c). *Proc. N. Central Branch Entomol. Soc. Am.* **20,** 20.

Callahan, P. S. (1965d). *Nature* **206,** 1172.

Casida, J. E., Coppel, H. C., and Watanabe, T. (1963). *J. Econ. Entomol.* **56,** 18.

Collins, C. W., and Potts, S. F. (1932). *U.S. Dept. Agr. Tech. Bull.* **336.**

Day, A. C., and Whiting, M. C. (1964). *Proc. Chem. Soc.* p. 368.

Dethier, V. G. (1957). *Surv. Biol. Progr.* **3,** 149.

Duane, J. P., and Tyler, J. E. (1950). *Interchem. Rev.* **9,** 25.

Eisner, T. E., and Kafatos, F. C. (1962). *Psyche* **69,** 53.

Evans, W. G. (1964). *Nature* **202,** 211.

Gara, R. I., Vité, J. P., and Cramer, H. H. (1965). *Contrib. Boyce Thompson Inst.* **23,** 55.

Gary, N. E. (1962). *Science* **136,** 773.

Gaston, L. K., and Shorey, H. H. (1964). *Ann. Entomol. Soc. Am.* **57,** 779.

Goeden, R. D., and Norris, D. M., Jr. (1964). *Ann. Entomol. Soc. Am.* **57,** 141.

Görnitz, K. (1949). *Anz. Schaedlingskunde* **22,** 145.

Götz, B. (1951). *Experientia* **7,** 406.

Graham, H. M., and Martin, D. F. (1963). *J. Econ. Entomol.* **56,** 901.

Henneberry, T. J., and Howland, A. F. (1966). *J. Econ. Entomol.* **59,** 623.

Henneberry, T. J., Shorey, H. H., and Kishaba, A. N. (1966). *J. Econ Entomol.* **59,** 573.

Holbrook, R. F., Beroza, M., and Burgess, E. D. (1960). *J. Econ. Entomol.* **53,** 751.

Ignoffo, C. M., Berger, R. S., Graham, H. M., and Martin, D. F. (1963). *Science* **141,** 902.

Jacobson, M. (1960). *J. Org. Chem.* **25,** 2074.

Jacobson, M. (1963). *Advan. Chem. Ser.* **41,** 1.

Jacobson, M. (1965). "Insect Sex Attractants." Wiley (Interscience), New York.

Jacobson, M., and Beroza, M. (1964). *Sci. Am.* **211,** 20.

Jacobson, M., and Jones, W. A. (1962). *J. Org. Chem.* **27,** 2523.

Jacobson, M., and Smalls, L. A. (1966). *J. Econ. Entomol.* **59,** 414.

Jacobson, M., Beroza, M., and Jones, W. A. (1960). *Science* **132,** 1011.

Jacobson, M., Beroza, M., and Yamamoto, R. T. (1963). *Science* **139,** 48.

Jander, R. (1963). *Ann. Rev. Entomol.* **8,** 95.

Jones, W. A., Jacobson, M., and Martin, D. F. (1966). *Science* **152,** 1516.

Karlson, P., and Butenandt, A. (1959). *Ann. Rev. Entomol.* **4,** 39.

Keller, J. C., Mitchell, E. B., McKibben, G., and Davich, T. B. (1964). *J. Econ. Entomol.* **57,** 609.

Kettlewell, H. B. D. (1942). *Entomol. Record* **54,** 62.

Kettlewell, H. B. D. (1943). *Entomol. Record* **55,** 107.

Kettlewell, H. B. D. (1946). *Entomologist* **79,** 8.

Kettlewell, H. B. D. (1961). *Entomologist* **94,** 59.

Kliefoth, R. A., Vité, J. P., and Pitman, G. B. (1964). *Contrib. Boyce Thompson Inst.* **22,** 283.

Kliewer, J. W., Miura, T., Husbands, R. C., and Hurst, C. H. (1966). *Ann. Entomol. Soc. Am.* **59,** 530.
Knipling, E. F. (1959). *Science* **130,** 902.
Knipling, E. F. (1960). *J. Econ. Entomol.* **53,** 415.
Knipling, E. F., and McGuire, J. U., Jr. (1966). *U.S. Dept. Agr. Inform. Bull.* **308,** 1.
Laithwaite, E. R. (1960). *Entomologist* **93,** 113, 133, 232.
Miles, W. R. (1966). *Science* **152,** 675.
Pitman, G. B., and Vité, J. P. (1963). *Contrib. Boyce Thompson Inst.* **22,** 221.
Rau, P., and Rau, N. L. (1929). *Trans. Acad. Sci. St. Louis* **26,** 83.
Rudinsky, J. A. (1963). *Contrib. Boyce Thompson Inst.* **22,** 23.
Rudinsky, J. A., and Daterman, G. E. (1964). *Z. Angew. Entomol.* **54,** 300.
Schneider, D. (1962). *J. Insect. Physiol.* **8,** 15.
Schwinck, I. (1953). *Z. Vergleich. Physiol.* **35,** 167.
Schwinck, I. (1954). *Z. Vergleich. Physiol.* **37,** 19.
Schwinck, I. (1955a). *Z. Vergleich. Physiol.* **37,** 439.
Schwinck, I. (1955b). *Z. Angew. Entomol.* **37,** 349.
Schwinck, I. (1958). *Proc. 10th Intern. Congr. Entomol., Montreal, 1956* **2,** 577.
Seitz, A. (1940). *Z. Tierpsychol.* **4,** 40.
Shorey, H. H. (1966). *Ann. Entomol. Soc. Am.* **59,** 502.
Shorey, H. H., and Gaston, L. K. (1964). *Ann. Entomol. Soc. Am.* **57,** 775.
Shorey, H. H., and Gaston, L. K. (1965a). *Ann. Entomol. Soc. Am.* **58,** 597.
Shorey, H. H., and Gaston, L. K. (1965b). *Ann. Entomol. Soc. Am.* **58,** 604.
Shorey, H. H., and Gaston, L. K. (1965c). *Ann. Entomol. Soc. Am.* **58,** 833.
Shorey, H. H., and Gaston, L. K. (1967). *Ann. Entomol. Soc. Am.* (in press).
Shorey, H. H., Gaston, L. K., and Fukuto, T. R. (1964). *J. Econ. Entomol.* **57,** 252.
Shorey, H. H., Gaston, L. K., and Roberts, J. S. (1965). *Ann. Entomol. Soc. Am.* **58,** 600.
Spieth, H. T. (1952). *Am. Museum Natl. Hist. Bull.* **99,** 397.
Steinbrecht, R. A. (1964). *Z. Vergleich. Physiol.* **48,** 341.
Steiner, G. (1953). *Naturwissenschaften* **40,** 514.
Steiner, G. (1954). *Naturwissenschaften* **41,** 287.
Teale, E. W. (1961). "The Insect World of J. Henri Fabre." Dodd, Mead, New York.
Vité, J. P., Gara, R. I., and von Scheller, H. D. (1964). *Contrib. Boyce Thompson Inst.* **22,** 461.
Waters, R. M., and Jacobson, M. (1965). *J. Econ. Entomol.* **58,** 370.
Wilson, E. O. (1963). *Sci. Am.* **208,** 100.
Wilson, E. O., and Bossert, W. H. (1963). *Recent Progr.* **19,** 673.
Wishart, G., and Riordan, D. F. (1959). *Can. Entomologist* **91,** 181.
Wood, D. L. (1962). *Pan-Pacific Entomologist* **38,** 141.
Wood, D. L., Browne, L. F., Silverstein, R. M., and Rodin, J. O. (1966). *J. Insect Physiol.* **12,** 523.
Wright, R. H. (1958). *Can. Entomologist* **90,** 81.
Wright, R. H. (1964a). *Science* **144,** 487.
Wright, R. H. (1964b). *Nature* **204,** 121.
Wright, R. H. (1964c). *Nature* **204,** 603.
Wright, R. H. (1965a). *Bull. At. Scientists* **21.**
Wright, R. H. (1965b). *Nature* **207,** 103.

7 REPELLENTS

Ruth R. Painter

AGRICULTURAL TOXICOLOGY AND RESIDUE RESEARCH LABORATORY
COLLEGE OF AGRICULTURE
UNIVERSITY OF CALIFORNIA
DAVIS, CALIFORNIA

I. PLACE IN ECOLOGY

When considering modern methods of pest control, the investigator and public alike often ignore insect repellents, the oldest and for many centuries the most widely used method of insect control. The successful use of DDT and other potent synthetic insecticides for controlling insect populations in both large and small areas and the ready availability of these products, especially in the "all purpose spray can for household use," undoubtedly has also led to a mistaken belief by many that repellents are no longer needed, Therefore, the use of repellents to control insects within a relatively small area seems of little importance. However, because of their past importance and continued widespread use by individuals, a short discussion of the present status of repellents is included.

Although one sees little large-scale advertising of repellents, the fact that they are readily available in drugstores, supermarkets, and sporting-goods stores shows that they are still in demand. It would be of interest to know if the sale of repellents has increased since the publication of Rachel Carson's "Silent Spring" (1962) and the more general awareness by the public of the dangers and problems in indiscriminate use of the potent synthetic organic insecticides. The development of resistance by insects to one compound after another in the last 10 years is another factor which has brought renewed efforts for the formulation of better repellents. From the standpoint of ecology repellents might well be the method of choice because the comfort and welfare of the subject is achieved while ecosystems are left relatively undisturbed (Egler, 1964; Rivnay, 1964; Tisdale, 1963).

II. WHAT IS A REPELLENT

The definition of a repellent as something that repels seems direct and straightforward; but it has proved to be ambiguous and at times even misleading. At best, the term is much less specific than the term attractant. Dethier (1947) has defined repellents as "those substances which as stimuli elicit avoiding reactions." He has further classified their action as physical or chemical.

A. Physical Repellents

Physical repellents include the usual swatting or pinching off of insects, and contact-stimuli repellents such as the surface texture of dusts, granules, water, oils, spines, or waxes. Visual and auditory repellents could also be included in this category. Visual stimuli are known to be attractants for insects, easily recognizable in the congregation of insects under street lamps. However, no reports on repellency brought about by visual stimulation of insects have been located in the literature. The yellow lamps for which commerical claims have been made offer merely minimal attraction (Hocking, 1963). Sound, on the other hand, can be either a repellent or attractant for insects. Amplified sound has been reported to be effective in repelling pyralid moths (Belton, 1962), sand fleas, and various species of mosquitoes (Kirkpatrick and Harein, 1965). Kirkpatrick and Harein investigated the possibility of using amplified sound to protect stored food from damage by the Indian meal moth (*Plodia interpunctella* Hübner).

B. Chemical Repellents

Chemical repellents are frequently confused with other chemical control agents. Much of the confusion with respect to whether a compound is a toxicant, antifeeding compound, arrestant, deterrant, and/or repellent can be avoided if chemical repellents are defined as chemicals which result primarily in the insect making directed movements away from the area. In an effort to define more precisely behavioral responses and the classification of compounds eliciting each responses, Dethier *et al.* (1960) proposed the scheme shown in the following tabulation:

Type of Chemicals in Terms of What They Do[a,b]

Movement		Example
(1) Stops or slows	Arrestant	Sugar, +odor
(2) Starts or speeds	Locomotor stimulant	DDT, pyrethrum
(3) Orients toward	Attractant	+Odor (geraniol)
(4) Orients away	Repellent	Odor (Indalone)[c]
Feeding, mating, and oviposition stimulant		
(1) Initiates or drives	Feeding or ovipositional stimulant	Sugar, +odor
(2) Inhibits	Deterrent	HCl, demissin, −odor

[a] Data from Dethier *et al.* (1960).

[b] To designate these actions the following terms are defined:

(1) *Arrestant.* A chemical which causes insects to aggregate in contact with it; the mechanism of aggregation is kinetic or has a kinetic component.

(2) *Locomotor stimulant.* A chemical which causes, by a kinetic mechanism, insects to disperse from a region more rapidly than if the area did not contain the chemical.

(3) *Attractant.* A chemical which causes insects to make oriented movements toward its source.

(4) *Repellent.* A chemical which causes insects to make oriented movements away from its source.

---

(1) *Feeding, mating, or ovipositional stimulant.* A chemical which elicits feeding or oviposition in insects. The term is synonymous with "phagostimulant," coined by Thorsteinson (1953, 1955).

(2) *Deterrent.* A Chemical which inhibits feeding or oviposition when present in a place where insects would, in its absence, feed or oviposit.

It is possible to make several generalizations about the use of these terms: (a) the same compound may have multiple effects on behavior; (b) any given effect may be elicited by chemicals which in other respects act differently; (c) movements involving orientation can be evoked by concentration gradients, or by currents carrying a chemical which may not necessarily be present in a gradient; there is also a theoretical possibility that solids dispersed in a gradient of density may elicit oriented responses.

[c] *n*-Butyl 6,6-dimethyl-5,6-dihydro-1,4-pyrone-2-carboxylate.

Although some of the definitions and generalizations might be debatable, the distinctions between repellents, arrestants, and deterrents should be recognized and used to distinguish the responses of insects to chemical stimuli. These responses are due to stimulation of the insect chemoreceptors.

1. *Chemoreception*

Chemoreception is the physiological process occurring in certain cells (chemoreceptors) as a result of their contact with certain chemicals (Hodgson, 1958). This process has been studied extensively in insects. Frings and Frings (1949) and Dethier and Chadwick (1948) reviewed the early empirical findings. During the next decade, basic research using electrophysical techniques contributed greatly to the understanding of the functions of the receptor cells. This work was reviewed by Hodgson in 1958 and again in 1964. He states that chemoreception triggers a variety of important behavior patterns, including feeding behavior, habitat selection, host–parasite responses and behavior integrating caste functions among social insects; chemoreceptors also mediate the response of insects to attractants, repellents, and some insecticides. The senses are classed (Dethier and Chadwick, 1948) as olfactory, gustatory, and the common chemical sense. Chemicals have likewise been categorized as olfactory and gustatory repellents, but because we are more concerned with the compounds causing stimulation than with insect chemoreception itself, we will use the classification of vapor and contact repellents (Dethier, 1953).

2. *Vapor Repellents*

Compounds which act in the vapor phase are most often stimuli of the olfactory receptors. But they may also act through the gustatory and common chemical sense, as demonstrated by Roys (1954) in a study on the effect of the vapor phase of compounds on the legs of cockroaches and by Slifer (1954) in a study with grasshoppers. The majority of synthetic chemicals found to be effective are repellent in the vapor phase. Since these compounds are usually detectable by other animals, including man, one of the most important criteria of an economically feasible chemical is that it be repellent to insects, while still acceptable to the animal species for whose protection it is intended. For the hunter—be he sportsman or soldier—the ideal repellent should be nondetectable by the hunted. One problem in the use of repellents that has received little attention is that the insect can become conditioned to the presence of the repellent, particularly in the vapor phase.

3. *Contact Repellents*

Contact repellents are those compounds with which the insect must come into direct contact and which act upon receptors not normally sensitive to vapors (Frings, 1946; Frings and Frings, 1949; Dethier and

Chadwick, 1950; Dethier, 1956). Such compounds frequently control feeding; the gustatory receptors on mouth parts and tarsi are generally involved. It is in this area that antifeeding compounds and stomach poisons are often confused with repellents. The proprietary substance Mitin FF* is considered to be a repellent for carpet beetles but is a stomach poison for clothes moths (Dethier, 1956). Most mothproofing preparations in general use are either insecticides or antifeeding compounds.

III. REPELLENTS—PAST AND PRESENT

Although no attempt will be made in this chapter to present a detailed history of the use of repellents and survey of the literature, the following sections describe briefly some of the historical background of their use and efforts to produce more efficient and longer-acting synthetic chemicals. More detailed information is available in "Chemical Insect Attractants and Repellents" (Dethier, 1947) and reviews by Dethier (1956), Busvine (1957), Shambaugh *et al.* (1957), and Vladimirova (1965). The natural repellents discharged by insects have been recently reviewed by Roth and Eisner (1962) and Jacobson (1966a,b).

A. The Use of Physical Repellents

Some of the oldest methods of repelling insects are still in use today. The simple flyswatter (where the fly is more often missed than hit), elaborate ceremonial feather dusters and fans in primitive cultures, and even horses' tails, are examples of physical means of repelling insects. The attendant depicted waving a large fan above the sleeping princess in ancient drawings obviously was using this method. This technique is still employed by mothers and nurses in various parts of the world.

While certain methods of physical control of insects cannot be classed as eliciting avoiding reactions, they are normally discussed under the heading of physical repellents. These include the mechanical removal by pinching, or picking off, of the undesired insect used by many animal species. Another means of mechanical control is the presence of a barrier, natural or man-made, between the insect and the attractive surface of host. Some plants have natural barriers, such as waxy or spiny coverings. Artificial barriers can be applied as dusts or oils. Roadside plants covered with dust are quite unattractive to insects (Dethier, 1947). Water barriers and tar or oil bands around tree trunks have been used for many centuries. Pimentel and Weiden (1959) recommend that the housewife employ the barrier technique using tightly sealed plastic coverings for the protection of stored woolens, rather than relying on so-called mothproof-

* *N*-3,4-dichlorophenyl *N'*-5-chloro-2-(2-sodium sulfonyl-4-chlorophenoxy)phenyl urea.

ing chemicals. Mesh screening has won popular acceptance and is so widely used in areas where it is readily available that its importance as a physical repellent to noxious insects is frequently overlooked in insect control programs.

B. Early Chemical Repellents

The first chemical repellent was undoubtedly discovered soon after man became acquainted with fire, when he found that a fire (especially a smoky, slow-burning fire) was a fairly efficient method of repelling insects. From smoke in general to smoke produced by burning certain plants or substances was a natural step. Campers, woodsmen, and even city patio enthusiasts still make extensive use of such fires and discuss at length the relative merits of various types of smudges or aromatic woods.

Probably the area most closely associated with folklore is the use of plants, plant extracts, or other substances because they are strong-smelling, odiferous, and pungent. These might be hung from the rafters, rubbed on the skin or clothing, or worn in a charm bag around the neck. Some are still in use today. Their use has led directly to the identification of several essential oils which are fairly efficient repellents, e.g., oil of citronella and oil of camphor. Included in the preparations which Howard recommended in 1911 was oil of citronella. Dethier (1956) states that oil of citronella was the most widely used mosquito repellent from about 1901 to 1938 and was the repellent against which new compounds were tested. Oil of citronella, which is extracted from *Andropogon nardus* (L.), contains geraniol as its primary component, with lesser amounts of citronellol, citronellal, boreneol, and terpenes. Citronellol and the corresponding aldehyde, citronellal, are considered to be the principal mosquito repellents of oil of citronella (Shambaugh *et al.*, 1957). On the other hand, geraniol has been identified as an attractant for the oriental fruit fly (Howlett, 1912, 1915, cited in Green *et al.*, 1960), and geraniol, citronellol, and citronellal for the Japanese beetle (Richmond, 1927). It should also be noted that certain compounds which are attractants at low concentrations are repellents at higher concentrations (Dethier, 1956).

Bordeaux mixture, a water suspension of copper sulfate and lime which has been used as a plant fungicide since about 1882, is often cited as the first synthetic chemical repellent in general use because agriculturists early observed that insects tended to avoid sprayed plants.

C. The Search for Synthetic Chemical Repellents

Research in chemical repellents for pest control between 1935 and 1955 was directed towards the production of synthetic chemicals for the protection of man, and to the screening of such compounds. Attempts were made to etablish reliable testing procedures so that the screening of compounds would be more meaningful. Although the majority of studies was directed towards the development of repellents for biting arthropods, some attention was given to the improvement of repellents for use against crawling and chewing pests.

Prior to World War II, four standard repellents for biting arthropods were in general use: the previously mentioned oil of citronella, dimethyl phthalate, Indalone, and Rutgers 612. Dimethyl phthlate, first patented in 1929 as a fly repellent (U.S. Patent 1,727,305), was among the compounds reported by Granett in 1942 (U.S. Patent 2,293,256) as insect repellents suitable for human use. Indalone (*n*-butyl 6,6-dimethyl-5,6-dihydro-1,4-pyrone-2-carboxylate) was patented as a repellent in 1937 (U.S. Patent 2,070,603). Rutgers 612 (2-ethyl-1,3-hexanediol) now called "612," was the most successful result of the extensive screening and testing program established by Granett in 1935, and a patent for its use was issued (U.S. Patent 2,407,205) (Granett and Hayes, 1945).

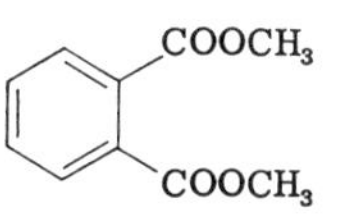

Dimethyl phthalate

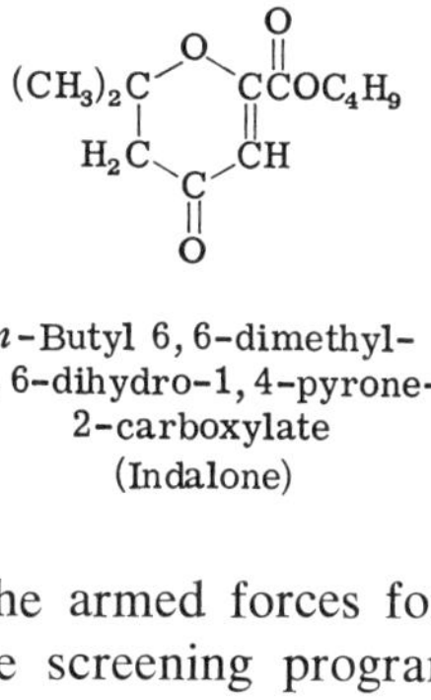

n-Butyl 6,6-dimethyl-5,6-dihydro-1,4-pyrone-2-carboxylate (Indalone)

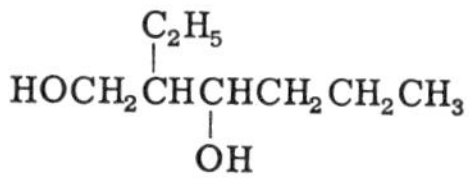

2-Ethyl-1,3-hexanediol

The requirements of the armed forces for repellents in World War II brought about large-scale screening programs of synthetic chemicals in the United States and Great Britain. The need was for more effective, longer-lasting, and cosmetically acceptable preparations for use on skin and clothing under tropical conditions of heat and sweat. Christophers (1947) reviewed the program carried on for the British Forces. King (1954) summarized the testing of some 11,000 compounds received by the Entomological Research Branch of the United States Dept. of Agriculture by 1952 (7000 prior to 1945, 4000 between 1945 and 1952). These compounds were obtained from three general sources: (1) Division of Insecticide Investigations of the Bureau of Entomology and Plant Quarantine; (2) other government agencies and universities under contract with the Office of Scientific Research and Development (OSRD); and (3) commercial sources. At first, the chemicals were tested for insecticidal and repellent action against body lice, mosquitoes, and chiggers.

In the later testing period, tests against houseflies, ticks, and fleas were included. The compounds were rated from 1 to 4A for effectiveness and only compounds of 3 or above were considered for further testing. Many of the promising repellents were found to attack one or more types of plastics and almost all had some effect on paint and varnish (Ihndris *et al.,* 1955).

The formulations recommended for use by the armed forces against a wide range of insects were (Anon., 1955):

For skin application:

1. M-250 or 6–2–2: 60% dimethyl phthalate, 20% 2-ethyl-1,3-hexanediol, 20% Indalone
2. 100% dimethyl phthalate
3. M-2020: 40% dimethyl phthalate, 30% 2-ethyl-1,3-hexanediol, 30% dimethyl carbate
4. M-2043 (improvement of M-2020): 40% dimethyl phthalate, 30% 2-ethyl-1,3-hexanediol, 30% propyl *N,N*-diethyl succinamate

For clothing treatment:

1. M-2086: 45% benzyl benzoate, 45% dibutyl phthalate, 10% emulsifier (Tween 80)
2. M-1960: 30% 2-butyl-2-ethyl-1,3-propandiol, 30% benzyl benzoate, 30% *N*-butylacetamide, 10% emulsifier (Tween 80)

Not all of the chemicals in the above formulations were made available for civilian use, but similar mixtures have proved useful.

Several thousand more chemicals were tested between 1952 and the date the wholesale screening program was phased out. Chemicals considered to have practical value as repellents summarized by Metcalf *et al.* (1962) are listed in Table I. Although several different compounds have been marketed for human use, the market now appears to be dominated by 2-ethyl-1,3-hexanediol (Rutgers 612) and *N,N*-diethyltoluamide. The latter compound, commonly called diethyltoluamide, is the best single repellent available at the present time (Hall *et al.,* 1957). It is effective against a wide range of insects, including mosquitoes, flies, chiggers, and biting flies, and was the result of the intensive research carried on by the Entomology Research Division of the United States Department of Agriculture in synthesizing and screening of potential repellents (Gilbert *et al.,* 1957a,b). Recently, the emphasis has been on new formulations to make the repellents more cosmetically acceptable and easier to apply. In addition to the original clear liquid (50%) solution of diethyltoluamide, the compound is now available in a cream case, a foam-type lotion, and

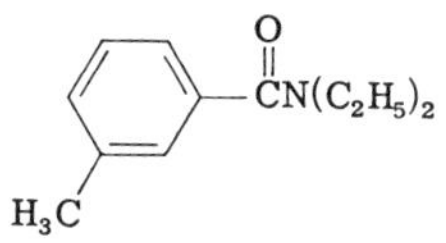

$$\text{H}_3\text{C}-\text{C}_6\text{H}_4-\overset{\text{O}}{\overset{\|}{\text{C}}}\text{N}(\text{C}_2\text{H}_5)_2$$

N,N-Diethyl-*m*-toluamide
(Diethyltoluamide, Delphene, or Deet)

a spray, and is effective when used on skin or clothing. Although they are easier to use, have a pleasant smell, and are not objectionable when applied to skin, repellents tend to dissolve paints (Ihndris *et al.,* 1955) and leave spots on some fabrics and paper. Many repellents are irritating to the mucous membranes (Hocking, 1963). One spray formulation has a particularly strong solvent action and dissolves paint, fingernail polish, and ball-point ink very readily. Many are also fairly combustible mixtures.

D. Repellents in the Bee Industry

The importance of repellents to the apiarist is well recognized. Smoke has long been used as a control measure, enabling the beekeeper to safely handle his bees. In this instance, smoke is a quieting influence as well as a repellent. This may well be an inherent reaction which has evolved through centuries of exposure to forest fires (Grout, 1963; Laidlaw, 1966).

Phenol has been used extensively to drive bees from the honeycomb at harvest time. In 1961 the Entomological Research Division of the USDA recommended that propionic acid be used because it is safer to handle than phenol (Anon., 1961). Townsend (1963) recommended benzaldehyde (artificial oil of almonds) for the same purpose. Woodrow *et al.* (1965) reports that propionic anhydride and propionic and acetic acids are all ideally suited for use as repellents to remove bees from the honeycomb because they are good repellents and do not contaminate the honey.

The loss of bees from insecticide poisoning has been a major problem of the bee industry since the advent of large-scale spraying with synthetic organic insecticides. Efforts to keep bees from potentially dangerous areas by incorporating a bee repellent in the insect control program have been tried repeatedly. However, no repellent effective enough to give dependable, long-lasting protection against the natural attractants has been reported. Recently, Woodrow *et al.* (1965) tested 195 compounds as bee attractants and repellents, using a modified olfactometer and field tests of the more promising compounds. Although 19 compounds were rated as moderately or strongly repellent by the olfactometer test, none

TABLE I

REPELLENTS FOR BLOODSUCKING INSECTS, MITES, AND TICKS[a]

Name	Formula	Properties	Toxicity, LD_{50} to rat, mg./kg.	Uses
Benzil	C_6H_5—C(=O)—C(=O)—C_6H_5	Solid, m.p. 95°C		Clothing impregnant for chiggers, mites
Benzyl benzoate	C_6H_5—C(=O)OCH_2—C_6H_5	Oily liquid, b.p. 323°C, m.p. 21°C, sp. gr. 1.12	1,900	Clothing impregnant for chiggers, mites
2,3,4,5-Bis-(Δ^2-butenylene)tetrahydrofurfural (MGK 11)	HC=CH, H_2C, CH, C, O, H_2 ring structure with HC=O	Liquid, sp. gr. 1.121	2,500	Repellent for cockroaches, biting flies on cattle
Butoxypolypropylene glycol	C_4H_9O—$(CH_2CHO)_n$—CH_2CH—OH with CH_3 side groups	Liquid, 400 and 800 molecular-weight fractions, d. 0.973-0.990	11,200	Fly repellent for cattle
N-Butylacetanilide	C_6H_5—N(C_4H_9)C(=O)CH_3	Liquid, b.p. 277-81°C, m.p. 22°C, sp. gr. 0.99	2,830	Clothing impregnant for ticks, fleas
n-Butyl-6,6-dimethyl-5,6-dihydro-1,4-pyrone-2-carboxylate (Indalone)	$(CH_3)_2C$, O, C(=O)OC_4H_9, H_2C, CH, C=O ring structure	Brownish liquid, b.p. 110-115°C/1mm., sp. gr. 1.06	7,800	Mosquito and fly repellent, used in mixtures
o-Chloro-*N*,*N*-diethylbenzamide	Cl—C_6H_4—C(=O)N$(C_2H_5)_2$			Mosquito repellent
Dibutyl adipate	$C_4H_9OC(=O)CH_2CH_2CH_2CH_2C(=O)OC_4H_9$	Liquid, b.p. 183°C/14 mm., sp. gr. 0.965	12,900	Tick repellent
Dibutyl phthalate	$C_6H_4(COOC_4H_9)_2$	Clear liquid, b.p. 340°C, sp. gr. 1.045	21,000	Clothing impregnant for chiggers, mites

TABLE I (*Continued*)

Name	Formula	Properties	Toxicity, LD_{50} to rat, (mg/kg)	Uses
Di-*n*-butyl succinate (Tabutrex)	$C_4H_9OC(=O)CH_2CH_2C(=O)OC_4H_9$	Liquid, b.p. 108°C/4 mm., m.p. -29°C	8,000	Repellent for cockroaches, biting flies on cattle
N,N-Diethyl-*m*-toluamide (Delphene, Deet)	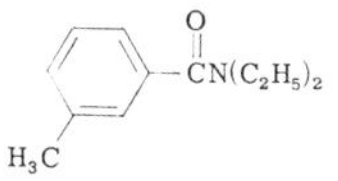 H_3C; $C(=O)N(C_2H_5)_2$	Clear liquid, b.p. 111°C/1mm., sp. gr. 0.996	2,000	General-purpose repellent
Dimethyl phthalate	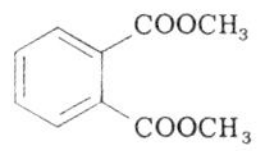 $COOCH_3$; $COOCH_3$	Clear liquid, b.p. 282°C, sp. gr. 1.189	8,200	General-purpose mosquito repellent
Di-*n*-propyl isocinchomeronate (MGK 326)	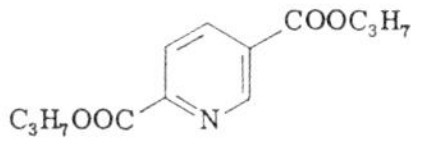 C_3H_7OOC; N; $COOC_3H_7$	Amber liquid, sp. gr. 1.08, b.p. 186°C/1mm.	6,230	Fly repellent for cattle
2-Ethyl-2-butyl-1,3-propanediol	$HOCH_2C(C_2H_5)(C_4H_9)CH_2OH$	Solid, m.p. 40-42°C, sp. gr. 0.931	5,040	General-purpose mosquito repellent
2-Ethyl-1,3-hexanediol (612)	$HOCH_2CH(C_2H_5)CH(OH)CH_2CH_2CH_3$	Clear liquid, b.p. 244°C, sp. gr. 0.94	2,400	General-purpose repellent for mosquitoes, flies, fleas, mites
5-Norbornene-*cis* 2,3 dimethylcarboxylate (Dimethyl carbate)	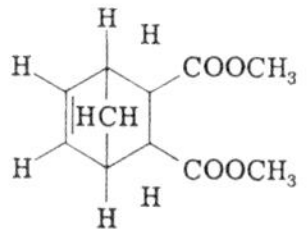 H; H; HCH; H; H; $COOCH_3$; $COOCH_3$; H; H	White solid, m.p. 40°C	1,000	Mosquito repellent used in mixtures
2-Phenylcyclohexanol	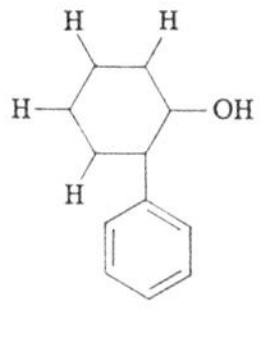 H; H; H; OH; H	Solid, m.p. 41°C		General-purpose repellent for mosquitoes, flies, mites, ticks
n-Propyl-*N,N*-diethylsuccinamate	$C_3H_7OC(=O)CH_2CH_2C(=O)N(C_2H_5)_2$	Sp. gr. 1.01	6,400	Mosquito repellent

[a] Data from "Destructive and Useful Insects" by C. L. Metcalf and W. P. Flint; rev. by R. L. Metcalf., McGraw-Hill Book Co., Inc., 1962. Used by permission of McGraw-Hill Book Co.

were recommended on the basis of the field experiments. However, the authors believe their work will be helpful in selecting and testing other formulations which may be of more practical use.

E. Present Areas of Research

1. *The Search for Oral or Systemic Repellents*

The possibility of an effective systemic repellent, preferably an oral repellent, is intriguing but the search for it has not yet been successful. Interest in oral repellents increased when thiamine chloride (vitamin B_1) was reported to be a repellent for mosquitoes and an aid in controlling itching when administered orally (Shannon, 1943). However, other workers (Wilson *et al.,* 1944; Lal *et al.,* 1963) have been unable to detect any repellent action for thiamine chloride. Kingscote (1958) introduced a wide variety of chemicals, including known repellents, vitamins, coffee, and many other types of organic chemicals into the stomachs of animals and men without preventing attacks by mosquitoes. Bar-Zeev and Smith (1959) found that in rabbits intervenous injections of diethyltoluamide did not repel *Aedes aegypti,* and they concluded that the compound was removed from the blood very rapidly after injection.

It would appear that if an effective oral repellent can be developed, it will have to be a compound not recognized as a repellent at the present time. Studies such as those of Khan and co-workers (1965) with mosquitoes, on the varying degrees of attractiveness of humans, and of Quigley (1965), on family differences in attractiveness of chickens to the chicken body louse, suggest that a difference in body chemistry may be responsible for the relative unattractiveness of some individuals and that a naturally occurring compound (or compounds) could be utilized as an oral repellent. In this connection, Khan and co-workers (Skinner *et al.,* 1965) also have reported that skin-surface lipids contain substances that are very repellent to female *Aedes aegypti.*

2. *Factors Influencing Repellency*

Recently, in an effort to better understand variations in repellency, some interesting experiments have been directed towards correlating the physical and physiological state of insects with their reaction to known repellents. Among other factors influencing repellency, Zolotarev and Gaverdovskii (1964) found that fleas are more sensitive to repellents in the spring than in the fall. Zolotarev and Elizarov (1964) stated that an increase in substrate temperature increased the sensitivity of mites to repellents.

3. *Improved Methods for Testing*

Research into better ways of testing compounds has included both (a) the development of improved olfactometers to determine the effectiveness of potential repellents (Section IV,C) and (b) the use of radioactive chemicals in the testing of repellency against the natural attraction of a host animal. Bar-Zeev and Schmidt (1959) used P^{32}-labeled phosphoric acid added to blood to study the probing action of mosquitoes in the presence of repellents. Lal *et al.* (1963) injected radioiodated serum albumen into repellent-treated mice and successfully used standard methods of scintillation counting to determine the number of mosquitoes biting the mice, thus measuring the repellency of the administered compounds.

4. *Repellency and Chemical Structure*

Although numerous attempts to explain repellency in terms of the physical and chemical properties or structure of known repellents have been made, no satisfactory correlation has been found. Dethier (1956) reviewed the literature and concluded that attempts to relate field performance to certain molecular characteristics were not fruitful. One could only say that repellents possess the proper balance of properties which make for inherent repellency.

An interesting theory has been advanced by Wright (Wright, 1956; Wright and Kellogg, 1962) that compounds which have a high infrared absorption (near 25 μ) are repellent to insects. However, since many compounds have this characteristic, it is difficult to believe that this is a primary criterion of repellency.

IV. SCREENING METHODS

A. Problems in Repellency Measurements

When efforts were made to set up standard testing procedures, many hitherto unrecognized variables were found to complicate the picture. A plea for well-designed experiments was voiced early by Christophers (1947). Dethier (1956) has an excellent discussion of the problems associated with repellency measurements. Among variables recognized at the present time are the concentration and toxicity of the compound itself, host attraction, physical and physiological condition of the test insect and of the host, and atmospheric conditions such as temperature, humidity, and wind velocity. Unfortunately, because such variables were unrecognized or hard to control, the literature on repellents is difficult to evaluate and the validity of many testing procedures is questionable.

Screening of potential compounds should logically proceed through four separate steps requiring the cooperation of several research disciplines, including such areas as biochemistry, chemistry, entomology, toxicology, and physiology. Consumer acceptance must also be considered. The four steps are: (1) laboratory screening for absolute repellancy involving olfactometer tests; (2) laboratory screening for "standard" or "host" repellency against natural or synthetic attractions similar to those against which the repellent would be competing in normal use; (3) screening of acceptability for use for which the repellent is intended; (4) screening under field conditions for final evaluation.

In the evaluation of repellency, clear definitions of different types of repellency measurements need to be added to our vocabulary. The following are proposed:

(1) Absolute or intrinsic repellency—repellent action measured in the absence of any attractant
(2) Standard or comparative repellency—repellent action measured against a standard quantity of a standard attractant
(3) Host repellency—repellent action measured against the attraction of the natural host

B. Tests against a Natural Host

While recognizing the difficulties in the evaluation of some studies because standardized methods of screening have not been adopted, the following section is included to direct the reader's attention to various types of methods used and to further sources of information. Shambaugh *et al.* (1957) discussed methods of testing repellents for biting arthropods, while Dethier (1956) considered the variables encountered in testing and the control and evaluation of variables. Busvine (1957) in his "A Critical Review of the Techniques for Testing Insecticides" has a chapter on the testing of insect repellents including cage tests against biting insects, tests against food, olfactometer tests, housefly bait tests, cockroach and tick repellent tests, and discussion of some of the variables involved. A detailed report of methods that have been used for screening repellents against mosquitoes, flies, fleas, ticks, chiggers, termites, bees, and stored-product damage by cockroaches, clothes moths, and silverfish has been compiled by Granett and Starnes (1960).

Tests for mosquito repellents usually consisted of applying the test compound to skin, or fabric over skin, and inserting the test areas into a cage of hungry female mosquitoes. Originally, human "volunteer" subjects were used almost exclusively and there were almost as many variations in quantity of repellent applied, method of application, and ways of measuring the repellent action as there were laboratories reporting. One of the first attempts to standardize testing was set up by Granett (1944) in which the amount of repellent applied, the skin area covered, and criterion of efficiency (time to first bite) was stated. The mass-screening program of the Entomological Research Branch of the USDA (King, 1954) used the paired-feeding technique in which the test chemical was applied to one limb and a chemical of known repellency was applied to the corresponding area of the other limb. Gilbert *et al.* (1957a,b) used a similar approach with a balanced, incomplete-block design, in which each compound was compared with all other compounds being evaluated.

Comparable results have been achieved when animals such as guinea pigs, mice, and rabbits have been the test subjects. The use of animals has enabled investigators to introduce refinements, such as that employed by Lal *et al.* (1963), in which a radioactive chemical was injected into a test animal and then the mosquitoes which had bitten the test animal were detected by counting the number of radioactive mosquitoes. Membranes over animal blood (Bar-Zeev and Smith, 1959) and over animal blood containing radioactive chemicals (Bar-Zeev and Schmidt, 1959) have been employed for laboratory testing.

Similar methods have been used for tests against fleas, ticks, and chiggers. The test compound is applied to one area of skin or stocking on one foot and an untreated or standard treated area marked out on the other foot. Both feet are then placed in pans containing the test insects and a visual count is made of the insects on each test area.

C. Olfactometer Studies

Insect response to vapors can be measured in an instrument known as an olfactometer. In its simplest form this instrument is a Y tube in which insects are given their choice of treated or untreated air. Originally, such instruments were used primarily for studying behavioral responses to chemical vapors. Modifications have made it possible to use these not only for qualitative repellency measurements, but also for quantitative measurements of the thresholds that insects can perceive. Dethier (1947) and Busvine (1957) reviewed the literature and described various designs that were useful in insect studies. Many recent modifications are based on the design of Willis (1947). Figure 1 is a schematic diagram of a modification designed by Wilson and Bean (1959) which was successfully used with several species of arthropods. Skinner *et al.* (1965) used a modified Willis olfactometer to determine the repellency of skin-surface lipids of humans to mosquitoes. The modified designs enable the investigator to determine the comparative repellency of compounds against specific attractants as well as the absolute (inherent) repellency. A simplified model

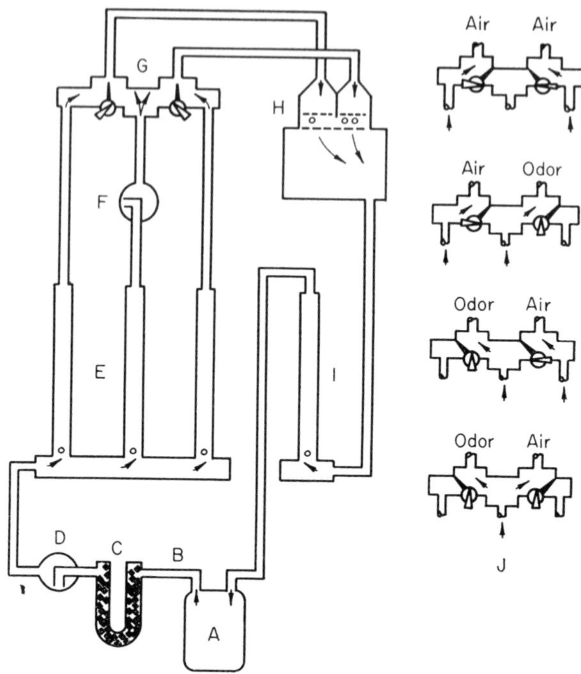

FIG. 1. Schematic diagram of olfactometer. A, continuous circulating air pump attached to ¼ horsepower electric motor; B, plastic (Pygon) tubing; C, U-tube with chunk bone-charcoal; D, humidity regulating vessel; E, influents flowmeter; F, odor vessel; G, valve mechanism for shunting airflow; H, observation chamber; I, exhaust flowmeter; J, diagrammatic valves showing possible airflow combinations (Wilson and Bean, 1959).

described by Howell and Goodhue (1965) was originally designed for use with cockroaches, but with slight modification proved useful for studies with *Drosophila,* honeybees, and houseflies.

V. GAS-CHROMATOGRAPHIC IDENTIFICATION OF INSECT REPELLENTS

One of the new scientific methods that has proved useful in the insecticide field is gas chromatography. Mixtures of 13 of the most important repellents have also been resolved on columns of Carbowax 20M, Tide, and Silicone 550, thus permitting their qualitative identification (Acree and Beroza, 1962; Bevenue, 1963). It would appear that quantitative methods using gas chromatography can be developed.

VI. THE FUTURE FOR REPELLENTS

In the present discussion we have indicated that there is a definite place for repellents in integrated pest-control programs. New scientific techniques should help with the screening of potential compounds. Except for the studies using radioactive tracers (Bar-Zeev and Schmidt, 1959; Lal *et al.,* 1963), all of the reports we found describing repellency measurements used either visual observation or simple manual photography to count the insects in the different test areas. Newer methods of photographic or electronic counting and recording can and should be adopted. Some scientists are working with these or equivalent methods which will increase the accuracy of repellency measurement (Kashin, 1966).

Repellents will probably be considered for greater use in ways other than the conventional application on the surface of the individual. The possibility of the use of space or area repellents was suggested as early as 1935 (Ginsburg), but has received little attention. Hocking (1963) in his discussion of the use of attractants and repellents in vector control, lists the possibility of space control of insects via air currents containing effective repellents which have been rejected because of irritation or other toxic effects when used on skin or clothing.

Renewed interest in the development of better repellents was stressed at the Symposium on Insects and Disease held during the 1965 convention of the American Medical Association (Smith, 1966). The need for more cosmetically acceptable and longer-lasting all-purpose repellents for clothing impregnation was emphasized (Gilbert, 1966; Gouck, 1966). The continuing search for oral repellents was presented by Sherman (1966), who also delineated the problems involved in finding an effective oral repellent.

In conclusion, we would like to suggest the possibility of greater use of repellents in integrated control programs of noxious insects. As already indicated, repellents might be used more widely for temporary control in limited areas, or they might be applied to some natural breeding grounds to direct insects to other areas for control by insecticides or chemosterilants. Maybe the "compleat" scientist, a disciple of Isaak Walton, who is trying to relax with a few days of trout fishing will come up with a solution because he is being driven from his favorable spot by a plague of mosquitoes or biting gnats.

REFERENCES

Acree, F., Jr., and Beroza, M. (1962). *J. Econ. Entomol.* **55,** 128.
Anonymous (1955). *U.S. Dept. Agr. Circ.* **977.**

Anonymous (1961). *U.S. Dept. Agr. Entomol. Res. Div. Corres. Aid* **33–16.**
Bar-Zeev, M., and Schmidt, C. H. (1959). *J. Econ. Entomol.* **52,** 268.
Bar-Zeev, M., and Smith, C. N. (1959). *J. Econ. Entomol.* **52,** 263.
Belton, P. (1962). *Nature* **196,** 1188.
Bevenue, A. (1963). *In* "Analytical Methods for Pesticides, Plant Growth Regulators and Food Additives" (G. Zweig, ed.), Vol. 1, pp. 189–225. Academic Press, New York.
Busvine, J. R. (1957). "A Critical Review of the Techniques for Testing Insecticides." Commonwealth Inst. Entomol., London.
Carson, R. (1962). "Silent Spring," Houghton, Boston, Massachusetts.
Christophers, S. R. (1947). *J. Hyg.* **45,** 176.
Dethier, V. G. (1947). "Chemical Insect Attractants and Repellents," McGraw-Hill (Blakiston), New York.
Dethier, V. G. (1953). *In* "Insect Physiology" (K. D. Roeder, ed.), pp. 544–576. Wiley, New York.
Dethier, V. G. (1956). *Ann. Rev. Entomol.* **1,** 181.
Dethier, V. G., and Chadwick, L. E. (1948). *Physiol. Rev.* **28,** 220.
Dethier, V. G., and Chadwick, L. E. (1950). *J. Gen. Physiol.* **35,** 589.
Dethier, V. G., Browne, L. B., and Smith, C. N. (1960). *J. Econ. Entomol.* **53,** 134.
Egler, F. E. (1964). *Am. Scientist* **52,** 110.
Frings, H. (1946). *J. Exptl. Zool.* **102,** 23.
Frings, H., and Frings, M. (1949). *Am. Midland Naturalist* **41,** 602.
Gilbert, I. H. (1966). *J. Am. Med. Assoc.* **196,** 253.
Gilbert, I. H., Gouck, H. K., and Smith, C. N. (1957a). *Soap Chem. Specialties* **33** (5), 115.
Gilbert, I. H., Gouck, H. K., and Smith, C. N. (1957b). *Soap Chem. Specialties* **33** (6), 95.
Ginsburg, J. M. (1935). *Science* **82,** 490.
Gouck, H. K. (1966). *Arch. Dermatol.* **93,** 112.
Granett, P. (1944). *Proc. New Jersey Mosquito Exterm. Assoc.* **31,** 173.
Granett, P., and Hayes, H. L. (1945). *J. Econ. Entomol.* **38,** 671.
Granett, P., and Starnes, E. B. (1960). *In* "Methods of Testing Chemicals on Insects" (H. H. Shepard, ed.), Vol. II, pp. 101–119. Burgess, Minneapolis, Minnesota.
Green, N., Beroza, M., and Hall, S. A. (1960). *Advan. Pest Control Res.* **3,** 129.
Grout, R. A. (1963). "The Hive and the Honey Bee." Dadant and Sons, Hamilton, Illinois.
Hall, S. A., Green, N., and Beroza, M. (1957). *J. Agr. Food Chem.* **5,** 663.
Hocking, B. (1963). *Bull. World Health Organ.* **29,** Suppl. 121.
Hodgson, E. S. (1958). *Ann. Rev. Entomol.* **3,** 19.
Hodgson, E. S. (1964). *In* "The Physiology of Insecta" (M. Rockstein, ed.), Vol. 1, p. 363. Academic Press, New York.
Howard, L. (1911). *U.S. Dept. Agr. Farmers' Bull.* **444.**
Howell, D. E., and Goodhue, L. D. (1965). *J. Econ. Entomol.* **58,** 1027.
Howlett, F. M. (1912). *Entomol. Soc. London, Trans. Pt. II* p. 412 (cited in Green *et al.*, 1960).
Howlett, F. M. (1915). *Bull. Entomol. Res.* **6,** 297 (cited in Green *et al.*, 1960).
Ihndris, R. W., Gouck, H. K., and Bowen, C. V. (1955). *U.S. Dept. Agr.*, **ARS 33–7.**
Jacobson, M. (1966a). *Ann. Rev. Entomol.* **11,** 403.

Jacobson, M. (1966b). *Advan. Chem. Ser.* **53,** 17.
Kashin, P. (1966). *J. Insect Physiol.* **12,** 281.
Khan, A. A., Maibach, H. I., Strauss, W. G., and Fenley, W. R. (1965). *J. Econ. Entomol.* **58,** 694.
King, W. V. (1954). *U.S. Dept. Agr. Agr. Handbook* **69.**
Kingscote, A. A. (1958). *Proc. 10th Intern. Congr. Entomol., Montreal, 1956* **3,** 799.
Kirkpatrick, R. L., and Harein, P. K. (1965). *J. Econ. Entomol.* **58,** 920.
Laidlaw, H. H., Jr. (1966). Personal communication.
Lal, H., Ginocchio, S., and Hawrylewicz, E. J. (1963) *Proc. Soc. Exptl. Biol. Med.* **113,** 770.
Metcalf, C. L., and Flint, W. P., Rev. by Metcalf, R. L. (1962). "Destructive and Useful Insects, Their Habits and Control." 4th Ed. McGraw-Hill, New York.
Pimentel, D., and Weiden, M. H. J. (1959). *J. Econ. Entomol.* **52,** 457.
Quigley, G. D. (1965). *J. Econ. Entomol.* **58,** 8.
Richmond, E. A. (1927). *Proc. Entomol. Soc. Wash.* **29,** 36.
Rivnay, E. (1964). *Ann. Rev. Entomol.* **9,** 41.
Roth, L. M., and Eisner, T. (1962). *Ann. Rev. Entomol.* **7,** 107.
Roys, C. (1954). *Ann. N.Y. Acad. Sci.* **58,** 250.
Shambaugh, G. F., Brown, R. F., and Pratt, J. J., Jr. (1957). *Advan. Pest. Control Res.* **1,** 277.
Shannon, W. (1943). *Minn. Med.* **26,** 799.
Sherman, J. L., Jr. (1966). *J. Am. Med. Assoc.* **196,** 256.
Skinner, W. A., Tong, H., Maibach, H. I., Khan, A. A., and Pearson, T. (1965). *Science* **149,** 305.
Slifer, E. H. (1954). *Proc. Roy. Entomol. Soc. London* **A29,** 177.
Smith, C. N. (1966). *J. Am. Med. Assoc.* **196,** 236.
Thorsteinson, A. J. (1953). *Can. J. Zool.* **31,** 52.
Thorsteinson, A. J. (1955). *Can. Entomologist* **87,** 49.
Tisdale, E. W. (1963). *Phi Kappa Phi J.* **43,** 32.
Townsend, G. F. (1963). *Gleanings Bee Culture* **91,** 464.
Vladimirova, V. V. (1965). *Med. Parazitol. i Parazitarn. Bolezni* **34,** 340.
Willis, E. R. (1947). *J. Econ. Entomol.* **40,** 769.
Wilson, C. S., Mathieson, D. R., and Jachowski, L. A. (1944). *Science* **100,** 147.
Wilson, L. F., and Bean, J. L. (1959). *J. Econ. Entomol.* **52,** 621.
Woodrow, A. W., Green, N., Tucker, H., Schonhorst, M. H., and Hamilton, K. C. (1965). *J. Econ. Entomol.* **58,** 1094.
Wright, R. H. (1956). *Nature* **178,** 638.
Wright, R. H., and Kellogg, F. E. (1962). *Science* **195,** 404.
Zolotarev, E. Kh., and Elizarov, Yu. A. (1964). *Zool. Zh.* **43,** 549.
Zolotarev, E. Kh., and Gaverdovskii, A. N. (1964). *Zool. Zh.* **43,** 1155.

8 ANTIFEEDANTS

Donald P. Wright, Jr.

AGRICULTURAL DIVISION
AMERICAN CYANAMID CO.
PRINCETON, NEW JERSEY

I. INTRODUCTION

Antifeedants represent a different approach to crop protection in that the elimination of the pest is not the primary objective. Rather than killing, repelling, or trapping the insect pests, we can inhibit their feeding and thus protect the crop or commodity. In laboratory tests, insects remain on treated foliage indefinitely and eventually starve to death without eating the leaves. In field tests where the insects are free to wander elsewhere seeking food, they find weed plants to eat or they die of exposure, predation, or starvation.

A. Definition

While the idea that a chemical could prevent feeding is not new, early workers usually referred to such materials as repellents. Antifeedants are not repellent in the true sense of the word, since insects are not driven away or kept away. Anorexient is not suitable, since the appetite of the insect is not affected and it will keep foraging for suitable food. Gustatory repellent has also been used to cover this action, but repellent still connotes a driving away, regardless of the adjective. Dethier *et al.* (1960) propose the term "feeding deterrent," which aptly fits the situation, as does possibly the term "rejectant" used by the British Research Committee on Toxic Chemicals (Frazer, 1965).

B. Early History

The first antifeedant used in agriculture was Z.I.P. (the zinc salt of dimethyldithiocarbamic acid compound with cyclohexylamine). This is still used to keep rodents and deer from feeding on the bark and twigs of trees in the winter. It is too phytotoxic to apply to foliage and is not effective against insects. Prior to this, a number of materials had been used as mothproofing agents, and Moncrieff (1950) attributes their action to the prevention of feeding of the moth larvae, rather than to killing or repellency. A number of these compounds were introduced from 1928 through 1939 and included such chemicals as chlorinated triphenylmethanes and many triarylphosphines, -stibines, -arsines, and -tins as well as a variety of triphenylphosphonium salts. The most widely used of these was perhaps Mitin FF, 5-chloro-2-[4-chloro-2-(3,4-dichlorophenylureido) phenoxy]benzenesulfonic acid.

While these early materials acted by preventing feeding, it was not until the introduction of Antifeeding Compound 24,055 by American Cyanamid in 1959 that such a mechanism was recognized as a potential method of insect control in agriculture. Since that time, a growing amount of work has been done with a variety of types of antifeedants, but none so far have shown sufficient activity to be of commercial interest.

II. CHEMICAL TYPES

A. Triazenes

This class of antifeedants includes Compound 24,055, which is 4′-(dimethyltriazeno)acetanilide. They are the only class of antifeedants extensively tested outside of the laboratory. Compound 24,055 has been reported (Wright, 1963) to be the most active member of a series of 30 some triazenes which were prepared and tested. It is an essentially odorless and tasteless solid, but breaks down rapidly under acidic conditions, as evidenced by a pronounced darkening of color.

Antifeeding Compound 24,055 is considered to be moderately toxic by ingestion in single doses and has an acute rat oral LD_{50} of 510 mg/kg. The acute rabbit dermal LD_{50} is 1400 mg/kg, and the compound is nonirritating to rabbit skin or eye. In chronic-feeding studies with rats fed diets containing up to 1250 ppm for 36 days, there was no mortality, and only insignificant effects were noted on food intake and weight gain. No phytotoxity has been noted with rates up to 8 lb/acre (Anon., 1959, 1964).

At practical rates, 24,055 is nontoxic to most insects, however it will inhibit the feeding of most surface-feeding chewing insects. Initial laboratory tests indicated that this material was most effective against chewing insects and subsequent field tests in 1959–1961 throughout the United States and Canada served only to reinforce the original conclusion, with modifications. Chewing insects that fed under the treated surface, such as miners, borers, earworms, and codling moth larvae, did not have to encounter the material after the first bite or so, and were not affected. On the other hand, small boring insects trying to penetrate treated containers were affected since they had to chew a relatively large hole through a bag or box impregnated with 24,055, which required many bites of treated material.

Compound 24,055 was found to be relatively nonspecific, and pests as diverse as caterpillars, beetles, weevils, flies, bedbugs, roaches, and snails were controlled, provided they were chewing surface-feeders. Such pests did not represent a large enough market to justify the costs involved in registration for use at the rates required, and no further work was done with 24,055 until 1964, when the development of resistance in the bollworm prompted a more thorough evaluation against this pest, but again the material was found to be only marginally effective at competitive rates.

B. Organotins

The only other class of antifeedants that has been more than superficially studied are the organotins. Much of this work was done by Ascher in Israel, beginning with his report that Brestan (triphenyltin acetate) when used as a fungicide inhibited the feeding of insects on the treated foliage (Ascher and Rones, 1964). Subsequent work has revealed that several triphenyltins have pronounced antifeedant activity. Since such materials readily dissociate in aqueous solutions to form $(C_6H_5)_3Sn^+$, the substituent should have relatively little effect on the antifeeding properties of triphenyltins in general, and this has been found to be the case (Ascher and Nissim, 1964).

They report these materials to be effective in the field and laboratory against the cotton leafworm, while Murbach and Corbaz (1963) report similar results with Brestan-treated potatoes, which are protected from attack by Colorado potato beetles. Triphenyltins have also found use against the larvae of the potato tuber moth and the caterpillar of *Agrotis ypsilon* (Ascher and Nissim, 1964).

Earlier workers often reported insect control or reduced feeding damage, which in the light of our present thoughts would be called antifeeding. Thus, grasshoppers were said to feed less on baits treated with $SnCl_2$ (Richardson and Seiferle, 1938) and earlier it was claimed that organotins were effective mothproofing materials in a number of patents issued to I. G. Farbenindustrie in 1929 and 1930.

C. Carbamates

While carbamates are primarily known for their insecticidal action, several workers have reported on their antifeeding properties. The first to do so was Guy (1937), who stated that foliar applications of several thiocarbamates inhibited the feeding of Mexican bean beetles, Colorado potato beetles, and Japanese beetles.

More recently, it has been shown that a large number of phenylcarbamates with alkyl or alkoxy substituents on the ring prevent feeding of the salt marsh caterpillar at dosages of about one-tenth the lethal rates (Georghiou and Metcalf, 1962). Similiar effects are reported with the boll weevil (Matteson *et al.,* 1963). Baygon, (*o*-isopropoxyphenyl)-*N*-methylcarbamate, in particular has been shown to be a systemic antifeedant against the boll weevil at rates from 40 to 100 ppm. At rates from 5 to 20 ppm, the material gives only partial protection (Matteson and Taft, 1963). This is one of the few systemic antifeedants known to be effective at reasonable doses.

D. Botanical Extracts

In the course of the years a great number of plant extracts have been reported to have repellent effects on insects. By modern standards some of these can be said to be or guessed to be antifeedants, although the majority of them are probably true repellents. Among the few cases that appear to be unequivocably antifeedants, pyrethrum is stated to be a "gustatory repellent" against biting flies *Glossina* and *Culicoides;* these insects would settle on residues but would not bite (Dethier, 1947). Hoo Soo and Fraenkel (1964) report finding a water-soluble antifeedant effective against the southern armyworm in extracts of the Boston fern.

E. Miscellaneous

Several unrelated types of chemicals have shown antifeedant activity in passing. Dethier (1947) states that copper stearate and copper resinate prevent the feeding of tent caterpillars, and mercuric chloride is claimed to be a "gustatory repellent."

Tahori *et al.* (1965) report that several plant-growth regulants at relatively high rates inhibited the feeding of the cotton leafworm by foliar applications and by immersion of the leaf petiole in solutions of the compounds. These were:

Phosfon: 2,4-dichlorobenzyltributylammonium chloride
Cycocel: (2-chloroethyl)trimethylammonium chloride
B-nine: *N,N*-dimethylaminosuccinamic acid
Carvadan: 3-isopropyl-4-dimethylamino-6-methylphenylpiperidine-1-carboxylate methochloride

Since Phosfon was the most effective, most of their work was done with that material, which gave 89% inhibition of feeding with a leaf dip in a 4000 ppm solution. A solution of 400 ppm gave only 34% inhibition of feeding of the leafworm, but 40 ppm in the water was enough to protect cut chrysanthemums from damage, and, in limited field tests with peppers and peanuts, artificial infestations of cotton leafworms were controlled with foliar sprays of 1200 and 400 ppm, respectively.

III. MODE OF ACTION

Insect feeding consists of several stages: (1) orientation or attraction, (2) biting, and (3) swallowing or sustained feeding (Dethier, 1954; Hamamura, 1959; Thorsteinson, 1960; Hamamura *et al.,* 1962; Loschiavo, 1965). When insects are given the choice between antifeedant-treated or -untreated plants, there is little or no difference in initial attractiveness and equal numbers will be found on both groups of plants. Biting takes place on both groups, but the difference lies in whether there is a continuance of biting or a sustained feeding response. Those on treated plants continue taking tiny bites from the leaf as they wander around, whereas those on untreated plants will feed in one spot unless disturbed.

In order to have sustained feeding, the insect requires: (1) the presence of a gustatory stimulus, (2) the absence of an inhibitory stimulus, and (3) it must be hungry. Evidence indicates the first of these as the point at which an antifeedant acts. Apparently antifeedants inhibit the taste receptors of the mouth region so that, lacking the proper gustatory stimulus, the insect fails to recognize the treated leaf as food and continues foraging.

Unless the antifeedant comes in contact with these receptors, there is no inhibition of feeding, as shown by a number of experiments in our laboratories. Compound 24,055 has been placed into the mouth cavity, thus bypassing the taste receptors, and there was no effect on feeding.

Injection of the material into the body cavity or dipping caterpillars into solutions (except for the head) again had no effect on feeding.

Waldbauer and Fraenkel (1961) and Waldbauer (1962) in their studies with maxillectomized larvae of the tobacco hornworm came to the conclusion that taste receptors on the maxillae regulate feeding by perception of feeding deterrents and also by means of a spontaneous input to the nervous system which inhibits feeding unless sufficiently stimulated by feeding stimulants.

IV. EVALUATION OF THE ANTIFEEDANT CONCEPT

A. ADVANTAGES

While most antifeedants have had very limited testing, there is still enough information available from field evaluations of 24,055 and others to draw some conclusions with respect to the utility of the concept of antifeedants for insect control. First and foremost, this is a method which does not harm the beneficial insects—the parasites, predators, and pollinators (Anon., 1964). Only those insects attacking the protected crop are affected, and even these are not necessarily killed, since they starve to death only if they can not go elsewhere for food. Not only are the parasites and predators not killed by the antifeedant, but the hosts are not killed, thus enabling the beneficials to survive more readily. This selectivity makes an antifeedant a perfect partner in integrated control programs in which the aim is the use of chemicals such that the beneficial insects will be harmed as little as possible.

In general, most antifeedants are not highly toxic. The triazenes, typified by 24,055, have rat oral LD_{50} values in the range of 500 mg/kg. The triphenyltins mentioned in Section II,B have rat oral LD_{50} values in the range of 150 mg/kg (Ascher and Nissim, 1964).

With adequate coverage, antifeedants will usually limit feeding damage more effectively than conventional insecticides, which require a certain amount of time before the toxicant can act, during which the insect can continue feeding.

B. DISADVANTAGES

On the other hand, there are some disadvantages found in the use of presently known antifeedants. Perhaps the greatest limitation is that only surface-feeding chewing insects can be controlled. Since antifeedants affect the sense of taste, those insects which do not feed on the treated surface, but pierce the surface to feed either by sucking or chewing internally are not affected. A good systemic antifeedant would be the answer here, and recent reports of sytemics are encouraging (Section II,C and II,E).

Another disadvantage is that good coverage is required and new growth is not protected. With antifeedants this is a particular problem because the insects have not been killed elsewhere on the plant. Here again a good systemic would help solve this problem.

V. SUMMARY

While efforts to find a commercially successful antifeedant have been to no avail, work is continuing in several laboratories around the world. Entomologists have become aware of the concept and are alert to the possibilities of an antifeedant. Furthermore, the extensive field evaluation of 24,055 has shown that the concept is acceptable to growers and is a practical method of insect control.

REFERENCES

Anonymous (1959). *Am. Cyanamid Co. Tech. Bull.*

Anonymous (1964). *Am. Cyanamid Co. Tech. Bull.*

Ascher, K. R. S., and Nissim, S. (1964). *World Rev. Pest Control* **3,** 188.

Ascher, K. R. S., and Rones, G. (1964). *Intern. Pest Control* **6,** 6.

Dethier, V. G. (1947). "Chemical Insect Attractants and Repellents." McGraw-Hill (Blakiston), New York.

Dethier, V. G. (1954). *Evolution* **8,** 33.

Dethier, V. G., Browne, L. B., and Smith, C. N. (1960). *J. Econ. Entomol.* **53,** 134.

Frazer, A. C. (1965). "Report of the Research Committee on Toxic Chemicals." Agr. Res. Council, London, 1964, *Rev. Appl. Entomol.* **A53,** 211.

Georghiou, G. P., and Metcalf, R. L. (1962). *J. Econ. Entomol* **55**, 125.

Guy, H. G. (1937). *Delaware, Univ. Agr. Expt. Sta. Bull.* **206.**

Hamamura, Y. (1959). *Nature* **183,** 1746.

Hamamura, Y., Hayashiya, K., Naito, K., Keiko, M., and Nishida, J. (1962). *Nature* **194,** 754.

Hoo Soo, C. F., and Fraenkel, G. (1964). *Ann. Entomol. Soc. Am.* **57,** 798.

Loschiavo, S. R. (1965). *Ann. Entomol. Soc. Am.* **58,** 576.

Matteson, J. W., and Taft, H. M. (1963). *J. Econ. Entomol.* **56,** 892.

Matteson, J. W., Taft, H. M., and Rainwater, C. F. (1963). *J. Econ. Entomol.* **56,** 189.

Moncrieff, R. W. (1950). "Mothproofing," pp. 69, 175. Hill, London.

Murbach, R., and Corbaz, R. (1963). *Phytopathol. Z.* **47,** 182.

Richardson, C. H., and Seiferle, E. F. (1938). *J. Agr. Res.* **24,** 1119.

Tahori, A. S., Zeidler, G., and Halevy, A. H. (1965). *J. Sci. Food Agr.* **16,** 570.

Thorsteinson, A. J. (1960). *Ann. Rev. Entomol.* **5,** 193.

Waldbauer, G. P. (1962). *Entomol. Exptl. Appl.* **5,** 147.

Waldbauer, G. P., and Fraenkel, G. (1961). *Ann. Entomol. Soc. Am.* **54,** 477.

Wright, D. P., Jr. (1963). *Advan. Chem. Ser.* **41,** 56.

9 INTEGRATED CONTROL

Ray F. Smith and Robert van den Bosch

DEPARTMENT OF ENTOMOLOGY AND PARASITOLOGY
UNIVERSITY OF CALIFORNIA
BERKELEY, CALIFORNIA

I. INTRODUCTION

The middle decades of the twentieth century have been years of revolution in the field of pest control. This revolution was triggered during World War II by the discovery of the insecticidal properties of DDT and the successful utilization of this synthetic organic material against a number of pests. After the war, research chemists and the chemical industry, largely spurred by the success of DDT, developed and brought into use a tremendous array of pesticides of various types. The products of this effort continue to proliferate to this very day. The benefit that the insecticide revolution has brought to mankind stands as a remarkable testimonial to human intelligence and technological prowess. Over vast areas, pest problems dating back beyond biblical times have been virtually eliminated. The benefits, measured in terms of lives saved, diminished suffering, and economic gain, are inestimable (Metcalf, 1965). Furthermore, the ever-burgeoning arsenal of synthetic organic chemical insecticides promises even greater benefit for mankind.

There can be no doubt that from the standpoint of human welfare the insecticide revolution has been an enormously fortunate event. But as is so often the case, the blessing is not an unmixed one. Problems have arisen, some quite serious, to detract from the benefit realized from the new insecticides (Bartlett, 1964c; Brown, 1951; Carson, 1962; Chichester, 1965; Egler, 1964a,b; Kuenen, 1949, 1961; Massee, 1954, 1958; Pickett, 1948, 1949; Ripper, 1956; Rudd, 1964; Schneider, 1955; Solomon, 1953; Stern *et al.,* 1959; van den Bosch and Stern, 1962; Wigglesworth, 1945). This is related in large measure to the fact that ecological considerations were essentially ignored in the development of these materials and chemical, toxicological, and economic criteria used instead. Pest control is largely an ecological matter (Burnett, 1960; Milne, 1965), consequently, modern insecticides, developed essentially without taking this into account, have with distressing frequency engendered serious problems through their disruptive impact on the ecosystems to which they have been applied.

It has become absolutely imperative that a fresh approach to pest control be undertaken and that this approach be essentially an ecological one. All persons connected with pest control must come to look upon artificial controls simply as tools to be fitted as unobtrusively as possible into the total environment to effect suppression of pest species at the times and places where they escape the repressive effects of natural control. All control pratices, whether they be chemical, cultural, physicals, or genetic, must be laced together with the existing components of the environment so as to be mutually augmentative and to bring about the most effective, least ecologically disruptive, pest control possible.

This in essence is the objective of integrated control, and it will be our purpose in the following pages to define this technique and describe its implementation. In this connection, our discussion will relate in large measure to agricultural pest problems. However, the same principles hold for control of pests of medical significance, those affecting domesticated animals, forests, etc. (Anderson and Poorbaugh, 1964; Laird, 1963; Stark, 1961; Voûte, 1964).

II. INTEGRATED CONTROL DEFINED

Integrated control is a pest population management system that utilizes all suitable techniques either to reduce pest populations and maintain them at levels below those causing economic injury or to so manipulate the populations that they are prevented from causing such injury. Integrated control achieves this ideal by harmonizing techniques in an organized way, by making the techniques compatible, and by blending them into a multifaceted, flexible system.

This definition of integrated control is much broader than that of Bartlett (1956) and Stern *et al.* (1959). Early considerations of integrated control stressed the importance of developing procedures where chemical and biological controls could be used compatibly. This was logical because these two approaches are our primary resources in the struggle to control arthropod pests and, as utilized in many instances, are in direct conflict with one another. Practical experience and logic have shown that we must, in the final analysis, integrate not only chemical and biological control, but all procedures and techniques into a single pattern aimed at profitable production and minimum disturbance of the common environment (Agricultural Research Council, 1964; Chant, 1964; Franz, 1961a,b; Smith and Reynolds, 1966; van den Bosch and Stern, 1962). We agree with Geier (1966) that integrated control must be more than a mere empirical superimposition or juxtaposition of chemical and biological control techniques. Neither must integrated control be built around chemical control; in fact, in some instances, an integrated control program can develop without involving any chemical control.

III. THE NECESSITY FOR AN INTEGRATED CONTROL APPROACH

The unilateral use of any control procedure can have unwanted and unintended side effects (Doutt, 1964). The application of a chemical to destroy an insect pest, or the planting of an insect-resistant variety of crop plant, or even the introduction of a new parasite to control a pest, may have surprising impact on other parts of the agricultural environment. A new variety of plant may be more susceptible to fungus diseases or may require modifications of harvesting procedures. This occurred in California with the introduction of alfalfa varieties resistant to *Therioaphis trifolii.* The reduction of a pest through a new biological agent may affect the other organisms that prey upon it and may otherwise result in modification of the environment. For example, it has been suggested that the elimination of screwworm may result in an increase in the deer and jackrabbit populations in the southeastern states and thereby produce increased competition for range vegetation (Blair, 1964). The reduction in Australia of the carrot aphid, *Cavariella aegopodii,* by an introduced *Aphidius* has resulted in the virtual disappearance of carrot motley dwarf virus (M. F. Day, *in litt.*). Anderson (1966) has shown how systemics used on livestock upset the ecological system providing for the decomposition of cattle dung in pastures.

Stark (1965) has discussed the diverse side effects on insect pests, diseases, and vertebrate damage when fertilizers are applied to forests. In the same way, the use of nitrogenous fertilizers tends to increase the level of phytophagous mites in orchards (Hamstead and Gould, 1957; Hukusima, 1958; Kuenen, 1949). However, where adequate predator population levels are maintained outbreaks do not occur even with fertilizer treatments (Lord and Stewart, 1961).

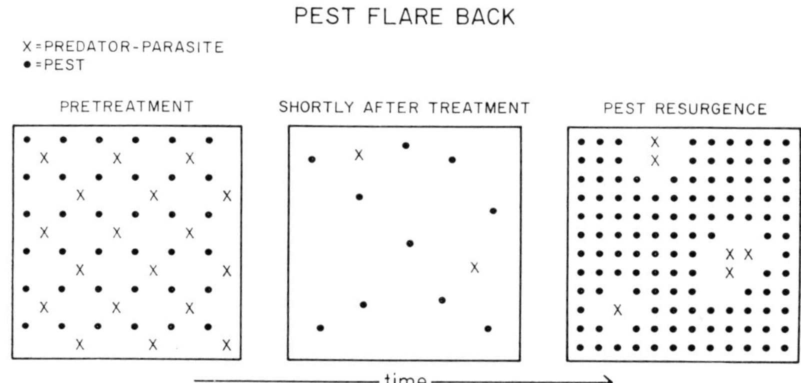

FIG. 1. Diagrammatic sketch of the influence of a chemical treatment on natural enemy–pest dispersion and resulting pest resurgence. The squares represent a field or orchard immediately before, immediately after, and some time after treatment with an insecticide for control of a pest species represented by the solid dots. The immediate effect of treatment is a strong reduction of the pest, but an even greater destruction of its natural enemy (enemies), represented by X's. The resulting unfavorable ratio and dispersion of hosts (pest individuals) to natural enemies permits a rapid resurgence of the former to damaging abundance.

The use of chemical pesticides without regard to the complexities of the agricultural environment has been in recent years a major cause of disruption. Often, the target pest species has become tolerant of the pesticides and no longer can be controlled economically with chemicals (*pesticide resistance*). The population of the target species may quickly recover from the pesticide action and for a variety of reasons may rise to new and higher levels (*pest resurgence*) (see Fig. 1). Other nontarget insects may, following the pesticide treatment, increase in numbers to damaging levels (*secondary pest outbreaks*) (see Fig. 2). The pesticide chemical may remain in or on the crop, or in the soil, drift to other nearby crop areas, flow into streams and drainages, and thereby create a hazard to man or his animals, or produce additional side effects (*residue problems*). The pesticide chemical may create hazards to pollinaters, wildlife, and other beneficial forms (*hazard to nontarget species*). Finally, product contamination or environmental pollution can lead to legal actions (*legal problems*).

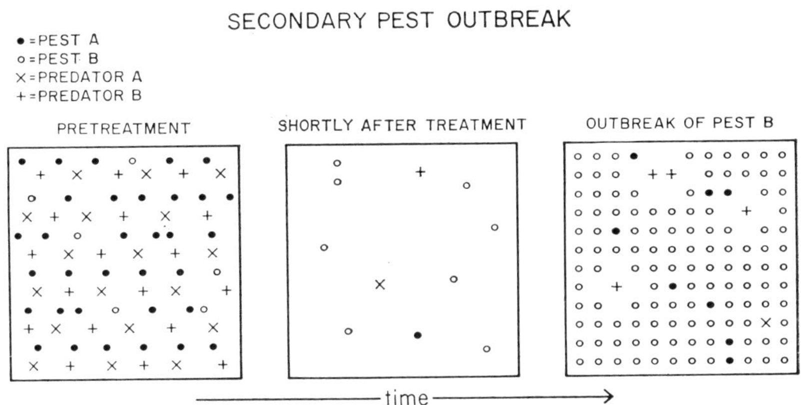

FIG. 2. Diagrammatic sketch of the influence of a chemical treatment on natural enemy–pest dispersion and the resulting secondary pest outbreak. The squares represent a field or orchard immediately before, immediately after, and some time after treatment with an insecticide for control of pest A, represented by (●). The chemical treatment effectively reduces pest A as well as its natural enemy (×), but has little or no effect on pest B (○) present in low numbers before treatment, but devastates the natural enemies (+) of pest B. Subsequently, because of its release from predation, pest B flares to damaging abundance.

Any modification of the environment may have extensive ecological implications. If we attempt to make the environment unfavorable to a particular pest organism, we may make it favorable to something else. The use of chemicals has dramatized this point in a most spectacular fashion and has also demonstrated that unilateral use of chemicals may end in disaster or excessive expense. The integrated control approach advocates the combined use of chemicals, other control methods, and naturally occurring mortality factors, and is based on a sound knowledge of the ecology of the pest organisms and the agro-ecosystem. The importance of better pest control in man's future and the increasing frequency of difficulties with a unilateral approach make obvious the need for integrated control.

IV. ECOLOGICAL BASIS OF INTEGRATED CONTROL

Integrated pest control is essentially the management of pest populations in man's agro-ecosystems. Bates (1964) discusses the "man-altered landscape" or "human ecosystem" that is rapidly coming to cover a large portion of the earth's temperate and tropical land areas. Agro-ecosystems are a part of this man-altered landscape modified and controlled for the production of agricultural and forest crops. Included are a great variety of agricultural enterprises, e.g., banana plantations, rice paddies, cornfields, steppe grazing lands, citrus groves, berry patches, woodlots, managed forests, apple orchards, fishponds, date gardens, and so forth. The stewardship of these agro-ecosystems may at times be inept and occasionally a complete failure. Nevertheless, man and his modern civilization are dependent upon reaping a harvest from these agro-ecosystems. Without this harvest in all its variety and abundance, the civilization we know today would perish and man would be thrown a long way back toward his former savage state of huntsman, root grubber, and berry picker.

Integrated control, developed in the context of the agro-ecosystem, does not necessarily have as its goal the preservation and maintenance of the system without change. Rather, the goal of integrated control is the manipulation of the agro-ecosystem to hold pests below economic levels and to avoid disruption of the system with its ensuing chaos.

As the first step in the discussion of the ecological basis of integrated control, we shall consider the components of an agro-ecosystem. Subsequently, we shall discuss their interactions and ways they can be manipulated to achieve better pest control.

A. Components of Agro-ecosystems

From the viewpoint of integrated pest management, the agro ecosystem is a unit composed of the total complex of organisms in the crop area together with the overall conditioning environment as modified by the various agricultural, industrial, social, and recreational activities of man. The major components include the crop plants, the soil substrate and its essential biota, the chemical and physical environment, an energy input from the sun, and man. In particular agro-ecosystems or at specific times, additional elements, such as weedy plant species, certain plant pathogens, or phytophagous arthropods, may become critical or dominant components in the system. Stated in oversimplified terms, man's objective in agricultural management is to hold the plant community at an early stage in ecological succession and to harvest the crop unmolested by the competing first-order consumers (Bates, 1963).

The ecosystem ecologist in his analysis is especially concerned with the measurement of energy flow in the ecosystem (Fig. 3) (Macfadyen, 1964). The same applies to some aspects of the analysis of an agro-ecosystem, but rarely if ever do we have an isolated system. However, the energy gains and losses are measured in the more familiar terms of fertilizer application, harvested crop, stocking and planting, migration, days of sunshine, etc.

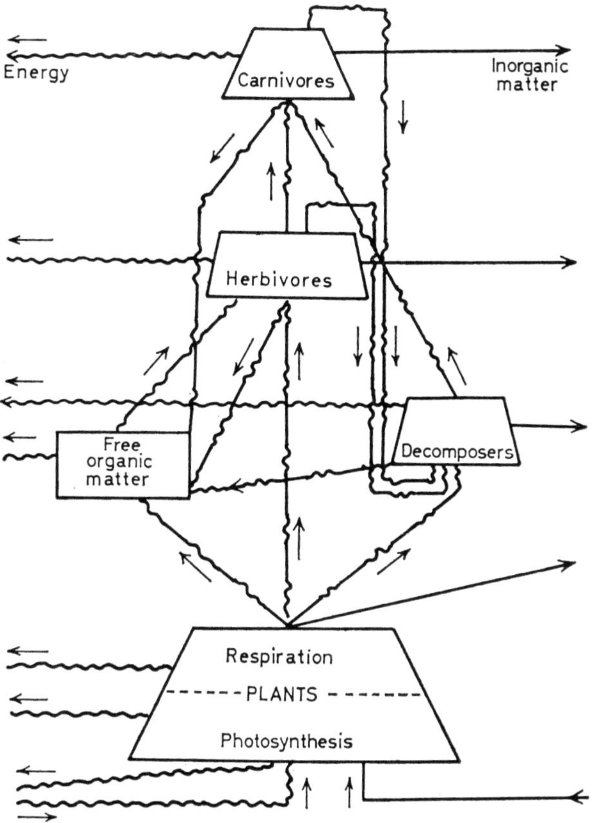

FIG. 3. Theoretical energy flow in an isolated ecosystem. Solar energy enters from the bottom: ~~~~~~~~, paths of energy flow; —~~——~~—, paths of organic matter; ————————, paths of inorganic matter (from MacFadyen, 1964).

In contemplating pest management in a given agro-ecosystem, the number of pest species, the abundance and distribution of the individual species, their biological and behavioral characteristics, competing forms, the organisms that prey upon the pests, the main and alternative food supplies of the pests, and the manner in which the other elements of the environment modify these must all be taken into consideration. The resources available for pest control research rarely, if ever, permit a total or undirected analysis of the agro-ecosystem. A convenient and practical point of orientation is the "life system" of Geier (1964, 1966) and Clark (1964). The life system is a subsystem comprising "only those components of an ecosystem which affect the population(s) of the subject species significantly" (see Fig. 4). For example, in the development

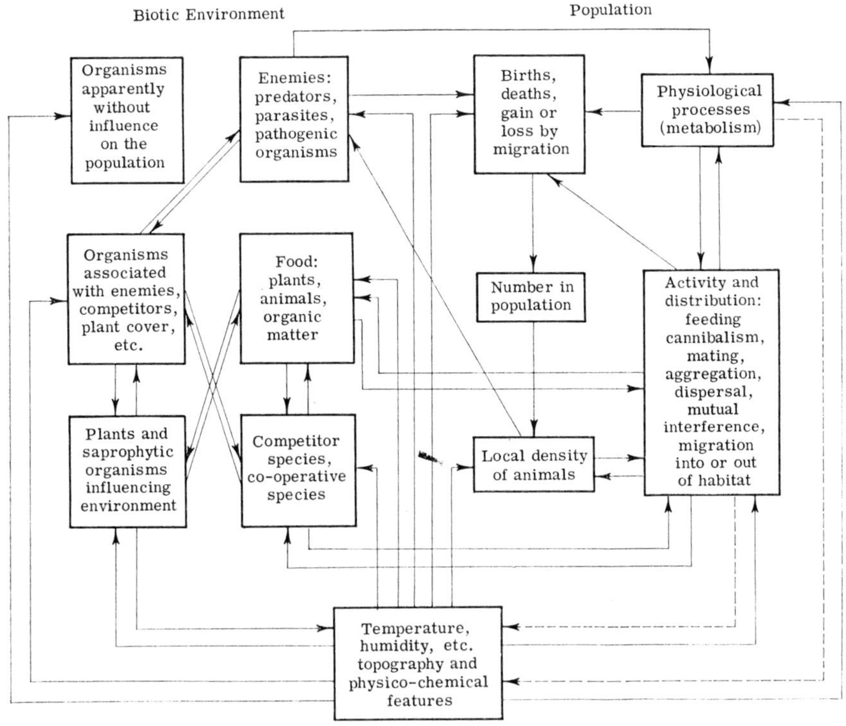

FIG. 4. Chief interrelationships of a population of one species with its environment (from Solomon, 1953).

of integrated control of alfalfa pests in California, all efforts were, out of practical necessity, first concentrated on the spotted alfalfa aphid. Later, consideration was given to the pea aphid, alfalfa caterpillar, alfalfa weevils, cutworms, and other pests. It is doubtful, however, that we will ever have a reasonably complete knowledge of the more than 1000 arthropod species associated with alfalfa in California. Even with the convenient and practical approach of the life system, the task of analysis and investigation is not easy; and eventually the overlapping life systems must be understood as parts of the total agro-ecosystem.

B. Stability of Agro-ecosystems and Complexity

Agricultural ecosystems vary widely in stability, complexity, and the area they occupy. Each of these factors has important effects on pest populations and consequently affects integrated control practices. The kinds of crops, agronomic practices, changes in land-use patterns, its total complexity and self-sufficiency, and weather are important elements affecting the degree of stability of an agro-ecosystem. All of these, except perhaps weather, are subject to considerable modification by man and therefore may be manipulated to influence pest management practices. Variations in stability from season to season and area to area are also greatly influenced by the length of life of the crops or dominant vegetation and by the length of the active season.

The level of migratory movements in and out of an agro-ecosystem is positively correlated with the degree of stability in the system (Southwood, 1962). Agricultural environments tend to be unstable because of cultural practices and ecological simplification. As a result, most major agricultural pests have a high degree of vagility and are highly adapted to colonizing, changing, or temporary agricultural environments.

Most ecologists will agree that as complexity increases, particularly with reference to the number and kinds of trophic interactions, the stability of the agro-ecosystem will increase (Burnett, 1960; Elton, 1958; Odum, 1964, Pimentel, 1961b; see Watt, 1965, for a contrary point of view). In fact, the emphasis is usually on the lack of stability in agro-ecosystems because of the low degree of complexity (Cole, 1964).

Voûte (1964) has properly emphasized the importance of the number of trophic interactions in affecting the stability of animal communities. He states: "We must realize that this stability within animal communities is the result not of the diversity of the species per se but of the number of interconnections between species, and with other components of the milieu . . . a rich community with only a few connections between species will show less stability than a poorer one with many connections." In this sense, the introduction of the spotted alfalfa aphid into the United States "enriched" the alfalfa community, but it was not until several parasite species were introduced, a fungus disease developed epizootically, and native coccinellid species adjusted their habits to the aphid populations that some degree of stability was regained. In the same way, the elimination of general predators by a nonselective chemical treatment can greatly reduce stability in an agro-ecosystem (Pimentel, 1961a).

The importance of "crop life" and "cropping practices" in agro-ecosystems is clearly evident by a comparison of the production methods for such agricultural commodities as radish (a short-term annual harvested at an early stage of development before flowering; and frequently several plantings per year in the same soil), cotton (a perennial plant grown as an annual row crop with harvest of the matured fruit 6 months or more after planting time), alfalfa (a perennial grown in solid plantings with a life of 2–20 years; harvested 2–8 times per year), apple (a tree crop grown in uniform orchard stands with a life of 20 or more years; fruit harvested annually), and permanent pasture (a mixture of legumes and grasses harvested intermittently by livestock). This array of crop types can, of course, be expanded and is complicated by consideration of utilization of the harvested crop, e.g., apples grown for cider or the fruit market, or grapes grown for wine pressing, drying, or for fresh consumption. There is a general tendency toward stability and complexity in perennial compared to annual crop plantings, and in mixed plantings or polycultures compared to monocultures. Likewise, there will be greater stability in uniform physical environments (for example, tropical environments) compared to those fluctuating markedly with the seasons, and in old undisturbed ecosystems compared to new disturbed agro-ecosystems.

Cole (1964) outlined the species diversity, resistance to invasion, susceptibility to pest outbreaks, and general stability of a tropical rainforest, the mixed hardwood forest of eastern United States, the boreal forest, the arctic tundra, and agro-ecosystems. Macdonald (1965) has stressed the ability of the boreal forest to recover from the severe stresses of wildfire, DDT treatments, or spruce budworm outbreaks. Voûte (1946, 1964) has compared the resistance to pest outbreaks of virgin forests, managed forests, and pure stands. It is clear that rarely do agro-ecosystems have the complexity, resistance to pest outbreaks, and resiliency of naturally occurring ecosystems. The necessary stresses imposed by man aggravate the inherent instability of agro-ecosystems. Again we must conclude that careful management is essential.

Change in complexity may become a tool for pest management. In integrated control programs, the fundamental importance of complexity should be recognized (Doutt, 1964). This does not mean that blind efforts should be made to preserve or increase complexity; rather it should be evaluated in each case and utilized as appropriate (Kennedy, 1965). Rudd (1964) advocates a reversal of the "trend towards environmental simplification" and the "conscious restoration of more complicated ecosystems." This may be appropriate as a general planning principle, especially in multiple-use areas; however, the application to specific agricultural situations should be made with extreme caution. Lewis (1965) has recently analyzed the effects of shelter belts, hedges, and uncultivated land on the ecology of adjacent crops. He concludes that such shelter often concentrates pests and leads to serious local damage. He states: "completely uncontrolled sheltering vegetation seems to be a dangerous reservoir of pests as well as bridging barriers to their further natural spread and encouraging concentrations on crops." He does not advocate the elimination of the shelter belts, but rather the control of the plant composition of the shelter areas and of pest infestations within them. van Emden (1965) has reviewed other relationships between uncultivated land and crop infestation (Fig. 5). He stresses the provision of alternative foods to both beneficial insects and pests. He points out that the practice of mowing vegetation along roadsides is detrimental to pest management because it removes the flower heads needed by the adult beneficial forms and leaves much of the foliage needed for food and shelter by the pest

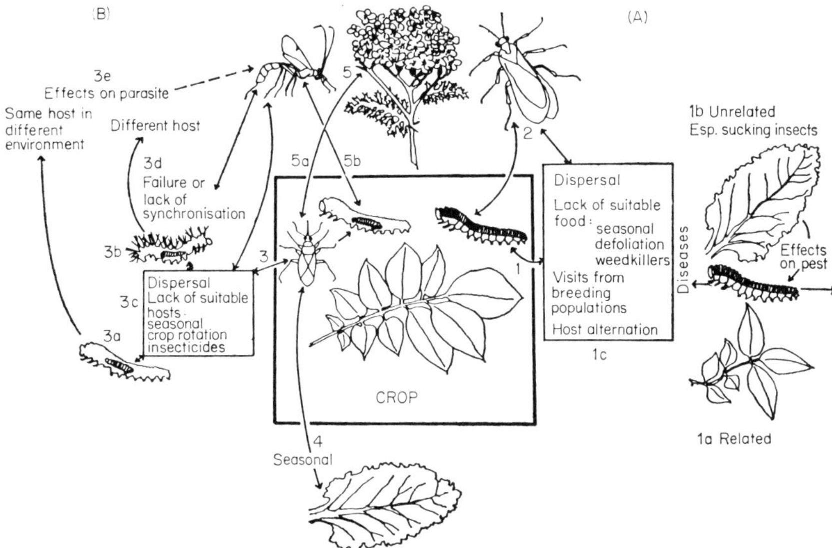

FIG. 5. Biological relationships between uncultivated land and crop pests (on the right) and beneficial insects (on the left). 1, injurious species feeding both on the crop and on plants on uncultivated land; 1a, wild plants related to the crop plant; 1b, wild plants not related to the crop plant; 1c, causes of pest movement between uncultivated land and crop; 2, the importance of flowers for pest insects; 3, the presence outside the crop of alternative prey; 3a, alternative prey related to the crop pest; 3b, alternative prey not related to the crop pest; 3c,d, reasons for the utilization of the alternative prey; 3e, effects on beneficial insects of attacking prey on uncultivated land; 4, beneficial insects feeding phytophagously on uncultivated land; 5, the importance of flowers for predators and parasites (from van Emden, 1965).

species. Bombosch (1966) also stresses the importance of a succession of flowering plants in field margins to support adult syrphids and other beneficial forms. Allen and Smith (1958) reported that agricultural areas which contained alternative food sources for adult *Apanteles medicaginis,* a parasite of *Colias eurytheme,* favored increased longevity and generally had a higher level of parasitization. van Emden concludes "There is yet insufficient evidence to establish whether, on balance, uncultivated land is beneficial or harmful to pest control." Cole (1964), on the other hand, indicates "We should try in every way to increase diversity."

Doutt (1965) and Doutt and Nakata (1965) have reported how increasing the complexity of vineyard agro-ecosystems has aided in the development of an integrated control program of grape pests. *Anagrus epos* was not an effective parasite of the grape leafhopper in most vineyards of the southern San Joaquin Valley. In this region, the parasite, unlike the grape leafhopper, cannot overwinter in the vineyard area, but must continue to breed through the winter on an alternate leafhopper, *Dikrella cruentata,* that exists on wild and commercial blackberries. Blackberries are now being planted on an experimental basis near the vineyards in these southern areas in hope of providing alternate winter hosts for *Anagrus.*

Another interesting relationship between cultivated and uncultivated land has been discussed recently by Riegert *et al.* (1965). Native grasslands do not support large populations of *Cannula pellucida* except on the margins adjacent to cultivated cereals. This supports their hypothesis that outbreaks occur when a favorable food habitat (cereals) and an oviposition habitat (short grass prairie) are readily available to the grasshoppers.

Another example of the utilization of crop diversity in pest management is the development of "strip-cutting" of alfalfa. Figure 6 shows the effect of the more stable (strip-cut) and less stable (solid-cut) environments on the relationship between the pea aphid and its hymenopterous parasite, *Aphidius smithi,* in California. In the more stable environment, both species persist at higher densities through the unfavorable (hot, dry) summer period and the parasite responds quickly to the increase in aphid density in the late summer. In the less stable environment, both aphid and parasite populations are reduced to very low levels in summer time and the low populations persist for several months. The aphid population rises steeply in early autumn (cool, humid weather) but significant response of the parasite is delayed until late autumn (from unpublished data of van den Bosch, Lagace, and Stern).

In general, in the development of an agro-ecosystem, man has tended to organize and simplify the environment to maximize the yield of a

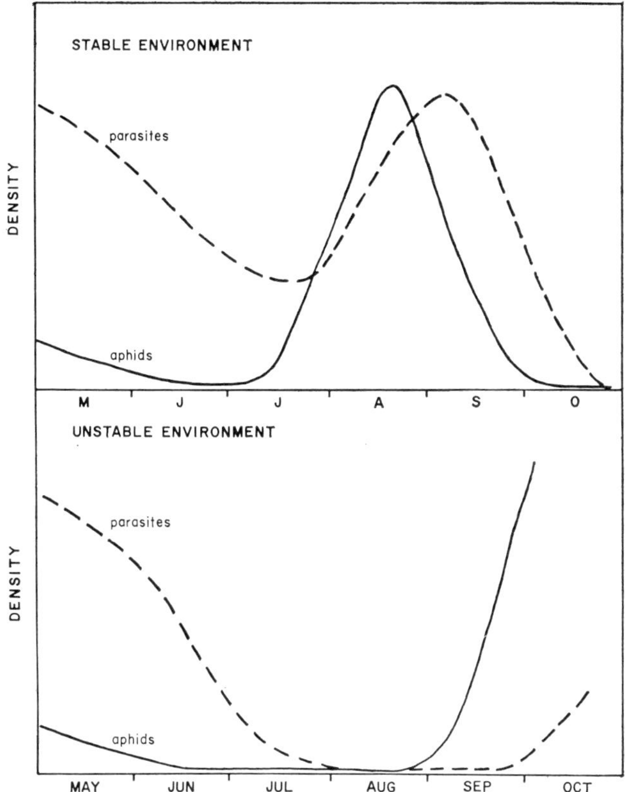

FIG. 6. Effect of more stable (strip cut) and less stable (solid cut) agro-ecosystems on the relationship between the pea aphid and its hymenopterous parasite (for discussion, see text).

special product, e.g., cotton fiber, beef, and apples. He eliminates competition through control of weeds, diseases, and arthropods. He changes complex food webs into simple, short food-chains. He attempts to hold ecological succession at an early seral stage. Consequently, we see an increasing trend toward huge monocultures and uniform stands of vegetation.

Cole (1964) concludes that modern agro-ecosystems are so simplified that we must expect a continued succession of outbreaks of pests. Nevertheless, in spite of their simplification, the complexity of the agro-ecosystems is much greater than would appear at first glance. LeRoux (1960) found 144 plant species in the undercover and surrounding hedges of an apple orchard in Quebec. These plants formed an important reservoir of predators and parasites of apple pests and contributed significantly to the "modified spray program." He also found 214 insect species present.

In Germany, Steiner (1962) classified 50% of all arthropods living in the crowns of untreated apple trees as "indifferent species," that is, they were neither directly noxious nor useful. In the analysis of an agro-ecosystem, these species are not really indifferent because they affect the stability of the total arthropod population. Schlinger and his associates (unpublished data) estimate about 1000 species of arthropods to be associated with alfalfa fields in southern California. It is this "unseen diversity" that must be evaluated and conserved, and not lost unwittingly through insecticide treatment or other agricultural mismanagement. van den Bosch and Hagen (1966) estimate that 300–350 arthropod species occur in irrigated California cotton fields, but that only about 20% are phytophagous. In a more diverse region, Whitcomb and Bell (1964) have found about 600 species of arthropod predators alone associated with nonirrigated cotton in Arkansas.

Another factor counteracting the instability of the agro-ecosystem brought on by simplification is the stabilizing of effects resulting from the regularity and uniformity of agronomic management practices (LeRoux, 1964).

Agro-ecosystems vary radically from area to area. This is, in large measure, due to differences in physical environments, but also results from changes in the composition of the biota and cultural practices. Such differences are quite obvious, for example, when comparing cotton ecosystems in the Mississippi Delta and in the Central Valley of California. Very often, more subtle differences occur within a rather small area. An outstanding example of such differences has been analyzed by DeBach and Sundby (DeBach, 1965; DeBach and Sundby, 1963). They showed differences in relative proportions of five imported species of parasites of the California red scale in 17 different sampling areas in the citrus-growing districts of southern California as the result of interspecific competition, and climatic adaptation of the parasites.

C. Limits of Agro-ecosystems

The physical limits of an agro-ecosystem are rarely precise. Very few are self-contained. For practical purposes in integrated control, the limits are determined by the nature of the pest problem. The area included must be large enough so that the important biotic components carry out their major activities within its limits. The agro-ecosystem will usually include a group of agricultural fields or orchards (rather than a single field) together with their marginal areas, and often certain other intermixed elements such as woods, streams, and weedy or uncultivated areas. With cultigens of long crop-life and especially where the pests are sessile (e.g., scale insects), the size of the unit that must be considered in a life-system analysis is small—perhaps a single tree or even a portion of a tree. With agricultural plants of short crop life which are affected by very vagile pests, the size of the unit that must be considered is often very large—frequently many square miles in extent. In our studies of aphids and *Colias* in alfalfa fields in California, we must consider several large adjacent fields and their associated marginal areas; a total of about a square mile (Smith *et al.,* 1949). Consideration of an armyworm, *Prodenia praefica,* had to take into account the interaction of populations in extensive grazing lands and the adjacent irrigated area. Since agricultural practice creates a large amount of disturbance, there is typically much movement of species within, into, and out of the agro-ecosystem. Those insects which become agricultural pests are usually well adapted to this disturbance (Southwood, 1962). They tend to be highly vagile forms, with alternate hosts in crop margins and uncultivated areas, and are adapted to locating the new crop planting. The alfalfa butterfly, *Colias eurytheme,* is such a vagile pest and seems to be perfectly preadapted to modern alfalfa culture. The coccinellids of the genus *Hippodamia* are also important in agricultural crops because of their adaptation to the disturbed conditions of agro-ecosystems (Hagen, 1962). These aspects of agro-ecosystem size and the dispersion and vagility of its components are critical elements in the analysis of an agro-ecosystem for integrated control. They affect the patterns of pest infestation and population trends, the level of internal contacts and population densities, and they can be managed and manipulated to advantage in an integrated control system.

D. Kinds of Pests

There are usually only one or two *key* pests in each agro-ecosystem. Key pests are serious, perennially occurring, persistent species that dominate control practices because in the absence of deliberate control by man, the pest populations usually remain above economic-injury levels [Class I pests of Chant (1964)]. The key pest is the focal point of analysis and management for integrated control. While the total number of potential pest species is high, the number of key pests involved is low, usually only one at a particular time. LeRoux (1960) in a 5-year study of an apple orchard found 92 arthropod species that had at some time or other caused important damage in apple orchards. These were associated with 122 entomophagous insects and 23 insectivorous birds. However, LeRoux considered only five species to be major pests. He classified 17 others as minor pests of primary importance and the other 70 as minor pests of secondary importance. Similarly, LeRoux (1964) states "In the majority of pest control problems with which I have been concerned in the past in Canada, as for example, pest infestations of such agricultural crops as apple, pear, peach, corn, onion, cabbage, radish, etc., infestations for a given crop were usually the work of one pest species and such pest species dominated the crop ecosystem for several generations—the period of the rise and fall of the infestation." In California, on grapes the key insect pest is the grape leafhopper; on cotton, the key pests are the lygus bug and the bollworm. These crops, however, are populated by many other insects that cause damage sporadically.

The classification of an insect as a key pest species is influenced by the general level of abundance of the damaging stage at the time the crop is susceptible, the kind of damage caused by the pest, and the damage potential of a single individual. If the species attacks the harvested unit (e.g., the fruit), it may be necessary to hold the populations at relatively low levels and it is likely to be a key pest. In the same way, a vector of a virulent virus has a high damage potential per individual and is likely to be a key pest.

In contrast to key pests, "occasional pests" [Class II pests of Chant (1964)] cause economic damage only in certain places or at certain times. Such pests are usually under adequate biological or environmental control; this is disrupted occasionally, thus permitting the pest population to increase to economic levels. These occasional pests are especially suited to integrated control techniques and it is against such pests that a system of corrective opposed to prophylactic treatment has its greatest impact. With these occasional pests, means of predicting the sporadic occurrence of damaging populations assume special importance.

A third group, "potential pests," cause no significant damage under the conditions currently prevailing in the agro-ecosystem [Class III pests of Chant (1964)]. In attempts to control key pests and occasional pests, care must be taken not to alter conditions through injudicious use of chemicals or cultural practices to the extent that the potential pests are permitted to reach damaging status. The red-banded leaf roller is a classical example of such a pest that has realized its potential as the result of chemical treatment and cultural practices.

Migrant pests [Class IV pests of Chant (1964)] form a fourth pest category requiring special consideration. They are nonresidents of the agro-ecosystem that enter periodically for short periods of time. Chant (1964) gives the example of dispersing lygus bugs in Ontario orchards. *Alabama argillacea* as a pest of cotton in southern United States and *Diabrotica undecimpunctata howardi* as a pest of corn in the Midwest are well-known examples of migrant pests. Locusts, armyworms, and similar highly mobile insects usually fall into the migrant pest category.

E. Management of Pest Populations

The determination of insect numbers is broadly under the influence of the total agro-ecosystem, and a background knowledge of the role of the principal elements is essential to integrated pest population management. A knowledge of the agro-ecosystem is necessary to harmonize the various control techniques and to prevent disruptive efforts. In the same way, an understanding of the agro-ecosystem permits an assessment of the stabilizing mechanisms operating on the pest and potential pest populations and suggests subsequent manipulations to reinforce and enhance their action.

The time comes in all practical pest problems for action. We can not wait for the complete analysis and dissection of the agro-ecosystem, but must move ahead with the knowledge available to protect the crop and the farmer's investment. Geier (1964) has provided a unique and rather comprehensive examination of the basic actions possible in pest control. He states: "The abundance of a population is a function of the fitness of the innate qualities of individuals in the population for the operative features of the environment, for example, climate and weather, and the supply and distribution of vital requisites such as food and shelter. This fitness can be reduced (a) by modifying the innate qualities which enable individuals to perform their life functions within the population and the environment; or (b) by altering essential features of the environment in such a way as to make it no longer able to support large numbers of a population . . . ."

Alternative (a) involves the autocidal methods that have given such spectacular results in certain instances in recent years. They are discussed in other chapters in this volume. The aim of alternative (b) can be achieved "in three ways: by increasing the frequency and severity of environmental hazards and thereby increasing the probability of premature death for individuals; by preventing individuals from locating vital requisites (such as food, shelter, mates, etc.) for the performance of their life functions; and by curtailing the supply of vital requisites to such an extent that former habitats become unsuitable for the pest species" (Geier, 1964).

V. EVOLUTION OF AN AGRO-ECOSYTEM

Agricultural ecosystems are the evolutionary product of man's attempt to control nature and to meet his need for food and other products. Although somewhat simplified compared to many less-disturbed ecosystems occurring in nonagricultural areas, these agro-ecosystems are, nevertheless, dynamic, multifaceted, evolving complexes and they have unique complexities of their own. The complexity and the interacting features of the agro-ecosystem are often not evident even to farmers who are a part of the agro-ecosystem on a day-to-day basis.

There is continual interaction between the populations of plants and animals in the agro-ecosystem and the unstable nonliving environment. This interaction and those among the living organisms, are continually evolving. This is not always evident in a particular place, but may become dramatic when we compare populations at the limits of the distribution of a species. For example, the populations of the alfalfa weevil, *Hypera postica,* occurring in the Sacramento Valley of California and at Lethbridge, Canada are quite different from each other and from the weevil population from which they originated (the one introduced into Utah near the turn of the century). They differ in behavior, cold resistance, temperature requirements, damage potential, and probably in other ways.

Similarly, Birch (1960) has traced the evolution of the Queensland fruit fly, *Dacus tryoni,* as it spread from its tropical home in the rain forests of Queensland in Australia southward into the states of New South Wales and Victoria, a distance of some 1500 miles. Originally living in the fruits of rain-forest trees, the fly has spread into cultivated fruits which are now grown in many districts along its route. But in addition to this, the spread south has involved an evolutionary adaptation of the fly to the progressively cooler latitudes. The adaptation of codling moth in California to walnut is another example of evolution of the feeding habit in a pest species. The development of industrial melanism in the pepper moth is a classical example of adaptation of a species to a changing environment (Kettlewell, 1961). Apart from the well-known development of insecticide resistance, there are undoubtedly numerous unidentified parallels in every agro-ecosystem.

Even without the deliberate manipulation of man, the agro-ecosystem is always in a state of evolution. This evolution is most often in the direction of increased complexity; new components are added by adaptation and introduction. Successful integrated pest control must be sensitive to these changes that occur both on the short-term dynamic basis and the long-term evolutionary basis. Perhaps we can clarify our concept of the agro-ecosystem and its evolution by sketching the highlights of the origin and evolution of the argo-ecosytems in California's San Joaquin Valley.

When, in 1772, the first Spanish explorers entered the San Joaquin Valley, they found remarkably large populations of Yokuts Indians existing on an abundance of elk, antelope, fish, tule roots, acorns, pine nuts, and other seeds (Smith, 1939; Cook, 1955). Three huge lakes and their surrounding tule marshes covered most of the lowlands, at least in the spring months. Large areas of grassland and oak savannah, together with smaller amounts of saltbush desert, chapparal, and riverine communities were essentially undisturbed by man (Piemeisel and Lawson, 1937). Some of the insects we know today as crop pests (*Prodenia praefica, Colias eurytheme, Lygus hesperus, Diabrotica undecimpunctata, Eurythroneura elegantula, Heliothis zea, Estigmene acraea,* several grasshoppers and a number of other species) occurred there, but were not pests because no agricultural crops existed in the valley. These insects were greatly influenced by the seasonal occurrence of rain and the limited distribution of native annual vegetation. None of the Mediterranean winter annual herbs and grasses, such as bur clover, filarees, wild oat, foxtail, and the like, were present. There was, of course, no cultivation of alfalfa.

It was not until 1836 that the first cattle ranch was established by the Spanish in the northwest fringes of the valley. During the next decade, numerous Spanish cattle ranches, including a few in the west-side area, began a precarious existence. The discovery of gold and the rapid influx of settlers from the eastern United States created a demand for beef, and a pastoral agriculture developed. The period of 1850–1870 was one of huge cattle holdings. The cattle were pastured on the lush grass of the valley during the winter and spring and were taken to the foothills and mountains in summer. Overgrazing began to take its toll, especially in drought years, and the introduction of Mediterranean grasses and forbs changed the composition of the range. The white man had now developed huge pastoral agro-ecosystems and the indigenous Indian had virtually disappeared from the scene.

The discovery in the 1850's that the winter and spring rains were sufficient to produce tremendous crops of wheat brought on the wheat era that lasted until about 1890. The new American settlers planted huge grainfields, displacing extensive areas of the native grasslands. The crops were sent to market first by wagon to the rivers, then by barges on the rivers, and later by rail. Each increment to the railroad system increased the grain area. The full development of this wheat era was also dependent upon the development of gang plows, harrows, endgate seeders, and better harvesting machinery. Thus, the development of these huge grain agro-ecosystems was the outgrowth of new settlers, market demands, transportation systems, and technical developments as well as the natural resources of the grassland areas and rainfall.

The introduction of a railroad system also permitted the development of general agriculture along the rivers where water was available. A few plantings of alfalfa were made in these "agricultural colonies" during the sixties, but it was not until the development of irrigation systems and a dairy industry that the acreage of alfalfa became extensive. The irrigation systems had small beginnings and were continually threatened by problems involving riparian rights to water, financing, land frauds, and state laws concerning water rights and irrigation districts. The dairy industry was dependent on the development of electrical power and refrigeration. Thus, again the social, legal, and technical complexities in the development of agro-ecosystems was manifested.

With the introduction of extensive irrigation systems, the grasslands and alkali deserts of the San Joaquin Valley were transformed into an intensive irrigated agriculture. Along with the grains and alfalfa came tree fruits (deciduous and citrus), grapes, cotton, melons, sugar beets, rice, and vegetables. A variety of native insects found the lush irrigated fields or orchards an ideal haven. An abundant food supply was available the year round and their period of increase was no longer confined to the spring months. To the native pest fauna has been added an array of immigrant species including the alfalfa weevil, spotted alfalfa aphid, pea aphid, green peach aphid, cotton aphid, codling moth, peach twig borer, Oriental fruit moth, citrus red scale, olive scale, and many others. Each addition brought about significant changes in the agro-ecosystems. At the same time, agronomic practice was improving. Better land-leveling equipment produced better seedbeds and improved water flow and distribution. This resulted in better crop stands and few weeds. The use of fertilizers and new plant varieties increased yields and reduced diseases. Mechanization sped up harvesting. Each of these changes had its impact on the insect populations in the agro-ecosystems and often changed the pest problems. For example, in alfalfa, the newer mowing machines leave a higher stubble, the new varieties have more leaves low on the stems, and better irrigation technique brings water into the field soon after harvest. As a result of all of these, the alfalfa field is not a barren desert following each harvest in summer; more insects—both beneficial and harmful—are able to survive from one cutting period to the next. This continuity of the insect populations favors stability and decreases the chances of pest outbreak.

One would think, with all the change that has occurred in the San Joaquin Valley, that a stable state has finally been reached. Today the Indians, the lakes, the tules, the elk and antelope, the oaks and salt bushes, and the native grasses and herbs are gone and in their places are crops and exotic weeds, domesticated livestock, wells and dams, irrigation and drainage systems, cities, industries, railroads, airports, highways, and hundreds of thousands of human beings (Dasmann, 1965). But the system continues to change. More people arrive, cities expand, industries proliferate, wastes accumulate, wells deepen, new crops are planted, agrochemicals evolve and diversify, new pests appear, water tables drop, agronomic and horticultural techniques change, and agromechanization becomes more sophisticated. Characteristically, each change has impact that ramifies broadly in the whole system. For example, replacement of alfalfa varieties susceptible to a pest such as spotted alfalfa aphid with resistant varieties conceivably had ramifications on pest problems in such diverse crops as citrus, grain, and melons.

On citrus, where the drift of organophosphate aphicides from treated alfalfa fields triggered large-scale insect outbreaks, such drift-induced outbreaks were eliminated when the resistant alfalfa obviated the need for chemical treatment.

The grain insect problem was probably affected through the influence of the resistant alfalfa varieties on coccinellid populations. It is a matter of record that during the spotted alfalfa aphid era in California enormous numbers of coccinellids were produced in the infested alfalfa fields. Movement of these coccinellids from alfalfa into grain fields added biotic pressures to the grain infesting aphids. With the advent of resistant alfalfas and the decline in coccinellid abundance there was unquestionably less pressure on the grain aphid populations as well as populations of aphids in other crops.

Similarly, melons are affected by pests that normally occur at noneconomic levels in alfalfa. Populations of these pests increased explosively in alfalfa following the wide-scale use of the broad-spectrum organophosphates against the spotted aphid. Particularly heavy and widespread infestations of spider mites and leaf miners, serious melon pests, developed in alfalfa during the first few years following the invasion of the spotted alfalfa aphid. There can be little doubt that large numbers moved into adjacent melon fields and contributed to aggravated problems on that crop. With the return to normal conditions in alfalfa following the cessation of extensive aphid control treatments, the spider mite and leaf miner problems ameliorated in the melon fields. The impact of pea aphid or lygus bug resistance in alfalfa varieties could have equal ramifications in the agro-ecosystems of the San Joaquin Valley when such varieties are introduced in the future.

VI. ECONOMIC-INJURY LEVELS

The determination of the levels of tolerable damage is an essential prerequisite to the development of integrated pest control programs (Stern *et al.*, 1959; Smith, 1962; Turnbull and Chant, 1961). The designation of this damage-tolerance level sets the goal of the pest management system. Hard experience has revealed the hazards and futility of attempts to maintain agricultural plantings insect-free. Not only are such attempts extremely costly, but they are ecologically unsound. Today, only the most uninformed can honestly advocate or practice such an unrealistic approach to agricultural pest control. Apart from the importance of parasites and predators, arthropods play an essential role in the decomposition of plant and animal residues, in the aeration and fertilization of the soil, and in the pollination of many plants. The goal of a pest management system should be stated in terms of specific damage levels, not insect numbers.

Insect pest numbers and crop damage are not always perfectly correlated (Johnson, 1965). A higher pest abundance can be tolerated under some conditions than under others without causing economic damage. For example, under cool weather conditions a higher level might be acceptable. At early stages of plant development, the crop may be more or less susceptible to damage. Frequently, the tolerable population of the pests drops as maturation of the crop nears. Differences in irrigation and fertilization practice may also affect the susceptibility of the crop to pest damage. Some crop varieties may be more susceptible than others. For example, the alfalfa variety *Africa* is quite tolerant of spotted alfalfa aphid populations which are damaging to the variety *Caliverde* (Davis *et al.,* 1957). From these examples, it can be seen that it may be necessary to qualify damage-tolerance levels in terms of a local climate, time of year, stage of plant development, plant variety, and cropping practices.

Many plants can suffer moderate loss of leaves, stems, fruits, or roots before the crop yield is affected. This is also true, in some instances, of the harvested fruits. For example, heavy tillering varieties of rice can tolerate moderate levels of rice stem borer during early stages of crop growth without loss in yield because lost tillers are rapidly replaced (Munakata and Okamoto, 1966). Also, with cotton, every lygus bug feeding-puncture that causes a cotton square (i.e., flower bud) to drop can be classified as insect injury—but this is not necessarily economic damage. The cotton plant has a limited capacity to set bolls depending upon its growing conditions and health. Those cotton squares in excess of this level will drop from the plant even if no lygus bugs or other square-feeding insects are present. It matters not at all whether the squares drop because of lygus bug feeding or because of the limited fruit-carrying capacity of the plant. Such insect-caused injury does not result in crop loss (see Fig. 7). It can not be called economic damage. On the other hand, when the insect feeding reduces boll load to a level below the potential of the cotton plant, a crop loss has occurred. At low levels of crop loss, it may not be economically sound to attempt to reduce the lygus population. The decision to attempt artificial control will depend upon the value of crop that can be recovered or protected balanced against the cost of the control measure. Ordish (1952) has called this the cost/potential benefit ratio.

The costs of the control procedures can not always be measured by such simple arithmetic as the addition of the price of the insecticide materials plus the cost of application. For example, if the control procedure employed to reduce lygus bugs or thrips in early season makes it necessary to treat several times in late season to protect the nearly mature cotton crop from bollworms, the cost of the subsequent treatments must also be considered in the economic evaluation of the initial treatment.

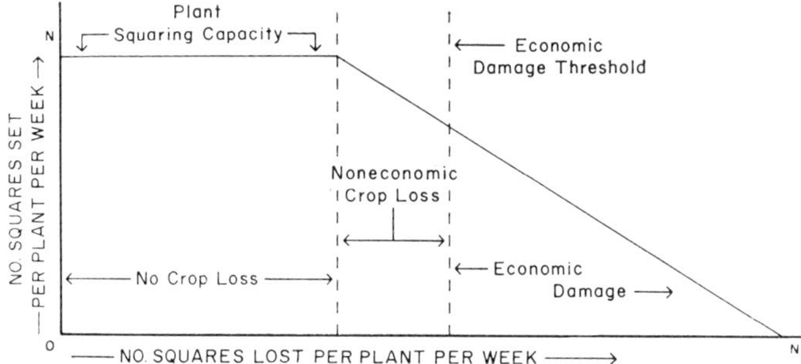

FIG. 7. Effect on cotton boll-set of plant capacity and insect injury.

This simplified analysis of lygus bug damage control implies that an increased yield will bring an increased economic return to the farmer. This is perhaps a valid assumption for the production of cotton fiber under present market conditions, but it may not be appropriate to the production of highly perishable crops, those with fixed or very limited market capacity, or others with highly variable market values.

The cost/potential benefit ratio is affected by market conditions, local conditions of grower economics, earlier investment in the crop, and sometimes the personal values of the people concerned. We must be concerned with economic levels of tolerable damage set under the prevailing economic conditions for that crop and the individual grower. Insect injury, no matter how severe, is not an economic consideration in a California field of melons that won't be harvested because the Chicago price for melons is insufficient to pay for harvesting and shipping.

Another aspect of decision making for pest control procedures involves an evaluation of the potential importance of low pest numbers serving as a food supply for entomophagous insects which will be needed later in the growth of the crop. An example of this aspect is discussed below (Section VII).

The ability to predict population trends makes it possible to set precise treatment times to avoid damage. Stern *et al.* (1959) have stressed the importance with some crops of establishing a "threshold damage level" or "treatment level" to provide time to initiate emergency procedures and prevent the pest population from reaching the economic damage threshold.

The preceding discussion of pest management emphasizes that the mere presence of an insect pest is not necessarily an indication of a threat of economic damage to the crop. In fact, we advocate the maintenance of subeconomic levels of the pests to support entomophagous forms. A pest species does not have to be eradicated (even locally) in order to be economically controlled. Economic control is achieved when economic damage is avoided no matter how dense the residual or continuing pest population may be. Achievement of economic control need not rely at all on killing the organisms; it may operate in other equally effective ways in preventing economic damage (Geier, 1966).

The philosophy of pest control based on eradication of the pest species is the antithesis of integrated pest control. Nevertheless, eradication is a legitimate goal under special circumstances. These circumstances do not prevail for most agricultural and forest pests. In most situations our goal should be to manage pest populations so as to eliminate them as pests but not to eradicate them. This philosophy was expressed by the PSAC panel on Cotton Insects (Watson *et al.* 1965). Cole (1964) also questions the eradication philosophy, on the basis that "we shall still be left with an inherently unstable biotic community, in which outbreaks of some new pest are to be expected."

VII. THE ROLE OF BIOLOGICAL CONTROL IN INTEGRATED CONTROL

Among the environmental factors affecting insect populations, entomophagous predators and parasites are the most highly susceptible to interference and disruption by the activities of man. One of the most disruptive of these interfering practices is chemical pest control. It is widely recognized that the interference of the modern insecticides with biological control has been a major cause of problems associated with the use of these materials (Bartlett, 1964c; Doutt, 1964).

There are three major viewpoints on the nature of biological control, and this difference of view has been a source of confusion in interpreting the role and importance of natural enemies. One viewpoint is utilitarian and considers biological control to be the use of organisms in the suppression of pest species (Fleschner, 1959; Franz, 1961b). This concept stems from classical biological control which involves the deliberate importation of exotic natural enemies to control pest species. (Figs. 8–10.) Man's manipulation of natural enemies is an essential aspect of this concept of biological control which usually minimizes the role of naturally occurring entomophagous species. Its adherents largely center their attention on applied problems.

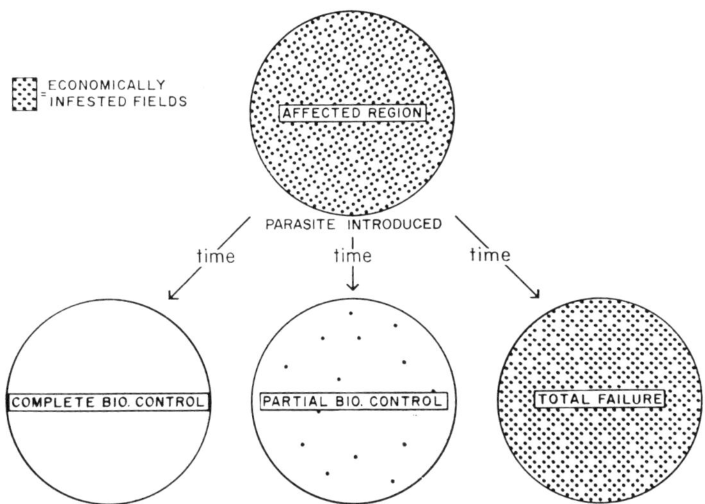

FIG. 8. Classical biological control as effected through natural enemy introduction. The top circle represents a hypothetical region (an island, an isolated valley, or even a continent) affected by a chronically epidemic-introduced pest insect species. The lower circles represent the three possible results of the natural enemy introduction (after a time lapse) as regards economic control of the pest. The left hand circle represents complete control (e.g., cottony cushion scale in California); the center circle partial control (e.g., citrus black scale in California); and the right hand circle total failure (e.g., Oriental fruit moth in Eastern United States) (examples from Clausen, 1956).

A second definition, like the first, emphasizes the utilitarian aspects of biological control. This view is broader in that it embraces biotic factors such as autosterilants, resistant strains of plants, and antibiotics, as well as entomophagy (Sweetman, 1958).

The third viewpoint of the nature and scope of biological control holds it to be the regulation of plant and animal population densities by predators, parasites, and diseases. It does not stipulate man's manipulation of

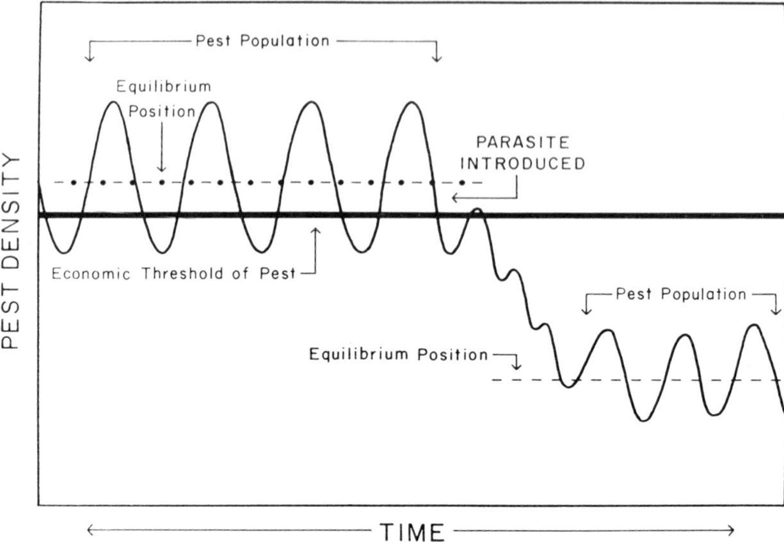

FIG. 9. Complete biological control (based on the economic criterion) of a pest by an introduced natural enemy. Note that it is not the economic threshold of the pest that is affected by the parasite, but rather its equilibrium position (e.g., long-term mean density).

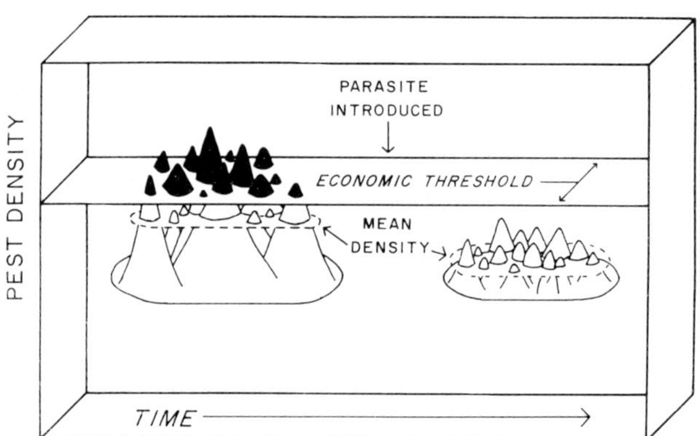

FIG. 10. Three-dimensional representation of biological control of a pest species by an introduced natural enemy. The two populations depict the distributional status of the pest species and its density at particular points in time before and after parasite introduction.

natural enemies, nor is it restricted to a consideration of pest species (Stern *et al.,* 1959; DeBach, 1964).

Argument over the definition of biological control has raged for years and a resolution of the matter can not be attempted here. However, a broader more naturalistic concept of biological control is an essential part of the development of integrated control programs. The native natural enemies play critical roles in many agro-ecosystems (Messenger, 1965). The basic ecological relationship of these native species to the pests and potential pests is not different from that of those introduced or otherwise manipulated by man.

Applied entomologists often equate the mechanics and objectives of biological control with those of chemical control. Only in the broadest view of pest control and management are they the same; in their operation and technique they are in most every respect different. With the rarest of exceptions, biological control involves permanent population regulation while chemical control almost invariably involves temporary destruction of localized populations. Only inundative colonization of parasites or predators for local pest suppression and the use of microbial insecticides in any way resemble chemical control, and even with these techniques there are fundamental differences. This is not always understood. Hence, when we judge the action of natural enemies in the same way we would evaluate a chemical control treatment, biological control is frequently seen to be inadequate. Indeed, the history of applied biological control is replete with such inadequacies and failures (Clausen, 1956). As a result, there is a widely held opinion among economic entomologists that biological control is too often inadequate or too slow in its action, and that chemical control by comparison is infinitely more efficient. This is most unfortunate because the full values of biological control must be assessed on a much broader, long-term basis. In fact the most significant role of biological control is its tremendously effective unaided action that helps us to exist on a reasonably equal footing with our arthropod competitors (see Fig. 11). Without this largely unnoticed action, it is questionable whether our modern civilization could have developed in the face of this competition for food. It is of critical importance that the full scope of "biological control" be generally understood—no matter where semantic debate may lead us.

With this broader role for biological control in mind, consideration should now be given to its place in the general scheme of integrated control. The logical place to begin is with natural control, of which biological control is a major component. It is generally recognized that tremendous pressures are exerted by environmental forces on plant and animal populations. Acting in concert, these forces determine the abundance and distribution of all living organisms. This in effect is natural control, which maintains a more or less fluctuating population density of an organism within certain upper and lower limits over a period of time by the combined action of abiotic and biotic environmental factors (Stern *et al.*, 1959).

The factors contributing to natural control can be simply classified into two major groups, namely, those that are density dependent and those that are density independent. This means that with one type of factor (density dependent) there is an increasing degree of impact with increasing density of the affected population, while with the other type (density

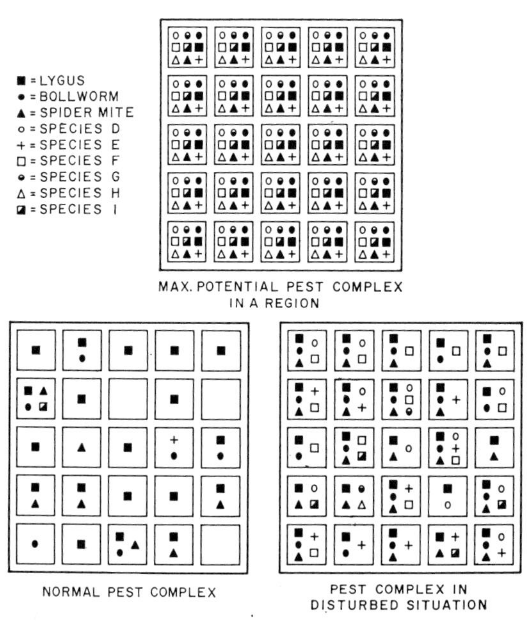

FIG. 11. Diagrammatic sketch indicating the role of naturally occurring biological control. The large squares represent a region (an island, an isolated valley, or even a continent). The small squares represent units of a given crop (cotton) in the region (units may be fields, acres, hectares, etc). The several symbols represent the phytophagous species occurring in the crop. The top square depicts the maximum potential pest complex in the region. The left-hand bottom square represents the normal situation where most of the pests are partially or totally under biological control by native natural enemies. The right hand bottom square represents a pesticide disturbed situation (similar to the Cañete Valley of Peru in the 1949–1956 era) where natural enemy activity has been interfered with so that several potential pests have reached economic status and the severity of the chronic pests has been increased.

independent) the degree of impact is *not* influenced by the density of the affected population.

Biological control agents are usually density dependent and what is even more important, they are often reciprocally density dependent (Huffaker and Messenger, 1964). That is, they not only respond with increasing intensity to increases in host density, but reciprocally their impact lessons as host density drops below a minimum threshold and continues to decline. [For a more thorough introduction into the intricacies of population dynamics see Huffaker and Messenger (1964).]

Thus, for example, the parasite wasp *Aphidius smithi* responds to rising density of *Acyrthosiphon pisum* with an increasingly heavy degree of partasitization and eventually causes the aphid population to crash. Then as the aphid population density declines below a certain minimum threshold, a progressively lower degree of parasitization is effected and the wasp, too, becomes scarce. Eventually parasitization drops to such a low level that the aphid is temporarily released from the controlling effect of the wasp and its numbers once more begin to rise. Ultimately as a certain aphid population density is reached, the wasp again responds and begins its trend towards an increasingly higher degree of parasitization on the rising aphid population (van den Bosch *et al.,* 1966).

This reciprocal density dependence of most biological control agents can be disrupted by agricultural practices, particularly insecticide application, directly through their lethal effects on the predators and parasites themselves and indirectly through effects on the host populations. Thus in California citrus groves, applications of DDT which were relatively harmless to the cottony cushion scale, *Icerya purchasi,* eliminated its predator, *Rodolia cardinalis.* This released the scale from effective biological control and the pest exploded to severely damaging abundance (DeBach, 1947). Cases of this sort have been commonplace in pest control during the past two decades (Bartlett, 1964c; Ripper, 1956; van den Bosch and Stern, 1962). Consequently, it is important in modern pest control to develop procedures that do not directly upset the activity of the biological control agents.

Direct interference with natural enemies is only one of two ways in which insecticides can disrupt the relationship between reciprocal density dependent agents and their hosts. Disruption of the natural enemies' food chains can also be important (Fig. 12). For example, if the aphid, spider mite, and thrips hosts of omnivorous predators such as *Chrysopa, Nabis, Geocoris,* and *Orius* are eliminated from cotton fields early in the season by chemical treatments, which in themselves are not significantly harmful to the predators, the latter will starve, emigrate, or cease reproducing. Later in the season when such strong-flying pest species as *Lygus hesperus* or *Heliothis zea* invade the fields, they are essentially free of predator attack and may reproduce explosively.

This is a major reason why in California we generally discourage early season chemical treatment of cotton for control of aphids, thrips, and certain caterpillars, even though they may cause some damage to the prefruiting plants. We consider this slight damage to be a minor sacrifice

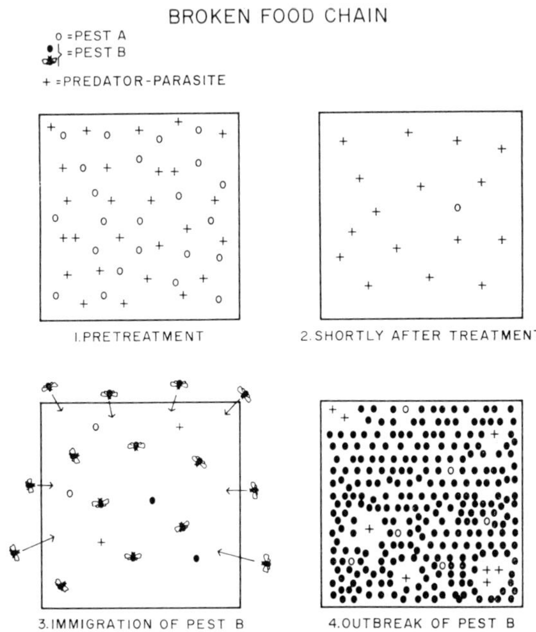

FIG. 12. Diagrammatic sketch of the influence of a chemical treatment on natural enemy–pest dispersion by disruption of food chain. The large squares represent an agricultural field or orchard: 1, pretreatment situation, pest A and predator abundant; 2, shortly after treatment, pest A is essentially eliminated but predator is still relatively abundant; 3, after time lapse, predator has starved or emigrated because of host scarcity and has itself become scarce, pest B immigrates; 4, pest B explodes to damaging abundance because of its freedom from predation.

when measured against the benefit derived from these pests because they sustain populations of entomophagous species. Additional examples could be cited. The important point is that since chemical insecticides kill insects, very large numbers of entomophagous species are destroyed when broad-spectrum materials are applied to control pest species; and even when the entomophagous species are not killed outright, elimination of their hosts causes them to starve, emigrate, or cease reproducing. Thus, biological control agents often are seriously interfered with by broad-spectrum synthetic insecticides, and this has been at the root of many of the insecticide-associated problems of recent years.

But insecticide application is not the only agricultural practice which adversely affects entomophagous species. Plowing, discing, cultivation, irrigation, fertilization, harvesting, pruning, weed control, and a number of other practices can all, directly or indirectly, affect the populations of predators and parasites (van den Bosch and Telford, 1964).

And so it is evident that we must not restrict our thoughts and efforts solely to the alteration of chemical control techniques in order to effect better utilization of biological control. We have to think in terms of the ecosystem and all the things that man does in it to affect arthropod populations.

VIII. SELECTIVE INSECTICIDES AND INTEGRATED CONTROL

Dependence on biological control alone does not offer a completely satisfactory alternative. In the first place, it is very doubtful that satisfactory biological control agents are available for all pest problems. Second, the fundamental nature of biological control is such that in many instances the control will not be perfect. At certain times the biotic regulating mechanisms may be short circuited and the pest populations will rise above tolerable levels. In other cases, the biotic agents provide a significant degree of control but this is not sufficient to keep the pest at tolerable levels. Indeed, the very fact that several thousand of the earth's insect species have attained major pest status is sufficient indication that biological control is not fully effective at all times and places. Nevertheless, most pest insects are affected by complexes of natural enemies and it is only because man has established arbitrary economic tolerances for these species that the bulk are regarded as pests. In the broad view, all pests are under natural control and, as indicated above, much of this is attributable to the action of entomophagous and entomogenous species. It is only logical, therefore, that the artificial control measures directed against insect pests be employed to augment the natural controls. And, of course, it follows from this that chemical control, the most widely used artificial control method, be employed in this augmentative manner.

This has been very difficult to do with available insecticides because the great majority of these are broadly toxic and therefore ecologically disruptive (Bartlett, 1963, 1964a,b). One of the greatest needs for the future development of integrated pest control is the development of selective or even specific insecticides (van den Bosch, 1965). In this case, we are not referring to differential selectivity between arthropods and mammals, but rather to selectivity within the Arthropoda. In other words, we need materials specific to such groups as locusts, lepidopterous larvae, weevils, pentatomid bugs, etc. Such materials should be sufficiently narrow in their toxic effects to have negligible impact on nontarget organisms and yet broad enough in their effects to make their (the chemical's) development and exploitation commercially feasible. Very few such materials are currently available and this will probably hold true for the foreseeable future (Lange *et al.,* 1965). Consequently, we must seek selectivity through the manipulation of the materials we now have in hand.

Modification of dosages, formulations, times of application, placement of materials, and other techniques can be utilized to increase the selectivity of chemical pesticides (Bartlett, 1964c; Ripper, 1956; van den Bosch and Stern, 1962). These all have the objective of lowering pest abundance to noneconomic levels while simultaneously sparing the ecosystem from highly disruptive impact.

Proof that selectivity in chemical insecticides can attain this highly desirable objective exists both in the historical and experimental record. For historical proof we have only to look back to the pre-World War II period. Insecticides in use during that time, by virtue of their methods of entry and chemical and physical characteristics, were largely selective in nature. As a result, they were relatively mild in their ecological impact and it is a matter of historical record that, during the era of their use, insecticide-associated problems were minor compared to the nightmarish ones of recent years.

Experimentation with selective controls has been carried out by a number of researchers in recent years.* Not all of these experiments have led to successful integrated-control programs, though some have; but in their totality they have clearly demonstrated the advantages of selective over broad-spectrum materials from the standpoint of ecological impact. The successful introduction and wide-scale use of a variety of selective acaracides illustrates the feasibility of future development of selectivity.

Thus we see from the historical record as well as from experimentation that selectivity in chemical pesticides does reduce problems attending their use. It therefore behooves pest control researchers to seek selectivity in

* In recent years there has been a voluminous literature on the impact of chemical insecticides on agro-ecosystems, on the testing of insecticides for selectivity, and on experimentation conducted to develop integrated control programs. These references are too numerous to cite in this chapter and the reader is referred to the journal *Entomophaga* (Publication de la Organisation Internationale de Lutte Biologique contre les animaux et les plantes nuisibles, Le François Editeur, Paris), in which a listing of the literature on integrated control has been compiled almost every year since 1956 by Franz, and Franz and Laux. When they occur, these listings are found in the fourth issue of each volume.

the currently available materials and where selective use of chemicals cannot be developed to seek alternative controls (Chichester, 1965).

Meanwhile, it has become quite clear that research chemists in the universities, governmental agencies, and industry should increasingly direct their efforts toward the development of selective chemicals, including hormones, pheromones, repellents, antifeeding compounds, and sterilants, as well as conventional materials.

IX. THE USE OF RESISTANT PLANT VARIETIES IN INTEGRATED CONTROL SYSTEMS

Very little experimentation has been devoted to utilization of insect-resistant plant varieties in integrated control programs (Kennedy, 1965; Beck, 1965). The development of resistant alfalfa varieties is a critical part of the final evolution of the integrated control program for the spotted alfalfa aphid in western United States. This program also includes the use of native predators, introduced parasites, several entomophagous fungi, chemical treatments, and early or strip-cuttings of the crop. A low level of plant resistance can increase the effectiveness of natural enemies where either alone is insufficient to maintain populations below the economic level (van Emden and Wearing, 1965) (Fig. 13). As resistant plant varieties become available there is usually little difficulty in integrating them with other control techniques. The comments of van Emden and Waring suggest that the plant breeder may not need to seek the "per-

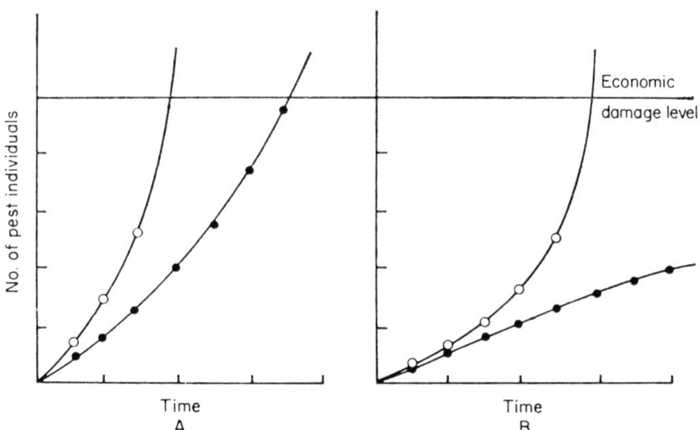

FIG. 13. The influence of a low level of plant resistance to pest attack on the effectiveness of natural enemies: A, susceptible plant (log pest multiplication rate: 2.25); B, resistant plant (log pest multiplication rate: 2.00); ○ without predators; ● with predators (from van Emden and Wearing, 1965).

fectly resistant" plant variety, and that the low levels of resistance provided by fertilizer treatment and level of water stress may have importance in integrated control. Way and Murdie (1965) have also suggested that "qualities should be sought which may not necessarily decrease damage by the pest but which alter its behaviour so that it is more readily controlled by other means . . ." and which "increase the plant attractiveness to the pest's natural enemies."

As a final point with respect to resistant varieties, it should be stressed that, just as with other control techniques, such varieties should be utilized only after taking into account their role and impact on other elements in the agro-ecosystem. For example, the new strawberry varieties that were introduced into California because of their resistance to certain pathogens lost much of this advantage because of their high susceptibility to cyclamen mite, a minor pest on the older strawberry varieties. Similar resistance is alfalfa to pea aphid might be of some economic benefit to the alfalfa industry in California. However, alfalfa can tolerate fairly heavy pea aphid populations without suffering economic damage and in California the crop is not severely damaged despite recurrent infestations of moderate density. In fact, there is reason to fear that elimination of these pea aphid populations through introduction of resistant alfalfa might well create a greater general economic loss than the savings effected by the reduction of the aphid as a pest of alfalfa. At present, the pea aphid populations in alfalfa serve as an important link in the food chains of coccinellids and other important predator species. The release of pests in crops associated with alfalfa from the biotic pressures produced by predators originating on pea aphid in alfalfa could significantly aggravate the pest problems in these associated crops.

X. PRACTICAL APPLICATION OF THE INTEGRATED CONTROL CONCEPT

Over the past 2 decades, a number of successful integrated-control programs have been developed in widely scattered areas (Boza-Barducci, 1965; Pickett and McPhee, 1965; Ripper, 1956; Stern *et al.,* 1959; van den Bosch, 1965).

It is most interesting that these programs have invariably been products of crises. In other words, growers and researchers driven to desperation by a variety of problems associated with largely unilateral use of chemical insecticides were forced to the ecological approach to solve these problems. This is perhaps one of the most significant developments in modern pest control, for out of chaos has come a whole array of imaginative techniques and ecologically selective materials which bid well to transform pest control at the applied level into a highly sophisticated technology.

The scattered successes have captured the interest and imagination of entomologists everywhere, and in recent years there has been a striking proliferation in the number of integrated control programs under experimental development. Certain of the programs that have been developed merit special discussion.

A. Apple and Pear Insects in Nova Scotia

The true landmark program was that developed in Nova Scotia by Pickett and his colleagues for control of apple and pear pests (Pickett and McPhee, 1965). This program had enormous impact on the field of pest control just at the time when the full significance of the pesticide dilemma was dawning on economic entomologists and the general public as well.

Thus while much of the entomological world neglected ecology and occupied itself in the unilateral exploitation of the new synthetic organic insecticides, Pickett and his colleagues working quietly, diligently, and in their far northern valley imaginatively evolved a brilliantly conceived, integrated control program. As a consequence, when the inadequacy of unilateral chemical control was finally generally recognized and alternatives were sought, the Nova Scotia program stood out like a beacon for all to use as a guide and model. Mistakes have been made by persons trying to adapt the program in its exact detail to different ecological conditions, but the principles that were demonstrated have remained fully valid and with proper adjustments are leading to significant developments in a number of areas (Geier, 1965; Hukusima, 1963; Mathys and Baggiolini, 1965; Scepetilnikova, 1965; Steiner, 1965; Wiakowski and Wiakowska, 1965; Wildbolz, 1963).

B. Cotton Insects in the Cañete Valley, Peru

A second integrated control program of outstanding significance was that developed for cotton insect control in the Cañete Valley of Peru (Boza-Barducci, 1965). This program was a product of extreme crisis and here too, an ecologically oriented entomologist, Wille (1951, 1952), very quickly grasped the basic nature of the problem and initiated studies leading to an intricately conceived integrated pest-control program.

The Cañete Valley is an essentially self-contained agro-ecosystem. It is one of a series of coastal valleys formed by streams running down from the high Andes to the Pacific Ocean. In their lower reaches, these valleys are cultivated to such crops as cotton, maize, potato, fruits, vegetables, etc. In their upper reaches, they connect with the inter-Andean valleys and are clothed in indigenous vegetation.

Most of the Peruvian coastal valleys have a unique climate characterized by little or no rainfall, low humidity, and abundant sunshine. All crops are under irrigation. Thus each valley is literally a self-contained agro-ecosystem isolated from the others by intervening vegetationally near sterile ridges and running in their upper reaches to wild vegetation.

The Cañete Valley is one of the most technologically advanced of these Peruvian agro-ecosystems. The growers in it are organized into an association (La Association de Hacendados de Cañete) and they have their own experimental station (La Estacion Experimental Agricola de Cañete) to serve their 22,000 hectares of cultivated land. Originally the valley was largely planted to sugar cane, but in the 1920's a shift was made to the cultivation of cotton. Currently about two thirds of the valley (approximately 15,000 hectares) is planted to that crop. The Cañete agricultural operations are the most highly mechanized in Peru. Other advanced agricultural, agronomic, horticultural, and plant-protection techniques have also been adapted to crop production in the valley.

It was the adoption of modern synthetic organic insecticides for cotton pest control which led the valley to the brink of economic disaster and ultimately to its salvation in integrated control.

The Cañete story stands as a classic example of the problems that can beset pest control which ignores ecology and relies on unilateral use of broad-spectrum insecticides (Table I). Once this approach to pest control was initiated, the Cañete was doomed to disaster because of certain factors peculiar to it. The most important of these was its status as an oasis in an otherwise arid and biologically impoverished land. Thus when synthetic organic insecticides were brought into use they were essentially applied as a blanket over the entire ecosystem. As a result, literally the entire biota of the valley was repeatedly exposed to these materials. This had two critical effects: (1) decimation of the parasite and predator fauna, which in turn relieved the key pest species as well as potential pests from biotic repression, and (2) rapid selection for insecticide resistance in the pest species.

Both developments occurred with great rapidity and with devastating effect. For example, though use of synthetic organic insecticides, principally DDT, BHC (1,2,3,4,5,6-hexachlorocyclohexane), and toxaphene, was not initiated until 1949, resistance to these materials had already developed by 1955. Where 5% DDT, 3% BHC, and a 3 : 5 : 40% BHC–DDT–sulfur mixture had proved effective in 1949, 15% DDT, 10% BHC, and a 5 : 10 : 40% BHC–DDT–sulfur mixture often applied at elevated dosages, were necessary for control by 1955. Furthermore, the interval between treatments was progressively shortened from a range of 8–15 days down to 3 days. Meanwhile, a whole complex of previously innocuous insects had risen to serious pest status. Included were *Argyrotaenia sphaleropa, Platynota* sp., *Pseudoplusia rogationis, Pococera atramenalis, Planococcus citri,* and *Bucculatrix thurberiella.*

TABLE I

PEST COMPLEX AND AVERAGE ANNUAL YIELD OF COTTON IN THE CAÑETE VALLEY, PERU DURING THREE PEST CONTROL REGIMES[a]

Pest complex		
Heavy metal and botanical insecticides, 1943–1949 (502.3 kg/hectare)	Synthetic organic chemical insecticides, 1949–1956 (603.8 kg/hectare)	Integrated control, 1956–1963 (789.1 kg/hectare)
Anthonomus vestitus	*A. vestitus*	*A. vestitus*
Anomis texana	*A. texana*	*A. texana*
Aphis gossypii	*A. gossypii*	*A. gossypii*
Heliothis virescens	*H. virescens*	*H. virescens*
Mescina peruella	*M. peruella*	*M. peruella*
Hemichionaspis minor	*H. minor*	*H. minor*
Dysdercus peruvianus	*D. peruvianus*	*D. peruvianus*
	Argyrotaenia sphaleropa	*P. atramenalis* (?)
	Platynota sp.	
	Pseudoplusia rogationis	
	Pococera atramenalis	
	Planococcus citri	
	Bucculatrix thurberiella	

[a] Data obtained from Boza-Barducci (1965).

Substitution of new insecticides, including organophosphorus compounds, for the older chlorinated hydrocarbons finally became necessary and then even these failed. Finally, as a result of pest resistance to the insecticides and the burgeoning complex of injurious species, the cotton yield plummeted alarmingly; the average yield per hectare in 1956 was the lowest in more than a decade.

With the situation bordering on disaster, the cotton growers of the Cañete Valley appealed to their experiment station to formulate a plan to bring about relief from their problem. This was quickly done, and the plan "Reglamentacion de Cultivo del Algodonero en el Valle de Cañete" was approved by the Peruvian Ministry of Agriculture under Ministerial Resolution No. 1427 (July, 1956) which stipulated a number of changes in pest control practices as well as the adoption of certain cultural practices pertaining to cotton production.

Among these were:

1. Elimination of cotton production in marginal lands.
2. Prohibition of ratooning in the more newly developed part of the valley.
3. Obligatory "dry cultivation" of that part of the valley where 2-year ratooning was permitted in order to cause maximum mortality to *Heliothis virescens* pupae.
4. Prohibition of the use of synthetic organic insecticides and a return to the old materials (i.e., calcium and lead arsenates and nicotine sulfate) except by dispensation of a special commission established to review exceptional cases. This commission usually recommends use of systemic materials at 25–50% of the dosages suggested by the manufacturers.
5. The repopulation of the valley with beneficial insects introduced from other Peruvian valleys or from foreign areas.
6. Establishment of planting deadlines and deadlines for plowing under or burning crop residues, pruning ratoon cotton, and application of the first irrigation.
7. Employment of a number of cultural practices leading to the establishment of a vigorous and uniform cotton stand.

As a result of this program, there was a rapid and striking reduction in the severity of the cotton pest problem in the Cañete Valley. The whole complex of formerly innocuous species, which cropped up in the organic insecticide period, reverted to their innocuous status. Furthermore, the intensity of the key pest problems also diminished and so there was an overall reduction in direct pest control costs. But what was most impressive was the sharp increase in cotton yield which attained its highest levels in the history of production in the Cañete Valley, averaging 789.1 kilos per hectare for the period 1957 through 1963 contrasted to the 603.8 kilo average for the 6-year organic insecticide era.

The Cañete Valley experience is especially significant because it compressed into a short period of time and a restricted area so many of the adverse factors that can plague a pest control program which is solely reliant on broad-spectrum insecticides. Throughout the world many situations have developed where pesticide resistance, pest flarebacks, or secondary pest outbreaks have occurred. But to have the combination so starkly evident and disastrously detrimental as in the Cañete Valley has been rare. This does not mean that other similar situations have not occurred or do not now exist. Indeed, if entomologists and growers were to take a hard look at a number of modern pest control programs they might well discover that in terms of cost, total pest complex, and even in quantity and quality of produce, matters are no better and perhaps even worse now than they were in the pre-World War II era.

But even if cases such as that in the Cañete Valley are not everywhere evident, it is important that entomologists and growers be made aware that such incidents can and do happen. They should also profit from the manner of solution of the Cañete problem which most significantly was not brought about solely by substitution of selective insecticides for more broadly toxic ones, but rather by a combination of practices which collectively produced a better ecological balance in the valley. It is of added significance that in those Peruvian Valleys where programs similar to that invoked in the Cañete had long been in force disaster of the sort that struck the Cañete never did occur (personal communication from J. Simon to our colleague L. E. Caltagirone).

C. Spotted Alfalfa Aphid in California

A third integrated control program of major significance was that developed for the control of the spotted alfalfa aphid, *Therioaphis trifolii,* in California (Stern *et al.,* 1959).

Therioaphis trifolii, an accidental immigrant into North America, was first detected in California in 1954. Within 2 years it had spread over all of the major alfalfa-producing areas of the state and had exploded to such devastating abundance that it threatened the very existence of the industry (Smith, 1959).

Initially, biological controls were inadequate to meet the aphid's onslaught and an emergency chemical control program involving broad-spectrum organophosphates was invoked as a stopgap measure. Meanwhile, native predators, particularly coccinellids, a fungus disease, and imported parasites began to take a heavy toll of the aphid (Davis *et al.,* 1957). But the broad-spectrum insecticides interfered with the predators and parasites so that they could not attain full efficiency. This led to extensive use of chemical control. When the aphid developed resistance to the organophosphates, control costs spiraled and aphid damage approached disastrous proportions.

The only solution seemed to be in the development of a selective chemical control that would augment and not interfere with biological control. Experimentation was directed toward this end and very quickly a selective chemical control (low dosages of demeton) was developed and put into widespread application. Within 1 year a general integrated control program was invoked, biological control augmented by selective chemical control came into full effect, and the crisis passed. Between 1957 and 1958 costs and losses attributable to the aphid dropped from approximately $9,705,000 to about $1,694,000 (DeBach, 1964). Subsequent development of aphid-resistant alfalfa varieties has all but eliminated *T. trifolii* as a pest of alfalfa in California.

The development of a selective chemical control for *T. trifolii* was significant in itself—but this was secondary to the impact that this dramatically effective program had on the general pattern of pest control in California and perhaps elsewhere too. Certainly in California it opened the eyes of research entomologists and agriculturalists to the potential of integrated control. Subsequently, several additional integrated control programs based on selective use of insecticides were developed. Currently, even more sophisticated programs involving cultural practices as well as selective use of insecticides are underway (van den Bosch, 1965).

XI. RETROSPECT AND PROSPECT

The remark is frequently made that there is nothing unique about integrated control; it is simply good pest control. One can hardly argue this point. Instead there simply appears to have been a lot of bad pest control during the past 2 decades, for what else can explain the manifold problems that developed. However, among entomologists there seems to be an increasing awareness of the mistakes that were made, and there now is a general trend towards an ecological approach to pest control. This indicates that good pest control is on the increase, which seemingly portends a bright future for integrated control (Smith and Reynolds, 1966).

Wigglesworth (1965) states: "If one looks back to the methods of countering insect pests in earlier days one always finds 'integrated control'; no holds were barred; a multiplicity of possible control measures were brought to bear. By and large, these control measures were ineffective." We cannot agree with Sir Vincent's conclusion that these early efforts comprised integrated control. By and large, pest control before 1945 did not involve a conscious effort to integrate the various elements of pest control into the agro-ecosystem. Instead, for the most part, control was applied piecemeal. Cultural methods, pest-resistant plant varieties, and biological and chemical controls were independently developed and exploited. They were not coordinated into pest management systems.

The insecticides that were then available were not very efficient, but neither were they particularly disruptive to the general environment. In fact, ecological impact was hardly of concern to the applied entomologists of that era. They were able to make relatively trouble-free use of insecticides, not because they tried to integrate them into the agro-ecosystems, but simply because these materials had little impact on the other elements of these systems.

The very fact that broad-spectrum organic insecticides were developed and brought into wide-scale use is proof in itself that entomologists in general were not thinking as ecologists. If they had done so, such materials would never have been used indiscriminately. If more entomologists and others involved in pest control had thought of the ecological and genetic implications, they would have known that unilateral use of broad-spectrum insecticides was doomed to failure. They clearly did not anticipate such complications.

The synthetic organic insecticides were developed on the basis of chemical, toxicological, and economic criteria. Essentially, no ecological considerations went into their creation. With rare exception, when these materials were handed to the economic entomologists, they unhesitatingly applied them to the current pest problems.

As late as the mid-fifties, an experienced entomologist counseled his colleagues, perhaps facetiously, to make collections of major pest species for deposition in museums since it would only be a matter of time before these pests would be eradicated by the new insecticides.

We feel that it is abundantly clear that 20 years ago ecological consideration was only rarely a factor in pest control. The general awakening to the ecological nature of pest control only occurred after the chaotic events of the organic insecticide era made it apparent. A few individuals (among others, Wigglesworth, 1945; Pickett, 1949) can point out that they held an ecological viewpoint, but their voices were lost in the wilderness. It is a matter of history that the ecological approach was virtually a lost concept during the first decade of the organic insecticide era.

Another very significant item that can be directly attributed to the synthetic organic pesticides was the enormous expansion in the field of economic entomology and pest control technology. Initially this activity was largely restricted to areas of pesticide chemistry, pesticide toxicology, and field exploitation of the vast array of new materials. As the full impact of these new materials came to be felt in the various agro-ecosystems, the significance of their ecological disruptiveness was manifested. This caused very great concern to all and indeed has created a furor in some quarters. A major outcome of this concern was a significant redirection of pest-control research emphasis to areas other than insecticide chemistry. As a result, in very rapid order unprecedented financial support and intellectual effort were directed into pest control. Most importantly, investigations were initiated along previously unexploited channels. Research progress on pheromones, attractants and repellents, chemosterilants, antimetabolites, hormones, insect pathology, host resistance, insect nutrition, biological control, and other areas has been truly staggering. Already, amazing and unprecedented pest control techniques have been made available and additional startling developments are just over the horizon.

Let us not again make the same mistakes with these new tools and techniques. If we seize them and unheedingly impose them on our agro-ecosystems, we will invite disasters of equal or of greater magnitude than those resulting from the unilateral use of the synthetic organic insecticides. Imagine the enormous ecological impact of juvenile hormone if such were ever to be used unilaterally or indiscriminately in the general environment. Similarly, if through autosterilization techniques we eradicate a key pest, we cannot be sure that this will not release from competitive suppression a second species which will then burgeon to economic abundance. Even the development of pest-resistant plant varieties, which has been extolled as the most desirable form of pest control, must be carefully considered in an ecological context.

The imposition of any pest control technique upon the agro-ecosystem can cause serious ecological problems. The very fact that we have available to us today a wider array of techniques of much greater sophistication and potency than ever before makes it extremely critical that we think of pest control in the broadest ecological terms. From this we can only conclude that the need for the integrated control approach is more critical than ever before.

The integrated control approach will not come about simply because we want it to. For the entomologist it entails intensive and well-balanced training, high competence, objectivity, ingenuity, cooperativeness, dedication, and perseverence. Nevertheless, even if these characteristics are generally inculcated into research entomologists, good pest control will not necessarily prevail if the grower, the chemical manufacturer, and the consumer do not comprehend its nature or desire it.

The grower wants quick, effective, uncomplicated, low-cost pest control. The chemical industry desires inexpensive, widely effective, highly competetive products that will be at an advantage in the battle for markets. The consumer wants high quality inexpensive produce unmarred by pest damage and uncontaminated by insect parts. All of these desires are understandable, and to a considerable degree they can be met. But there must be compromise, for as they stand now they are too rigid to permit a workable ecological approach to pest control. Integrated control is not simple, it is not always inexpensive, and it often will not give the astonishingly rapid and seemingly complete kills which we have come to expect from the powerful insecticides. The grower, the pesticide manufacturer, and the consumer must understand this. It will not be an easy task to bring about such understanding, but it must be done if integrated pest control is to become a widespread reality.

This is the challenge that confronts the field of applied entomology. With careful, imaginative experimentation and working cooperation among the persons involved in research as well as effective communication with those other groups concerned with pest control, this goal can be attained.

REFERENCES

Agr. Res. Council (1964). "Report of the Research Committee on Toxic Chemicals," 38 pp. London.

Allen, W. W., and Smith, R. F. (1958). *Hilgardia* **28,** 1.

Anderson, J. R. (1966). *Bull Entomol. Soc. Am.* **12,** 342.

Anderson, J. R., and Poorbaugh, J. H. (1964). *J. Med. Entomol.* **1,** 131.

Bartlett, B. R. (1956). *Agr. Chem.* **11,** 42.

Bartlett, B. R. (1963). *J. Econ. Entomol.* **56,** 694.

Bartlett, B. R. (1964a). *J. Econ. Entomol.* **57,** 366.

Bartlett, B. R. (1964b). *J. Econ. Entomol.* **57,** 559.

Bartlett, B. R. (1964c). *In* "Biological Control of Insect Pests and Weeds" (P. DeBach, ed.), Chapt. 17, pp. 489–511. Reinhold, New York and Chapman & Hall, London.

Bates, M. (1963). "Man in Nature." Prentice-Hall, Englewood Cliffs, New Jersey.

Bates, M. (1964). *Bull. Entomol. Soc. Am.* **10,** 67.

Beck, S. D. (1965). *Ann. Rev. Entomol.* **10,** 207.

Birch, L. C. (1960). *Ann. Rev. Entomol.* **5,** 239.

Blair, W. F. (1964). *Bull. Entomol. Soc. Am.* **10,** 225.

Bombosch, S. (1966). *Proc. FAO Symp. Integrated Pest Control, October, 1965, Rome.*

Boza-Barducci, T. (1965). Experiencas sobre el empleo del control biologico y de los metodos de control integrado de las plagas del algodonero en el Valle de Cañete—Peru, S.A. Trabajo presentado por el autor en el seminario organizado por el comité de Investigacion de la Producción del Algodon (Washington, D.C., 18–19 de Mayo 1965) bajo los auspicios del Comité Consultivo Internacional del Algodon, con motivo de su XXIV Reunion Planaria (Washington, D.C., U.S.A., Mayo de 1965). 25 pp., mimeo.

Brown, A. W. A. (1951). "Insect Control by Chemicals," 817 pp. Wiley, New York.

Burnett, T. (1960). *Federation Proc.* **19,** 557.

Carson, R. (1962). "Silent Spring." Houghton, Boston, Massachusetts.

Chant, D. A. (1964). *Can. Entomologist* **96,** 182.

Chichester, C. O., ed. (1965). "Research in Pesticides." Academic Press, New York.

Clark, L. R. (1964). *Australian J. Zool.* **12,** 362.

Clausen, C. P. (1956). *U.S. Dept. Agr. Tech. Bull.* **1139.**

Cole, L. C. (1964). *Am. J. Public Health* **54,** 24.

Cook, S. F. (1955). *Anthropol. Records* **16,** 31.

Dasmann, R. F. (1965). "The Destruction of California." Macmillan, New York.

Davis, C. S., *et al.* (1957). "The spotted alfalfa aphid and its control in California," 43 pp. Calif. Agr. Extension Serv. Pam., Berkeley, California.

DeBach, P. (1947). *Calif. Citrograph* **32,** 406.

DeBach, P. (1964). *In* "Biological Control of Insect Pests and Weeds" (P. DeBach, ed.). Reinhold, New York and Chapman & Hall, London.

DeBach, P. (1965). *In* "The Genetics of Colonizing Species" (H. Baker and G. L. Stebbins, eds.), pp. 287–306. Academic Press, New York.

DeBach, P., and Sundby, R. A. (1963). *Hilgardia* **34,** 105.

Doutt, R. L. (1964). *Bull. Entomol. Soc. Am.* **10,** 83.

Doutt, R. L. (1965). *In* "Research in Pesticides" (C. O. Chichester, ed., pp. 257–64. Academic Press, New York.

Doutt, R. L., and Nakata, J. (1965). *Calif. Agr.* **19** (4), 3.

Egler, F. E. (1964a). *Am. Scientist* **52,** 110.

Egler, F. E (1964b). *BioScience* **14,** 29.

Elton, C. S. (1958). "The Ecology of Invasions by Animals and Plants." Methuen, London.

Fleschner, C. A. (1959). *Science* **129,** 537.

Franz, J. (1961a). *IUCN Symp., Warsaw, July 1960,* p. 93.

Franz, J. (1961b). *Z. Pflanzenkrankh. Pflanzenschutz* **68,** 321.

Geier, P. W. (1964). *Australian J. Zool.* **12,** 381.

Geier, P. W. (1965). *Proc. 12th Intern. Congr. Entomol., London, 1964* p. 598.

Geier, P. W. (1966). *Ann. Rev. Entomol.* **11,** 471.

Hagen, K. S. (1962). *Ann. Rev. Entomol.* **7,** 289.

Hamstead, E. O., and Gould, E. (1957). *J. Econ. Entomol.* **50,** 109.

Huffaker, C. B., and Messenger, P. S. (1964). *In* "Biological Control of Insect Pests and Weeds" (P. DeBach, ed.), pp. 45–117. Reinhold, New York and Chapman & Hall, London.

Hukusima, S. (1958). *Hirosaki Daigaku Nogakuku Gakujutsu Hokoku* **4,** 72.

Hukusima, S. (1963). *Gifu Daigaku Nogakubu Kenkyu Hokoku* **18,** 61.

Johnson, C. G. (1965). Entomol. Dept. *Rept. Rothamsted Expt. Sta., 1964,* 177.

Kennedy, J. S. (1965). *Ann. Appl. Biol.* **56,** 317.

Kettlewell, H. B. D. (1961). *Ann. Rev. Entomol.* **6,** 245.

Kuenen, D. J. (1949). *Tijdschr. Entomol.* **91,** 83.

Kuenen, D. J., ed. (1961). *IUCN Symp., Warsaw, July, 1960.*

Laird, M. (1963). *Bull. World Health Organ.* **29,** 147.

Lange, W. H., Feichtmeir, E. F., Persing, C. O., Gardner, L. P., Jeppson, L. R., Lyon, R. L., Keith, J. O., and French, W. L. (1965). *Bull. Entomol. Soc. Am.* **11,** 71.

LeRoux, E. J. (1960). *Ann. Entomol. Soc. Quebec* **6,** 87.

LeRoux, E. J. (1964). *Bull. Entomol. Soc. Am.* **10,** 70.

Lewis, T. (1965). *Sci. Hort.* **17,** 74.

Lord, F. T., and Stewart, D. K. R. (1961). *Can. Entomologist* **93,** 924.

Macdonald, D. R. (1965). *Brookhaven Symp. Biol.* [**BNL 917** (C 43)].

Macfadyen, A. (1964). *In* "Grazing in Terrestrial and Marine Environments" (D. J. Crisp, ed.), p. 336. Blackwell, Oxford.

Massee, A. M. (1954). *Rept. Commonwealth Entomol. Conf. 6th, London, 1954* p. 53.

Massee, A. M. (1958). *Proc. 10th Intern. Congr. Entomol., Montreal, 1956* **3,** 163.

Mathys, G., and Baggiolini, M. (1965). *Deut. Pflanz.-Tag. Mitt. Biol. Bundesrep. Land-Forstwirt.* **115,** 21.

Messenger, P. S. (1965). *Ann. Appl. Biol.* **56,** 328.

Metcalf, R. L. (1965). *In* "Research in Pesticides" (C. O. Chichester, ed.), pp. 17–29. Academic Press, New York.
Milne, A. (1965). *Ann. Appl. Biol.* **56,** 338.
Munakata, K., and Okamoto, D. (1966). "Symposium on the Major Insect Pests of Rice." Johns Hopkins Press, Baltimore, Maryland.
Odum, E. P. (1964). *BioScience* **14,** 14.
Ordish, G. (1952). "Untaken Harvest." Constable Press, London.
Pickett, A. D. (1948). *Ann. Rept. Entomol. Soc. Ontario* **79,** 5.
Pickett, A. D. (1949). *Can. Entomologist* **81,** 67.
Pickett, A. D., and McPhee, A. W. (1965). *Proc. 12th Intern. Congr. Entomol., London, 1964* p. 597.
Piemeisel, R. L., and Lawson, F. R. (1937). *U.S. Dept. Agr. Tech. Bull.* **557.**
Pimentel, D. (1961a). *J. Econ. Entomol.* **54,** 108.
Pimentel, D. (1961b). *Ann. Entomol. Soc. Am.* **54,** 76.
Riegert, P. W., Pickford, R., and Putnam, L. G. (1965). *Can. Entomologist* **97,** 508.
Ripper, W. E. (1956). *Ann. Rev. Entomol.* **1,** 403.
Rudd, R. L. (1964). "Pesticides and the Living Landscape." Univ. of Wisconsin Press, Madison, Wisconsin.
Scepetilnikova, V. (1965). *Proc. 12th Intern. Congr. Entomol., London, 1964* p. 600.
Schneider, F. (1955). *Verhandl. Deut. Ges. Angew. Entomol.* **13,** 18.
Smith, R. F. (1959). *Hilgardia* **28,** 647.
Smith, R. F. (1962). *Proc. N. Central Branch Entomol. Soc. Am.* **17,** 71.
Smith, R. F., and Reynolds, H. T. (1966). *FAO Symp. Integrated Pest Control Rome. October, 1965.*
Smith, R. F., Bryan, D. E., and Allen, W. W. (1949). *Ecology* **30,** 288.
Smith, W. (1939). "Garden of the Sun." Lyman House, Los Angeles, California.
Solomon, M. E. (1953). *Chem. Ind. (London)* p. 1143.
Southwood, T. R. (1962). *Proc. 11th Intern. Congr. Entomol., Vienna, 1960* **3,** 54.
Stark, R. W. (1961). *J. Forestry* **59,** 493.
Stark, R. W. (1965). *Ann. Rev. Entomol.* **10,** 303.
Steiner, H. (1962). *Entomophaga* **7,** 207.
Steiner, H. (1965). *Proc. 12th Intern. Congr. Entomol., London, 1964* p. 599.
Stern, V. M., Smith, R. F., van den Bosch, R., and Hagen, K. S. (1959). *Hilgardia* **29,** 81.
Sweetman, H. L. (1958). "The Principles of Biological Control." W. C. Brown, Dubuque, Iowa.
Turnbull, A. L., and Chant, D. A. (1961). *Can. J. Zool.* **39,** 697.
van den Bosch, R. (1965). *In* "Research in Pesticides" (C. O. Chichester, ed.), pp. 297–301. Academic Press, New York.
van den Bosch, R., and Hagen, K. S. (1966). *Calif. Agr. Expt. Sta. Bull.* **820.**
van den Bosch, R., and Stern, V. M. (1962). *Ann. Rev. Entomol.* **7,** 367.
van den Bosch, R., and Telford, A. D. (1964). *In* "Biological Control of Insect Pests and Weeds" (P. DeBach, ed.), pp. 459–488. Reinhold, New York and Chapman & Hall, London.
van den Bosch, R., Schlinger, E. I., Lagace, C. F., and Hall, J. C. (1966). *Ecology* **47,** 1049.
van Emden, H. F. (1965). *Sci. Hort.* **17,** 126.
van Emden, H. F., and Wearing, C. H. (1965). *Ann. Appl. Biol.* **56,** 323.

Voûte, A. D. (1946). *Arch. Neerl. Zool.* **7,** 435.

Voûte, A. D. (1964). *Intern. Rev. Forestry Res.* **1,** 325.

Watson, J., Giles, W. L., Johnston, H. G., Lovvorn, R., Newsom, L. D., Roeder, K., and Smith, R. F. (1965). "Cotton Insects." President's Sci. Advisory Comm., Washington, D. C.

Watt, K. E. F. (1965). *Can. Entomologist* **97,** 887.

Way, M. J., and Murdie, G. (1965). *Ann. Appl. Biol.* **56,** 326.

Whitcomb, W. H., and Bell, K. (1964). *Arkansas, Univ.* (*Fayetteville*) *Agr. Expt. Sta. Bull.* **690.**

Wiakowski, S. K., and Wiakowska, I. (1965). *Proc. 12 Intern. Congr. Entomol., London, 1964* p. 601.

Wigglesworth, V. B. (1945). *Atlantic Monthly* **6,** 107.

Wigglesworth, V. B. (1965). *Ann. Appl. Biol.* **56,** 315.

Wildbolz, T. (1963). *Schweiz. Z. Obst-Weinbau* **72,** 630.

Wille, J. E. (1951). *J. Econ. Entomol.* **44,** 13.

Wille, J. E. (1952). *In* "Entomologia Agricola del Peru" (Junta Sanidad Vegetal Direccion Gen. Agr. Ministerio Agr. eds.), 2nd Ed., Revised and Enlarged, 543 pp. Lima, Peru.

Part II

VERTEBRATE PESTS

10 BIOCONTROL AND CHEMOSTERILANTS

Walter E. Howard

DEPARTMENT OF ANIMAL PHYSIOLOGY
UNIVERSITY OF CALIFORNIA
DAVIS, CALIFORNIA

I. INTRODUCTION TO VERTEBRATE CONTROL

To obtain enough food, clear water, and clean air and to satisfy man's needs for leisure, recreation, and aesthetic experience necessitates a sound ecological understanding of the problems of managing vertebrate pests. And because of the continuum of complex interactions that exists among living things, this requirement is not easily satisfied. Unfortunately, man's "progress" by definition means alteration; and whenever the new balance or equilibrium (which animal populations establish following any manipulation of the original environments) either threatens the existence of a desired environment or becomes a direct pest to man, some form of reductional control or management then becomes necessary to alleviate the difficulty.

Curiously obscured in an era of technological domination is the fact that one living thing depends for its existence on another. The goal of crop, livestock, timber, and wildlife managers is the effective channeling of energy from living things into useful products. Pest control at its simplest ensures the well-being of man by protecting his person and by inhibiting those living competitors that would take from him what he wants and needs (Rudd, 1964).

The interrelationships of man and animals have become increasingly complex as human populations have increased. Man's demand for additional food, fiber, and timber have required more intensive use of the lands and waters. Most habitats for wild creatures have become so altered that many forms have suffered large population reductions. Other animals, finding these alterations to their liking, have established new balances and substantially increased in density, often creating problems that adversely affect man's interests and welfare. That some vertebrate animals, just as with insects, may at times become pests is a self-evident platitude: "In point of fact, there are numerous situations where control of predators, rodents, and even some birds is essential to protect agricultural and pastoral interests or human health and safety" (Leopold *et al.,* 1964).

Since times and social values are changing, even many "pest" animals are now assuming recreational significance and these new aesthetic values must be recognized. The citizenry is becoming more deeply concerned, and rightly so, with the changing environments and patterns of life in the world in which they live. The growing reaction against unwarranted killing of any animal, which in the eyes of many is not actually creating a problem, can no longer be ignored in our present affluent society. It is paramount that control programs fully recognize this sliding scale of values and that all animal control activities be limited strictly to the troublesome species and preferably to the specific individuals concerned whenever possible.

The use of poisons to cope with vertebrate pest problems has developed rapidly and without the desired degree of concomitant research concerning the evaluation of these toxicants, determination of dosages to use, and methods of application. The primary reason for this situation is that these logical counterbalances have been grossly underfinanced. But our concern in this chapter is with the use of biological, chemical, and other nontoxic means of "artificially" regulating the densities of vertebrate populations and of preventing damage by these animals.

My objective in this chapter is first to introduce the field of vertebrate pest control and then to evaluate and analyze critically various aspects of the little-known fields of biocontrol and chemosterilant use as a means of preventing animal damage either by controlling local populations of unwanted mammals, birds, and other vertebrates or by preventing the individuals from doing damage. Even though there is little knowledge presently available regarding the utility of these nonconventional methods of regulating undesirable population densities of vertebrates or of preventing damage by other means, they fortunately show some promise, especially with integrated control. Integrated control places the primary dependence upon natural regulating forces and sparingly intercalates other means of reductional control of the target species when they are necessary. The need for a keener insight into these nonpesticide approaches to the control of certain mammals, birds, and other vertebrate pests in order to help create self-regulatory ecological units is, obviously, of paramount importance.

Since discussions of the control of vertebrate animals frequently become emotional, I would like to dispel all parochial feelings among the affected parties at the onset in hopes that the contents herein then can be perused dispassionately. This report is not intended to be a philosophical treatise of the emotional aspects of this subject; they have been well-covered by Carson (1962) and Rudd (1964). Even though the multiplicity of vertebrate pest problems may not always have been fully represented by all authors, and even though certain liberties have been taken in statements of some facts and in the predictions of some eventualities that have not materialized, such authors have presented, nevertheless, a compendium of needed data to document and expose the real, the potential, and the incipient hazards in the use of pesticides in regulating the densities of vertebrate pests. The need for ecological concomitance—the living with natural forces—and the lucid appeals by many authors for more sophisticated ecological approaches that will advance the field of integrated control, is well-taken by all, I am sure. There is no question that some people who are responsible for vertebrate control programs are inclined to minimize the injurious effects of the methods they use, while some of those who are primarily interested in the conservation of wildlife are equally inclined to maximize the intrinsic value of any animals lost due to control operations. However, there is no rebuttal to the statement that the use of toxic chemicals must be kept to the absolute minimum in the control of vertebrate animals. Vertebrate pest control is applied ecology, i.e., it is the management of the behavior of individual animals and the regulation of population levels, not the destruction of individuals. All animal control must be based on a prudent translation of the ecological laws of nature into an effective management policy.

The primary objective of vertebrate control should always be the alleviation of damage and not the destruction of vertebrates. And, fortunately, some vertebrate pest problems do not require that the density of a population be reduced either by destroying individuals or by utilizing biocontrol or chemosterilant procedures. One such method is the use of sound as a deterrent, which is discussed in Chapter 11. Other methods not discussed herein include chemical repellents, traps, snares, wetting agents that reduce an animal's resistance to cold and moisture, stupefacients and anesthetics, electric shock, shooting, exclusion by fences and other barriers, and many types of frightening devices (shape, color, light, odor, and sound). Bounty payments seldom work, but they are always in vogue with the uninitiated. The bounty system seldom stimulates effective destruction of the breeding stock; if the payment is increased to make it worthwhile to bring about really effective control, then fraud and other cheating practices result. Instead of paying bounties, it usually costs less and better control is achieved when people are hired for the task (Howard, 1966).

A. Definition of a Problem Vertebrate

It is much simpler to coin a definition of a vertebrate pest than it is to establish the tenets and doctrines of a vertebrate pest control program. By definition, a vertebrate pest is any native or introduced species (wild or feral) of vertebrate animal that is currently troublesome locally, or over a wide area, to one or more persons, either by being a health hazard, a general nuisance, or by destroying food, fiber, or natural resources (Howard, 1962). Also, it is important to recognize that any animal that may currently be a pest to one or more persons, may at the same time be considered either desirable or of neutral value to someone else. Judgment with respect to the propriety of controlling vertebrate pests is a relative matter. Most animals cannot be classified as either good or bad. Whether an animal is beneficial or undesirable depends entirely upon one's relationship with it. The same is true with plants, and that is why the more avid a person is as a gardener, the stronger will be his aversion to weeds.

A homeowner usually will not tolerate the presence of a single rodent, snake, or other animal that he may consider a pest, whereas a farmer usually does not object to most of these same species, unless they become so numerous as to cause him economic loss. Most state fish and game organizations have a policy of actively trying to protect nearly all kinds of wild animals, including predators; yet, at game farms, wildlife refuges, fish hatcheries, and for removing rough fish in lakes and streams, we find that fish and game managers too have well-organized programs, using poisons, traps, and other control measures, to cope with their vertebrate pests. Few naturalists enjoy watching moles make runways in their lawns or gophers feed on their prize flowers. For these reasons, both a clear understanding of and a degree of tolerance for other people's relation to the situation are required in judging someone else's decision with respect to what he may think is a pest.

With the ever-increasing demand for more exploitation of our natural resources due to today's expanding human population, it is no wonder that we have produced a controversial land-use quandary. To live in harmony with wild and feral vertebrates in the face of a growing human population is not simply a matter of preserving some primeval conditions; rather, and unfortunately on a much larger scale, it is to learn how to live harmoniously in seminaturalistic situations, under intensive agricultural development, and with an expanding urbanization.

We should have no illusions that some day we might be free of vertebrate pests, for by definition any animal in competition with another organism is then a pest to that individual. The balance of nature is the dynamic adjustments—the survival of the fittest—which occur between organisms of the ecosystem. Man is merely a tamed or partially domesticated animal, and it is an axiom of nature that all life has other forms of pestiferous life to contend with. What is really striven for in vertebrate pest control is to reduce the troublesome populations to a tolerable level.

But who decides what is a tolerable level, the affected party or a professional diagnostician who presumably does his best to operate within the prescribed parameters of the law or his list of regulations? This, of course, is the underlying problem with vertebrate pest control. There usually is a greater conflict of views when debating whether or not the control of a particular vertebrate animal or population is needed than there is in deciding what methods are to be used.

In man's world, pest control—artificially, if it cannot be accomplished by habitat manipulation—is an integral segment of land harmony. Man cannot use land to any extent and expect to preserve the original vegetation and its associated fauna. In fact, to ignore necessary control measures to keep certain pests in check when the habitat has been changed is to invite ecological disharmony of land.

Land harmony, or more specifically in this discussion, the healthfulness of the environment, relates to the impact of man's activities on his surroundings and the concomitant effects of these environmental changes on the animal populations. But before one can engage in an intelligent discussion of environmental health or land harmony, he must first come to an understanding of what is really meant by the term, "balance of nature." Does it involve sharing our apples with the codling moth and our lawns with the mole? Just what do we mean when we say this or that will upset the balance of nature, that is, will alter the natural scheme of things? And, why do we unconsciously imply that to do this is always bad? Has not man survived and flourished in proportion to the extent that he has gained control of nature and manipulated its balance to his advantage? The only means of reversing this trend is to reduce the number of humans living on this earth, because only after the human population growth has been halted will man then have a really favorable situation for effectively reducing the present and potential dangers arising from environmental contamination as a result of air and water pollution, sewage and industrial wastes, radioactive substances, food contamination, occupational hazards, and pesticides.

B. Examples of Vertebrate Pests

The need to control a great variety of vertebrates to improve man's biological advantage over other animals and to ameliorate the problems they cause him in public health, crop protection, forestry, etc., is universal and legendary. Of the various kinds of nuisance, destructive, disease-carrying, or predatory mammalian pests, rodents are the main targets of control. There are several hundred kinds of rodents in North America. Next in importance are the carnivorous forms, such as coyotes and foxes, which kill and harass livestock (especially sheep) and poultry: "That losses to individual livestock growers from coyotes are sometimes severe is beyond question" (Rudd, 1964). Poultry ranches are subject to severe predator losses at times from a variety of animals, including particularly bobcats, coyotes, and raccoons. Some of the smaller carnivores such as skunks are important as disease carriers, notably of rabies. Of lesser importance are rabbits and ungulates—deer, goats, and antelope—which cause local plant damage or compete with livestock for forage. And, of course, a large assortment of other kinds of mammals (opossums, moles, bats, beaver, weasels, vagrant house cats, etc.) occasionally require control measures.

Some of the most difficult control problems concern birds in cities, at airports, around homes, and those which gather in great flocks and cause damage to cereal grain crops, animal feedlots, truck and fruit crops, at roosting sites and in reforestation, or are a general nuisance. Examples of troublesome species include the horned larks, house sparrows, feral pigeons, finches (including the linnet), various species of blackbirds, starlings, magpies, jays, crows, game birds, swallows, robins, woodpeckers, and many others (Howard *et al.*, 1961). Many types of snakes, other reptiles, and even certain amphibians become pestiferous at times. Certain fish are also considered to be pests when their presence conflicts with man's desires.

C. Ecological Implications

A new approach is required to fill a scientific void which exists in most applied ecological investigations of vertebrates. Even though there is a great volume of data accumulated on animal control, the area is underdeveloped with respect to the basic principles and theories of both autecology and synecology which are needed to provide a clearer insight into the mechanics of the related ecological processes.

The ecosystem is the basic unit of structure and function that must ultimately be dealt with, but it is kaleidoscopic in nature, and such changing scenes and patterns are not easy to deal with. There is some merit to setting up conceptual models from those constituent parts that are explicable in physiochemical terms in hopes that their combination will imitate or reproduce the properties of the whole component. However, it often is difficult to test the validity of conceptual models concerned with the ecology and behavior of vertebrate populations, even when the performance of the model appears identical with that of the actual biological system, because of the danger of analogy of other and obscure mechanisms having the same relationship between input and output. Nevertheless, even though the relationship between stimulus and response in the behavior of vertebrates is dynamic and highly complicated, the existing knowledge of the parts should be scrutinized carefully, and then an attempt made to resynthesize this information into appropriate concepts.

Obviously, a keener insight is needed into the factors that regulate the productivity and stability of vertebrate communities. It appears that the essential logic of applied ecology, in this instance the control of vertebrate pests, will be determined in the future from new concepts of the complex biological entities involved and their resolution—not the quantification of biological measurements. The complexity of most ecosystems largely prevents elegant explanatory constructs for the time being, and this restricts effective use of sophisticated computerized language because any theoretical formulation or model should have one or more testable field consequences. Vertebrate control is a specialized field of applied ecology, and its primary research objective is to discern the correct entities, for we are now in an era of conceptualization. Even though measurements are essential to permit classification in logical order, they are no longer very profound by themselves. Quantification per se is no longer as important as it used to be, except in facilitating the publication of research. Unfortunately, sometimes statistical analyses of research data also can make an apparently paradoxical set of measurements or observations intuitively acceptable to an editor, especially if he happens to be impressed by formulas he does not understand.

The balance of nature is the complex interplay of the life (birth to death) of all organisms within any area. Neither it nor harmony imply stability. If you analyze the unlimited number of times the balance of nature has been adjusted, modified, or altered intentionally, you will note that the overwhelming majority of recent modifications are on the favorable side of the ledger. Of course, this does not mean we are licensed to be reckless and make costly mistakes with our resources, as has happened, by trying to wrench productivity from the land with the expectation that technology (often at government expense) can or will automatically come to the rescue. Rather we need more knowledge to learn how to modify or alter nature to man's advantage in a way that will ensure harmonious land use.

Man has survived and improved his standard of living in direct proportion to his ability to gain control of nature and to manipulate the environment to his advantage. We do not desire wildfires and floods just because they both are natural events. And pest control is not just treating the symptoms while ignoring the disease. Since man is part of nature, it is axiomatic that he will come into conflict with other phases of the environment around him, just as all other animals also have numerous species that are pests to them. There is never, of course, a place for unwarranted destruction of animals or unwise use of control materials that upset the dynamics of beneficial species. The primary objective of all control methods should be to accomplish the desired effect with a maximum of safety to man and to forms of life useful to him.

It should not be forgotten that artificial regulation of the density of many animal species is not only common sense but also good conservation of natural resources. To try to protect all species of vertebrate "pests" in the interest of conservation may actually be working against the very principle striven for. Man's use of a resource may inadvertently provide suitable homesites for certain vertebrates, which then become too numerous and cause soil erosion, prevent the resource from renewing itself, or conflict with man's interest in other ways. The balance of nature is a natural phenomenon which often does not result in the conservation of natural resources.

The frontiersman and his grazing animals are not the only ones who have wasted away land's healthy flesh (its topsoil) to expose its rocky bones beneath. In much of the desert country of western and southwestern

United States, for example, native animals have also been responsible for making the land sick. Wherever weather or soil conditions do not favor a rank growth of vegetation, many kinds of rodents, rabbits, harvester ants, and other invertebrates, which appear to thrive best in a relatively bare habitat, are capable of keeping these areas "naturally" bare. In other instances, such as with ground squirrels in California, they take advantage of even moderate grazing by livestock "to get on top of" the potential grass growth and further deplete the vegetative cover, not only greatly slowing soil buildup, but sometimes accelerating erosion as well. Indians warned our early pioneers to be wary in crossing the great plains because about every 7 years the grasshoppers would denude the land, leaving no feed for horses.

To attempt to control weeds, insects, and vertebrate pests is not just "treating the symptoms while ignoring the disease." Landuse is habitat alteration, and when the habitat is changed man must expect some undesirable effects along with the good. To fulfill the nation's food and fiber requirements, habitats must and will be altered drastically. Nowadays this is usually, though not always, done in a way that is harmonious with the principles of good land management. For example, the establishment of alfalfa or other irrigated pastures creates a favorable habitat for pocket gophers and meadow mice as well as certain insects, plant diseases, and weeds, but this does not mean that the land is abused. At the same time that this altered habitat becomes more suitable to a few species of vertebrates and other potential pests, it also becomes inhospitable to a large number of other kinds of plant and animal life. Land is sick due to man-made causes only when its nutrients are being depleted or when soil is eroding away and being destroyed more rapidly than it is being replaced. Then man is not living in harmony with his land. But the zealous desire of some to preserve too many feral burros, ground squirrels, certain forest rodents, or other animals on grazed rangelands or logged forests in the interest of conservation may produce consequences that actually work against the very effect striven for.

Conservation of vertebrate animals centers primarily on maintenance of stable habitats as suitable places for these communities of animals and plants to live. The maintenance of inviolate sanctuaries and refuges is not the answer to the perpetuation of animal communities. This too often leads to an unbalanced ecosystem where the species to be protected end up destroying their own habitats. Furthermore, man must adequately appraise the effects on wild animals of the inevitable changes man must inflict on the environment. He must consider not only how habitat modification may create vertebrate pest problems, but also examine how these habitat alterations may create problems for the animals. It is nice to be wanted, and that is why game species in many localities have profited from habitat improvement investments in their behalf, making them better off than if they were not hunted. The problems of reducing animal damage are far more complex, however, than the relatively straightforward issue of attempting to produce more fish and game. However, since there are many biological interactions affecting both fields of endeavor, we must avoid developing a meaningless dichotomy of interest. It is important to recognize the need for a greater insight into the factors that regulate productivity and stability in modified ecological communities.

Major breakthroughs in the development of new principles and theories of vertebrate control are likely to result, I believe, by investigating simultaneously the complex behavioral, physiological, and ecological processes. By working within such a broad framework, animal control policies can then be derived from a clearer insight into both animal behavior and various sophisticated ecological principles—the new science of biotic control—rather than using control policies based on the more restrictive research directed to animal destruction. Effective animal control is the translation of ecology into policy. We must develop the ecological management of each vertebrate pest in a way that promotes a new but tolerable natural balance, i.e., self-regulatory ecological units (Howard, 1964b). Another way of saying this is to state that we need to learn how to orient certain innate behavioral traits of a species against its own welfare in order to achieve endogenous control.

II. BIOCONTROL

A. Objectives

Since vertebrate species that have become pests have done so simply because they are so well adapted to the prevailing habitat conditions, biological control, in particular the making of a habitat unsuitable to the species in question when such can be accomplished, is generally considered to be a much more desirable and effective procedure for controlling animals than just attempting to destroy or otherwise remove the troublesome individuals. Also, as will be brought out later in the discussion of predators, the destruction of individuals by artificial means (shooting, poisoning, etc.) or by natural predation may have only a temporary effect and in the long run may result in stimulating the offending population of animals to increase to a level of density that is greater than it would have reached if no control had been attempted.

"Man has enormously, and often recklessly, modified ecosystems through interruption of ecological succession and substitution of simple (agricultural) plant communities for natural diversity, through disruption of energy balance of communities by assisting (through artificial feeding) his herbivorous domesticates to escape the checks and balances that normally exist between primary producers and their consumers" (Blair, 1964). Since intensive agricultural practices consist of planting, fertilizing, and irrigating new and highly palatable varieties of plants, it is no wonder that such monocultural practices, which produce these simplified biological environments, find certain species of vertebrates responding to become pests.

Biocontrol of vertebrates, at least in agricultural situations, probably has its greatest potential when there is conscious harmonizing of natural control with chemical or other direct control methods. This is the concept of integrated control, where two kinds of control are combined against the pest species, and the two kinds of control are integrated in a sequence best suited to the crop ecosystem. "With these as operating premises, it is clear that any chemical used in an integrated control program must disrupt natural equilibria as little as possible and should survive no longer than necessary. The general purpose of chemical agents in such a program is to reduce pest populations to levels at which antagonistic organisms take over the function of control. Needless to say, the chemicals must do little or no harm to these competing organisms" (Rudd, 1964). Biotic control also has considerable merit because carefully planned ecological dislocations in nature are considered far more palatable to most people than the repugnant and hazardous aspects of the elimination of nuisance vertebrates with poisonous materials.

"The search for bird-resistant varieties of cereal grains is a field of biological study likely to yield long-lasting results, even through early results have been somewhat disappointing . . . . Cultural practices and habitat manipulation may also be relied upon to alleviate bird damage in some situations" (Besser, 1962). One reason why biological control of birds has been difficult is the great ecological versatility of those species which have become pests.

Some of the noxious forms of vertebrates, especially those in urban areas, can be controlled by carefully regulated practices that cause a minimum of adverse side effects (undue hazards to health, injury to nontarget species, or residues on agricultural commodities), but control practices against those species which attack agricultural crops, foods in storage, rangelands, and forest trees present more difficult problems (Swift, 1964). A great deal of additional ecological knowledge is required in these instances. What is needed for biocontrol of vertebrates to be effective is some self-accelerating method of control that forces populations down by eroding their homeostatic capability (Watt, 1964). The basic method of biological control is the enrichment of species representation to reduce the number of individuals of the offending species.

Even though the biological control of vertebrates has not been developed to the same degree of sophistication and effectiveness as that for certain insects and mites, it still behooves us all to exploit every ounce of its potential effectiveness now. Actually, biocontrol of vertebrates is practiced considerably more than most people are aware of. To appreciate this, merely visualize how many cultural and agricultural practices could easily be altered to create more serious pest problems, and even create new pests from species that are now quite innocuous.

B. Dynamics of Vertebrate Populations

Before attempting to exploit the full potential of biocontrol of vertebrates, it is necessary to have some knowledge of the dynamics of vertebrate populations and to understand fully both the advantages and consequences of intentionally creating biotic imbalances. Any change in the composition of habitats can lead to other problems, such as a widened distribution of the troublesome species into new niches. All local biota have vacant niches; the irregular pattern of success and failure of vertebrate introductions throughout the world bears out this. Vertebrates are acclimatized often without any apparent reduction in the densities of other species of vertebrates. The exploratory drives or innate dispersal traits (Howard, 1960) within vertebrates tend to accentuate this spreading when habitats are disturbed. The wider the tolerance of a new arrival, the greater will be the number of new but suitable niches available to insure its survival; no immediate genetic differentiation is required. Research with rodents indicates that their principal dispersal movements are made about the time these animals attain puberty, and it strongly suggests that many of the observed dispersal patterns are governed by the laws of heredity (Dice and Howard, 1951; Howard, 1949, 1960; Howard and Childs, 1959). The "purposiveness" of *innate* dispersal—the extensive dispersal movements made by nonphilopatric individuals—is not for the individual's welfare; rather, in spite of the high rate of mortality of innate dispersers, it seems to have distinct survival value for the species. But instinctive behavior is not necessarily advantageous to the individual in a social species.

"The density of any particular species of mammal or of the total vertebrate components of the biomass at any particular moment, often appears not to be influenced intrinsically as much by a fluctuation in the amount of primary production as it is by variously inherited behavioral traits of the vertebrates involved and by various intraspecific stresses [psychological, competition for food or mates, territoriality, weather, disease, or other vicissitudes of life] . . . . Animal populations have considerable powers of self-limitation which prevent severe over-populations that otherwise would destroy the species. Self-limitation counteracts their innate ability to produce a surplus of offspring. This compensatory mortality tends to increase, percentagewise, as density increases above a certain equilibrium point, and to decrease as density falls below this point (Nicholson, 1954; Morris, 1963)" (Howard, 1965b). Even though members of a species become a brake counteracting their own great reproductive potential, the upper density limits may be raised or lowered within certain ill-defined parameters whenever man modifies the environment. Individual animals often starve to death, die of disease, or are killed by storms, but populations of wild vertebrates do not completely exhaust all of the food over a sizeable area so that all of the occupants must then starve. And, similarly, if provided food *ad lib*, the population will stop increasing when a certain equilibrium density is reached. It is this level of density that somehow triggers operation of the complex, self-limiting controls. This is nature's way of preserving the species.

Faunal diversity provides a limited degree of natural control, i.e., it increases the capacity for continuous self-regulation of species with respect to their environment. When there is considerable diversity of fauna and flora, living things and systems then appear to be endowed with a self-regulating feedback mechanism which guarantees their sustenance, adaptiveness, and perpetuation in a dynamic continuum. Just how the social behavior and bioenergetics of species control population density is not well understood, but individual animals as well as populations are dynamic in structure and behavior; they are not static units. The best example of diversity providing biological stability can be found in the tropics, where insect epidemics and extensive defoliation of vegetation is a rare phenomenon. By way of contrast, compare this situation with the extreme cyclic irruptions of herbiverous lemmings in the arctic tundra, and how these rodents periodically denude their habitats. There is no question that biological diversification results in greater stabilization of the environment and increased biological control. The reason diversification cannot be the sole objective is that reduced diversity, even monoculture as an extreme example, also leads to greater control of nature for the benefit of man.

Biocontrol of vertebrates, in particular the manipulation of habitat conditions, should not be employed *a priori,* because if done improperly the treatment can create more problems than it cures. In fact, it can cause more problems than the use of poisons. With chronic poisoning of vertebrate pests, there is concern about the possibility of subtle and undesirable physiological and behavioral responses, or of carcinogenic and mutagenic effects on both the target species and on nontarget populations, although none have been recognized so far. But when a habitat is modified, there is little doubt that it will produce more pronounced interactions with other species of animals than would usually result as a consequence of population reductions by means of either chemosterilants or toxicants. "Observations indicate that natural biomes have a well-established, stable, animal–soil–vegetation complex which is *not* delicately balanced. A natural change (e.g., by disease) or man-caused change (e.g., by shooting), in the density of a native species of browsing, grazing, seed-eating, or predatory mammal does not precipitate a dramatic 'balance-of-nature' type chain reaction of responses by other components of the community. Such chain reactions usually are the consequences of the introduction of alien plants or animals, farming, grazing, logging, man's use of fire, or natural catastrophic events" (Howard, 1965b), all of which are a form of habitat modification. And the main way that our native biotas have been degraded and fragmented is by the alteration of habitats. Take the coyote as an example. The coyote started out as a lonely predator of prairie dogs and rabbits in the Great Plains and other western parts of the United States, excluding Alaska. Then man came along with lambs, and chickens, and while he spread his civilization into the coyote's habitat, the coyote backtracked along man's trail to the point that now the coyote can make a living almost anywhere from California to Maine and also Alaska.

The best example I know of where a stable, balanced biome was drastically disrupted by the introduction of mammals is New Zealand. These introduced mammals have upset the natural stability of the habitats over large areas by destroying vegetation, thus also causing extensive erosion. The principal reasons for the destructiveness of the exotic big game animals, fur bearers, and wild pigs and goats to certain habitats in New Zealand are: (1) some of the soils arc highly susceptible to erosion; (2) the mountainous country often gets high intensity rainfall; and (3), of great significance, many of the endemic plants which evolved without browsing mammals have little innate resistance to heavy, selective grazing or browsing. As a result of these introductions, an irreversible change in the composition of the vegetation has occurred throughout many of the mountain ranges. However, where enough soil has remained, and where browse-resistant and unpalatable plants have adequately replaced those destroyed by the browsing mammals, a new and stable equilibrium of the animal–vegetation–soil complex has developed. But on some mountains both the A and B soil horizons have been lost (Howard, 1964a, 1965a). It is also interesting to note that even in this instance where such drastic ecological changes are occurring that it is still difficult to obtain sufficient evidence, or even empirical answers, with respect to what are the various homeostatic reactions the introduced animals undertake in developing a harmonious relationship with their new environment.

C. Predators

1. *Introduction*

For man to utilize natural enemies (the predators) to help him control pest populations of vertebrates is not a simple procedure. The role played by vertebrate predators as enemies of pest vertebrates (e.g., rodents, rabbits and birds) is not a phenomenon easily interpreted empirically. Before we try to discuss intelligently any potential methods of employing any or all of the predators of pest forms of mammals and birds as a means of biocontrol, it seems appropriate to first analyze some of the basic predator–prey interactions to learn to what extent vertebrate predators are determinants of the population density of prey species.

It has been my observation that the combined predation pressure by hawks, owls, snakes, and carnivores usually perpetuates a greater, not lesser, seasonal and annual density of species of vertebrate prey than would otherwise exist. The suitability of other aspects of the habitat and self-limitation resulting from intraspecific stresses used in the broad sense—not interspecific relationships between predators and their prey—largely determine the magnitude of the natural vertebrate densities that have been in existence for long periods of time over large areas.

Another weak point concerning the effectiveness of predators in controlling pestivorous mammals and birds is that the predators are not host specific, and in general are usually opportunists taking what is most readily available. One way they stimulate their prey to become more abundant is by feeding on the weak and unfit, thus tending to increase the vigor and adaptiveness of their prey. Predators are also self-limiting, usually reproduce much more slowly than their prey, and often prey on other predators. Furthermore, even in situations where predators are thought to be depressing the size of a vertebrate prey population, this still does not necessarily mean that the predators are actually "controlling" the pest species by reducing it to a density acceptable to man's needs.

It should be pointed out that in this discussion we are not concerned with the need to control predators in order to increase the density of a vertebrate prey which man intends to prey upon (harvest), such as lambs, chickens, fish hatcheries, game farms, or any other situation where man makes such heavy demands on the prey species that he does not wish to share many individuals with natural predators. Man, of course, is the ultimate in predators, and through his intelligence he has devised means whereby he can easily reduce the populations of many vertebrate prey species to a very low level or even extinction in a few cases. In his blind predaciousness and ecological ignorance, he has sometimes done just that.

Densities of animal populations vary within relatively narrow limits in a particular habitat and unknown regulatory mechanisms prevent further population increase. Factors that might act to limit populations of vertebrates include not only predation but also emigration, shelter, food, social interaction, and other vicissitudes of life which operate as stress factors on populations. And any one of these forces, including predation, can play the dominant density-regulatory role at certain times under special conditions. Some of these situations will be discussed below.

The hypothesis by Howard (1965b) that vertebrate predators usually do more to increase population densities of field rodents than they do to depress them implies that, without predation, self-limitation stress factors come into play at lower density levels, and that these forces operate more drastically as population controls than does predation. This predator–prey theory also implies that natural selection has enabled native predators to stimulate their natural prey populations to exist at the ecologically optimum density. Optimum density here refers to the total number of individuals that the habitat can adequately support during the most favorable season on a sustained yield basis.

This concept is not new. As Scott (1958) pointed out, Forbes (1880), in writing about birds, asserted that "annihilation of all the established 'enemies' of a species would, as a rule, have no effect to increase its final average numbers," and that excessive populations are "in one way or another, self-limiting" (Forbes, 1882). According to Huffaker (1958), "density-induced autoinhibition or intraspecific competition in the broad sense is the only true governing or equilibrating mechanism." Specific mortality factors such as predation seldom persist long enough to have an appreciable effect on the overall density of prey populations; instead, the mortality level may precipitate natural population responses tending to offset it (Errington, 1956). For the most part, predators feed on prey that already have a poor life expectancy, and even excessive predation can be compensated for by accelerated reproduction. Accrued evidence indicates that much predation, even when conspicuous and severe, may operate in an incidental fashion rather than as a true depressant (Errington, 1946). Rodents, hares, and grouse all decreased in numbers at the very time when 22 species of their avian and ground predators were thought to be controlled (Crissey and Darrow, 1949). Pocket gopher numbers showed no correlation with the presence or absence of coyotes (Robinson and Harris, 1960). "Voles probably exemplify a general law that all species are capable of limiting their own population densities without either destroying the food resources to which they are adapted, or depending upon enemies or climatic accidents to prevent them from doing so" (Chitty, 1960). "Instead of competing directly for food, animals compete for conventional substitutes, e.g., territory or social position, which are capable of imposing a ceiling density at the optimum level, and can prevent it from rising to the starvation level which would endanger future resources" (Wynne-Edwards, 1959). It seems obvious that the predators of lemmings in Alaska, as reported by Pitelka (1958), largely were opportunists exploiting prey that could not have survived anyway; the predators only delayed, hence perhaps magnified, an already inevitable rodent population crash. Predation may be considered a beneficial service for most prey species, and for some species even may be important to survival (Latham, 1951). The rabbit plague in West Wales leaves no doubt that, wherever "skimming" a population by trapping was introduced, "the rabbit population increased by leaps and bounds" (Hume, 1958).

The concept that predators usually do more to stimulate an increase in numbers of vertebrate prey than they do to depress them requires a proper appreciation of both the time interval and the size of area involved. Unless otherwise stated, the time period is a long one, including many generations of the species; it does not apply just to part of the normal lifetime of specific individual vertebrates. Likewise, the size of each area involved is sufficiently large to support entire populations of both the predator and prey species; the hypothesis is not concerned with small areas containing only a few individual territories or home ranges.

Natural selection has seen to it that both the vertebrate prey and their predators require each other to exist in nature in optimum densities. It is the author's hypothesis that predators usually cannot depress vertebrate population densities below the environmental capacity permanently; instead, they usually increase the vigor and reproductive success of the prey sufficiently to eventually more than replace any individuals the predators may destroy. Since predators have evolved under natural selection, predation in some way must favor keeping their food supply (the prey species) at the maximum density inherently permissible under self-limitation, and which remains within the thresholds of security of the ecosystem. It seems likely that not only would the average population density be less, but that the cyclic peak densities (irruptions) of most species also would be less, not more, if there were no natural predators. Without predation as a continuous or periodic stimulus for prey population to increase in numbers, the self-limiting factors of population controls apparently come into play before the density of such populations have attained what the habitat could support under the influence of sustained predation.

One way in which predators may bring about a higher density is by maintaining a younger age class within the prey population. Younger but reproductively mature individuals of at least some species may have smaller home ranges and less-defended territories, yet reproduce as rapidly as older animals. Many types of stimuli can prompt vertebrates to increase their numbers. With laboratory mice, Petrusewicz (1957) demonstrated a significant increase in reproduction and an increased rate of survival of weaned mice whenever the population was transferred to a new cage, regardless of whether the new cage was smaller or larger.

Game managers have been proclaiming for some time that the proper intensities of fishing and hunting can result in an increase in fish and game. If a deer herd is protected from all natural predators as well as from hunters, it will "become to some extent self-limiting after it develops an over-population," and its productivity and population density then will decline (O'Roke and Hamerstrom, 1948). Darling (1937) points out how "anti-social behavior" limits red deer. The Wisconsin Conservation Department has shown that liberal hunting regulations increase the deer herd (Dahlberg and Guettinger, 1956). In the western United States there is a positive correlation between the development of overpopulated big-game herds and an increase in total number of hunters (Rasmussen and Doman, 1947). Control of predators was once credited with a pronounced population increase and subsequent die-off in the early 1920's in the protected deer of the Kaibab forest in northern Arizona, but a closer look at the factors responsible seems to indicate otherwise. According to Lauckhart (1961), "Game men are now convinced that the removal of cougar from the Kaibab had nothing to do with the boom and bust of the deer herd. The deer increase apparently was the aftermath of some habitat changes."

Without natural predation, as is the case with deer in New Zealand, various intraspecific stress agents may be able to exert more adaptive pressures than would be possible if predators also were modifying the demographic structure of the population. This needs to be verified. The 120 or so men employed each year in New Zealand as government deer cullers are predators of deer. In the government's earlier deer control operations, problem areas were abandoned and the men moved to new sites just as soon as deer numbers were reduced appreciably. "Investigations generally into government shooting operations have led to the conclusion that undesirable population surges have been induced in many of New Zealand's wild animal herds as a direct result of the application of spasmodic and fluctuating shooting pressures" (Anderson and Henderson, 1961). But much more knowledge about the social behavior and bioenergetics of deer will need to be available before we can postulate the effect that absence of predators may have on the ecogenetic homeostasis of deer herds. In fact, many questions on vertebrate predator–prey interrelationships require better answers before the full significance of predation can be determined.

Many bird and mammal predators are attracted to areas that support concentrations of palatable prey species; but once the surplus or excess prey disappears, the remaining prey is thus largely inaccessible to predators living in that area. As a consequence, those predators that are highly mobile may well leave such areas, if they are not too involved in breeding responsibilities; and they may become pestiferous predators in their new habitats. It is common knowledge that insectivorous birds often disappear from areas following insect control operations as a consequence of loss of food supply (e.g., see Couch, 1946; DeWitt and George, 1960; Rudd and Genelly, 1956). Less mobile animals, such as fish in a stream, may succumb if the insect fauna is artificially depleted. Even though it is virtually impossible to separate analytically all influencing factors in these situations, a number of studies have shown that loss of food supply following spraying operations has contributed materially to fish die-offs (Cope, 1960; Cope and Springer, 1958; Graham and Scott, 1959; Kerswill, 1958). A similar phenomenon occurs when rodents are removed. Their predators, the hawks, owls, snakes, and carnivores, must then move to new localities or suffer the consequences of malnutrition (Howard, 1953).

2. *Natural Predators*

The previous introduction makes a fairly strong case against natural predators as effective population regulators of vertebrate prey species. However, the answer is not that simple, and there undoubtedly will be found bona fide exceptions to the aforementioned hypothesis. And the science of bio-control is primarily concerned with creating such exceptions. Nevertheless, it is apparent that in analyzing the regulatory mechanisms involved in the control of vertebrate prey species by natural enemies, much more insight is needed into the basic behavior, ecology, and host–parasite and predator–prey interactions.

Until proven otherwise, however, I think that natural enemies should always be given the benefit of any doubt with respect to their importance in preventing the population level of troublesome vertebrates from rising still higher. But, it should also be recognized that this hypothetical degree of control, by itself, is usually grossly inadequate in preventing noxious vertebrates from becoming economic pests. Such species are pests because they are so versatile and well adapted to the existing habitat conditions, and there is little evidence to indicate that their existing population levels could be altered much more than what might be called just an academic degree with or without the presence of natural predators. Most vertebrate pests are troublesome even when their populations are so low that they are not vulnerable to natural predation.

Perhaps the most thorough study of natural predation by hawks and owls upon the rodent population of one township was conducted by Craighead and Craighead (1956). They "conclude that the total weight of food required by a raptor population and the number of prey animals killed during a year are of such magnitude that raptor predation must be recognized as an effective biological control; furthermore, that the way in which raptor predation acts on collective prey throughout the year to effect this prey reduction strongly suggests a precise regulatory force . . . . In the case of mice, predation is usually considered of value, while in the case of predation on pheasants, it is usually considered injurious. The actual good or bad that results should not be interpreted solely in terms of man's self-interest; predation must be viewed as a biological function tending to keep prey populations balanced within the limits of their vulnerability." Because the raptors consumed such vast numbers of rodents, they concluded that predation regulates the population densities of prey more precisely than other stressing forces. These views are shared by many people. But, unfortunately, we have no way of knowing what densities the prey species would obtain if there were no raptors, and whether without raptors other mortality factors would then become even more effective than the predators. If other factors were not inhibiting the potential reproductive capacity of the rodents, the raptors could not begin to cope with the populations, which would then irrupt to the point that they would eat themselves out of house and home; this occurs every 3 or 4 years with lemmings in the arctic. As a matter of semantics, I am not sure they are correct in calling raptor predation "an inexpensive form of control" of noxious prey species and also contend that "under natural conditions no one prey species can draw enough predation pressure to keep its population at a dangerously low level," because before most rodent pests can be considered controlled, they must be reduced to such a low level that the population cannot promptly recover.

One must be careful how he appraises the pros and cons of the value of vertebrate predators. For example, one author (whom we will not cite since his views have probably changed) reasoned that the destruction of 110,495 coyotes during 1 year in the western United States was a serious mistake. He said that if they had not been taken through control then the coyotes would have justified their existence, theoretically, by killing and consuming the entire mouse population from 33,000,000 acres, and that the removal of the mice by the coyotes would have saved considerable amounts of forage. The fallacy to this line of reasoning, of course, is that both the time factor and the reproductive potential of rodents have been overlooked. The rodents would have been reproducing during the entire year, quickly compensating for those eaten by the coyotes. The coyotes also would have been having pups during the period when they were destroyed. No matter how many coyotes are present, they do not seem to be able to decrease, let alone eradicate, their food supply. Natural selection has evolved a situation where predators are favored most when they can stimulate the development of the maximum possible prey population. At best, predators act only on the symptoms of most vertebrate pest problems; they cannot treat the disease, which is the condition of the habitat.

It seems conceivable that natural predation, in certain situations, can be sufficiently effective to be considered of economic significance. Such a situation might be one in which the habitat conditions are marginal for the prey. Another would be one in which the distribution of the prey is restricted to a localized area, while the predators involved are highly mobile, ranging over a much larger area. In these situations, even though the predators usually cannot reduce the level of vertebrate pests low enough so that individuals are no longer a pest, it does seem reasonable to assume that there may be instances when they can effectively prevent the prey species from becoming even more noxious. But, in general, natural predation probably inherently does more to increase the vigor and general healthfulness of its prey species by deleting the less fit, rather than permanently suppressing prey densities.

"Birds are common targets of predaceous mammals and carnivorous birds . . . . In the best study of its kind, Tinbergen (1946) showed that European sparrow hawks took a large percentage of house sparrows during a given year (perhaps 50 per cent). Nevertheless, he could not conclude that the population was importantly impaired from year to year. Moreover, in other areas of England and the Low Countries where sparrow hawks have been assiduously removed, there seems to be no compensating increase in house sparrow populations . . . . Irrespective of a bewildering welter of apparently conflicting data and views, it seems likely that predators are not a dominant influence in the control of most mammalian populations . . . . Some mammals, like some birds, can be extraordinarily effective hunters . . . . But even among these skilled hunters, the victims seem most frequently to be the handicapped—the immature, the wanderer, the ill-adapted. The removal of such prey may conceivably result in more vigorous prey populations rather than less" (Rudd, 1964).

Many agricultural crops can tolerate a limited number of pest vertebrates. And, conceivably, there may be situations in these instances where the habitat of the predators of the pest species (rodents, rabbits, or birds) might be improved sufficiently to attract additional predation effort, especially at certain seasons. Ways of encouraging predators to spend more of their foraging time in localized areas are designed to improve the shelter and roosting places of the predators. An even more effective means, which usually is not practical to employ, is to provide supplemental food for the predators in order to entice them to spend more time in that locality. If the pest species are troublesome only at specific stages in crop development or if the winter carry-over breeding population is the weak link, then perhaps the supplemental feeding of the predators can be reduced sufficiently at these specific time periods to force the predators to temporarily feed more intensively on the prey species. However, most vertebrate predators are quite mobile, and they cannot be starved too much or they will then leave the area. Also, since the predators concerned are not specific in their prey selection, any increase in predators may result in more preying upon beneficial, nontarget species.

In conclusion regarding natural predators, it should be pointed out that the unquestionably high esthetic value of natural predators cannot be emphasized too strongly. Also, since natural predators are a dynamic component of ecosystems, their numbers should never intentionally be artificially manipulated, either up or down, without first employing considerable ecological wisdom in the process. And we must all be quick to confess that many significant questions concerning the importance of natural predation in controlling troublesome species of vertebrates are yet to be resolved.

3. *Introduced Predators*

It has been demonstrated that one or more introduced predators in a localized situation, e.g., house cats, sometimes can depress, and keep depressed below environmental capacity, a confined population of rats or house mice for as long as the diet of the cats is supplemented periodically with other food. When house cats living about a farmhouse or barn are forced to subsist entirely upon what is available to them in the form of wildlife, they can do so only if the wild animal habitats are sufficiently favorable so that the cats are unable to destroy the populations of mice and rats, which are their food supply; otherwise the cats also would die or be forced to disperse. According to Elton (1953), cats have a great advantage over human beings when controlling rat populations of low density: "they do not seem to get bored! The psychological problem of such maintenance work may to that extent be helped . . . . From these and certain other instances, it is concluded that if a sufficient number of cats (say, four) is introduced after complete rat extermination has been done, and if part of their food is supplied as milk, they will maintain the immediate area of the farm buildings rat-free. They will not necessarily clear a farm of an existing rat infestation. The quantity of cats is probably more important than their quality. To keep cats on this scale is certainly more expensive in human food (used for the cats), than if human servicing was used for rat clearance, but it supplies a useful and efficient source of additional labour, which has the important attribute of maintaining the efficiency of control at very low rat densities" (Elton, 1953). It should also be recognized that having cats around a household or farm means that certain species of birds will not be able to nest successfully in the same area.

Predators introduced for biological or entomophagous control of insects occasionally can reduce the density of a localized population of a species of prey that evolved in the absence of that predator, although most purposely introduced vertebrate predators usually have not been as effective in reducing permanently the population densities of their newfound species of prey as was anticipated by those who introduced them.

"During the period 1910–1930 an area of 600 hectares of the Frisian island of Terschelling was planted with young trees. Much damage was done to these plantations by water voles, *Arvicola terrestris terrestris* (L.). For the biological control of this pest 102 weasels, *Mustela nivalis* L., and 9 ermines, *Mustela erminea* L., were introduced in 1931. The weasels disappeared within 3 years, the ermines on the other hand increased strongly and had to be controlled in turn. They exterminated the water voles within 5 years and reduced the population of the rabbit, *Oryctolagus cuniculus* (L.), on the island to an extremely low level" (Van Wijngaarden and Bruijns, 1961). Soon the ermine commenced feeding on sparrows, starlings, terns, shellducks, curlews, other waders, poultry, tame ducks, and even turkeys. "After 1939, however, a state of natural balance seems to have established itself, the ermines, though still common, are by no means a pest anymore."

Dogs, either tethered or running free, can be utilized to frighten away coyotes and smaller predators from poultry, deer from small paddocks, etc. At several Canadian and European airfields, trained falcons have recently been employed to reduce the numbers of birds that present a potential hazard to aircraft operation.

The intentional introduction of predators to control troublesome species of vertebrates obviously should not be undertaken until all potential ecological consequences have been carefully scrutinized. The introduction of a predator onto a very small island or other isolated locality might result in the complete extermination of a certain kind of vertebrate prey, but in most situations the introduction of alien predators into a new ecosystem not only may be perilous, but may prove to be catastrophic, because the predators of vertebrate pests are not host-specific. Tragic examples include the introduction of the mongoose in Hawaii to control rats, the fox into Australia to check the rabbits, and New Zealand's introduction of weasels, stoats, and ferrets in the mistaken belief that they would control the rabbit. All of these introduced predators not only failed to accomplish their mission, but they have become troublesome predators as well.

D. Habitat Manipulation

Before biocontrol of vertebrates can come into full bloom, solutions to some difficult problems must be obtained. For example, it will be necessary to learn what are the interacting factors or forces that enable each species living in a community, which is composed of dynamic populations of many species, to share the components of the complex environment in which they live and to produce a relatively stable biota. But, out of necessity, to advance the understanding of basic ecological principles it also will be essential to limit such investigations to a manageable number of factors. An ecosystem has many hierarchical levels and is like a quasiorganism, with its *modus operandi* consisting of constant changes in the production, storage, modification, utilization, and loss of energy; hence perhaps its determining factors are irreducible wholes which necessitate a holistic analysis in the development of the needed concepts.

Whenever food supply, shelter, or other factors of the habitat are changed, the animals will reflect this change in some way. This is the basis for biological control: to discover a means of modifying existing biotic conditions in order to effect reductions in specific pest populations by causing either their emigration, or their starvation, or some type of debilitation, resulting in lowered natality and/or increased mortality.

Habitat selection is an important intrinsic factor that favors the localization of populations. As was pointed out earlier, usually the most important factor determining the presence or absence of animals in a given locality is the suitability of the local habitat conditions to the species in question. And any artificial manipulation of habitats will alter the entire ecosystem more drastically than if members of specific species of animals were removed by some selective control method. Since the key to the sustenance of vigorous animal populations is the maintenance of suitable habitats, it is an axiom that habitat modification is also an effective way, although not always the most desirable procedure, of controlling many vertebrate pests.

Animal numbers will not keep increasing indefinitely, even if surplus food and the best of conditions are provided. By providing surplus food, Bendell (1959) established an insular population of white-footed mice (*Peromyscus leucopus noveboracensis*) that did not survive in trials without extra food. The artificially supplied food gave a measurable increase in survival of young to approximately 1 month of age; but even in the presence of surplus food the population did not continue to increase indefinitely or even become very abundant.

By providing surplus food on two 1-acre fields plots of sagebrush in northeastern California for 1 year (1957–1958), I obtained a threefold increase of rodents (from 4 to 12 individuals per acre) of all species present compared to control plots. This is not an impressive increase, however, for in the same general area the total density of all species of rodents fluctuated naturally from 20.5 rodents per 100 trap nights in 1958 to only 1.5 in 1960, nearly a fourteenfold change in density, although this represents the extreme differences observed during a period of 5 years (1955 and 1960). A fivefold seasonal change in density of *Peromyscus maniculatus bairdii* on a 300-acre plot in Michigan was considered normal (Howard, 1949).

When a farmer replaces native vegetation with nonnative plants that have been developed by breeding experiments in which such factors as natural selection by native animals have been ignored, he alters the habitat to such an extent that sometimes the native wild animals can no longer exist there. In other instances, the alien forage or crop may stimulate certain species of native fauna to become so numerous that locally they may result in destruction of certain types of native vegetation. The probability that introduced mammals or birds will disrupt the natural stability of their new habitats depends upon many factors. If native animals closely related to those introduced are present, the chances that the stability of the habitats will be weakened are much less. This is why many species of big-game ungulates introduced into the United States have not upset the stability of habitats and become serious pests, as they have in New Zealand.

In some situations, there is every reason to believe that the intensity of undesirable browsing of young conifers by deer might be substantially reduced if the amount and availability of alternate and more preferred species of browse were increased. But increasing the food supply in this manner does not necessarily mean that then there will be a corresponding increase in deer numbers. In general, the most logical way of managing temporary deer problems related to forest regeneration is to liberalize both the bag limit and the hunting season. Where agriculture abuts extensive deer range, "buffer strips of palatable deer forage are sometimes planted to relieve the pressure on agricultural croplands" (Eadie, 1954).

"The use of effective bird-resistant varieties of crops, while not at present a reality, may yet prove to be one of the more promising means of combatting bird depredations in the future," according to De Grazio

(1964), who cites the following research results. Three blackbird-resistant varieties of grain sorgum (Northrup King 120, Northrup King 125, and Adkins-Phelps 614) planted alongside heavily damaged cornfields had less than 1% damage. However, a tight-husked hybrid variety of corn thought to be bird resistant sustained heavier damage than did the surrounding corn, because it was slow in maturing and was in the vulnerable dough stage after the other varieties began to harden. It is more difficult to repel birds from a crop if other palatable foods are not readily available. Efforts have also been made to develop a high-tannin milo that will repel blackbirds.

If one is willing to accept the ecological consequences, many kinds of vertebrate pest problems can be largely alleviated by modifying the habitat. Orchards or other trees near a vineyard invite an increase in bird depredations. The removal of oak trees and cottonwoods adjacent to walnut and almond orchards reduces losses from crows, magpies, woodpeckers, and jays. Woodlots in close proximity to cultivated fields often increase pest problems to agriculture, regardless of whether hawks and owls take up residence in the trees. "Clean farming that eliminates cover along fence rows and field margins is generally frowned upon by conservationists. However, it is a practice that pays off where bird attack upon rice is prevalent" (Neff and Meanley, 1957). Restricting the storage of grain in field shocks to a minimum prevents rodents from building up to pest proportions. It hardly needs elaboration here that the removal of the nesting, shelter, or roosting places of many species of birds in particular, but also of other vertebrates, can significantly reduce many types of vertebrate pest problems.

A dramatic example of how important providing troublesome vertebrates with an alternate or buffer food can be in helping to reduce crop damage can be found in the solution to waterfowl depredations. Ducks and geese can be very destructive to many kinds of crops (Biehn, 1951), and the conventional means of frightening them away from valuable grains, vegetables, and pastureland become workable only if waterfowl refuges, which provide adequate resting areas and sufficient food to hold the birds until the crops are harvested, are available in the general vicinity. Otherwise, the effectiveness of the various herding and frightening devices would be much less.

Indirect mole control, through the reduction of food supply, is a comparatively expensive method; but it is useful on lawns and golf turfs. "Several insecticides are capable of reducing the population of earthworms and soil insects to a point where the soil no longer provides sufficient food to fulfill the mole's daily requirements. The effect on the moles cannot be expected for several weeks following treatment. This method of control is most suitable for turf areas and will often serve a two-fold purpose by also ridding the lawn of harmful insects which are found in lawns" (Marsh, 1962). The same technique can be utilized to discourage skunks from digging in lawns for grubs and other insects (Merrill, 1962).

An example of how habitat modification can reduce a vertebrate pest to a low level is provided by Keith *et al.* (1959). Aerial spraying of weedy, mountain rangeland in western Colorado with an herbicide (2,4-D) resulted in the following changes 1 year after treatment: the pocket gopher population was reduced 87%, the production of perennial forbs was reduced 83%, and grass production was increased 37%. This decrease in gophers was probably due to the decrease in forbs available to them. Gophers do better on forbs than grass, and as a result of the spraying their diet was changed from 82 to 50% forbs and from 18 to 50% grass. In unpublished United States Department of Interior Fish and Wildlife Service Reports of 1946–1948, M. W. Cummings found that up to 90% of the gopher population on Grand Mesa, Colorado, were removed by weed control with herbicidal sprays (Cummings, 1962).

There are many more examples of how habitat modification either creates or controls various species of field rodents; but only a few illustrations will be cited here. Land use is habitat alteration, and when the habitat is changed man must expect some undesirable effects along with the good. The establishment of alfalfa or other irrigated pastures creates a favorable habitat for pocket gophers and meadow mice; but at the same time this altered habitat becomes inhospitable to a large number of other kinds of rodents. Even light grazing by cattle on California annual rangeland is enough to make the habitat more suitable to ground squirrels. New Zealand has probably controlled more potential European rabbits with their extensive applications of lime and superphosphate fertilizers than they regularly control with toxicants, because their rainfall is equally distributed throughout the year and the vegetative response to fertilization produces such a rank growth of grass that such habitats then become unsuitable to the rabbit (Howard, 1958). If a dense stand of herbaceous vegetation is created, it favors meadow mice; but if it is mowed or grazed by livestock, it no longer is a favorable habitat for these mice. Cultural practices in apple orchards require a sod type of ground cover with the addition of heavy mulches around tree bases, and this favors meadow mice. The piling of prunings around the base of the trees as a buffer food often minimizes mouse damage except during severe winters (Fitzwater, 1962). The bare habitat produced temporarily by the range-improvement practice of converting brushlands into grasslands greatly favors some species of rodents, which then become a serious pest for 1–2 years, until the herbaceous vegetation is well established. Cutting timber improves the habitat for certain rodents, deer, and rabbits, thus creating serious silvicultural problems until new seedlings get established and grow out of reach of the troublesome species of vertebrates.

E. Diseases

Disease can be looked upon as the result of the forces of ecology, for epizootics that from time to time locally decimate populations of some kinds of vertebrates do so as a result of a dynamic relationship between three principal factors. They are the effect on the host species, the agent of disease, and the current environmental conditions. The way disease affects the dynamics of vertebrate populations is largely ecological and not a mere interplay between the host and the agent. It is because of these ecological forces that disease becomes a potential biological controlling factor. "Under any situation, an epidemic is only a temporary phase, and it can only be understood when it is related to preceding and succeeding events" (Jones, 1964). I am sure even natural diseases may someday become a more important means of helping to control vertebrate pests, but first we must obtain more knowledge concerning both the complete ecology of such epizootics and the inherent hazards associated with the artificial manipulation of the dynamics of such agents. A commercial bacterial rodenticide called Ratin, which has a short shelf life, has had only limited success.

"In searching for natural foci of disease agents in wildlife, it is logical to study habitat types which have a large and relatively stable wildlife population. In such foci one does not expect to observe diseases in the reservoir wildlife hosts, but if the viruses, rickettsia, bacteria, or fungi set up chains of infection in aberrant hosts, this may result in epidemics of disease, sometimes having a high mortality" (Johnson, 1964). Peak populations of wildlife are reached in altered environments and during successional changes in the vegetative cover. It is during these situations that disease agents in wildlife are apt to spread to new hosts and start epidemics. But, of course, the intentional introduction of exotic species of animals or plants into an environment just to precipitate epizootics among the native pest vertebrates is not a procedure to initiate recklessly. Yet, someday there may be sophisticated ways developed for modifying the habitat to increase the incidence of certain fairly specific pathological mortality factors in pest vertebrates.

The number of different kinds of significant debilitating diseases which may cause numbers of wildlife to succumb is quite large. And the number of kinds of parasites and other pathogens involved is of course much greater. The specific diseases of man known to be derived from wildlife are many. For example, in just the state of California we find: "rabies, Western encephalitis, Colorado tick fever, Rocky Mountain spotted fever, relapsing fever, Q fever, plague, tularemia, murine typhus, lymphocytic choriomeningitis, psittacosis, leptospirosis, salmonellosis and toxoplasmosis. A variety of bacterial infections may be contracted from wild animals, notably those caused by *Pasteurella pseudotuberculosis, Pasteurella multocida, Bacillus anthracis, Erysipelothrix rhusiopathiae, Clostridium tetani,* and *Listerella monocytogenes*. Certain of the fungus diseases such as coccidioidomycosis and histoplasmosis are derived from exposure to wildlife habitats. California virus has been isolated from arthropods and Rio Bravo virus from bats collected in California. It has been shown by serological tests that these viruses may produce infection in man. In the course of field studies of arthropods, small mammals and birds in California and Oregon, several viruses have been isolated which may prove to be of importance as disease agents, that is, Modoc virus, Turlock virus, Kern Canyon virus, Hart Park virus and two yet unnamed viruses isolated from *Microtus montanus* meadow mice" (Johnson, 1964) The public health importance of field rodents, rabbits, and hares is discussed by Weinburgh (1964). Herman (1964) discusses disease as a factor in control:

> There is a tremendous volume of published data on the occurrence of potential disease-causing organisms in wild birds. Most of it relates to discovery of new parasites and their taxonomy. Case reports, histories of prevalence or frequency of specific diseases in limited areas or summaries of such reports are rare. Details of pathogenesis or pathology, or a clarification of life cycles, that tell of disease potential or mode of infection are even rarer. Epizootics occasionally have been recognized in wild bird populations, but the causes often remain unidentified. In contrast, there are many case reports for single individuals; these only suggest a potential for losses and are not evidence of actual epizootics . . . . Epizootics in which large numbers of birds die of a diagnosed cause have been recognized in very few cases. The most dramatic is botulism which causes extensive losses among birds, particularly waterfowl. Development of the disease is related to habitat contamination; a toxin is produced by the growing botulinus bacterium and the birds get sick from consuming this toxin. The bacteria grow best in the absence of oxygen and thus the disease occurs in association with decaying animal or plant matter which produce conditions ideal for such growth. Outbreaks have been reported primarily among water-fowl, shorebirds, pheasants and poultry . . . . Bacteria of the *Salmonella* group are among the chief causes of disease losses among captive birds, such as poultry. These bacteria are pathogens of the intestines and cause disease, often fatal in a wide variety of animal hosts . . . *Salmonella* infections have been reported from starling, rusty blackbirds and cowbirds in New Jersey . . . . Potentially they could cause extensive losses among wild birds but, to date, no severe outbreaks have been noted in North America. Encephalitis is a virus-caused disease which has had much publicity in recent years . . . . English

sparrows and pheasants are the only species of our wild birds known to have died from encephalitis . . . Pox, another virus infection frequently recognized in birds, is manifested by the development of small tumors (up to the size of a pea), usually on the beak or feet . . . . The so-called Roux sarcoma virus is another potential source of bird losses . . . . The Roux sarcoma virus appears to be connected in some way with the occurrence of leukemia, a disease which takes a large toll in poultry and is known to occur in wild birds as well. In a survey we have been conducting on blackbirds, we have uncovered at least 65 species of parasites, either by our own examinations or from reports in the literature. All of these parasites must be considered to be potential pathogens, even though we have not yet uncovered evidence of disease that can be attributed to any of them. Each must be studied experimentally to determine mode of infection and the circumstances under which it can be harmful or fatal to the host. The main point I wish to bring out here is that none of these parasites are host-specific to the blackbirds and they can be expected to be found in at least a variety of passeriform birds if not in most species of birds. Parasites are likely to be more narrowly host specific than bacteria or viruses, but even with parasites the range of infective hosts can usually be expected to include most of the passeriform species.

Quoting further from Herman:

The classic example of a disease agent used to control a wild animal population is provided by the story of myxomatosis and rabbits in Australia. Prior to release of the myxoma virus (causative agent of myxomatosis) many experiments had been performed to test its control potential. It had been demonstrated that the virus is nearly always lethal to wild and domesticated forms of the European rabbit (*Oryctolagus cuniculus*). Further, it had been demonstrated that all common domesticated animals, as well as representatives of the native fauna (mostly marsupials) and the introduced hare (*Lepus europaeus*), were refractory.

Early experiments with myxoma virus did not portend the later success . . . . The importance of mosquitoes as vectors of the disease was not known when these initial field tests were made. Transmission of the virus was reported to be mechanical. Australian investigators have shown that a number of bloodsucking arthropods are capable of transmitting the infection but that mosquitoes are the chief vector. The vector has been referred to as a 'flying pin'; in other words, its mouth parts become directly contaminated with the virus rather than by the virus developing or multiplying within its body. Thus any arthropod which would feed on a lesion on an infected rabbit and then bite a susceptible host would transmit the virus by contamination.

An interesting series of events also occurred in Europe. A French doctor, desiring to reduce the native rabbit population on his walled estate in France, released the virus and dramatically reduced his local population. Since this virus can be transmitted by free-flying arthropods, the wall around his estate was no barrier and the disease spread through much of Europe and also to the British Isles. The kill of the native rabbits was as dramatic as it was in Australia. The gains to agricultural interests have been great, but the sportsman lost his most important trophy, and the numerous people who kept a few rabbits in the backyard as a source of food lost this supply of supplementary protein.

For completeness of the myxoma story, I must point out that for several decades outbreaks of myxomatosis have occurred among commercial rabbitries in California. Our native cottontails are presumably the reservoir of infection. The cottontails are susceptible to the virus but develop no characteristic lesions, manifestations of the disease, or fatalities.

Davis and Jensen (1952) reported on experimental attempts to introduce an epizootic among wild rats living naturally on a farm in Maryland. They inoculated a bacterium, *Salmonella enteritidis,* into this population which preliminary studies indicated was free of any *Salmonella.* This *Salmonella* is considered highly pathogenic to rats. It causes extensive intestinal involvement and is transmitted in contaminated feces from infected animals.

Herman's (1964) analysis of this experiment follows:

The rats lived in four buildings on a farm and had all the characteristics of a wild population. They were trapped alive in box traps, marked, bled and swabbed, and released at the place of capture. During the 2 years of the study, about 2,000 individual rats were caught and there were about 3,000 recaptures. In February 1950, 20 rats in one building were inoculated with the culture of *Salmonella.* Then in October 1950, another group in the same building was inoculated along with a group in another building. As pointed out by these authors, the determination and comparison of mortality rates in a wild population is complex and difficult. Mortality rate is a rather loose term used to indicate a proportion of deaths. In analysis of their results they substituted 'probability of surviving' as their criteria for interpretation. This probability was derived from a statistical analysis of retrapping data; they also determined agglutinin levels of sera obtained by heart puncture and cultured cloacal swabs and feces. They recognized that their procedures gave only an index of infection, although their data does show evidence of spread of the infection within the population. Greatest spread apparently occurred during May 1951, when there were radical changes in food supply and shelter that apparently caused considerable movement of the rats. There was no more spread during the summer and fall of 1951, and the number of positives gradually declined. The infection was at a low level when the study was terminated in October 1951.

Although Davis and Jensen did not fully explore the population changes, it was certain that the population doubled during the interim of their study, and thus the induced *Salmonella* could not be considered effective in lowering the population. They point out that their data show clearly that an organism of potential pathogenicity may have no measurable effect on population size, mortality or reproduction. They emphasize that their data indicate the complexity of disease phenomena and warn against hasty conclusions about the role of pathogens in population management.

There undoubtedly have been many unreported attempts to control wildlife populations with disease-causing agents. It is known that as early as the 1880's Pasteur recommended the introduction of a bacterial pathogen to reduce the rabbit population in Australia; in fact he sent one of his assistants to Australia with cultures of the organism. However, cautious government officials vetoed the project.

A number of years ago I was told of a 100% successful project to eliminate the wild pig population on a privately owned island off the coast of California by the introduction of hog cholera virus. But to my knowledge this event was never documented. More recently, similar attempts . . . to eradicate native swine on another island off the California coast with this virus yielded disappointing results. While it was demonstrated that the disease was well established in a few animals there was little spread and the investigators concluded that the use of live cholera virus in depopulating wild swine is not satisfactory. Because of the repercussions that might occur from potential dangers of introducing a disease into a population, it is only subsequent events, such as resulted from the release of myxoma virus by the French doctor, that bring these attempts to the attention of the public or even the scientific community.

An epidemic (involving man) or epizootic (involving animals) is a complex phenomenon. Its full understanding requires a thorough knowledge of the biology of the causative agent and associated organisms, of the definitive host and vectors; and of the transmitters or intermediate hosts if they are involved. It also involves a knowledge of the inter-relations of various hosts that may become a part of the complex, plus data on ecology, environment, behavior, food supply, immunology, pathology, and more."

Herman concludes with a few imperative rules that must be basic to any consideration of introducing a potential disease-causing organism into a wildlife population as a method of controlling that population.

1. The applicant organism must be demonstrated to be highly pathogenic to the prospective subject species. Usually a disease which normally occurs in the subject species is not a potential applicant or it would already be doing an adequate job. Therefore the applicant is more likely to be an organism exotic to the subject species.

2. The potential killing power, residual duration, and ultimate resistance must be anticipated. One should strive for as complete knowledge as possible concerning the long range consequences to the total population and survival of the subject species.

3. The applicant organism must be host specific. We cannot introduce a disease into blackbirds that would be a threat to other birds, livestock, or man.

4. The applicant organism must be available. Not only is it necessary to be able to provide a sufficient supply of infective material for the initial implant, but the natural environment must be favorable for its perpetuation to provide the impact desired. If a vector or intermediate host is essential, it must be present in the environment.

5. If initiated, the control program should be monitored in every detail to insure its progress in the direction anticipated without adverse, detrimental side events not anticipated.

I do not wish to leave you with the impression that control of wildlife populations by implantation of a disease-causing organism is an impossibility. On the contrary, it has much merit if the criteria outlined above can be met. The events that followed application of myxoma virus in rabbits demonstrates this. However, be aware that this is a complex problem. We may have an acceptable applicant organism tomorrow and, again, one may not be discovered during our lifetime.

III. CHEMOSTERILANTS

A. Introduction

There is a very great need for a suitable means of artificially regulating the birth rate of those populations of wild vertebrates that live in seminaturalistic situations and have become troublesome to man. This would be a desirable procedure for equalizing some of man's conflicting interactions with the ecological biota, and it would help manage a healthy environment in perpetuity. Even though at the moment a great deal more insight into chemosterilants is required before antifertility methods can be widely applied, I believe it is both a challenging and promising field of research with mammals, birds, reptiles, and fish. Especially do I see great hope for the application of chemosterilants in integrated control. By using a combination of sterilants and conventional control methods, it will sometimes be possible to obtain synergistic effects, where the degree of control achieved will greatly exceed the sum of the independent effects of each method alone.

"Birth control should not be viewed as the panacea for pests, however. By definition vertebrate pests are weeds—organisms with high tolerances, low requirements, and quantitative resilience. The vertebrates that we classify as pests have very steep population growth curves. Populations of many of the vertebrate pest species are probably at or near the leveled off top of the classical sigmoid growth curve. To artificially push them off the plateau, by biogenetic control or any other means, onto the precipitous slope of the sigmoid growth curve is of no avail unless the biogenetic control effort be unrelenting or else it be accompanied by a concurrent control of the biotic requirements that nurture the growth curve and survival rate. A half sterilized bacterial culture doesn't remain sterile for long, neither does a reproductively inhibited population of dump fed rats" (Wetherbee, 1964).

Many scientists have postulated that the presence of a given number of sterile individuals in a population exerts a much greater biological control pressure on that population than if the same number of fertile individuals were removed. This concept that the sterile individuals which remain not only fail to contribute to the next generation but meanwhile also compete for space, food, and social order is correct biologically, since it is based on the population principles of density dependence and the sigmoid growth curve; but it is of little importance to the farmer or other individual who may be troubled by the vertebrate pests in question. Since vertebrates are so long-lived, in contrast to insects, it usually is not feasible to wait out their life span before the noxious individuals are removed. Therefore, antifertility agents probably will prove to be most helpful in maintaining populations of vertebrates at a reduced level after the populations have already been suppressed by other means.

When evaluating the effectiveness of chemosterilants on wildlife populations, it is important to consider the field of ethology in order to determine whether the sterile individuals behave socially in the same manner as fertile individuals. Should they lose all aggressiveness and become submissive individuals, it is conceivable that then they might be ignored in the normal establishments of territories and peck orders. If this should happen, the population threshold then might actually be increased due to the presence of sterile individuals that do not become active participants in establishing the social hierarchy.

In the search for antifertility agents, it would seem more advantageous to have a female reproductive inhibitor than a male one unless, of course, it is possible to make both sexes sterile. If a choice must be made, biological economy places more value on the ovum than on the spermatozoan, regardless of whether the species is monogamous or polygamous.

In the endocrine area, there are several theoretical possibilities for achieving sterility. There is an intimate interrelationship between the hypothalamic area, the anterior pituitary, and the gonads. The production of gonadotrophic hormones and the production of ova and sperm depend upon stimulation from the anterior pituitary. The gonadotropic hormones are dependent upon releasing factors from the hypothalamic area. There is also an inhibitory action at the hypothalamic area, plus other involvements. What is needed to control reproduction is some means of interfering with the complete cycle that produces sterility. A sufficiently high concentration of estrogens or androgens can cause inhibition of the formation of ova and of spermatogenesis; also, the production of steroid hormones (estrogens and androgens) is markedly reduced. In adult animals, this response is not so marked.

According to Wetherbee (1964):

> Spermatocidal drugs and chemosterilants offer the greatest potential threat to male fecundity. The numbers of compounds tested in human birth control research are legion; those that have been tested in wildlife species and found practical are few. Jackson *et al.* (1961) orally administered a number of simple alkane sulfonic esters to rats and reported extraordinary results. Cumulative doses produced predictable periods of reversible sterility according to dose rate. No effects upon libido were noted nor any toxic side effects.
>
> Experimentally spermatogenesis has been shown to be inhibited by reduction of light, deficiency of vitamin E, altered hormonal balance, hybridization, shortwave irradiation, and the action of drugs. The reduction of light obviously is impractical for operational use in the natural habitat . . . . The use of most nutritive deficiences (vitamin E) in inhibiting spermatogenesis is impractical, as the diets in wild populations cannot be restricted where natural foods are available *ad libitum*. The genetic isolation conferred upon the first generation of hybrid crosses is well known . . . but who can enforce hybridization in nature, and (in contrast to work with insects) it is impracticable to release large numbers of vertebrates propagated from the laboratory . . . . The effects of ionizing radiation usually come to mind when one begins to search for a weapon of sterilization . . . . It would seem that whole-body irradiation sufficient to cause irreversible sterility is too near the lethal level in vertebrates. Radionuclide ingestion might well be used against spermatogenesis but the problems of environmental contamination are

probably insuperable. High energy microwave (radar) failed to sterilize birds or eggs under field conditions.

Paralleling the familiar medical immunity response that develops from vaccination in the practice of disease prevention, vertebrates can be made immune to fertilization by inoculation with a vaccine containing gonadal or germinal tissues or fluids. Immunologic procedures include the formation of antigonadotrophic substances, the formation of antibodies in the agglutination process, the formation of antibodies against spermatogenesis, and even the formation of antihormone antibodies . . . . Until an *orally* effective immunity can be induced this line of approach will have little but theoretical value to pest control of large wildlife populations. Inoculation of elk by means of propelled syringes, however, may be a possible contemporary use of the injection technique.

Interference with the spontaneous process of embryological development or interference with the biological conditions that sustain the life of the embryo is the most certain kind of biogenetic control . . . . Selective control is made easier by distinguishing between a placenta and an egg shell, between a zonary placenta and a discoidal placenta, between holoblastic and meroblastic yolk cleavage, between parental dependence and independence.

B. Birds

Davis (1959, 1961, 1962) found that triethylenemelamine (TEM) inhibited testicular recrudescence in starlings, causing the testes to become merely an interstitial organ and concluded that the use of gametocide promises to add sensitivity to control measures. In a field test on red-winged blackbirds, this cytocide inhibitor of meiosis did cause a measurable reduction in hatchability and number of nestlings produced per nest (Vandenbergh and Davis, 1962). Of great significance is their observation that TEM had no discernible effect on pairing or other mating behavior in the blackbirds. Behavior changes also did not appear in rats administered TEM (Bock and Jackson, 1957). One difficulty in using TEM on the red-winged blackbirds is that it had to be made continuously available to the birds during the breeding season. "While TEM on the basis of Davis' work presently has the highest candidacy for operational use against *male* birds (in spite of negative findings that we have experienced when the compound is used on sexually *active* male birds), Enheptin (2-amino-5-nitrothiazole) also has potential for special purposes. This compound tested at the Massachusetts [Co-operative Wildlife Research] unit has no effect upon rats but has differential potencies among bird species" (Wetherbee, 1964).

In Elder's (1964) 4-year search for a practical oral contraceptive for controlling nuisance birds, viz., pigeons, most substances that were effective in inhibiting ovulation in other animals proved to have little effect on pigeons, even in nearly lethal doses. Compounds he tested included tranquilizers, gametocides, antithyroid compounds, hypophyseal inhibitors, insecticides, fungicides, and coccidostats. "Practical results were obtained with the anticholesterol compound SC-12937 (22,25-diazacholestanol dihydrochloride). When this compound constituted 0.1 percent of the diet for 10 days in early November, no eggs were laid for 3 months, full fertility among some birds was not reached for 6 months, and some remained anovulatory for 12 months. Following spring feeding, ovulation was almost completely inhibited for 3 months, and after 6 months remained 75 percent inhibited." Provera at 0.1% or more in the diet and arasan at 0.35% inhibited ovulation without severe debilitation of the birds, but the effect was lost as soon as the materials were withdrawn from the diets. Nichols and Balloun (1962) also have shown that anticholesterol compounds can reduce egg laying. According to Wetherbee (1964), "These hypocholesterolemic agents [SC-12937 and SC-11952] seem to be the most potent female gametocides available to date for the control of over-populations of birds. They are new, and more research is needed in their possible side-effects not to say in the economics of their industrial production and techniques of selective administration."

Six valuable review papers, focused upon avian sterility either as reproductive failure deliberately induced, reproductive failure incidentally discovered, or sterility mechanisms postulated, are the following in "Recent Findings in the Inhibition of Avian": (1) Sperm Sustentation by R. D. Crawford; (2) Ova Sustentation by P. F. Consuesra; (3) Embryogenesis by M. J. Landy; (4) Oogenesis by R. G. Somes, Jr.; (5) Spermatogenesis by B. C. Wentworth; and (6) Embryo Sustentation by R. P. Coppinger (Wetherbee *et al.,* 1962). In a subsequent paper on Vertebrate Pest Control by Biological Means, Wetherbee (1964) reviews some of the significant pest control points brought out in the six seminar papers.

The transovarian deposition of colored dyes into yolks of hens' eggs has been recognized for a long time by poultry scientists (Denton, 1940). Recently, however, Wetherbee *et al.* (1964) showed that when an oil-soluble dye, Sudan Black B, was fed to laying adult female birds, that it not only labeled the yolks with discrete layers of black so that positive identification was available that the female had fed on the treated bait, but also it adversely affected the hatchability of such eggs when fed in low concentrations. After conducting extensive tests with the small Japanese quail (*Coturnix coturnix*), they found that almost all fertile eggs laid for the ensuing 10 days by females fed 500 mg/kg in a single acute dose of Sudan Black B failed to hatch. When laying females were fed levels lower than 20 mg/kg, the yolks of all eggs laid for about 1 week were discolored; hatchability was adversely affected only at levels of 167 mg/kg and above of Sudan Black B. They report a very broad margin of safety between effective dose in inhibiting hatch and lethal dose (only

25% mortality occurred at 16,000 mg/kg dosage). "The likelihood of secondary effects on predators eating the adults or the dyed eggs is remote, as only a small fraction of the low dose ingested finds its way to the yolk; most passes out the digestive tract" (Wetherbee, 1964).

Quoting further from Wetherbee (1964):

> The hormonal balance which regulated the formation of spermatozoa in the testis has been a natural site of attack for reproduction prohibitionists. The administration of prolactin is antagonistic to gametogenesis and also inhibits gonadotropic hormone secretion (Bates *et al.,* 1937). However, proteinaceous hormones cannot be administered orally, and it is impractical to capture and inject animals from any appreciable fraction of a wild population . . . . Diethylstilbestrol (DES)-induced capons that used to be available on the poultry market bear testimony to the effectiveness of the hormonal approach to sex reversal, but those capons were produced by implantation of long-lasting hormonal pellets under the skin. The ingestion of hormones (the ethinylated steroids are more effective orally than the non-ethinylated) is ineffective unless continued over an extended period and most of these hormones are prohibitively expensive.
>
> In chickens a great many drugs and feed contaminants have a statistically noticeable effect on hatchability of eggs. Gossypol, found in crude cottonseed oil or meal, was found to reduce hatchability by Bird (1956). The oil of the fava olive and cyclopropene fatty acids found in many mavalaceous plants (*Sterculi foetida*) suppress hatchability . . . . Any of the nutritional or pharmacological agents that affect quality of yolk, albumen or shell tend to depress hatchability. We have already mentioned the fungicide Arasan causing the production of soft-shelled eggs, that will not sustain embryos. For one reason or another scores of these candidate compounds have been eliminated from our screening tests with the quail (*C. coturnix*); either they are not available commercially, or they are too expensive to produce, or they are apparently detoxified, or as with Arasan have a taste disagreeable to the bird (Arasan is actually used as a bird repellent!). Elder (1964) working with the pigeon reports parallel experience.
>
> Emulsified oils, sprayed over the eggs of the herring gull in Maine and Massachusetts was used by the United States Fish and Wildlife Service from 1934 to 1953. This pioneer embryocidal programme was a move in the right direction, but economical and sophisticated methods and tools had not been developed at that time.

C. Mammals

According to Wetherbee (1964), "the antioestrogen U-11, 55A, a diphenylindene derivative made by the Upjohn Company . . . and related compounds, except for frank oestrogenic agents, were the most potent oral mammalian antifertility agents reported up to 1963 (Duncan and Lyster, 1963)."

A small concentration of estrogen or a larger amount of androgen can exert an action at the hypothalamic–pituitary areas in newborn rodents (up to 1 week or 10 days of age) in such a way that their future ability to produce ova or sperm and sex hormones is essentially lost. Barraclough

(1961) thinks the physiological mode of action works through the crippling of pituitary growth. This is the area where the highly potent estrogen, mestranol, and other estrogens are most effective in creating sterility in rodents. Estrogens in the rat can also cause abortion and in other ways interfere with pregnancy (in contrast to little effect with humans). "Mestranol (developed by Syntex Corporation) is a highly effective sterilizing agent in rats and mice. When administered to the very young rodent [up to about 10 days of age] either by subcutaneous injection, gavage, or through the milk while nursing, both males and females are irreversibly sterilized throughout life. When the steroid is administered to normal adult females in minute quantities, serious impairment in ovulation, fertilzation, and implantation follows" (Anon.). At this time, mestranol appears to me to be a promising oral mammalian antifertility agent.

As pointed out by Linhart and Enders (1964), diethylstilbestrol is inexpensive, fat-soluble, stable under extremes of temperature, and importantly to vertebrate control, is effective when taken orally. Therefore, estrogen placed in effective baits could be distributed in the field during the breeding season of foxes and other pest species. By inhibiting reproduction, it should result in lowered densities of the species in question. Diethylstilbestrol has been shown to adversely affect reproduction in dogs (Jackson, 1953), mink (Travis and Schaible, 1962), and other species.

Biweekly testicular biopsy failed to reveal any adverse effects on spermatogenesis when 50 mg of diethylstilbestrol were given to male foxes (Linhart and Enders, 1964). However, subsequent biopsies did not measure any possible loss of libido of retardation of spermatic development of less than 2 weeks duration for the males dosed with this synthetic estrogen. As Jackson (1959) points out, since antifertility effects may be produced without obvious histological damage, alterations in fertility should be the primary concern, using testicular histology as an ancillary investigation.

When silver fox vixens were force fed single doses of 50 mg of a synthetic estrogen, diethylstilbestrol, those given the estrogen any time ranging from not more than 9 days before mating to not more than 10 days after mating failed to produce offspring (Linhart and Enders, 1964). The diethylstilbestrol was dissolved in a few drops of ethyl alcohol and mixed with 10 ml of melted tallow. The mixture was then force fed to anesthetized (ether) vixens by a syringe and an 8-in. copper tube inserted into the esophagus.

Field tests of the acceptance by wild red and gray foxes (*Vulpes fulva* and *Urocyon cinereoargenteus*) of baits for administering antifertility agents were conducted in New York in 1961–1963 (Linhart, 1964). Foxes, dogs, and crows, in that order, most frequently consumed the baits.

Other wild and domestic species took baits only occasionally. "The results suggest that the possibility for finding a 'superbait' is not promising," but additional trials were recommended. "Development of a bait which would repel dogs but be readily taken by foxes does not seem likely because of the close kinship between the two . . . . If control of fox abundance through baits containing an antifertility agent is to be practical, a high proportion of a fox population must consume the baits to achieve a significant reduction in productivity." This study also pointed out the importance of having the subjects accept the bait at the proper phase of their breeding season.

Intensive research on the use of antifertility agents in management of mammalian predator populations has been conducted by Balser (1964a,b) and his associates at the Denver Wildlife Research Center of the United States Fish and Wildlife Service. Balser points out that not only will a wide variety of agents be required, but more importantly, detailed knowledge is needed on a variety of techniques of application, proper timing, dosage, and the dispersal of baits. In many instances, he thinks the problems of application far outweigh the development of a suitable drug.

Balser (1964a) points out that:

> Applying antifertility agents in a bait to species having one litter per year is much simpler than application to those which produce several litters a year. The latter may require drugs that produce permanent sterility to make baiting practical; otherwise bait would have to be repeatedly applied or continuously available. For species breeding once a year, whether monestrus or polyestrus, agents must be selected that will block reproduction for the entire breeding season rather than result in delayed breeding after the effects of a drug wear off. Blocking reproduction in species such as the raccoon (*Procyon lotor*) and striped skunk (*Mephitis mephitis*), which go into dormancy or hibernation, precludes the use of agents that interfere with follicle development. In the north, these animals are believed to breed shortly after movement begins in the spring. These animals must either be treated during breeding and gestation, or some way must be found to interfere with reproduction before they go into hibernation.
>
> Animals that do not normally breed their first year (partly true in the case of the coyote) have a lower reproductive potential which increases the vulnerability to antifertility agents and prolongs the effects on the population. Where population reduction of the black bear (*Ursus americanus*) is necessary and an adequate hunting harvest is not possible, summer or fall baiting with antifertility agents may prove practicable. This situation exists in the Pacific Northwest where extensive bear damage occurs on timber.
>
> The breeding season of coyotes in most cases is not concurrent with the smaller carnivores but precedes them. This adds greatly to the selectivity of this method. Selection of bait carrier, dispersal of baits, and choice of baiting sites are all expected to minimize the effects on associated species and increase selectivity for the target species. An added advantage is the non-lethal effect on domestic dogs that frequent control areas.

Concerning the effect of stilbestrol on future reproduction, foxes, mink, and dogs have produced successful litters the year following administration of this drug. Indications are that stilbestrol would not have lasting effects except in cases where an extreme overdose causes damage to ovarian tissue.

A further question is raised about possible delayed breeding that would negate the effects of antifertility agents. This is true of progestational agents as used in dogs and humans and may be true of stilbestrol when administered prior to ovulation during follicle development. Two of our penned coyotes ovulated 30–45 days after administration of a single 100 mg oral dose of stilbestrol when given about 3 weeks before the normal peak of estrus. However, in coyotes, when stilbestrol is administered after ovulation to interfere with implantation or gestation, new follicles cannot be raised until corpora of either pregnancy or ovulation become non-functional. This prevents the coyote from recycling before the next breeding season.

Results of the field trial indicate that a wild coyote population can be successfully treated. Whether similar results can be obtained under a wide variety of conditions will be determined in future field trials.

IV. SUMMARY

Vertebrate pest control is applied ecology, i.e., it is the regulation of population levels—not necessarily the destruction of individuals. All animal control should be based on a prudent translation of the ecological laws of nature into an effective management policy. Since the control of vertebrates is quite different than insect control, this chapter on Biocontrol and Chemosterilants includes a discussion of both the field of vertebrate control and the ecological implications related to any effective employment of either biocontrol or chemosterilant procedures in managing troublesome vertebrates.

Since vertebrates are long-lived and the troublesome individuals usually cannot be tolerated until they die of natural causes, the greatest hope for both biocontrol and chemosterilants in vertebrate control is to use them in conjunction with integrated control procedures, where some other technique is employed to initially reduce the density of the vertebrates. Then various combinations of biological control and antifertility agents can be used to maintain the vertebrates at a tolerable density level.

The potential role of natural and introduced predators of vertebrate pests is discussed in some length. Introduced predators, such as house cats supplemented with milk, appear to have more utility in vertebrate control than do the naturally existing predators, which are more likely to stimulate their prey species to increase in density than depress them. Since natural predators are a dynamic component of ecosystems, their numbers should never intentionally be artificially manipulated, either up or down, without first obtaining considerable ecological knowledge about the consequences.

When one is willing to accept the ecological consequences, many kinds of vertebrate pest problems can be largely alleviated by modifying the conditions of the habitat. This statement is supported by the fact that the most important factor determining the presence of animals in a given locality is the suitability of the habitat to the species in question. However, it must be recognized before habitats are intentionally manipulated to control vertebrates that any artificial alteration of the physical conditions of habitats will alter the entire ecosystem more drastically than could result if members of specific species of animals were removed by some host-specific control method.

Before the implantation of disease-causing organisms can become an important tool in the control of vertebrates, it is essential that we have a keener insight into the complete ecology of such epizootics and into the inherent hazards associated with the artificial manipulation of the dynamics of such agents. The complexity of disease phenomena should serve as a warning against any hasty assumptions about the role of pathogens in population management. At the present, various frightening devices, repellents, exclusion by barriers, electric shock, stupefacients, and anesthetics show much greater utility in the field of vertebrate pest control.

Even though at the moment a great deal more insight into chemosterilants is required before antifertility methods can be widely applied to control the birthrate of troublesome populations of vertebrates; I believe it to be a challenging and promising field of research with mammals, birds, reptiles, and fish. By using a combination of sterilants and conventional control methods, it probably will be possible to obtain synergistic effects, where the degree of control will greatly exceed the sum of the independent effects of each method. A variety of promising birth control materials and procedures for birds and mammals are discussed.

REFERENCES

Anderson, J. A., and Henderson, J. B. (1961). *New Zealand Deerstalkers Assoc. (Spec. Publ.)* **2,** 37 pp.

Anonymous. *Pennsalt Chem. Corp., Agr. Chem. Div., Develop. Bull.* **36.**

Balser, D. S. (1964a). *J. Wildlife Management* **28,** 352.

Balser, D. S. (1964b). *Proc. 2nd Vertebrate Pest Control Conf., Anaheim, Calif., 1964* (M. Cummings, ed.), Univ. Calif. Davis Agr. Exten. Serv., Davis, California. pp. 133–137.

Barraclough, C. A. (1961). *Endocrinology* **68,** 62.

Bates, R. W., Riddle, O., and Lahr, E. L. (1937). *Am. J. Physiol.* **119,** 610.

Bendell, J. F. (1959). *Can. J. Zool.* **37,** 173.

Besser, J. F. (1962). *Proc. Natl. Bird Control Seminar* 4 pp. Bowling Green State Univ., Bowling Green, Ohio.

Biehn, E. R. (1951). *Calif. Fish Game, Game Bull.* **5,** 71 pp.

Bird, H. R. (1956). *Proc. Cornell Nutr. Conf. Manufacturers* p. 24.

Blair, W. F. (1964). *BioScience* **14,** 17.
Bock, M., and Jackson, H. (1957), *Brit. J. Pharmacol.* **12,** 1.
Carson, Rachel (1962). "Silent Spring." Houghton, Boston, Massachusetts.
Chitty, D. (1960). *Can. J. Zool.* **38,** 99.
Cope, O. B. (1960). *Trans. 2nd Seminar Biol. Problems Water Pollution, 1959* p. 72.
Cope, O. B., and Springer, P. F. (1958). *Bull. Entomol. Soc. Am.* **4,** 52.
Couch, L. K. (1946). *Trans. 11th N. Am. Wildlife Conf.* p. 323.
Craighead, J. J., and Craighead, F. C., Jr. (1956). "Hawks, Owls and Wildlife." Stackpole, Harrisburg, Pennsylvania.
Crissey, W. F., and Darrow, R. W. (1949). *N.Y. State Conserv. Dept. Div. Fish Game Res. Ser.* **1,** 28 pp.
Cummings, M. W. (1962). *Proc. Vertebrate Pest Control Conf., Sacramento, Calif., 1962,* pp. 113–125. Natl. Pest Control Assoc., Elizabeth, New Jersey.
Dahlberg, B. L., and Guettinger, R. C. (1956). *Wisconsin Conserv. Dept. Tech. Wildlife Bull.* **14,** 282 pp.
Darling, F. F. (1937). "A Herd of Red Deer." Oxford Univ. Press, London and New York.
Davis, D. E. (1959). *Anat. Record* **134,** 549.
Davis, D. E. (1961). *Trans. N. Am. Wildlife Conf.* **26,** 160.
Davis, D. E. (1962). *Anat. Record* **142,** 353.
Davis, D. E., and Jensen, W. L. (1952). *Trans. N. Am. Wildlife Conf.* **17,** 151.
De Grazio, J. W. (1964). *Proc. 2nd Vertebrate Pest Control Conf., Anaheim; Calif., 1964* (M. Cummings, ed.), pp. 43–49. Univ. Calif. Davis Agr. Exten. Serv., Davis, California.
Denton, C. A. (1940). *Poultry Sci.* **19,** 281.
DeWitt, J. B., and George, J. L. (1960). *U.S. Fish Wildlife Serv. Circ.* **84** revised, 36 pp.
Dice, L. R., and Howard, W. E. (1951). *Univ. Mich. Contrib. Lab. Vertebrate Biol.* **50,** 15 pp.
Duncan, G. W., and Lyster, C. (1963). *Fertility Sterility* **14,** 565.
Eadie, W. R. (1954). "Animal Control in Field, Farm, and Forest." Macmillan, New York.
Elder, W. H. (1964). *J. Wildlife Management* **28,** 556.
Elton, C. S. (1953). *Brit. J. Animal Behaviour* **1,** 151.
Errington, P. L. (1946). *Quart. Rev. Biol.* **21,** 144.
Errington, P. L. (1956). *Science* **124,** 304.
Fitzwater, W. D. (1962). *Proc. Vertebrate Pest Control Conf., Sacramento, Calif., 1962,* pp. 67–78. Natl. Pest Control Assoc., Elizabeth, New Jersey.
Forbes, S. A. (1880). *Illinois Natl. Hist. Surv. Bull.* **1,** 3.
Forbes, S. A. (1882). *Illinois Hort. Soc. Trans. 1881* [*N.S.*] **15,** 120.
Graham, R. J., and Scott, D. O. (1959). *Mont. Fish Game Dept., U.S. Forest Serv. Rept.* 35 pp.
Herman, C. M. (1964). *Proc. 2nd Bird Control Seminar* pp. 112–121. Bowling Green State Univ., Bowling Green, Ohio.
Howard, W. E. (1949). *Univ. Mich. Contrib. Lab. Vertebrate Biol.* **43,** 50 pp.
Howard, W. E. (1953). *J. Range Management* **6,** 423.
Howard, W. E. (1958). *New Zealand Dept. Sci. Ind. Res. Inform. Ser. Bull.* **16,** 47 pp.
Howard, W. E. (1960). *Am. Midland Naturalist* **63,** 152.

Howard, W. E. (1962). *Trans. N. Am. Wildlife Conf.* **27,** 139.
Howard, W. E. (1964a). *J. Wildlife Management* **28,** 421.
Howard, W. E. (1964b). *Proc. 2nd Bird Control Seminar* pp. 126–137. Bowling Green State Univ., Bowling Green, Ohio.
Howard, W. E. (1965a). *New Zealand Dept. Sci. Ind. Res., Inform. Ser. Bull.* **45,** 96 pp.
Howard, W. E. (1965b). *In* "The Genetics of Colonizing Species" (H. G. Baker and G. L. Stebbins, eds.), pp. 461–484. Academic Press, New York.
Howard, W. E. (1966). *Proc. Congr. Protection Tropical Cultivations, Marseilles, 1965,* pp. 627–629.
Howard, W. E., and Childs, H. E., Jr. (1959). *Hilgardia* **29,** 277.
Howard, W. E., Cummings, M. W., and Zajanc, A. (1961). *Calif. Vector Views* **8,** 14.
Huffaker, C. B. (1958). *Proc. 10th Intern. Congr. Entomol., Montreal, 1956* **2,** 625.
Hume, C. W. (1958). *UFAW Courier* **15,** 1.
Jackson, H. (1959). *Pharmacol. Rev.* **11,** 135.
Jackson, H., Fox, B. W., and Craig, W. (1961). *J. Reprod. Fertility* **2,** 447.
Jackson, W. F. (1953). *Calif. Vet. Nov.-Dec.,* 22.
Johnson, H. N. (1964). *Proc. 2nd Vertebrate Pest Control Conf., Anaheim, Calif., 1964* (M. Cummings, ed.), pp. 138–142. Univ. Calif. Davis Agr. Exten. Serv., Davis, California.
Jones, D. O. (1964). *Proc. 2nd Bird Control Seminar* pp. 4–9. Bowling Green State Univ., Bowling Green, Ohio.
Keith, J. O., Hansen, R. M., and Ward, A. L. (1959). *J. Wildlife Management* **23,** 137.
Kerswill, C. J. (1958). *Atlantic Advocate* **48,** 65.
Latham, R. M. (1951). *Penn. Game News* (*Spec. Issue*) **5,** 96 pp.
Lauckhart, J. B. (1961). *Proc. 41st Ann. Conf. Western Assoc. State Game Fish Commissioners, Sante Fe, New Mexico, 1961.*
Leopold, A. S., Cain, S. A., Cottam, C. M., Gabrielson, I. N., and Kimball, T. L. (1964). *Trans. N. Am. Wildlife Conf.* **29,** 27.
Linhart, S. B. (1964). *N.Y. Fish Game J.* **11,** 69.
Linhart, S. B., and Enders, R. K. (1964). *J. Wildlife Management* **28,** 358.
Marsh, R. E. (1962). *Proc. Vertebrate Pest Control Conf., Sacramento, Calif., 1962,* pp. 98–107. Natl. Pest Control Assoc., Elizabeth, New Jersey.
Merrill, H. A. (1962). *Proc. Vertebrate Pest Control Conf., Sacramento, Calif., 1962,* pp. 79–97. Natl. Pest Control Assoc., Elizabeth, New Jersey.
Morris, R. F. (1963). *Entomol. Soc. Can. Mem.* **31,** 311.
Neff, J. A., and Meanley, B. (1957). *Arkansas Univ.* (*Fayetteville*) *Agr. Expt. Sta. Bull.* **584,** 89 pp.
Nichols, E. L., and Balloun, S. L. (1962). *Poultry Sci.* **41,** 1982.
Nicholson, A. J. (1954). *Australian J. Zool.* **2,** 9.
O'Roke, E. C., and Hamerstrom, F. N., Jr. (1948). *J. Wildlife Management* **12,** 78.
Petrusewicz, K. (1957). *Ekolog. Polska* **A5,** 281.
Pitelka, F. A. (1958). *Proc. 15th Intern. Congr. Zool., London, 1958* Sect. 10, Paper 5, 3 pp.
Rasmussen, D. I., and Doman, E. R. (1947). *Trans. N. Am. Wildlife Conf.* **12,** 204.
Robinson, W. B., and Harris, V. T. (1960). *Am. Cattle Producer* **42,** 2 pp.

Rudd, R. L. (1964). "Pesticides and the Living Landscape." Univ. of Wisconsin Press, Madison, Wisconsin.

Rudd, R. L., and Genelly, R. E. (1956). *Calif. Dept. Fish Game, Game Bull.* **7,** 209 pp.

Scott, T. G. (1958). *Illinois Nat. Hist. Surv. Bull.* **27,** 179 pp.

Swift, J. E. (1964). *Proc. 2nd Vertebrate Pest Control Conf., Anaheim, Calif., 1964* (M. Cummings, ed.), pp. 71–76. Univ. Calif. Davis Agr. Exten. Serv., Davis, California.

Tinbergen, L. (1946). *Ardea* **34,** 1.

Travis, H. F., and Schaible, P. J. (1962). *Am. J. Vet. Res.* **23,** 359.

Vandenbergh, J. G., and Davis, D. E. (1962). *J. Wildlife Management* **26,** 366.

Van Wijngaarden, A., and Bruijns, M. F. M. (1961). *Lutra* **3,** 35.

Watt, K. E. F. (1964). *Proc. 2nd Vertebrate Pest Control Conf., Anaheim, Calif., 1964* (M. Cummings, ed.), pp. 24–28. Univ. Calif. Davis Agr. Exten. Serv., Davis, California.

Weinburgh, H. B. (1964). *U.S. Dept. Health Educ. Welfare* **PHS-CDC** 87 pp. Atlanta, Georgia.

Wetherbee, D. K. (1964). *Proc. AAAS Symp. Pest Control, Montreal, 1964.*

Wetherbee, D. K., Crawford, R. D., Consuegra, P. F., Coppinger, R. P., Landy, M. J., Somes, R. G., Jr., and Wentworth, B. C. (1962). *U.S. Fish Wildlife Serv. Spec. Sci. Rept. Wildlife* **67,** 97 pp.

Wetherbee, D. K., Coppinger, R. P., Wentworth, B. C., and Walsh, R. E. (1964). *Univ. Mass. Agr. Expt. Sta. Bull.* **543,** 16 pp.

Wynne-Edwards, V. C. (1959). *Ibis* **101,** 436.

11 BEHAVIORAL MANIPULATION (VISUAL, MECHANICAL, AND ACOUSTICAL)

Hubert Frings and Mable Frings**

DEPARTMENT OF ZOOLOGY
UNIVERSITY OF HAWAII, HONOLULU, HAWAII

Man has used his knowledge of animal behavior to aid him in protecting his food and property from pests since prehistoric times. Frightening devices or noises have been his main tools. With modern studies of animal behavior yielding a harvest of precise information on reactions of animals —both approach and withdrawal—to a wide variety of stimuli, other aspects of behavior can be exploited for pest control. We shall deal in this chapter not only with means for frightening vetebrates, chiefly birds, but also with behavioral manipulations not involving fright. The most important of these are reactions of animals to their own communication signals. We shall restrict this discussion to the use of visual, mechanical, and acoustical signals for control of pest vertebrates.

I. FRIGHTENING DEVICES

The use of visual and acoustical scaring methods to chase pest mammals and birds from homes or crops is as old as man's competition with

* Present address: University of Oklahoma, Norman, Oklahoma.

these animals. In primitive societies, and even in advanced civilizations, these obvious methods are still used. The farmer who dashes from his home to scare marauding deer or blackbirds is matched, more formally, by the "bird-boys" of Africa, who patrol grain fields waving branches, swinging rattles, or yelling. Man, always seeking methods that consume less manpower, has tried scarecrows, flashing strings, dangling ropes, bright mobiles, and various noise producers, such as wind-rattles, chimes, etc. With increasing sophistication, he turns to mechanically driven scarecrows and pyrotechnic devices. Where these can be used for relatively short times to achieve the desired effects, results are often satisfactory. Where long-term controls are needed, however, results are far from satisfactory. Animals soon learn to ignore objects and sounds that portend no danger. Most modern studies in this field have been made to find some means to reduce or circumvent this habituation. Progress has been made, but most of the problems still remain.

A. Visual Frightening Devices

1. *Moving Bio-objects* (*Scarecrows*)

Scarecrows represent the oldest of all devices used by man as substitutes for his presence. All cultures have their own patterns, and all find rather soon that birds and mammals rapidly habituate to the presence of these foreign objects. There follows, then, efforts to increase effectiveness. Addition of strips of blowing metal to the scarecrows, or moving the objects about (Neff and Meanley, 1957) may slow down habituation, sometimes sufficiently to allow their use, but rarely are these completely satisfactory. Two-surfaced owl dummies (LaFond, 1961) which turn in the wind have been used with some success to repel gulls from towers. Machinery to move scarecrows, and loudspeakers to broadcast sounds from them are modern modifications, but so far these have proved either too expensive or relatively ineffective.

There has been some work on the use of models of hawks and other birds of prey to protect fields from bird attack. These have been hung from balloons (Kuchly, 1961; Schmitt, 1962a), suspended from poles (Busnel and Giban, 1960, pp. 193–196), or swung about from poles (Hardenberg, 1960). The results have been various. For instance, in Holland, a model of a goshawk swung about over a field was effective in 1957, but not in 1958 (Hardenberg, 1960). These models often depend for their supposed effectiveness on the theory that short-necked flying "birds" terrify others on the ground, while long-necked fliers do not (Tinbergen, 1951). This idea is based on studies on reactions of fowl to shapes moved above them. Using a birdlike dummy with a long projection on one side of simulated wings and a short projection on the other, Lorenz and Tinbergen found that the fowl exhibited fear reactions when the model moved above with the short projection forward, but remained unperturbed when the long projection was forward. They interpreted this to mean that the birds distinguished hawks as "short-necks," and geese and ducks as "long-necks."

Schleidt (1961) has retested this idea, using turkeys as the test birds and a variety of shapes moving overhead. He concludes that the short-neck, long-neck idea is incorrect, finding that the degree of response depends mainly upon the size and speed of movement of the object overhead. Circular disks are just as effective as those with projections. The birds react to strange objects overhead, but there is rapid habituation to continued exposure. Thus, he thinks that, because in the earlier studies the test animals were birds that were conditioned to having flying ducks and geese overhead, they did not respond to an object looking like them, but did to the same shape "flying" in reverse. From the standpoint of the possible usefulness of these rather expensive mechanisms for bird repellency, the rapid habituation suggests that they may be of limited value.

Among bio-objects, albeit not moving, that have been used for scaring birds are dead specimens of the same species. Along the coast of Maine, for instance, fishermen generally believe that hanging a dead gull on the mast of a ship will keep other gulls away, thus saving the catch from being taken. Likewise fish drying companies kill gulls and hang their bodies up near the drying racks. Models of the birds seem to be almost ineffective. Kuchly (1961) reports this also for Starlings (*Sturnus vulgaris*) in France. Hardenberg (1965) and Wright (1965) report the use of dead gulls along airport runways in Europe.

As Wright (1965) notes, however, there are no scientific tests of the effectiveness, or even of the reactions of birds to the sight of dead fellows. With Herring Gulls (*Larus argentatus*) (Frings *et al.,* 1955b,c) it was observed that birds in the air emit a characteristic alarm call when they sight a hanging dead gull. A recording of this call proved just as repellent as the dead bird, suggesting that the mechanism of action involves the well-developed acoustical communication system of these animals. Obviously birds that see a dead gull must recognize it by sight, and it would be interesting to know what the signal-releasing factors in this situation are. Certainly, a careful study of the reactions of various species to captive and dead fellows is in order. In practical terms, habituation seems to be fairly rapid, and Hardenberg (1965) points out that to continue to be effective the bodies must be moved about, and above all else be kept in orderly, recognizable condition. The last suggests that models would have to be quite lifelike if they were to produce results, and Wright (1965) reports that attempts to use dummies near airports were without success.

In Europe, efforts have been made to repel birds from airfields by the use of trained falcons (Brown and Sugg, 1961; Van der Heyde, 1965; Wright, 1965). The British workers report success with the method, but training of falcons is costly, losses are often appreciable when the birds are released, and falcons are becoming scarce. In Holland, falcons did not prove successful, for they were not, even when trained, aggressive enough against larger birds. The effectiveness of trained falcons probably depends upon two main factors—their typical appearance recognized by the birds, and since habituation to this occurs rapidly, attacks of the falcons on birds that fail to respond, thus reinforcing the effect. There might also be reactions to the sounds of the predators. If the first two alone are responsible, it is hard to see how any artificial arrangement could be set up to replace the living birds. If the third is effective, recordings of the calls might be useful, if rebroadcast to resting birds. This will be discussed later.

2. *Scintillating Objects, Flags, etc.*

A tremendous variety of flashing, moving, flickering, and waving objects has been used with variable success, usually depending upon length of time in use. Floodlights and moving beacons have been used in attempts to disturb roosting Pigeons (*Columba livia*) and Starlings, usually with little or no success (United States Fish and Wildlife Service, 1948). Whirling or flashing bits of metal, or windwheels suspended over fields or on buildings have also been used, with somewhat more success (Neff, 1945; Kuchly, 1961, among others). For long times or large areas, habituation generally occurs, and the expense is large. Most of the work with these has been done without rigid standards of experimentation—particularly the limits of vision of the birds have been ignored—and a careful behavioral study of reactions of birds to specific patterns and colors might develop some interesting leads. The unpredictability of results in this field may be illustrated by the results of Keil (1962), who found that ordinary large glass balls atop poles repelled birds of prey, and other birds too, for long periods of time.

Flags hung from wires or poles are often used to scare birds. Different people favor various designs for the flags, almost totally unsupported by critical scientific data. Bruns (1959b) reports that black and white striping greatly increases the effectiveness of flags used in Starling control and recommends against red, which was commonly used in Germany. In Europe, various colored, or black and white, nylon, dangling-thread masses, usually with trade names (Kuchly, 1961; Bruns, 1961; Przygodda, 1962; Schmitt, 1962a) are used above vineyards, etc. These seem to be effective for at least short times, if sufficiently close together, sufficiently bright, and moved by the wind. Habituation occurs, but often not soon enough to prevent harvesting of fruits before appreciable bird damage can occur. It would appear that the effective stimuli here are the movements of strange objects near the feeding places of the birds. Certainly some comparative studies should be made on reactions to colors, shapes, degree of mobility, etc. Perhaps if these streamers were hung on a grid sufficiently closely spaced to prevent easy flight (McAtee and Piper, 1936), the method would function effectively.

An illustration of a special flag-type repellent has been reported for Laysan Albatrosses (*Diomedea immutabilis*) on Midway Island (Frings and Frings, 1959a). These birds do not fly under overhanging solid objects, such as canopies on airplane hangars. They turn aside whenever such sizeable, overhanging surfaces appear before them. With this as a basis, a piece of flag cloth, 3 ft wide and about 12 ft long, was hung lengthwise over a regular flight path of these birds. The birds now turned aside, even though there was a perfectly clear flyway about 15 ft wide and 30 ft high beneath. When the flag was lowered to the ground, the birds flew between the poles on which it had been hoisted, at altitudes down to a few feet. This suggests that streamers hung at an altitude of about 50 ft along the edges of the runway, particularly at danger areas, might turn the birds back from traversing the runway and thus creating a hazard. Near runways, of course, aviators do not wish to have hanging objects, unless these would be easily destroyed if hit by a plane. So it was suggested that the flags be made of very thin materials and suspended from a cord between fragile balloons. This would probably have remained only a curiosity, however, for the wind at Midway Island is strong, and it is doubtful whether balloons could be kept in place.

Salted herrings suspended from strings above vineyards or orchard trees have been used in Europe for some years to repel Starlings (Bruns, 1959c; Kuchly, 1961; Przygodda, 1962; Schmitt, 1958). The literature on this is scanty, even though these fish are apparently widely used by farmers. In 1958, at an international colloquium on bird control, Schmitt (1960b) expressed the belief that the smell of the herrings was the effective agent. This was doubted by others, however, because of the widely held belief that birds have a poorly developed olfactory sense. The other alternative seems to be, that their flashing in the sunlight is the effective feature. In 1961, the matter was again discussed, briefly at a second colloquium (Giban, 1962b, p. 26), and a few tests were reported. These showed that dummies cut from bright metal did not repel the birds, while pieces of cloth soaked in fluid from the herrings did; this seems to support the olfactory hypothesis. This was further supported by a report at the earlier colloquium that chopped salted herrings scattered near the hutches of domestic rabbits are used in some parts of France to repel predators. The results with the herrings are variable, from seemingly fully effective repellency for appreciable periods of time to no effect at all. These are included here among visual devices, not because there is proof that the effects are visual, but because the olfactory hypothesis has not been confirmed beyond reasonable doubt.

While most fireworks depend for their frightening effects on sound, many also produce bright flashes, smoke puffs, or similar visual accompaniments, and some, such as Roman candles, are purely visual in operation. Many kinds of pyrotechnics are used for scaring birds, and these have been recommended for a long time (cf. Kalmbach, 1928; Neff and Meanley, 1957). Reviews of their use and degree of effectiveness have been published by Bruns (1959a,b, 1960, 1961) and Schmitt (1962a) in Germany, and by Brown and Sugg (1961), Brown *et al.* (1962), Kuhring (1965a,b), and Wright (1965) in Britain and Canada. It is generally assumed that the major repelling factor in fireworks is the sound, and this will be discussed below, but Kuhring emphasizes the increased effectiveness if sounds are accompanied by smoke puffs, smoke trails, etc. Many people use smoke puffs, nonexploding rockets, and bright lights as repellents for birds. Generally the reports are that these are more effective than continuously present scintillating objects; the irregularity of appearance facilitates the effects. Habituation does occur, but can be circumvented by shifting from one type to another. Research is now mainly directed toward discovering the most suitable times of application, particularly minimum times, for fireworks have an appreciable cost. So far, no generally accepted rules emerge from this research, except the need for variety in timing and materials.

B. Mechanical and Acoustical Frightening Devices

1. *Biosounds*

These are sounds produced by a species other than the pest species that might influence the behavior of the pest species—for instance, the cry of a predator. There are many anecdotal reports of the responses of hunted mammals to the sounds of predators, but, to our knowledge, no critical studies of their possible use in pest control. With birds, Miller (1952) reports reactions, usually mobbing, by many small species to his imitations of calls of hawks and owls. Since these birds did not respond to other sounds, and he found specific responses, he concludes that there is auditory recognition of predators by birds. For those interested in practical bird control, there might be some utility in exploring a wide gamut of hawk and owl calls under a variety of situations.

Boudreau (1965) found that a protest call of the Sparrow Hawk (*Falco sparverius*), when recorded and broadcast to feeding English Sparrows (*Passer domesticus*) was strongly repellent, with no habituation to the broadcasts after 6 days of treatment. This reaction is the reverse of that found by Miller, and raises some interesting questions: does it result from predator recognition, or does the hawk call resemble some call of the sparrow (cf. Busnel, 1965)? Further research, using the methods now fairly well developed for studies on communication signals should yield interesting data. It has been reported that sounds produced by crickets attract Starlings, but laboratory tests of this hypothesis (Nelson and Seubert, 1964) give no support to it.

2. *Loud Noises*

The use of sudden loud sounds, usually noises, produced by a wide variety of methods matches the use of scarecrows in antiquity. From the housewife who claps her hands at a marauding cat to the agriculturist who hangs an exploding acetylene bird-chaser in his orchard the methods are basically the same. These have been used for scaring mammals, such as deer or mice, but the major studies have been with birds.

Almost all animals, from protozoans to man, respond to sudden loud sounds or vibrations with startle reactions or withdrawal. Unfortunately, all animals generally soon habituate to loud sounds, so that continued effective use is usually impossible. For birds, sudden noises are often the first, and sometimes the only, attempted control means. For short times, with alterations in rhythm and nature, they may be all that is needed. Where bird control requires long times of treatment, or extended areas must be treated, however, these usually fail to live up to their early promise.

There has been little scientific work on the use of loud noises to control mammals. Certainly, deer, foxes, squirrels, rats, mice, etc., respond to sudden clicks or other noises with startle reactions. Calhoun (1963), for instance, found that rats scurried for cover with any clicking, even that made by a movie camera. For short-term control, these sounds may be quite effective, but mammals soon learn to ignore noises, unless something happens.

It has been suggested that sharks might be kept from swimmers by intense underwater vibrations or sounds. So far, the results have been disappointing. Possibly with increased knowledge of the sensory capacities of sharks, practical ideas may be formulated, but right now, too little is known about the matter (Backus, 1963; Moulton, 1963; Tavolga, 1960, 1965; Winn, 1964). The hearing ranges of fish in general, and sharks in particular, are narrower than those of most other vertebrates and centered at low (below 1000 Hz) frequencies (Kleerekoper and Chagnon, 1954; Dijkgraaf, 1963b; Wisby *et al.,* 1964). Intense underwater sounds in these ranges are hard to produce, although they carry well once produced. The recent interest in near-field effects (Van Bergeijk, 1964), which could be received through the lateralis system of fish (Dijkgraaf, 1963a,b) might lead to some new ideas in acoustic fish repellency. Fortunately, except for sharks and a few other species, fish are not considered pestiferous by man, so the matter has little import here. Interest in guiding fish around dams, etc. (cf. Burner and Moore, 1953), or in attracting them for catching, will probably lead to further discoveries about acoustical behavior of fish, and thus might lead to ideas for acoustical shark control.

Two main groups of methods have been used to produce loud sounds for driving birds from areas of depredations. First, there are literally hundreds of methods for mechanically producing noises—banging together pieces of wood or metal, rattling bamboo or wooden pieces, windwheels with ratchets, shouting of large groups, cymbals, drums and other percussion instruments, etc. Second, there are many types of firecrackers—rockets, bombs, etc.—which produce loud sounds as well as visual stimuli.

In the first category, there are no really modern devices, in the sense that they employ new principles. Their use has been reviewed by Büttiker (1962) and Bruns (1959b) among others. Generally the cost is low, but they may or may not be effective, depending upon conditions. The sound must be loud enough to startle the birds, and many problems arise in getting sufficiently intense sounds. Other problems arise in getting repetition without employing human beings, and in varying the rhythm of application. Even with repeated changes of equipment and placement, habituation is always rapid; this is the major problem. For short times and restricted areas, some method can usually be found for a particular bird problem. Otherwise, mechanically produced sounds soon become ineffective.

Among recently tried techniques is the use of a helicopter to drive Starlings from nocturnal roosts (Haag, 1958; Schmitt, 1960a,c). While undoubtedly the sight of the aircraft and wind from it were involved, it would seem probable that the noise was the dominant factor. By having the helicopter fly low over the roost for short times on two or three evenings, the roost was cleared. Obviously habituation was avoided here, because the time of application was short. Schmitt feels, however, that this is not a practical solution, for it is very expensive and involves an element of danger.

Another case in which habituation seems to have been obviated because of short exposure time—in this case of individuals, not the whole population—is reported by Frings and Boudreau (1964). In an attempt to keep Laysan Albatrosses on Midway Island from crossing the runways and colliding with airplanes, they tried the use of very high intensity noises broadcast into the flight path of the birds. With intensities of 125 decibels (dB) at 1 meter from the speakers, recorded sounds of noisy propeller-driven aircraft and recordings of other interrupted noises were found to turn back the flying birds at distances up to 600 ft, sufficient to allow the sounds, broadcast from the sides of the runway, to turn birds back at either side. Since the sound was used only during takeoff and landing of aircraft—about 5 minutes each time—and since different birds were in the air at different times, habituation did not set in, even after 3 months, during which traversals of the runways were reduced by about 35%, and aircraft strikes reduced accordingly.

At present, much interest centers on the use of pyrotechnics for the production of loud noises. Single firecrackers, double crackers (one of which throws the other for some distance), various rockets and explosive charges, shell crackers fired from shotguns, and a variety of projectiles and blanks for use in pistols and other types of firearms are being used. The selection of types depends upon expense, distance to be covered, and other factors. As with all noises, variety may be important, otherwise habituation is rapid. Earlier uses of pyrotechnics are reviewed in Kalmbach (1928, 1945) and Neff (1945). Bruns (1959a,b, 1962) reviews the extensive German tests with rockets constructed to explode at set distances from the firing points. These are set off selectively by a man stationed on a tower in an orchard or grape vineyard to follow birds with rockets exploding behind, thus forcing them to fly in one direction, usually that of their migration. The German workers have an elaborate battery of pyrotechnics available, and use them with precision.

Brown and Sugg (1961), Cooke-Smith (1965), Kuhring (1965a,b), and Wright (1965) review the use of fireworks to repel gulls and other birds on British and Canadian airfields. This is their chief weapon against these hazards to aircraft. As with the Germans, they use a wide variety of firecrackers, shells, and rockets. They find that the presence of human beings gives them added effectiveness. Seubert (1965) and Nelson and Seubert (1964) also report favorably on the use of pyrotechnics, chiefly shell crackers. Habituation is slow if irregular times of application are used, and changes in types of fireworks may overcome it.

In the United States, there has been considerable use of so-called rope firecrackers for bird repellency (Neff and Mitchell, 1956). These are made by inserting fuses of firecrackers into a piece of rope, hanging this in a field, and igniting the rope. The rope burns slowly, detonating the firecrackers at intervals. The time between explosions can be determined by the spacing of the firecrackers along the rope; thus it is possible to set up a long string that will detonate periodically for hours. Obviously, one must use enough of these to have sufficient explosions throughout an area to affect the birds. Because of the loud report and somewhat irregular times of explosion, these devices have proved quite effective for short times, and in some cases for long times (Neff and Meanley, 1957; Nelson and Seubert, 1964).

All these pyrotechnic devices have two potentially serious drawbacks. First, they are relatively expensive, for they are destroyed; second, they may be fire hazards. The first must be balanced against the gain in harvested crop, and so may or may not be a factor. The second varies with weather conditions, etc., and must be faced. At present, it would seem that fireworks are the most widely used means for scaring birds from airfields and agricultural areas in the United States, and are used to a high degree also in Canada, Britain, and Germany, but there they are usually part of a multiple approach. Variability in effectiveness is great, and habituation is a problem. Reports in the literature tend to deal with special treatments, and studies on fundamentals are scarce.

Acetylene exploders are another pyrotechnic means for producing loud, interrupted sounds. These were introduced some years ago for bird control in orchards, and are now used, mainly in the United States, at airfields, in orchards, etc. They are variously constructed, mostly commercially, using carbide to produce acetylene, which is exploded at various intervals determined by adjustment of the flow of water onto the carbide. The exploders are hung in the orchard or field to be protected, and once set, operate with little care. Van der Heyde (1965), however, reports that in Holland they are unsatisfactory, because they are sensitive to temperature changes, etc. Cardinell (1937), who reported some time ago on their use for preventing damage to cherries by American Robins, found no habituation. Later workers have not had such complete success. Brown and Sugg (1961) and Wright (1965), reporting on their use near airport runways, say that they have a maximum effectiveness of 2 weeks. They are quite expensive, but are less a fire hazard than fireworks.

A major problem with all noise producers is habituation, and this has been much discussed, without much enlightenment. Studies of this process under controlled conditions in the laboratory and field might yield valuable information. For every one of the noise-producing devices, there are experts offering it as the best available for bird control, and at the same time others finding it of low value. As always, the importance of habituation varies with the nature of the problem and time of control needed. Thus, with breeding birds habituation is rapid, with feeding birds less rapid. If one must protect a crop from depredations over long periods of time, habituation becomes a major problem with all noise devices. If the time of protection is short, habituation may not be a consideration. A factor in habituation to noise producers that has not been critically studied is its relationship to the intensities and patterns of sounds produced by the devices. Careful acoustical studies of these devices are long overdue.

This leads to an important consideration. Since loud sound is the operative feature in all of these, the sound could just as well be applied by recording and broadcasting as by the wasteful procedures now used. Thus, a loop of tape of variable length with recorded loud reports broadcast through speakers placed appropriately could, without fire hazard and at lower expense, produce the same noise. In view of the fact that tape recorders and other acoustical equipment have been available for some time, one wonders why more has not been done along this line. As long ago as 1950, Reich reported on the use of recorded sounds for chasing Starlings. In tests with the Starling distress call (Frings *et al.,* 1955a), which is broadcast in this manner, recorded loud noises were often used as reinforcement. The work with Albatrosses using recorded noises has been recounted above.

Actually, it is not even necessary to have firecrackers or guns for recording. If one overloads the microphone by some sudden loud sound, such as hand-clapping, an excellent artificial pistol shot or cannon-cracker explosion results. A tremendous variety of sounds can be made by using pieces of wood banged together, pieces of metal struck variously, handclaps directly in front of the microphone, drum smashes, popguns, etc. These can then be indefinitely reproduced, arranged by cutting and splicing the tape in any desired time sequence (so that, for instance, every day's program can be different), and can be broadcast through appropriate amplifiers and speakers at any desired intensity. Original costs may be more, but the equipment is durable unlike fireworks. Actually, unless one wants to do experimental work, even a tape recorder is not needed, for loud clicks, easily amplified into bangs, can be readily produced by short-circuiting wires leading to the amplifier. Certainly, electronic methods for producing sounds—while not solving the habituation problem—seem preferable to potentially hazardous pyrotechnic methods. Experimentally, electronics allows production of sounds of known composition and time intervals, and thus permits critical studies on the mechanisms of action of loud sounds on pest species. The last are sorely needed.

3. *Ultrasonics, etc.*

Ultrasonic (formerly called supersonic) sounds are those whose frequencies are so high that the human ear does not receive them. Physically, ultrasound is defined as sound with frequencies higher than 15,000 cycles per second (Hertz, Hz). Physiologically, with reference to man, it is sound above human hearing limits, about 20,000 Hz. Other species have other ultrasonic limits.

Earlier biological studies on ultrasonics dealt mainly with the destructive effects of underwater ultrasound on tissues and small organisms. This, coupled with the seeming mystery of sounds that could not be heard, led the popular press, and even some scientists, to regard ultrasound as a somehow mysterious form of energy. Actually, all the destructive effects of ultrasound can be duplicated with ordinary sound. Unfortunately, this mystique still surrounds ultrasonics in many peoples' minds, and almost any studies on damaging effects of sounds on animals are said to be ultrasonic research. The most flagrant and long-continued example of this muddled thinking is the continual reference in the popular press, and even by many scientists, to the control of birds by broadcasts of the quite-sonic recorded distress signals as ultrasonic control. Ultrasound is not entirely like other sound, for the higher frequencies have different transmission properties than do lower frequencies, but there is nothing mysterious or particularly effective about ultrasonics for pest control. The only obvious advantage that ultrasounds have over regular sounds is that humans cannot hear them and thus be disturbed along with the pests.

Shortly after World War II, concern was felt for the possible damaging effects of the loud sounds produced by jet aircraft upon human beings. Since many jet planes produce high proportions of ultrasonic sound, which would not be noticed by a man entering the sound field, the question arose whether high intensity airborne sound, at ultrasonic frequencies, could produce damage to organisms as did underwater ultrasound. Allen *et al.* (1948), Frings *et al.* (1948), Frings and Senkovits (1951), and Danner *et al.* (1954) studied the physical effects of high intensity sounds, partly ultrasonic and partly sonic, on mice and other animals and found that these sounds, if sufficiently intense, could injure or kill the organisms. For mice, however, the intensity was exceedingly high—1 watt per square centimeter (W/cm^2)—sufficient to boil water if absorbed by a piece of glass wool surrounding a container of water.

These studies showed that the major, if not the only, deleterious effect of irradiation by high intensity sound, in mammals at least, is heating caused by absorption of sound in the fur. With hairless mice, about 100 times greater sound intensities are needed to produce heating than with furred mice. For so large an animal as man, without dense fur cover, the level needed for damage would be well above 1 W/cm^2, a level obtainable only with extremely high-powered acoustic equipment.

These reports of damage to mice and insects led naturally to the question of possible use in pest control. This has been discussed previously (Frings, 1948; Frings and Frings 1960, 1963), and it has been pointed out that sound fields of such high intensity are extremely limited in size, very expensive to produce, and in general economically impractical. If one can capture and retain a mouse, for instance, in a small parcel of high intensity sound, he could much more easily kill it by many cheaper means. There seems to be no reason to change this opinion today. Unless someone develops a cheap means for producing very high intensity sounds and finds some way whereby these can be transmitted through air over some distance (there are special problems of rapid loss of intensity with high intensity sounds), pest control by sonic destruction is not practical.

However, even if ultrasounds could not be used for destruction, they might be used for scaring pest mammals or birds. Considering first of all birds, there have been suggestions, mostly in the popular press, that birds can be chased by ultrasonic beams. Furthermore, some commercial organizations have claimed successes against Pigeons, Starlings, and English Sparrows with ultrasonic sound producers. All attempts so far, to get critical data from these claimants have proved fruitless. The usual return is a sales circular stating that the device, which usually also produces plenty of sonic sound, chased birds in some town or on some building, presumably checked only by the customers. All scientific tests, so far, in which prechecking and checking during tests have been made, have shown no scaring of birds by ultrasonic sounds (United States Fish and Wildlife Service, 1948; Kalmbach, 1954; Brown and Sugg, 1961; Schmitt, 1957; Van der Heyde, 1965; and Wright, 1965, for example). In short, up to now all carefully conducted tests have indicated no responses by pest birds to ultrasonic sounds, unless accompanied by sonic components. Claims of commercial operators are not buttressed by scientifically acceptable studies.

Actually, all the studies on hearing ranges of birds have shown them to have an ultrasonic limit no higher than that of man, and generally lower (cf. Schwartzkopff, 1955a,b; Frings and Slocum, 1958; Frings and Cook, 1964; Spurlock, 1962). Since many smaller birds have songs with ultrasonic, or near ultrasonic, frequencies of sound, biologists have been led to believe that they hear ultrasounds. Thorpe and Griffin (1962) have shown that the notes of smaller birds do have ultrasonic components, but that these are overtones, and at such low intensity that, coupled with their poor carrying power in air, they almost certainly are negligible. If birds cannot hear ultrasonic sounds, obviously they cannot react to them through their sense organs, and, as we have seen, physical effects require tremendously intense sounds. There is no doubt that an ultrasonic scaring device for birds, particularly where it is to be used near man, would be preferable to the usual noise producers. However, unless someone develops a method for getting the sounds through to the birds, the possibility seems remote.

To be fair, we should note that there are two ways in which ultrasonic sounds might prove to be significant in bird control. The first is in the determination of the waveform of sonic signals. The presence of ultrasonic frequencies in the notes of birds alters the aggregate waveform. Removal of these frequencies, for instance, in recording with a recorder not sensitive to ultrasonics, results in waveforms that differ from the original. Usually this is not appreciable enough to affect the tone quality, but in some cases it may be. Thus, the lack of hearing capabilities of birds for ultrasounds does not mean that they may not distinguish recorded sounds from originals. We shall discuss this later. Second, although ultrasonic signals, as such, may not be audible to birds, combinations of these signals producing beat frequencies might be found that could stimulate particular receptors—ears, vibration receptors, etc.—of birds, but not of man. If such were found, the desired scaring of birds by sounds inaudible to man might be achieved. So far, however, there are no critical studies on this possibility.

With small mammals—dogs, cats, mice, rats, etc.—the situation is quite different. These animals do indeed hear frequencies above those heard by man—their ultrasonic limit is higher. Dogs and cats hear at least one octave higher than man, and rodents hear two or more octaves higher (Prosser and Brown, 1961; Berlin, 1963). This means that studies on the reactions of these animals to ultrasonic noises might have value (Frings, 1948; Frings and Frings, 1963). There has been very little practical work, and much of it has been done by persons who do not report their results in the open literature, because they wish to commercialize their findings. So, reports pass by word of mouth or in advertising copy about successes in chasing rats or mice with ultrasonic sounds, but the nature of the sounds, the intensities used, and the bases for the claims are unclear, or even concealed.

There is no doubt that rodents respond to clicks. For instance, Calhoun

(1963) reports on the excessively fine sensitivity to clicks in Norway rats, making it impossible to take movies without some masking sounds to cover up the clicking of the camera. While these are sonic, they are just as effective if ultrasonic, and there is plenty of evidence that it is the ultrasonic frequencies (for man, of course) in the clicks that are most effective. There would seem to be little reason, therefore, to doubt that rodents at least would react by flight to ultrasonic pulses or noises, but as with similar sounds in the sonic range, habituation would probably be rapid and use for long-term control would be impossible (cf. Marsh *et al.,* 1962). Perhaps as with other noise producers, changes of position, intensity, and rate of emission would make these effective, at least for some days.

While certainly not ultrasonics or even sound, the possible use of radar for pest control might be briefly mentioned here. Knorr (1954) and others report reactions of flying birds to radar beams and suggest that the flight patterns of birds might be influenced by radar. This, could it be done, might offer a means for driving pest birds from feeding areas, or for keeping birds away from airports. The situation seems to be far from clear. Busnel *et al.* (1956), Schwartzkopff (1959a), Kuhring (1965a), and Drury (in Busnel and Giban, 1965, p. 100) report no effects of radar on birds under ordinary conditions. Brown and Sugg (1961) also note that radar control would be very expensive, if radar affected birds at all. The reports of abnormal behavior of birds near radar or high-powered radio installations are few now, in spite of the increasing number of tracking stations. Pigeons and Starlings have been reported roosting on radio antennae. It would seem that ordinarily radar or high-intensity radio waves do not affect bird behavior. Yet the earlier reports in the literature, and other anecdotal reports, suggest that a wide-ranging study might be in order.

On the other end of the spectrum from radar waves are heat rays, and there is a preliminary report by Kuhring (1965a) that heat can attract gulls to land surfaces—indeed, that this may be why these birds congregate on airstrips during the winter. He suggests that possibly furnishing the birds with warm strips away from the airports might lead them to desert the runways in favor of these proffered resting places. One can immediately think of a number of difficulties in using this as a practical means of control. How is the decoy airstrip to be made more attractive than the real one? How are the birds to find out that this area exists? Perhaps one could use the food-finding call of gulls, an attractive call to be described below, to draw them away from the real airstrip. The important thing about this observation, it seems, is that it emphasizes the fact that heat as a stimulus influencing the behavior of animals has not been exploited for control of pest vertebrates. Perhaps mammals other than man and birds might be able to receive infrared rays, and this reception might elicit certain behavior patterns. This would be an interesting study.

4. *Audiogenic Seizures*

Small mammals, particularly laboratory strains of rats and mice, when subjected to certain sounds, suffer audiogenic seizures. The occurrence of these seizures in rodents leads one to believe that, with the right sounds, these animals might be driven from places where they are not wanted. Audiogenic seizures roughly resemble epileptic seizures in man, and some workers believe that they are basically the same. The animal first enters a wild running state, which terminates in uncoordinated spasms or convulsions. Rats do not die in the seizures, but mice often do. These seizures are induced by sounds, which must be sufficiently intense and generally within certain frequency ranges. Certainly it would seem that, if a mouse or rat were to be thrown into such a convulsion by sound, it would, if free, leave the area in which the sound occurs. Since habituation is slow and usually takes place less rapidly the greater the exposure, it would seem that the animals should not become inured to these sounds in any reasonable time. The frequencies most effective in inducing these seizures are ultrasonic, 15–30 kHz, and the intensities needed in this range not exceptionally high, 80–90 dB (Frings and Frings, 1952). The possibilities seem good that some form of control for rodents based on these seizures might be evolved.

The literature on audiogenic seizures is extensive, dealing almost entirely with laboratory bred rats and mice. The studies were generally designed to elucidate the physiological, genetic, or behavioral mechanisms involved. Reviews may be found in Finger (1947), Busnel (1963a), and Lehmann and Busnel (1963); these also have excellent bibliographies. There is thus a good fund of basic information about seizures, but an almost total absence of data on wild rodents. That which exists leads one to think that, although wild rodents do not react as do laboratory strains, they might have some form of intense reaction to high frequency sounds. Except for the report of Marsh *et al.* (1962), there seems to be no report of practical attempts to use seizure-inducing sounds. Some companies manufacturing ultrasonic pest-chasers have suggested that their devices act through this principle, but as noted, critically controlled tests are not reported, and generally even the nature of the sounds is not stated. Thus it is impossible to judge the claims made for them.

The possibility of exploiting the seizure susceptibility of rodents for rodent control seems good enough that critical studies are in order. For instance, wild rat and mouse colonies could be established for observation and experimentation, such as described by Calhoun (1963), and broadcasts of sounds of known frequencies and intensities made while the animals were observed. Infested buildings could be selected and equipped with infrared or television cameras to monitor the actions of the rodents before and during sound emissions. At least one commercial group claims to have done something like the latter with a device they produce that emits ultrasonic noises supposedly correctly structured. But they do not state what the sounds are, nor detail the experiments; without clearly stated test data, evaluation is impossible.

Even if one were to assume that ultrasonic sounds of some sort could, through their irritating effects on rodents, be used for rat and mouse control, the planning of specific controls would still not generally be a simple matter. Obviously, the sound must reach the rodents at sufficient intensity to be irritating. Ultrasonic frequencies offer some special problems here. They are more rapidly attenuated in air than are ordinary sounds, so that higher initial intensities are needed or shorter distances are covered. They behave more like light than do lower frequencies. Thus, they do not bend around corners as readily. So sound shadows can be produced by objects in the path. These could be particularly vexing in places such as warehouses, where there are many objects standing about. This means also that these sounds do not penetrate into the burrows of rats or nests of mice. Pouring sound down a rathole would be as fruitless as pouring water. Furthermore, ultrasonic frequencies are much more readily absorbed by soft or porous objects than are sonic frequencies. All of these would be problems of application, and their solution would require knowledge and skill on the part of the applicator to determine the correct number and placement of sound sources. The correct timing of treatments and exploitation of the existing behavior patterns of the animals would also be important. In short, any device using ultrasound for irritation of rodents, even if designed from extensive laboratory experimentation, would require, at least at first, individual consideration for each individual case. Claims that a device can merely be placed inside a building and that all the rats and mice will depart must be regarded as overdrawn on the face of it, and open the claimants to suspicion of lack of appreciation of the nature of sound and of pest control. Practical rodent control by exploitation of the susceptibility to irritation of these animals seems not outside the realm of possibility. It will, however, require carefully controlled studies on rodent behavior and on the acoustical problems sure to arise when laboratory tests are converted to practical operations.

II. COMMUNICATION SIGNALS

Communication signals are chemical or physical productions, usually patterned in time or space, of one individual that, upon being received by a second individual of the same species, influence the behavior of the latter (Frings and Frings, 1964). We shall be concerned here only with visual, mechanical, and acoustical signals; chemical signals will be considered elsewhere in this book.

Experiments with communication signals for pest control have been done almost entirely with the acoustical signals of birds. Possible uses of visual or tactile signals are virtually untested. The development of the tape recorder has made possible the recording, storage, and delivery of sounds produced by animals, thus accounting for the recent interest in this field.

Among mammals, studies on communication signals have been numerous (Tembrock, 1963), but none with practical objectives foremost. Strangely, there are few studies on wild rats and mice, and no reported attempts to use their signals for practical control. This would seem to be a ripe field for future investigation. Similarly with fish, acoustical signaling systems have been studied for some species (Moulton, 1963; Tavolga, 1960, 1965; Winn, 1964), but so far sharks, the pests in this group, have not been found to produce acoustical signals (Backus, 1963), so no practical tests have been made.

Acoustical communication signals have three advantages over noises for practical pest control. (1) They are effective at intensities as low as 3 dB above ambient noise levels (Boudreau, 1964); (2) habituation is usually slow, for the reactions are generally instinctive; and (3) the signals may be specific, thus allowing one to drive away only one species, or nonspecific, allowing a wide variety of species to be chased. The first use of recorded communication signals to control birds was made in 1953 (Frings and Jumber, 1954). Since then a sizeable literature has grown up on the subject. Three international colloquia devoted mainly to this subject have been held, and the published proceedings of these (Busnel and Giban, 1960, 1965; Giban, 1962b) form continuing surveys of progress.

A. Types of Communication Signals Possibly Useful in Pest Control

There is no generally accepted classification system for acoustical communication signals of animals, even for birds (cf. Brémond, 1963). For practical pest control, only a few types have been studied. Thorpe (1961) classifies the call notes of birds as follows: (1) pleasure calls; (2) distress calls; (3) territorial-defense calls; (4) flight calls; (5) feeding calls; (6) nest calls; (7) flock calls; (8) aggressive calls; (9) general alarm calls; (10) specialized alarm calls.

Broughton (1963) develops a more elaborate system. For our purpose, only alarm calls, distress calls, and feeding calls have significance.

1. *Attractive and Identifying Signals*

Relatively few of these are pertinent to the present discussion, for only a few have been suggested for possible practical use. Herring Gulls have a call that they give on sighting food which is attractive to other members of the species. This has been called the food-finding call (Frings *et al.,* 1955b,c). When recorded and later broadcast to gulls in the wild, it attracts them from considerable distances—up to a few miles. A different type of call, but with the same effect, is the assembly call of the Common Crow (*Corvus brachyrhynchos*), emitted when the birds sight an owl or cat. This causes mobbing of the predator. When broadcast to the birds in the field, it attracts numbers of them within 5–10 minutes from distances up to 1 mile (Frings and Frings, 1957). Most of the calls used in reproduction, particularly bird songs, are attractive to birds, but no attempts to use these for control have been made. An interesting type of call, involved in flock cohesion and species identification, is reported by Hamilton (1962). He found that migrating birds flying at night emit calls causing other members of the species on the ground to fly up and join the flock. It would be interesting to know whether this could cause birds to move from places where they are not desired.

Actually none of these calls have been seriously tested for practical bird control. With gulls, for instance, one might think that the food-finding call, when broadcast at some place other than where the gulls were resting or feeding, would draw the birds away. However, some grave difficulties arise. If the gulls were feeding, it is almost certain that no call would attract them, and they do not remain for more than about 1 hour near the source of the call, unless food becomes available. It would seem unlikely, therefore, that birds could be drawn from a source of food by a food-finding call. They might temporarily be attracted away from resting places, however. With mobbing calls or assembly calls, the matter might be worthy of investigation, but once more, unless some visual stimulus is present, the birds do not remain after responding. A stuffed owl near a speaker emitting the calls enhances the effect and keeps crows occupied for a time, but not for more than 1 hour. Considering also that attractive calls must compete with real signals present in the environment, it seems that only special circumstances would permit their use for bird control.

2. *Repellent Signals*

a. Distress Signals. These have been, by all odds, the most used of any communication signals for practical bird control. They are calls emitted by birds when held in the hand or otherwise restrained. There has been some imprecision in terms here; alarm and distress calls have been confused. It is often uncertain, in reading reports of practical work, whether the writers used distress calls, alarm calls, or some other types. There has been a tendency to use the term distress calls for almost any repellent call.

This is not the only source of confusion, however. Thorpe (1961), whose classification is given above, does not mean by "distress calls" the calls described here. Instead, he refers to the calls of nestlings when disturbed, hungry, isolated, etc. Broughton (1963) attempts to resolve the problem by suggesting that distress calls, as here defined, be called desperation calls, while the calls of baby birds be designated fretting calls. However, in the same book with Broughton's suggestion (Busnel, 1963b), two other authors (Busnel, 1963b; Brémond, 1963) repeatedly use the term distress calls for those calls of adults. Giban (1965), in a book edited with Busnel, who is the editor of the book containing Broughton's suggestion, writes that distress call is the "terme qui par convention, proposée par Frings et adopté internationalement, sert à designer les séquences de cris que pousse un Oiseau tenu en mains par un Homme." We shall, therefore, use the term, distress call, to refer to the call emitted by a captive bird; alarm call will be defined below. Busnel and his colleagues earlier used a second term, cry of agony, to refer to the call of a bird caught by a hawk or other predator. This call, however, seems to be so similar to the distress call that they have more recently used only distress call.

Most species of birds, when held securely, emit a typical distress call, usually a set of piercing shrieks. The Starling distress calls used in the original tests (Frings and Jumber, 1954) were obtained by holding the birds upright by the legs, with no injury. Kuhring (1965b) states that Herring Gulls do not give a distress call easily [Frings *et al.* (1955b) report the same], and so he shocked the birds to induce them to emit the call. Pigeons, he finds, also require shocking to give distress calls. Müller (1960) states that when Starlings are held upside down causing the blood to flow to the head, their calls are more excited and effective than when the birds are held upright. Variations from bird to bird are so great that it might be difficult to put this to a critical test. The matter, however, is of no large importance in practice. In our work, we found considerable difference between the distress calls of male and female

Starlings, but made no critical studies to compare the reactions of birds to them. By alternating one with the other in broadcasting to the birds, a sharp change in sound was secured; this seemed to add to the effectiveness.

Distress calls, particularly those of French corvids (*Corvus corone, C. frugilegus,* and *C. monedula*), have been subjected to detailed acoustical analyses by Busnel and Giban (1960), Andrieu (1965), and Brémond (1965). In general, they are broad-band noises, with sharp discontinuities and abrupt shifts in intensity (transients). Among French corvids the distress calls have many features in common. Busnel and his colleagues are particularly interested in determining the effective parameters of the calls, hoping thus to be able to select the special features which cause the reactions and maybe thus to develop supernormal calls. So far, their results have not allowed them to select these features with precision. They have found that they can play the calls of corvids backwards with only slightly diminished effectiveness. But, as will be discussed below, the species with which they are working seem to have developed a broad spectrum of response to communication sounds. At any rate, analyses of these calls have not proceeded far enough to allow many generalizations to be made.

The reactions to distress calls vary with the species. Starlings are generally repelled by broadcasts of their distress call (Frings and Jumber, 1954). If flying, flocks are dispersed; if settled, the birds fly up and directly away from the sound source; in fact, one can thus "herd" them, as is done with cattle, driving them along selected routes. French corvids, on the other hand, show complex reactions, described in great detail by Busnel and Giban (1960), Giban (1962a, 1965), Gramet (1962, 1965), and Brémond (1965). Usually, groups of these birds are first attracted to the sound source and later disperse separately. This reaction is so precise that it has allowed the French workers to make critical studies of the semantic value of the calls. They have elaborated a special system for designating the types and phases of the reaction, which they call positive phonotaxis, adopting the nomenclature of students of tropistic behavior. They also recognize phonoresponses, in which the birds respond to broadcasts by calling, and phonokinesis in which there is an increase in activity without special orientation. They have analyzed in depth the patterns and variations with varying environmental, acoustical, and physiological conditions.

Gulls of a number of European species (*Larus argentatus, L. canus, L. fuscus, L. marinus, L. ridibundus*) likewise show this type of response (Busnel, 1965; Hardenberg, 1965), and Magpies (*Pica pica*) have a complex pattern of aggregation and dispersion when subjected to broadcast of their distress call (Brémond, 1962). Some birds, e.g., Oyster-catchers (*Haematopus ostralegus*) in England (Brown and Sugg, 1961), are attracted by distress calls and do not disperse. Laysan Albatrosses have a clearly defined distress call (Frings and Frings, 1959a), but do not respond to it by any movements. It may therefore be stated that, while generally distress calls alert birds of the same species and cause many to disperse, often after first aggregating, the reactions reported are still various enough that predictions are hazardous. The calls must be tested under field conditions to determine the effects.

Distress calls of mammals are known, but reactions to them are still relatively unstudied. Altmann (1952) found that the distress call of a young elk (*Cervus canadensis*) quickly attracts either the herd or a parent. Distress calls of small mammals, such as rabbits or rodents, are known to attract predators, but their effects on the same species are virtually unknown. There are many anecdotal reports of rats being driven from buildings by maltreatment of a rat on the premises which causes it to scream. Scientific verification, however, is still lacking. The most persistent of these stories recounts the repellent effects produced by burning a rat slowly, its screams of agony supposedly driving away its fellows, the effects lasting for at least a few months.

b. Alarm Signals. Alarm calls of birds are those produced when birds, which themselves are free, sight a predator or other alarming circumstance. The last phrase is necessary, if seemingly anthropomorphic, because gulls, for instance, emit their alarm call when they sight a person holding a captive or dead gull (Frings *et al.,* 1955b). This term actually may cover a family of calls (Marler, 1957), different for ground predators and for hawks. Boudreau (1965) has found different alarm calls in the same bird when confronted with humans, cats, or avian predators. Generally alarm calls are sharp, staccatto notes, variously repeated, often with differentiable sections, e.g., in the Herring Gull. Thus, birds that are merely disturbed may emit only low intensity, slow-paced calls, while the same birds, under severe threat, shriek as if in a paroxysm of excitement. For Brewer's Blackbirds (*Euphagus cyanocephalus*) and Bicolored Red-Winged Blackbirds (*Agelaius phoenicius californica*) for instance, Boudreau (1965) reports a series of alarm signals varying with the distance of an enemy from the colony or nest, "changing from a one-note 'chuck' to a high-pitched 'squee' as one approaches the nest or colony."

There have been few analyses of alarm calls of birds (Busnel and Giban, 1960; Brémond, 1965). In general, these calls are more restricted in frequencies than are distress calls, tend to be much less continuous in emission, and are usually clearly patterned in time. They vary from low to high frequencies (cf. Marler, 1955, 1957), depending upon species and type of threat. Wide variations in patterns exist, even with the same bird under different circumstances, and often, if more than one type of alarm call is used, there are intermediates of all grades. Andrew (1961), for instance, reports for Blackbirds (*Turdus merula*) a graded series of alarm notes, represented by him as "duck," "tix," and "ziep" in order of intensity. Since reactions of fellows to these different levels vary widely, possible use of alarm calls in bird control necessitates clear knowledge of the types used.

The reactions of various species to alarm calls are, in some ways, less predictable than to distress calls. Thus Starlings (Boudreau, 1962; Schmitt, 1962a,b; Keil, 1965; Pfeifer, 1965) are repelled, as with distress calls, but more effectively. Herring Gulls, on the other hand (Frings *et al.,* 1955b), behave toward their alarm call as they do toward their distress call, that is by approaching first and then dispersing, the latter generally only after the sound is turned off. The Common Crow (Frings and Frings, 1957), like the Starling, is repelled.

Many species of birds apparently have visual alarm signals, involving wing or tail flashing, or even silent and sudden flight (e.g., Common Crow). In the Quelea (*Quelea quelea*), sudden whirring of the wings in flight, thus involving both visual and acoustical cues, is a potent alarm signal (Crook, 1960); these birds also have an alarm squeal. Mammals, likewise, may have both visual and acoustical alarm signals, but so for studies are few and none were designed to test these signals for control.

c. History of Tests with Distress and Alarm Calls for Bird Control. Before proceeding with details of the use of recorded distress and alarm calls in practical bird control, we shall summarize briefly the history of development of this field. This has been done previously by Busnel and Giban (1960) and Schmitt (1962b). Pfeifer (1965) has briefly reviewed mainly studies in Germany, Hardenberg (1965) studies in Holland, and Brough (1965) and Sugg (1965) those in Britain.

The first practical tests using the distress call of Starlings to drive these birds from tree roosts were made in late summer, 1953, by Frings and Jumber (1954). This paper reports tests in two towns in central Pennsylvania. A second report (Frings *et al.,* 1955a) deals with further tests in these towns and in six more cities. By late 1954, therefore, tests of this method for urban problems seemed promising. In May, 1955 the suggestion was made that distress calls might be valuable also for bird control in orchards, but no tests were made (Frings, 1955).

Starting in 1955, the French scientists, Busnel and Giban, instituted what were to become the most extensive studies of the method, demonstrating in October of that year that corvids could be kept from food areas and driven from roosts with their distress call (Busnel and Giban, 1960;

Busnel *et al.*, 1955; Gramet, 1959, 1961, 1962; Schmitt, 1962b). In the summer of 1954, in the United States, studies on the alarm call of Herring Gulls were carried out, and clearance of dumps and fish processing plants by broadcasts of this call were demonstrated (Frings *et al.*, 1954, 1955a,b). In 1955, paralleling the French tests, studies were made with the alarm call and assembly call of the Common Crow in the United States, but no practical tests were made (Frings and Frings, 1957). This led to a series of cooperative studies between the American and French workers, involving exchanges of recordings and tests with calls of corvids and larids of the two countries (Busnel *et al.*, 1957; Frings *et al.*, 1958; Frings and Frings, 1959b; Busnel and Giban, 1960). These revealed the existence of dialects in the language of birds, that is, differences among geographically isolated populations of the same species. This explained the lack of reactions observed when recordings of gull calls made in Maine were sent to workers in Holland for testing with the same species.

In Germany, the first tests with distress calls for Starling control were made in 1955, and were unsuccessful (Reich, 1955; Bruns, 1956; Ehlgen, 1958). This result, contrasting with the results of the American and French workers, was puzzling. Later, however, Creutz (1956), Bruns (1959a–d), Gross *et al.* (1959), Schmitt (1958, 1959a,b, 1960a–c), and Schwartzkopff (1959b) showed that the earlier failures were the result of poor fidelity in the recording and playback equipment, and they obtained excellent controls in urban and agricultural problem areas. Since 1959, these men and others in Germany have developed distress call and later alarm call broadcasts into practical control for Starlings, particularly in vineyards.

In Switzerland, Müller and Gerig (1957) successfully chased Starlings and Blackbirds from areas where they were pests by broadcasts of their distress calls, and later Buchmann and Müller (1957) and Müller (1960) applied the technique in olive orchards in Tunisia. In Holland, the year 1958 was one of disappointment, but 1959 saw the development, by Hardenberg (1960) of control of gulls on airports by broadcasts of their distress call. In Britain, likewise (Brown, 1962; Brown and Sugg, 1961), the first attempts at control of gulls at airports were unsuccessful, but later studies (Brough, 1965; Cooke-Smith, 1965) led to some degree of success, by no means, however, that achieved in Holland and France.

In late November, 1958, while the results of broadcasts of starling, gull, and crow distress calls seemed promising but unpredictable, a colloquium was held in France at the Laboratoire de Physiologie Acoustique. This event spurred much further research in Europe and led to the exchange of ideas that made 1959 a year of progress with the method. When much the same group of workers met again, in October, 1961, at

Versailles, France studies on acoustical bird control, chiefly by distress calls, were the main item of discussion, and progress in solving problems of application and evaluating types of calls was striking. In 1963, a third colloquium, at Nice, France considered the problems of birds at airports, again with major emphasis on acoustical controls.

In the United States, meanwhile, little progress was made, chiefly because no group of biologists or governmental agency became committed to intensive research. There are many reasons for this, not pertinent here, covered in an earlier report (Frings, 1964). In 1960, G. Boudreau, an engineer in Phoenix, Arizona, began studies on agriculturally pestiferous birds, chiefly Prairie Horned Larks (*Eremophila alpestris*), Starlings, House Sparrows (*Passer domesticus*), and House Finches (*Carpodacus mexicanus*), and the development of economically feasible methods for their control, using distress and alarm calls, as well as other biosounds (Frings, 1962, 1963). The United States Fish and Wildlife Service in 1961 instituted a program of study on acoustical control of birds (Seubert, 1965). The rising interest in these studies in the United States is clearly reflected when one compares the *Proceedings of the First Vertebrate Pest Control Conference* in 1962 with those of the second conference in 1964 (Cummings, 1964). In the former, almost no mention was made of acoustical controls. In the latter, almost every report on bird control covered some tests with distress calls.

A summary of studies on distress and alarm calls for control of depredations by birds is presented in Tables I and II. These show clearly the preponderance of work on the Starling, with secondary emphasis on larids and corvids. Certainly these are the most widespread of birds whose activities disturb man, but many important species still remain to be studied, e.g., Quelea, Bobolink, American Robin, and many species of blackbirds, rice birds, ducks, and geese. Furthermore, the tables show clearly the tendency toward limitation of these studies to the use of distress calls, even though alarm calls, where tested (e.g., Frings *et al.*, 1955b; Schmitt, 1962a,b; Pfeifer, 1965; Boudreau, 1963, 1964), have proved to be more effective. The work has just begun, and the problems are just being visualized; really practical usefulness in the hands of inexperienced persons has yet to be developed, if it is possible at all.

B. Equipment for Recording and Broadcasting Acoustical Communication Signals

1. *Recording Equipment*

The development of tape recording since World War II has been a major factor in making possible acoustical bird control, as well, of course, as studies on bird song. It would take us too far afield to attempt a review

TABLE I

Studies in Which Distress Calls Were Tested for Possible Practical Use in Bird Control

Situation	Reference
	Laysan Albatross (*Diomedea immutabilis*) and Black-Footed Albatross (*D. nigripes*)
Experimental	Frings and Frings (1959a)
	Heron (*Ardea cinerea*)
Experimental	Hardenberg (1960)
	Oystercatcher (*Haematopus ostralegus*)
Airfields	Brown (1962); Brown and Sugg (1961); Brown *et al.* (1962); Hardenberg (1965)
	Lapwing (*Vanellus vanellus*)
Airfields	Brown (1962); Brown and Sugg (1961); Brown *et al.* (1962); Cooke-Smith (1965); Hardenberg (1965); Keil (1965)
	Gulls (*Laridae*)[a]
Airfields	Brough (1965) acr; Brown (1962) r; Brown and Sugg (1961) acr; Brown *et al.* (1962) acr; Cooke-Smith (1965) acr; Hardenberg (1962, 1965) acr; Keil (1965) ar; Kuhring (1965a,b) ad; Sugg (1965) acr
Experimental	Brown and Sugg (1961) r; Busnel (1965) amr; Giban (1965) afmr; Gramet and Hanoteau (1965) afmr; Seubert (1965) a
	Pigeon (*Columba livia*)
Roosting	Kuhring (1965b)
	"Crows" (*Corvidae*)[b]
Roosting or feeding	Busnel and Giban (1960) cfm; Busnel *et al.* (1955) cfm; Gramet (1959, 1961, 1962, 1965) cfm; Gramet and Hanoteau (1965) cfm; Hardenberg (1960) f
Airfields	Brown and Sugg (1961) fm; Brown *et al.* (1962) fm; Cooke-Smith (1965) m; Keil (1965) cf; Sugg (1965) fm
Experimental	Brémond (1962) gp; Busnel (1965) cfm; Busnel and Giban (1960) cfmgp; Busnel *et al.* (1955, 1957) bcfm; Frings *et al.* (1958) bcfm; Giban (1965) cfm; Gramet (1959, 1961, 1962, 1965) cfm; Hardenberg (1960) f
	Blackbird (*Turdus merula*)
Roosting and feeding	Müller and Gerig (1957)
	Starling (*Sturnus vulgaris*)
Roosting (urban)	Brough (1965); Brown and Sugg (1961); Busnel and Giban (1960); Frings (1954, 1962); Frings and Jumber (1954); Frings *et al.* (1955a); Hardenberg (1960); Müller (1960); Müller and Gerig (1957); Schwartzkopff (1959b)

TABLE I (*Continued*)

Situation	Reference
Feeding (agric.)	Boudreau (1960); Brough (1965); Bruns (1956, 1959a,b,c,d, 1960, 1961, 1962); Buchmann and Müller (1957); Creutz (1956); Ehlgen (1958); Müller (1960); Müller and Gerig (1957); Reich (1955); Schmitt (1959a,b, 1960*a*–c, 1962a)
Airfields	Brown and Sugg (1961); Brown *et al.* (1962); Cooke-Smith (1965); Keil (1965); Sugg (1965)
	Red-Winged Blackbird (*Agelaius phoenicius*)
Feeding (agric.)	Seubert (1965)
	Western House Finch (*Carpodacus mexicanus*)
Feeding	Boudreau (1962, 1963)

[a] Species studied (key letters used in table): *Larus argentatus* (a); *L. canus* (c); *L. delawarensis* (d); *L. fuscus* (f); *L. marinus* (m); *L. ridibundus* (r).

[b] Species studied (key letters used in table): *Corvus brachyrhynchos* (b); *C. corone* (c); *C. frugilegus* (f); *C. monedula* (m); *Pica pica* (p); *Garrulus glandarius* (g).

of types of tape recorders, with evaluation of the various types for particular purposes. Generally, the more portable the recorder the lower the fidelity, although mere bulk and weight do not in themselves confer fidelity. The slower the tape speed, in general, the poorer the frequency and transient response, but electronic circuitry is important here, and tape recorders with tape speeds of, let us say, 3¾ in. per second may respond to higher frequencies than do others running at 7½ in. per second. This however is not usually the case. Similarly with microphones, there is a tremendous variety available, each with special features. Some are particularly sensitive, some directional, others nondirectional, and so on. Except to say that the microphone should be matched to the recorder, with fidelity equal to or better than that of the recorder, little can be stated in general.

Andrieu (1965) lists what he considers to be necessary specifications for recording equipment, based on studies showing reduced effectiveness with lower fidelity. He believes that the recording chain—i.e., microphone, preamplifier, and recorder—should be capable of recording frequencies of at least 50–40,000 Hz, have high signal-to-noise ratio, and be very sensitive to transients. He suggests that a tape speed of 76 cm per second (30 in. per second) be used. The specification for frequency response up to 40,000 Hz may be based on Thorpe and Griffin's (1962) report of ultrasonic frequencies in bird songs. Otherwise, it would be unrealistic, for, as we have already noted, the hearing range of birds does not surpass that of man. Since tape speed such as this is not standard

TABLE II

STUDIES IN WHICH ALARM CALLS OR OTHER SPECIFIED CALLS WERE TESTED FOR PRACTICAL USE IN BIRD CONTROL

Situation	References
	Gulls (*Laridae*)
Experimental	Frings (1954) a; Frings *et al.* (1955b,c, 1958) am (also food-finding call)
	Prairie Horned Lark (*Eremophila alpestris*)
Feeding (agric.)	Boudreau (1964)
	"Crows" (*Corvidae*)
Experimental	Busnel and Giban (1960) bcfm; Busnel *et al.* (1957) bcfm; Frings and Frings (1957, 1959b) b; Frings *et al.* (1958) bcfm (also assembly call)
	American Robin (*Turdus migratorius*)
Feeding (agric.)	Boudreau (1963)
	Water Pipit (*Anthus spinoletta*)
Feeding (agric.)	Boudreau (1964)
	Starling (*Sturnus vulgaris*)
Roosting (urban)	Pfeifer (1965)
Feeding (agric.)	Bruns (1959d, 1960); Schmitt (1962a)
Airfields	Keil (1965)
	Western House Finch (*Carpodacus mexicanus*)
Feeding (agric.)	Boudreau (1962, 1963, 1965)
	House Sparrow (*Passer domesticus*)
Feeding (agric.)	Boudreau (1965)

commercially, and the mechanical and electronic equipment needed to produce this degree of response are costly, such recording equipment would be expensive. One must always balance expense against presumed effectiveness in practical control.

Actually, this seems to be too high a level of fidelity. Busnel and Giban (1960) and Busnel (1965) describe the equipment used by their laboratory in their extensive and fruitful tests, and it is far removed from this. Likewise, Boudreau (1960, 1962) and we (Frings *et al.*, 1955a,b, 1958) have obtained excellent responses from birds using equipment of much lower fidelity. As Busnel (1963c, p. 86) states: "It is to be noted that in all these field experiments, the quality of the broadcast was not very good, since considerable amplification was used. This did not result in a very faithful rendering of the signal. But, in spite of distortion and the increase in background noises, the reactions were very good."

Frings *et al.* (1958) state similarly: "With field equipment, fidelity of reproduction is at times much reduced and to the human ear at least the sounds broadcast to the birds are distorted or partially masked with background noise or hum. Yet they seem to be 'understood' quite readily by the birds. It may be that the physical parameters determining the effectiveness for birds are not the same as those determining fidelity for the human ear. The similarity of the ears of birds and mammals however, makes this seem not too likely. More likely, these communication signals are capable of being distinguished by the birds over high levels of background interference."

On the other hand, research workers in Germany (cf. Bruns, 1959b,c, 1960; Schmitt, 1958, 1962b) found at first that their attempts to chase Starlings with recorded sounds failed entirely, and that the results remained that way until they secured recording and broadcasting equipment of much higher fidelity. In our early work (Frings *et al.,* 1955a), the distress call of the Starling, as it finally left the speakers, was sometimes projected with such poor fidelity that we thought that it might be acting just as noise. Only by reduction in intensity, and later by comparison with attractive calls, were we convinced that the call itself was specific. Even in more recent experiments, the German workers continue to need high fidelity recording equipment to secure results, and Bruns (1961) states that, if birds fail to respond, invariably the fidelity of the broadcast is poor. Andrieu (1965) and Brémond (1965) also found, in controlled tests, that low fidelity equipment resulted in poor response by the birds. Pfeifer (1965) reports also that repeated copying of tape recordings reduces their effectiveness. Perhaps European birds other than Starlings have similar signals, and so the Starlings in Germany have learned to respond only to their own specific calls, while in the United States no such differentiation is needed and consequently the birds can respond to general features of the signals; this is believed to occur in corvids (Frings *et al.,* 1958).

At any rate, it is obvious that sufficient fidelity is needed to record the signals clearly. Since fidelity and tape speed are powerful price determinants, one must often select equipment of lower standards than those set by Andrieu. Briefly, it would seem advisable to use the best possible recording equipment consistent with cost, portability, etc. Fidelity may be lost before the sound is ultimately broadcast, but it can never be gained if not there in the first place.

For recording distress calls of birds, which can be done in the laboratory, Andrieu gives as a rule that one should have the bird at least 1 meter from the microphone in a nonreverberant room, preferably an anechoic chamber, and that the sound pressure at the microphone should never exceed 100 dB, otherwise the microphone overloads and distorts the signals. Furthermore, the preamplifier and recording unit should not be operated at peak capacity. He also states that one should immediately check the recordings to correct for deficiences. All these rules are certainly valid.

For recording alarm calls, which must be done in the field, other suggestions can be made. Obviously one cannot have anechoic conditions. More of a problem arises from stray noises. One method used to overcome this is to use a parabolic reflector behind the microphone. If this is done, one should keep in mind that the reflector itself alters the signals variably. For instance, the reflectors ordinarily used act as frequency cutoff devices, reducing frequencies below 1000 Hz (Little, 1964). This may or may not be important, but should be considered.

We generally prefer to place the microphone near enough to the point at which the call is likely to be given that the gain on the recorder can be set low, and thus background noise is not recorded at high levels. This requires more time, for one must observe the birds first to determine their patterns of movement so that the microphone can be placed correctly. Then some patience may be necessary and some tape used up while the birds do other than give the desired call. However, no distortion is introduced other than that in the basic system.

An important cause of distortion in recording is failure to realize that the devices used to indicate overloading on commercial tape recorders—electric eye, VU meters—respond much more slowly to changes in sound pressure than these changes occur in bird calls. Thus a sudden peak intensity in a call may last for so short a time that the pointer of the VU meter is merely started toward the true register, but fails to reach it, because the sound pressure has dropped off long before the pointer has reached peak. So recordings of bird calls made with peak settings of 0 on a VU meter are almost invariably overloaded at the peaks. If a VU meter is used with bird calls, the pointer should never be allowed to exceed about —5. The signal can be amplified later to any degree necessary, for the intensity relationship of signal to background noise is not changed by recording at lower levels. It is astounding to hear some of the poor quality recordings of bird calls distorted by overloading that are made with high fidelity equipment; almost invariably this VU meter error is the cause.

In summary, then, for recording animal communication signals, one should get the highest fidelity microphone and recorder he can obtain consistent with his economic resources. For different animals the necessary fidelity is different. Extreme cases of special need for fidelity would be for recording bat echolocation sounds or insect songs. The recording should be carried out carefully, with due regard for the fact that any extra piece of apparatus introduced into a recording system contributes some potential change in fidelity. Above all else, particularly with calls of birds, one should avoid recording at too high average levels, so that sharp peaks do not overload the equipment.

2. *Projecting Equipment*

In general, much the same can be said for the chain of equipment used for broadcasting sounds to birds in the field. The higher the fidelity and power the better, but high fidelity and power are expensive, particularly in work such as this, where often one is trying to send recognizable signals over long distances. Generally, the projection chain consists of a tape player, amplifier, and loudspeaker or speakers. Each of these affords a place for choice of fidelity of reproduction.

In field operations, the power supply needed to drive the tape recorder and amplifier may be a source of hum and distortion. This was particularly true in the earlier years of this work, when the usual power supplies were battery-driven DC–AC converters. These often introduced 60-cycle hum or other interfering sounds. With modern use of transistorized equipment, this problem has been almost overcome.

As with recorders and microphones, amplifiers and loudspeakers come in a bewildering array of types. For biologists, it is often most convenient to enlist the aid of acoustics specialists or sound engineers. Since these specialists often work with sounds that are quite differently structured than are animal communication signals, it is well to call their attention to the special nature of the material to be projected. As with the recording side of the chain, the weakest link determines the quality.

Again luckily, with most birds considerable distortion is tolerated—but not too much. As before, the German workers have found problems with fidelity that others have not; the fidelity necessary may depend upon the nature of local populations of the birds. There is no doubt, in some Starling tests, particularly those made by amateurs, that the equipment could not project sounds that would be recognizable, and no responses resulted. Given the proper signals, correctly recorded, as Brémond (1965) notes, the reactions seem to be almost independent of external conditions, such as other sounds, weather, etc.

C. Information Needed for Planning Control Programs

In one of the earliest papers published on the use of the distress call for Starling control (Frings *et al.,* 1955a), the statement was made: "It is hoped that this information on our tests will put to rest belief that this method is simple and fool-proof. Certainly the use of broadcast recorded calls is safe and relatively cheap. But the clearance of an objectionable roost requires attention to equipment, knowledge of the habits of Starlings, coordination during the treatments, and persistence in the work. No one expects to cure a disease with one easy treatment, nor to eliminate pest insects or rodents without continued effort. No one should expect to clear Starlings from objectionable roosts without intelligent and persistent work."

This is still as true as it was then. Even if one were to have a call so repellent to pests that not one that heard it remained in place, mere broadcasts of the call would not constitute control. If this seems so elementary as to be trite, one need only review some of the "tests" in the United States, with their "successes" in a few cases and much publicized "failures" in more, to realize the need for stating it clearly. In this section, we shall consider the biological and environmental knowledge that one must have to plan effective controls, and in the next section, the methods of application that have been used.

Gramet (1961), for agricultural problems, notes that planning a control program requires knowledge of: (1) the birds (foods, reactions to sounds, range of signals, etc.); (2) the place (vegetation, topography, alternate foods, etc.); (3) the crops (time of harvest, duration of harvest, value of crops, etc.); and (4) the climate and weather. Boudreau (1965) lists the factors influencing control, other than audio factors and the signals used, as follows: species present; feeding habits and foods; habitat available; weather, particularly wind; movements of birds, particularly patterns of leaving and arriving; persistence of birds; background noises; and human disturbance, such as farm activities. We shall reorganize these under six headings: (1) identification of species involved; (2) nature of problem; (3) population size and structure; (4) environmental conditions; (5) behavior patterns of animals involved; and (6) possible factors interfering with control operations. These can be discussed only in the most general terms, for kaleidoscopic variations in all are found in practice.

1. *Identification of Pest and Other Species*

This is so obvious that it needs little comment. It is surprising however, to find in how many cases untrained persons misidentify the species, or feel that this is unimportant. If one is using noises, or other indiscriminate scaring devices, identification may not be critical. But, when using communication signals—which are often specific—identification is of first importance. For biologists doing experimental work this should be no problem, but when control methods are released to the public, even pest control operators, identification may be a serious problem (Cooke-Smith, 1965). As Jackson (1964) has pointed out for pest control operators, there are many species of birds that are involved, and identification may be difficult; yet it is essential. For instance, in mixed roosts, with Starlings, Cowbirds, and American Robins, it was found (Frings *et al.,* 1955a) that the latter two species, not affected by broadcasts of the Starling distress call, remained in the roost and decoyed the Starlings back. Where the roosts were composed almost entirely of Starlings, this did not occur. In practical terms, also, one should note that citizens of towns with Starling problems may not recognize the difference between these birds and others; finding some birds still in trees, they believe that the Starlings have not been removed.

2. *Nature of the Problem*

Except for cases in which objection is raised to the mere presence of an animal, such as squirrels in attics, there are three problem areas with vertebrates: (1) roosting in large numbers on buildings or trees; (2) resting or loafing where man has critical traffic; and (3) feeding on foods destined for human use. Each of these poses special problems in control.

A number of species of birds—chiefly Starlings, Pigeons, English Sparrows, Mynahs, various species of corvids, Quelea, etc.—spend the night, during all or part of the year, in communal roosts. Where these roosts are in trees in urban areas, on facades of buildings, or inside open structures, the birds may be considered pests. Where this behavior pattern is seasonal, for instance, with Starlings, the seasonal occurrence will dictate time of control. The fixation of birds to roosts usually varies with time of year, availability of other roosting areas, length of time the roost has been used, etc. All these will affect planning for control. With sounds, the nature of the roost may create special acoustic problems. For instance, fully leaved trees are good sound absorbers; facades of buildings reflect sound and may have decorations that produce sound shadows; sheds or other buildings form reverberant chambers, producing distortion of the signals.

Some birds—chiefly, various species of gulls, Lapwings (*Vanellus vanellus*), Pigeons, Oystercatchers, plovers, corvids, etc.—at certain times of the year aggregate in specific areas, apparently for resting, since no food is available and they do not sleep there. The most important problems of this sort occur at airports, where birds settle near or along the runways. The attractive features in these cases are not always clear, although usually they involve preferred habitat, and apparently in winter, warmth of the runways (Kuhring, 1965a). A major problem posed by this type of situation arises from the large distances involved and the levels of background noise. Generally, the fixation of individual birds may be small, but the area remains attractive to others if those present are driven away.

A tremendous number of species of birds become agriculturally undesirable, because they feed on growing crops or stored foods; even a partial list would be extensive. Important factors affecting control include the season involved, the type of food and possible alternative foods available, the nature of the damage, potential economic value of the crops, etc. Acoustic problems vary with the situation; usually, the only critical problems are possible sound shadows in orchards, vineyards, etc., and the large areas to be treated. The tenacity of the birds is variable, depending chiefly upon alternative foods and length of time during which they have used the area.

In all these situations, as Neff and Meanley (1957) and Boudreau (1960, 1963) have pointed out, birds are much easier to dislodge early in the period of infestation than they are if allowed to become fixed in the place. Thus, a determination whether the pests are regular or incidental is of great importance, particularly in evaluating results.

3. *Population Size and Structure*

The size of the population may determine the area that must be treated. Small flocks of roosting birds may occupy one tree; large flocks may occupy many trees along many adjoining streets. In agricultural situations, however, even small flocks may move about over large areas.

Besides determining the sheer size of the problem, the population may also influence the reactions of the birds. Thus, Frings *et al.* (1955a), Boudreau (1963, 1964), Brémond (1962, 1965), and Busnel and Giban (1960) found that large flocks are more easily driven from roosts or foods than are small flocks or individual birds. There seems to be evidence of the type of chain fright reaction described by Calhoun (1963) for rats. Flight of one individual triggers off mass flight in all, and it is the rare individual that remains. However, Bruns (1959b,d, 1960) and Buchmann and Müller (1957) report that in Germany and Tunisia Starlings in small groups are more easily repelled than in large groups. Regardless of the reasons for this discrepancy, obviously the size of the groups may be important in determining reactions.

The composition of the population is also of great importance. Mixed roosts pose special problems, for if one uses only the calls of one species, other birds may remain and act as decoys, causing the pest birds to return. In tests in Rochester and Buffalo, New York (Frings, 1954; Frings *et al.,* 1955a), evidence was found to suggest that where Starlings formed the majority of the roosting birds, the other species fly out with them when the Starling distress call is broadcast, but where Starlings are in the minority, only the Starlings fly out, and later return. Furthermore, it seemed that if young Starlings were present in large numbers, their failure to respond—possibly they had not shifted from the freezing response of nestlings—left them in the roosts as decoys (cf. also Schmitt, 1960b). Boudreau has also noted this in birds feeding in vineyards and thus has had to use a series of distress and alarm calls. Much more study is needed on this phase of control, particularly at airports, where often a number of species are involved.

4. *Environmental Conditions*

The most important of these are topography and weather. In agricultural application, the placement of speakers may be dictated by terrain features that block sounds coming from only one source. In urban situations, the nature of the buildings and the flight paths of birds among the buildings determine placement of speakers or deployment of sound trucks.

Climate and weather, beside determining movements of the birds, may in themselves be important. Drury and Keith (1962), for instance, point out the importance of weather conditions in changing the course of migration of birds. Such changes could bring birds not usually found in a region to that region. Busnel and Giban (1960) have discussed the importance of weather conditions in determining sound distribution, and Boudreau (1965) and Brémond (1965) emphasize the particular importance of wind in determining the expected range of sounds. Micrometeorological conditions near the ground (Geiger, 1965) may create peculiar sound propagative circumstances. Thus, Schilling *et al.* (1946, 1947) have shown that there may be, from day to day at the same spot, variations in range of sounds up to two times. Negative temperature gradients in the lower atmosphere can, through diffraction, cause sound beams to be deflected upward, leaving sound shadows. Acoustic "mirages" may thus be produced as a result of temperature variations in the lower atmosphere. In some agricultural situations, wind may limit the carrying power; this has been shown for foghorns over open water (Wiener, 1961), in which sound transmission upwind may be severely limited. Conversely, with reduced or no wind, temperature irregularities are common, and these may result in irregular patterns of sound distribution. So far, there have been no studies on the direct effects of these factors on the use of sounds for pest control, but they certainly will be of importance.

5. *Behavior Patterns of Animals*

Obviously, the repertory of signals produced by an animal limits the selection of signals that can be used for possible control. And, related to that, the behavior patterns of the animal as it receives the communication signals are important. Thus, attraction or repulsion would be used differently, and mixed patterns such as in French corvids might have special value, or create special problems.

Besides these direct behavioral items, migrations and movements are factors for consideration. There are few birds that are nuisances all year round, particularly in temperate climates. Mostly, they move from place to place in annual, seasonal, and daily cycles, and it is during some phases of these that they become pests. Frings (1962) has diagrammed the annual movements of the Starling in the eastern United States. The birds are generally objectionable only in summer, when they form large communal roosts in trees along city streets and attack crop plants, and in winter, when they roost on buildings. Thus, attempted clearance of tree roosts in the autumn could lead one to conclude that the operation had been a success, when it only preceded the normal departure time.

Knowledge of the expected dates of depredations enables one to be prepared, so that treatment can be made early, before the birds establish themselves. The dispersal of roosts before fully established may actually bring about clearance lasting for some time, without further application of sound. In State College, Pennsylvania, for instance, the Starlings had for years moved from a few scattered summer roosts into a concentrated single autumn roost. When treatments of the summer roosts in 1953 dispersed them and drove the birds out of town, the autumn roost did not form. This roost did not form subsequently, even though no further sound treatments were made (Frings *et al.,* 1955a). Bruns (1962) points out that clearance of Starlings in southern Germany is easier than in northern Germany, because the time of treatment in southern Germany coincides with autumn migration, and so the birds, when chased, move away completely rather than remaining nearby.

Knowledge of diurnal rhythms of movement is also important in planning control operations. Thus, the biological studies of Jumber (1956), who traced the flight paths of Starlings into and out of roosts and determined that light intensity, and secondarily weather, determine the time of arrival of birds at roosts, were used by Frings and Jumber (1954) as the basis for timing applications of broadcasted distress calls to drive the birds from their roosts. Jumber found that Starlings arrive at roosts in the evening by a series of steps: to tall trees within about a quarter of a mile of the ultimate roost, then variously to the final definitive roosts. These were called by him the pre-roost and the definitive roost; Hasse (1963), who found a similar situation for crows, named them secondary roost and primary roost. In control, early stages of treatment were directed only at the definitive roost, for the Starlings moved easily from the pre-roosts to other alternative pre-roosts and thus necessitated large movements of sound trucks. Only when the birds had been disturbed for two or three evenings at the main roost were the pre-roosts treated as the birds arrived. The sound was then directed so that the birds were driven away from the definitive roost and, if possible, away from the center of town. This exploitation of already existing flight paths as dispersion routes for birds has also been behind the success of Boudreau (1960) in vineyards in Arizona. It is much easier to move a group of birds along accustomed flight paths than to move them at random, and the flight paths are followed much farther away from the starting point.

One of the most important aspects of behavior in planning control is the relationship of the bird to the roosting, resting, or feeding situation. Neff and Meanley (1957) stress the importance, for control of blackbirds, of early treatment regardless of method selected. Boudreau (1960, 1963), likewise, has found the difficulty of dislodgement directly related to the time that the birds have been allowed to feed in a given spot. Long-established roosts are more difficult to break than are newly formed roosts. Roosting and feeding flocks are more easily dispersed early in their seasonal occupation than later. Breeding birds, as one might expect, are extremely difficult to dislodge. Not all are so site-tenacious as the albatrosses of Midway Island (Frings and Frings, 1959a). which remain on the nest in spite of almost any stimulation, even when run over by trucks. Brémond (1965) and Gramet and Hanoteau (1965) note that gulls and corvids on breeding grounds move grudgingly, if at all, when subjected to broadcasts of their distress calls.

6. *Possible Interfering Factors*

In many situations a problem in broadcasting the calls of birds is the presence of background noises. As already noted, if the fidelity and power of the broadcasts are sufficient, this may be minor. But, if long distances are involved, or if the terrain to be covered is such that it creates acoustical problems, the presence of high levels of background noise can be troublesome. Using the alarm call of the Herring Gull in an attempt to chase the birds from the roof of a sardine canning factory, we (Frings *et al.,* 1955b) found that the call was effective up to about 200 ft until the factory started operation. Then the noise masked the rather low-intensity sound output, and the birds moved onto the far end of the roof. In these cases, of course, it would be sufficient to raise the signal level, so that the signal-to-noise ratio is proper, but this may require more expensive equipment, produce distortion, or make the sound near the source of inappropriate intensity. Gramet (1965), among others, emphasizes the need to have the sound intensity, *at the birds,* correct. This may mean subtle adjustment to get range yet not lose near-source effectiveness.

Boudreau (1963) in tests with English Sparrows near a major highway found it necessary to move speakers about as the noise level varied, or to use more than one set of speakers to cover the field properly. Calhoun (1963), in studies on rat behavior utilized the principle of masking with noise to enable him to take moving pictures without having the rats disturbed by the clicking of the camera. He played a continuous recording of noise at sufficient intensity to mask the clicks. He reports, however, that this apparently did not mask the communication signals of the rats. Unfortunately, however, high levels of background noise may mask signals in practical tests.

Certain physical factors of treated structures or areas may interfere with sound propagation and thus with correct coverage. For instance, buildings used as roosts by Starlings and Pigeons may create severe problems, because of their shape, necessitating the use of many speakers or sound trucks. If a building is part of a complex of buildings, reverberations and echoes may be set up, and these may mask the communication signals. Thus, in clearing Starlings from City Hall in Philadelphia (Frings, 1962), we found that a series of speakers was necessary to reach all parts, and these were supplemented with speakers on sound trucks that could be dispatched to trouble areas. Inside buildings such as hangars or sheds, complex sound fields may be set up if one tries to use one speaker at high intensity, and the signal is distorted and does not reach the birds with fidelity. This necessitates the installation of a series of speakers near the roosting birds, so that each may be driven at relatively low output to deliver the sound at correct intensity and fidelity to the birds. For instance, at LaGuardia Airport in New York, we attempted clearance of a hangar infested with roosting Starlings. At first, one speaker was used on the floor of the tall shed, but no response was obtained. It was obvious, even to the human ear, that reverberation in this gigantic echo chamber was resulting in an almost continuous mass of disorganized sound when the speaker was driven at high intensity. Accordingly, the speaker was installed at the ceiling level near the roost and driven at medium intensity; the birds were then driven away.

The most disturbing factors to the research scientist in acoustical bird control are those that arise from the activities and attitudes of people involved in or living near the experiment. Boudreau (1965) reports the disturbing influence of farm workers, whose necessary presence in the fields makes evaluation of results difficult or impossible. We have many times had overeager "helpers" while we were testing a call or a method of application who clashed together potlids or smote the trees with poles. Conversely, in one town where we were attempting clearance of tree roosts on residential streets, the police, with motives of their own and no advance warning, refused to let sound trucks drive other than one way on some streets. If these were merely harmless signs of irritation, they might be negligible, but when they make scientific tests uncritical or even impossible, they must be considered as a major interference.

When equipment is given to ordinary persons in an effort to determine its ultimate practical utility, one may encounter the situation noted by Neff and Meanley (1957). They point out that if a farmer wants to succeed in chasing blackbirds, he must really work at it and wish success. This, they found, many were unwilling to do. The same is true here, and half-hearted efforts are often made, with poor results, yet reported as if they were really true tests. Because of this, as well as the fact that some technical training is necessary to apply the sounds correctly, the French and German workers have not seen fit to release acoustical methods for general public use (Bruns, 1962; Schmitt, 1959b).

Another problem is that involving publicity. This need not be serious in agricultural and similar trials, but one cannot avoid at least some publicity in urban operations. This should seem to matter little, but it is not so simple as that. We have previously detailed some of the problems (Frings 1962, 1964). Even under good circumstances, it is difficult to have newspaper reporters resist the temptation to make the operation—and the operators—look funny. Under the worst circumstances, as in our studies in Philadelphia in 1960–1961, the editorial staff of a newspaper, wishing to embarrass the city administration, used the Starling tests as a weapon in their war, totally ignoring truth and ultimately triumphing over reality to bring about cessation of the tests. Since these tests were begun with the understanding that they were only an experiment on removal of Starlings from the City Hall (acoustically one of the most challenging buildings to be found), yet succeeded completely in the first phases, this editorial irresponsibility is disheartening. Other similar cases could be cited, all far removed from science, but impeding the enlargement of our knowledge of bird control, not just because the experimenter becomes irritated, but more directly because tests are cancelled or changes made that have no scientific validity. In the United States, we would list this as the major cause for lack of intensive research in accoustical bird control.

D. Methods of Application

As with insecticides or drugs, recorded communication signals of animals are only tools to be used in achieving practical control or management. Depending upon all facets of the situation, these tools must be used in a coordinated program to achieve the objectives. As with drugs, when used clinically, the patient is not usually satisfied with failure just to discover some new information. An agriculturist generally wants his crops protected, not a demonstration of how not to do the job. A town council wants the birds cleared from the trees or buildings, not data on sound intensities too low to do this. Unfortunately, at least in the United States, there are no experimental bird infestations maintained just for testing purposes. Thus, while scientifically one should keep all factors constant, practically he seldom can, and a constant feature of tests of acoustical bird control is changing the conditions as data accumulate. Consequently, reports of tests tend to be long and drearily detailed, hard to summarize briefly, often raising more questions than they answer. We shall try here to generalize as much as seems justified, then to cite a few specific tests in detail to illustrate the problems and the methods used.

1. *Selection and Deployment of Equipment*

We have already noted the importance of using the correct equipment for this work. Assuming that the original recordings are of highest possible fidelity, one wishes to apply them with comparable fidelity. However, equipment suitable for field use under various conditions usually does not have the fidelity characteristics of studio equipment. Some sort of compromise is usually necessary. A tape recorder can, of course, be used as a tape player, and this has been done in many experiments. However, a high fidelity recorder may be too valuable or too delicate to be exposed to winds, rains, or the bouncing of sound trucks. Furthermore, most tape recorders are not easy to operate while one may be driving a truck. Consequently, most workers now use a tape player, usually a cartridge type with a continuous tape loop that repeats the message at fixed intervals. There are a number of commercial cartridge tape players, the characteristics of which are given by the manufacturers. As with recorders, one must match his requirements for ruggedness and fidelity with the amount of money that can be spent on the project.

Amplifiers are likewise various. Under agricultural circumstances, it may be necessary to have weatherproofing not ordinarily needed for amplifiers, and, if amplifiers are to be placed in fields where sunlight is intense, some form of forced-draft ventilation may be necessary. Both player and amplifier require power, and 110-volt, 60-cycle lines can be used if available, or DC–AC converters with batteries, or, by use of transistors, batteries directly.

The choice of speakers presents similar problems (cf. Sugg, 1965). It is probable that the speakers used for sound emission may be the weakest link in most acoustical bird control experiments. Generally one tries to reduce the number, to save wiring and minimize his power needs. This often means driving speakers beyond their best response limits. Generally to obtain low-frequency sound output, large, heavy horns must be used, creating problems of installation and raising expense. Luckily, as already noted, the fidelity needed is usually moderate, so that even mediocre speakers suffice. Figures 1–4 show the setups used by the French experimenters and by Boudreau, illustrating typical playback chains.

Deployment of the speakers is not a biological, but rather an acoustical problem. The objective ideally is to have the signals reach all the birds at the correct intensity. As already noted, with many species, this may be very low (down to 3 dB above ambient noise levels), but with some species, e.g., Starlings, it is rather high. "Birds must be exposed to signals just as insects must be exposed to insecticides. One does not treat a field by spraying insecticide from only one end; coverage must be complete. This applies to sound coverage also. The most practical and efficient method for providing this coverage, considering all conditions, is the method to use" (Boudreau, 1965).

A sound-level meter is very valuable for determining sound coverage. One can place the speakers, if he is using a fixed installation and can then determine the sound pressures throughout the area, detecting possible areas of low intensity. These may be caused by features of terrain, wind conditions, or other factors. Often slight shifts in positions of fixed speakers will correct the situation. In other cases, it may be necessary to add a speaker or to change the height of the speakers from the ground. If one does not have, or under the conditions cannot practically use, a sound-level meter—for instance, in treating an extended tree roost—the reactions of the birds themselves must give the clues.

For urban applications, buildings or trees, sound trucks are usually most practical (Frings *et al.,* 1955a). These allow free movement during the critical minutes when the birds are arriving, and they economize on speakers and amplifiers. They also allow one to change the direction of application, which sometimes causes a reaction in a group that otherwise does not move. For airport applications, Brown *et al.* (1962) in England and Kuhring (1965a,b) in Canada found mobile units superior

FIG. 1. Fixed installation for broadcasting recorded communication signals of birds in a vineyard in California. The playback unit ("Amplitape") is a combined amplifier and tape player using a cartridge loop of prerecorded tape, with built-in timer. This is battery powered and activates the speaker, here mounted well above the ground to deliver sound across the grape vines (photo courtesy G. Boudreau).

to fixed speakers, although Hardenberg (1962, 1965) in Holland has used fixed speakers. Buchmann and Müller (1957) argue that, for large-scale agricultural applications, a sound truck is better than multi-speaker fixed installations. Since any mobile unit requires a driver, the ultimate objective should be, if possible, the use of fixed, automatically operated speakers. Unless one has had considerable experience with the

FIG. 2. Mobile unit in use in vineyard in California. This has the same "Amplitape" unit as in Fig. 1. The speaker is oscillated by hand while the truck moves at about 5 mph, thus giving broad coverage and eliminating sound shadows (photo courtesy G. Boudreau).

method and with acoustical problems, it is best to start with movable speakers to determine the correct placement.

2. *Selection of Signals*

So far, there has been little research using attractive signals for practical control of animal depredations. As already noted, there are a number of theoretical objections to this, and the possibilities of success have apparently, in most cases, seemed too remote for the time and money to be invested.

By far the major effort has been made on the use of distress calls, and collections of these from many species have been made in a few laboratories. Usually these calls are fairly specific, so it is necessary to get calls for each species involved. Hardenberg (1965), for instance, with Lapwings and three species of gulls on airports in Holland, uses a recording with distress calls of each serially arranged. The French workers find (Brémond, 1965; Busnel, 1965) with their species of gulls and crows

FIG. 3. Equipment used in France for broadcasting recorded communication signals of birds. At left, weatherproof box containing two 12-volt storage batteries to power the tape player. In the middle, tape player and amplifier units with photoelectric cell in lid of box to turn on at sunrise and off at sunset. At right, bank of loud speakers (photo courtesy of J. Giban).

however, that there is generally a reaction by all species to the calls of any one; thus they use only one distress call for all species. We shall discuss this below.

It has been our experience that alarm calls are generally more effective than distress calls. Bruns (1959d, 1960), Keil (1965), Pfeifer (1965), and Schmitt (1962a,b) report that in Germany they no longer use the distress call of the Starling, instead they use what they call the *Alarmruf* or *Warnruf,* and the *Kommandoruf,* emitted when the birds sight cats, hawks, or other enemies. It seems surprising that, even granting the successes achieved in France and Holland, students of the problems of gulls at airports have used distress calls, when alarm calls are more easily obtained and of striking repellency. This is particularly surprising in Britain and in the United States, where distress calls have not achieved the desired objectives. Yet reports from Britain and the United States show tests only with the distress call and no attempts to secure the variety of alarm calls produced by the gulls, and reported (Frings *et al.,* 1955b) as repellent.

While these two classes of calls are of known effectiveness, in some

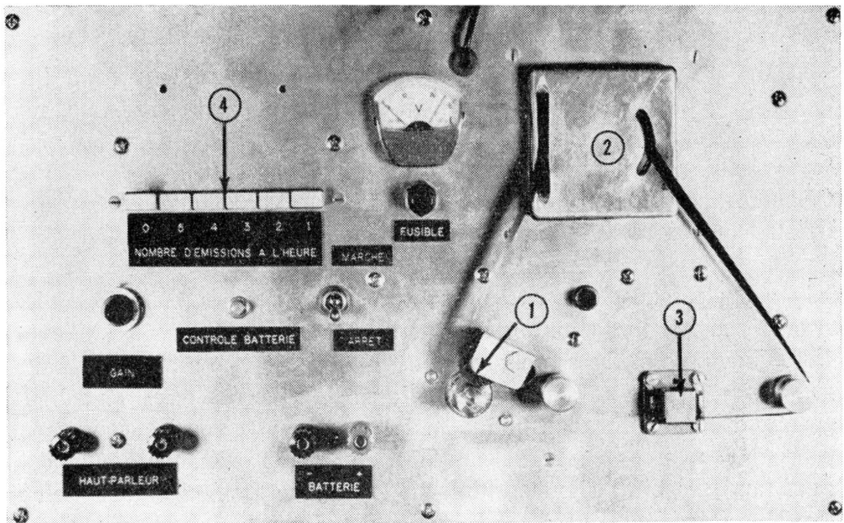

FIG. 4. Tape deck of equipment used in France. Cartridge (2) stores continuous loop of recorded tape, which is drawn past the playing head (3) by motor driven capstan (1). Leads from the playing head go to a preamplifier and 12-volt amplifier beneath, driving the speakers. The wire at the middle above leads to the photoelectric cell in the lid of the box. The number of emissions per hour can be programmed by the keys (4); the emissions are irregularly spaced in time during each hour. Battery voltage is determined by depressing the button (*Controle Batterie*) and reading the meter. To the right of this button is the On–Off switch. Output intensities can be varied by an attenuator (Gain) (photo courtesy J. Giban).

species there are other call notes that deserve study. Boudreau (1962, 1963, 1964) particularly, has found that combinations of various distress and alarm calls and notes, often with other sounds, such as hawk calls or even simulated shots, are much more effective than are individual calls. Also many workers find that shifting the calls, e.g., from one distress call recording to another, increases the effectiveness.

Busnel and his co-workers (cf. Busnel and Giban, 1960; Busnel, 1965) are trying to delineate the reactogenic parameters of the calls they are studying, in the hope that they can select and use just those items of the sound that are important. This might create a supernormal signal, in the sense of Tinbergen (1951), as sought also by Brown and Sugg (1961). These developments if possible would increase the effectiveness of the tools to be used, and such research has tremendous fundamental value in our understanding of bird behavior. They would, however, still leave the techniques of application of prime importance. It is too much to hope that any signal, no matter how supernormal, will be a foolproof control.

3. *Effects of Types of Problem*

At the present time, three basic types of bird problems have been studied in testing communication signals for control: viz., roosting, resting, and feeding. Each presents particular aspects that influence deployment of equipment, time of application, etc.

Generally roosts are fixed, occupied only at night, and in most cases tenanted only at certain times of the year. Thus, time of application is generally short—as the birds come into the roost, or after they have settled—and clearance is desired only for a particular season. A specialized form of roost is the communal nesting areas of some corvids and other birds. They present the same conditions, however. If the roost is of limited extent—for instance, a small woodlot, a small clump of trees, or a building—fixed installations can be made. If the roost is extensive, it is generally simpler to use sound trucks. Where pre-roosts exist, these should be found, and the birds should ultimately be driven from them, as well as from the definitive roost.

If Starlings or corvids are involved, the sound is broadcast at the roost as the birds try to come into it for the night. Generally the sound is turned on only as long as the birds try to enter the roost. This may mean that the total time of sound application is only 10–20 minutes in the course of about 1 hour when the birds would otherwise enter the roost without disturbance. Generally, three or four evenings are required to achieve a reduction of 80–90% in the population, and often one or more evenings after that to drive out most of the remainder. If the roost is large and well established, there seems to be always a small residue of birds that remain after all the others have deserted. The reasons for this are not known, but it has been suggested (Frings *et al.*, 1955a) that these may be the lower members of the peck-order setup in the flock taking advantage of their new freedom. The results of leaving these in the roost vary: in some cases, these birds soon join the others wherever they are; in others they remain and may decoy the flock back. Merely broadcasting the call to them after the fourth or fifth night, however, is useless.

With communally nesting aggregations of corvids, Giban and colleagues have found that the birds are hard to move during the day, usually returning quickly to their nests. However, when chased from their eggs in the middle of the night, they may not find their way back, and the eggs fail to hatch. Thus, these scientists have shown that the acoustic method of scaring birds may be useful in reducing bird populations, as well as in shifting their location.

Aggregations of resting birds are most publicized as hazards on airports. They are usually gulls and corvids, although other species may be involved (cf. Brown and Sugg, 1961; Busnel and Giban, 1965; Seubert, 1964, 1965). In some cases, the problem exists only during daylight hours, in others during both daylight and dark. The distances involved are large and background noise levels are high, creating special acoustical problems, discussed by Andrieu (1965), Gramet (1965), Brown *et al.* (1962), and Sugg (1965). Briefly, the problem is to get coverage of the extensive runways with minimum number of speakers. In some cases, the hazard is seasonal, in others year round. In the latter case, habituation to sounds may become serious, as it may not with other types of problems. Generally it is not practical to clear an airfield permanently or for long times, and so control is aimed mainly at keeping the birds away from aircraft during landing and takeoff. Thus, the sound need be on for just a few minutes before each plane movement (Hardenberg, 1962, 1965). If the airplane movements are relatively few, and the bird population shifts as birds leave to feed, etc., habituation may not take place. If the airplane movements are almost continuous, and the bird population is relatively constant, habituation can be a problem. Change of calls and change of direction of the source seem to aid (Brown and Sugg, 1961). After some efforts to use fixed speakers of various types (Sugg, 1965), these workers have gone over to the use of sound trucks. They also supplement the broadcasts with various fireworks, etc. (Cooke-Smith, 1965; Wright, 1965), as does Kuhring (1965a,b).

Feeding problems vary greatly in location and nature, occurring in open fields, vineyards and orchards, cattle feedlots, and food storage sheds, to mention only a few. With crop plants, generally the danger of bird damage exists only at critical times such as sprouting, harvesting, etc. In these cases, the time of application is rather limited. Usually, feeding occurs only in daylight, and feeding pressure varies with the time of day; it is greatest in the early morning and before sunset, and least shortly after midday. The equipment must be available for sound production during all the daylight hours, but not necessarily at night. Because of the rather limited overall time for which control is needed, habituation may not be serious. The major problem arises in determining the duration and rhythm of application for the sound. This will be discussed in the next section. Distances to be protected and acoustical properties of the environment vary with the crops: open fields are like airports, and orchards or vineyards like complex, absorbing buildings. Fixed installations are generally cheapest, for they do not take continuous manpower. Sound trucks may save on original cost, but require drivers, which may be expensive. However, in a large field, one sound truck can do what would require 15–20 fixed speakers. The problems involved in placement of fixed speakers have been discussed above and need not be repeated here.

In feedlots, depredations may go on all year or may vary with season. If birds use the feedlot as a winter food source, pressure on them to feed there when alternative foods are scarce is very great. If they use the lot as a supplementary food source, the fixation may be much less. Generally distances are not great, and acoustical problems are minor, but may be specialized. If the sound must be applied all year round, habituation may be serious. Shifts of the calls or positions of the speakers assist here. In storage depots or sheds, the same problems arise as with roosting birds in these places—reverberation, sound shadows, etc. Actually, these seem to be like extended roosting situations, and often the birds roost in or near the feeding area. Generally they are year-round problems, and thus habituation may develop, but may be obviated as stated above. Speaker placement is essentially an acoustic matter, and if appropriate rhythms of emission of the sounds can be found, the installation may become automatic, but must be watched for habituation by the birds.

4. *Timing of Sound Emissions*

Annual, seasonal, and diurnal movements of the pest species determine the basic times of application. Obviously, there is no sense in applying sounds when the birds are elsewhere; thus, roosting birds need to be subjected to the broadcasts only while coming into or being in the roost, feeding birds while attempting to feed, and so on.

More important, and more difficult to be specific about, however, is the duration of single emissions of calls, and intervals between these emissions if they are not continuous. It is generally believed by workers with these calls, but with little experimental evidence, that continuous use of the sounds results in the birds' becoming habituated. Consequently, the general rule has been to use the sound discontinuously, and as little as possible. In driving Starlings from tree roosts, for instance, the sound may be turned on, during the approximately 1 hour that the birds attempt to enter the roost, for only a few minutes in total, and usually only in bursts of 15–30 seconds each time. The German workers, however, suggest that this precaution is not necessary, for they report effectiveness with long-continued application. The matter needs careful, critical studies. In the meantime, the sounds are generally applied intermittently, each emission being relatively short.

The recommended rhythm of emission varies greatly with species, investigator, etc. The simplest situation results from having an observer to turn the sound on and off as the birds come into and leave the defended area. This is the method used for urban tree roosts, where the time of treatment is short. A modification of this is the use of a sound truck moving about through an extended area with the sound on continuously. German vineyards are usually protected by a man stationed on a watch-tower, armed with various means for bird control, which he uses as his experience teaches him—distress calls to flush or turn birds and exploding flash crackers or rockets to scare them in the air and to drive them directionally. If the control is being attempted with a species on which no data exist, this may be the only reasonable way to approach the matter. Generally, the person applying the sound should record his observations carefully, for these may give information from which automated programs can be planned. This planning may be difficult, however, for, as Gramet (1965) notes, the sound emissions may have to be timed to suit the "pressure" of the birds, that is their persistence in trying to return. Thus, he found that with the same species at the same location emissions from 1 to 10 times per hour were needed, with 1–2 minutes for each emission. At airports, where persons are already stationed, manual operation is most suitable and involves no major expense. Boudreau (1963, 1965) has developed a time sequence suitable for Western Robins, English Sparrows, and House Finches—10 seconds of sound delivered every 2 minutes during the day for about 3 days, after which less frequent emissions suffice. The sound applying device is turned on and off, morning and evening, by an electric eye. Thus, it is fully automated. He finds that habituation does not occur, in fact after a few weeks few or no birds even try to enter the protected area.

5. *Specific Uses*

The most usual use of distress or alarm calls is to chase birds from places where they are undesirable. We have already discussed the general principles involved, and here would like to illustrate the nature of specific problems encountered in carrying out control programs under practical conditions. Most of the reports in the literature give only the broadest information about the methods and results. This is often all that is necessary. However, at this stage of development, much may be learned by sharing day-to-day experiences.

The following is a report (Frings *et al.,* 1955a) of the earliest tests, in 1953, with the Starling distress call in State College, Pennsylvania. The call had been tested in a preliminary fashion, using only a tape recorder broadcasting under trees. It had been found that as darkness came on the birds were increasingly hard to move and that the cries of caged Starlings were ineffective against wild birds. First, a description of the infested areas:

In Aug. 1953, when the first tests with the distress call were made, the starlings were in two major areas and one minor area. The largest concentration of birds was in the area that we called the Pugh Street Area. About 15,000 starlings were in Norway maple trees along Pugh St. for about ¼ mile and along four consecutive cross-streets (Foster, Nittany, Fairmount, and Prospect Aves.) for about 50 yards on each side of Pugh St.

The second major area, that we called the Gill-College Area, was about ¾ mile away and roughly L-shaped. It included one block (about 75 yds.) on Gill St. and about 75 yds. on College Ave. that crossed Gill St. There was a smaller subsidiary to this area about ⅒ mile away on S. Barnard St., that crossed College Ave. next to Gill St., and on the other side of College Ave. from the Gill St. roost. The infested trees were Norway maples, and the break between the main area and the subsidiary was due to the fact that no trees were present along College Ave. and S. Barnard St. between them. The Gill-College Area had about 5,000–7,000 starlings.

The only minor area occupied at the time of the first tests was at Beaver Ave. and Garner St., in a grove of Norway maples around a house, about ¼ mile from the Pugh Street Area. There were about 1,000 starlings there.

Now, to the actual tests:

It was decided to try to clear a small area within a large roosting area. The spot selected was the population center of the Pugh St. Area: the corner of Nittany Ave. and Pugh St. A distance of 50–75 yards on each side of this on both streets was to be treated, forming a +-shaped test area. In this region there were about 5,000–7,000 roosting birds. The starlings all around were to be left to give a heavy reinfestation pressure.

A sound-truck, with a 15 W. amplifier and 2 trumpet-type, movable loud-speakers was used, with power supplied by batteries through converters. The peak sound pressure at 1 meter from the horns was about 120 db. The sound was applied as the truck moved away from the center of the roosting area and was off while the truck moved toward the center. Observers were stationed on the roof of a three-story apartment building within about 25 yds. of the corner; from there they had an excellent view of the whole area.

On the first evening (Aug. 25, 1953), the sound truck was started when about half of the starlings were in. The birds flew out immediately. From then until dark (65 min.), the truck made 14 treatments, each lasting about 2 min. There were estimated to be fewer than 1,000 starlings in the trees at night-fall.

On the second evening (Aug. 26), about 1,000–2,000 tried to return. These were driven out as before by periodic trips through the area with the sound-truck. The number of runs was 6, each again of about 2 min. duration. The total treatment time was 40 min., with one trip after dark to survey. At night-fall there were no more than 200 birds in the treated area, while one block away the starlings were as numerous as ever.

On the third evening (Aug. 27), only about 500 birds tried to return. These were easily dislodged by 5 runs, each of about 1 min., over a period of 40 min. On the second and third nights of these tests the reluctance of starlings to enter trees from which their fellows had been driven by the sound was obvious. Flocks of starlings flew over the tree-tops skimming them as if to land, but veering off and flying away without alighting. It seemed almost as if there were a transparent cover

over the tops of the trees. This was noted many times later, but it still remains without adequate explanation. It was this behavior that led us to postulate a possible warning odor (Frings and Jumber, 1954) that starlings might give off when thus scared from their roosts.

On the fourth evening (Aug. 28), very few starlings tried to alight in the test area. When the starlings were well settled in the surrounding untreated area, the sound-truck was driven through the test area with sound on, but only a few flew up from near the edges. A treatment just at night-fall showed none at all in the trees.

In spite of the fact that thousands of starlings were left all around the cleared region, reinfestation was slow. By Sept. 5, when a careful check of the whole area was made, only about 500 birds were in the treated area, mostly near the edges, and many of these were grackles and cowbirds. After this, the whole Pugh Street Area was cleared; thus the maximum time of clearance was not determined under these conditions.

Shortly after the experiment previously described, the clearance of a small town, Millheim, Penna., was carried out successfully. This showed that one could probably clear the whole of the Pugh Street Area in State College, and this was attempted. The decision was further motivated by complaints from persons nearby that the starlings had been left near them, and they did not appreciate being in the position of experimental controls. At the time of these tests, there were about 10,000 starlings in the area. The same sound equipment and procedures were used as before, and observers recorded the results from the apartment building.

On the first evening (Sept. 6), the truck was used on Pugh St. only, the speakers being directed down the side-streets as it passed. The total treatment time was about 60 min., with the sound on almost continuously. By night-fall the trees on Pugh St. were essentially free of starlings, but those on the side-streets were infested at distances beyond about 25 yds. from the corners. It was decided that, on succeeding nights, the sound-truck should be driven along the side-streets also.

On the second evening (Sept. 7) most of the starlings tried to return. The sound-truck was driven slowly through the whole region for about 60 min. before darkness, with the sound on continuously. At night-fall there were very few starlings left. On the third evening (Sept. 8), only about ⅓ of the original population of starlings tried to enter and they were easily driven away. The total treatment time was 40 min. On the fourth evening (Sept. 9), only about 300–400 starlings came near the area, and these were at the extreme edges. The sound-truck was driven through to assay the results and to drive these out. Treatment time was about 30 min.

On Sept. 10 there were only about 200 birds in the area, and these were mostly grackles, robins and cowbirds. Later checks, on Sept. 15 and Oct. 3, showed the area to be clear of starlings, and this was confirmed by persons nearby. Actually the results were only significant to Oct. 1, because at that time the birds normally moved into the autumn roost.

By Sept. 20, the starlings in State College were in two major roosts and two minor roosts. One major roost was the Gill-College Area, previously described. The second was one which usually was the annex to this first, on S. Barnard St. about ⅒ mile away. This had grown until it had become L-shaped also, including about 100 yds. on S. Barnard St. and 50 yds. on Foster Ave., around a corner. The one minor roost was the area at the corner of Beaver Ave. and Garner St., already described, and the other was about ¼ mile from the Gill-College area on W. Nittany Ave. All of these, except the Beaver-Garner Area, were on the southwestern side of town. The Beaver-Garner Area was about 1 mile from the others on the southeastern side. There were 10,000–12,000 starlings in town, about 1,500 in the minor roosts, the rest in the major roosts.

In view of the fact that the Pugh Street Area had remained clear, it seemed of interest to try to clear all of State College. Two sound-trucks, like the one described before, were used. One treated the two major roosts, following a fixed driving pattern that brought it back to any point in about 6–7 min. Once started the truck ran continuously in the pattern. The second truck treated the two minor roosts and places where starlings tried to settle as they left the treated areas. A likely possibility was that the starlings would try either to resettle in the Pugh Street Area or to enter the autumn roost on S. Burrowes St. Two sets of observers followed the movements of the starlings, one atop an apartment building near the major roosts, the other atop the apartment building near the Pugh Street Area.

On the first evening (Sept. 28), the truck with the fixed pattern of coverage was in action about 60 min. just before dark with the sound on continuously. The second truck struck at the two minor areas. The feared invasion of the Pugh Street Area did not materialize. Starlings flew over the area, after being driven from the major areas, and skimmed the tops of the trees. As before, however, it was as if there were an invisible barrier preventing their landing. The feared invasion of the autumn roost also failed to occur. The second truck, therefore, spent most of the time treating the minor roosts—actually in driving back and forth between them. At night-fall, it is likely that most of the starlings were still in State College, but they were in many unusual places—in evergreen trees on untreated streets, even in dwarf fruit trees and hedges near the one major area.

On the second evening (Sept. 29), the truck with the fixed route extended this to include the minor area about ¼ mile away. The cruising truck treated the pre-roosting assembly areas, about ½ mile from the major roosting areas, as the birds first started to come into town. It was thought that driving them from the pre-roosting areas, when the light intensity was still relatively high, would give the birds opportunity to find other suitable quarters. The truck patrolling the main areas began shortly after the treatment of pre-roosts and continued till dark, about 60 min. When the pre-roosting areas were cleared, the cruising truck treated the minor area at Beaver and Garner. Only a few starlings had tried to return there, however, and these quickly left. This truck, therefore, was used to survey, testing possible infested areas with bursts of sound. Only a few birds were found trying to roost here and there, and these were driven away easily. When darkness came, there were probably no more than 1,000–2,000 starlings in town, although it was impossible to make an exact census, because of the scattering which had occurred.

The treatments on the third night (Sept. 30) were carried out as on the second, with the fixed route truck active for about 45 min. and the cruising truck used to treat pre-roosts and to check for possible infested areas later. There were only about 1,000 starlings which came to the pre-roosts and these were driven away as soon as they arrived. Only a few hundred tried to enter the major roosting areas, and these were driven away by the truck driving on the fixed course. The Beaver-Garner roost was clear. Extra attention was given by the second truck to the W. Nittany Ave. minor roost, and it was cleared before dark. There were only about 300 starlings in scattered trees near the former major roosts at nightfall.

After allowing a few days for possible reorganization of roosts, a complete check of State College was made on the evening of October 3. This was supplemented by newspaper and radio requests to residents to report any concentrations of starlings. Only about 50–75 starlings were found in one tree about ¼ mile from the two major roosts and the same number in three trees about one block away. One report was received from a resident in the northern part of town, where occasional pre-roosting had occurred but no permanent roost. A check revealed that the offenders in this case were about 50 Purple Grackles. The town remained thus until leaf-fall, about Oct. 20. It is quite usual, in clearing operations, to find small groups of starlings which do not go with the main group. If there are only a scattered few, they constitute no problem.

The clearances achieved may be summarized as follows. The earliest treatment—on Foster Ave.—was two months before the usual fall departure of the birds. This remained clear for most of that time, but later treatments in the area make this difficult to evaluate. The central region of the Pugh Street Area remained clear after Aug. 25–28, even under pressure of later treatments. If Oct. 20 be taken as the approximate date of departure, this would be approximately 7 weeks. The Pugh Street Area as a whole was clear for approximately 6 weeks, and the whole of State College for 3 weeks. At the end of these periods of time there was no evidence of reinfestation, so that these do not necessarily represent maximum possible clearance times that might be achieved. The usual autumn roost of N. Burrowes St. did not form; this had existed previously for at least 15 years.

The second report is of an unsatisfactory test in Philadelphia; unsatisfactory, that is, in that the birds did not leave, but excellent as a training experience. Later in 1960, this same building was treated with success (Frings, 1962).

City Hall, in down-town Philadelphia, was infested with starlings for at least 15 years before 1954. This was a winter roost building up in population in early Nov. and being deserted in early April. The building is about 1 city block square, seven stories high (about 100 ft.), with a central courtyard about ⅓ city block square. At the middle of each of the south, east and west faces there is a large cupola extending about two stories higher. In the center of the north face, there is a tall tower. The building is highly decorated with sculptures, ledges and ornamental trim. It is made-to-order for starlings, and they show their appreciation by roosting on it in immense numbers—a minimum estimate of 40,000, probably more nearly 50,000. These birds use the cupolas, tower and a few surrounding buildings for pre-roosting. As darkness falls, they flock into the courtyard and onto the faces. An observer on the roof in their midst has the impression of being in a black blizzard. The major part of the flock roosts in the courtyard, and the rest chiefly on the south and east faces of the building away from the cold northwest winds.

The decision to try to clear this building was motivated by the surprising successes in 1953. While it was unlikely that so large a building with so many hiding places for the birds could be cleared, it seemed worth trying. Accordingly, permission was granted by city officials for a test to be made early in 1954. Preliminary observations of behavior and population counts were made in Dec., 1953, and for two days preceding the tests in Jan., 1954.

On the first evening of treatments (Jan. 29), three sound-trucks were used: one with 8 short, trumpet-type speakers driven by a 30 W. amplifier, and two others like those used previously. The last two were placed at opposite sides of the building in the street about 20 ft. from the sides of the building, and the drivers were instructed to circle it slowly with the sound at full intensity. The other truck was stationed in the court-yard with the intensity also maximum. The birds were allowed to become somewhat settled before the sound-trucks were given the signal to start. The total treatment time was 40 min., with the sound on all the time.

Even to the human ear, the sound quality left much to be desired. The equipment was being driven at full intensity, and there was considerable distortion. One of the trucks circling the Hall had much higher quality than the other, and the difference in repellency between the two was striking. Inside the courtyard the din was terrific, but the sound, as a result of overdriving of the amplifier and reverberation from the walls, was noise and not recognizably the distress call. The birds on the upper levels in the courtyard were somewhat repelled, but those near the speakers were less affected. At night-fall most of the starlings were in their usual places. For the first time, it was clear that noise alone was ineffective; the sound had to be recognizable.

On the second evening (Jan. 30), the courtyard was treated with a 125 W. amplifier driving two bull horns with four 30 W. drivers on each. This could be run at about ½ of capacity and achieve requisite volume without distortion. The speakers were directed by hand toward the various sides of the courtyard. Outside the two sound-trucks continued their circling.

The sound was started as the first large groups of birds tried to enter the courtyard, and was continued until nearly dark, a total of 65 min. The results in the courtyard were strikingly better. The birds were very excited and flew wildly away. They stayed out of the courtyard perching on the cupolas until it was almost dark, then some drove in rapidly even against the sound. At night-fall there were about 5,000 fewer birds in the courtyard than on the previous evening. Most of these birds were huddled in recesses near the large tower, where the horns could not be directed. On the outside, there was a corresponding increase in numbers, the trucks having little effect. The quality of sound in the courtyard, however still left much to be desired, for reverberation from the walls produced considerable distortion.

The third day being Sunday (Jan. 31), no changes in equipment or arrangements were possible. The treatments were the same as the night before, starting a little earlier, with total treatment time about 70 min. At night-fall, there were only about half the number of starlings in the courtyard as at the beginning. Those that had left were either on the tower, much higher than they usually roosted, or on the outside of the building.

On the basis of the observations up to this time, we decided to place the large horns on the roof of the building to lay down a wall of sound around it, thus preventing the birds from pre-roosting on the cupolas. Accordingly, during the fourth day, wires were prepared on the roof and the bull horns and 125 W. amplifier were taken there.

On the fourth evening (Feb. 1), the bull horns were on the roof, one sound truck was left to circle the outside, and the second was stationed in the courtyard. Sound was turned on as soon as the birds started to arrive. The quality was excellent, with little reverberation, and the results were strikingly better. Typical winter-time compact flocks were scattered as they approached 300–400 ft. above the building. The flocks seemingly exploded as the sound was turned on them. The starlings were held away from the building until it was almost dark. At that time, they started to drive in. Unfortunately the horns could be moved for only short distances, and the cupolas and tower on City Hall blocked the sound from wide areas. The birds flew in these sound-shadows, dodged around the tower or cupolas and into the courtyard. They were successfully repelled only from the sides nearest the horns. The total treatment time was 60 min., with the sound on most of the time. When night had come, there were about 25% of the original population in the courtyard, almost all where the sound could not reach. This was the first evening on which there was a significant effect.

It was obvious that horns were needed to reach parts of the building which were shielded from the sound, and that more powerful equipment would be needed outside of the building, if the background traffic noise were to be overcome while the sound had good fidelity. This necessitated changes that would have required about three days. There was no time for such an extended stay, and so the tests were discontinued with this evening.

On the next evening (Feb. 2), therefore, no treatment was made. The birds were wary in approaching the building, and many tended to avoid the cupolas which had been the pre-roosts. But, as darkness fell, they drove in, and at nightfall the appearance of the building was as if no treatment had been made. Obviously, there was no residual effect here, as had been observed in the summer experiments.

Two facts stand out in these tests. First, it is very important to treat early, while the birds can find other quarters. Once the starlings have landed, they are extremely difficult to dislodge. It seems as if their reaction, once in their established roosts, is to freeze in place on hearing the distress call rather than to fly away. Second, the sound must be recognizable. Reverberation and overdriving of sound equipment to cover large distances and to overcome background noise pose real problems in situations such as this. There was evidence from the work of the summer that the sound was relatively specific for starlings and was not just a noise, but the results here showed this to be true beyond reasonable doubt. Obviously, the best approach is to place the loud speakers so that the sound is directed into an essentially free field.

The possible use of recorded communication signals for population reduction, instead of population shifting, is still to be tried on a large scale. The experiments of Busnel and Giban (1960) have already been mentioned. These suggest a method for population diminution with birds that nest communally. The treatment of many scattered individual nests is probably impractical. It is possible to use broadcasts of the distress call of Starlings to herd the birds along selected paths. Thus, the call might be used to drive birds into traps, such as the giant light trap developed by researchers of the Fish and Wildlife Service (Seubert, 1963).

Totally untried is the possibility of using the jamming of communication systems suggested for insects by Wright (1964). His idea, discussed by him only for chemical communication in insects, is to flood the environment with artificial signals and thus mask the natural ones. It might be possible to flood an area with recorded territorial or courtship songs of birds and thus render it difficult for females to find males, or to flood the environment with communication sounds of parents and young, and thus reduce this system to uselessness. These seem, at first sight, almost certainly to be economically impractical, but perhaps with communally nesting birds or under special circumstances, the idea would be worth trying.

E. Problems and Advantages

1. *Habituation*

The well-known habituation of birds to visual and noisemaking devices leads one immediately to ask whether habituation to recorded communication signals might occur. As already noted, in all earlier work it was assumed that this was a serious problem and so very short times of treatment were used. This is still widely practiced. However, tests in Germany, and by Boudreau in the United States, have indicated that continuous operation may not be objectionable. It seems that birds do not habituate readily, if at all, to their own signals. This makes biological sense, for danger signals of small birds may be emitted many times daily, and the need to respond is always there.

By far the majority of workers with distress or alarm calls of birds report no habituation, even after extended periods of use. Thus, Gramet (1959, 1961, 1962, 1965), with corvids in France, found none after up to 59 days of continuous application in feeding situations. Where habituation did occur, in a few of his tests, it was only necessary to shift the call to another effective one, and the birds responded as at first. Boudreau (1960, 1962, 1964) likewise reports no habituation with Starlings, House Finches, English Sparrows, American Robins (*Turdus migratorius*), etc., after more than 2 weeks of treatment. In fact, he stresses the cumulative effects, finding that the first few days of application in crop areas may seem to have little effect, requiring repeated broadcasts, but after a week or so the birds flush readily and stay away for long times. Bruns (1959b,d, 1960, 1961) had previously reported this for Starlings in Germany with distress and alarm calls. Müller (1960) found no habituation by Starlings after 3 years in olive groves in Tunisia, and Schmitt (1962a) reports none in vineyards in Germany. Schmitt reports continuous operation in vineyards up to 40 days with increased effectiveness. Hardenberg (1965), working with gulls on airports found no habituation after some years of operation.

On the other hand, Brown *et al.* (1961, 1962) and Brown (1962) report habituation after a few weeks with distress calls of gulls, and this has been reported also for airports in the United States (Nelson and Seubert, 1964). Brown and Sugg (1961) found that changing the call, or supplementing the broadcasts with shooting or fireworks, restored the effectiveness. This difference between the gulls in Britain and Germany may be the result of different population structures (Seubert, in Busnel and Giban, 1965, p. 285) or possibly reflects some equipment problems, such as gradual deterioration. In spite of these results, however, habituation seems to be slow in many cases, or nonexistent, and changes in recordings seem to afford a means for overcoming it.

2. *Movements of Pests to New Problem Areas*

It is obvious that scaring pests from objectionable roosting or feeding areas does not kill them. The population remains, and the animals may become pests somewhere else. This is true with any repellent. There is no question that in many cases animals have achieved population levels that are incompatible with their own best interests, as well as those of man. In these cases, population reduction is essential. Acoustic methods offer no direct approach to this, except as tried with the corvids in France. Also Buchmann and Müller (1957) report that their chasing the birds from olive groves brought some to below minimal food-intake levels and resulted in large numbers of deaths. Sounds might be used to increase the effectiveness of decoy traps (Seubert, 1963) or light traps, but this is a minor contribution. Hardenberg (1965) feels correctly that the airport problem is not solved merely by broadcasting distress calls, even when these work as well as they do in Holland. Population reduction and management of cover and food near airports is the ultimate solution, as he sees it, with distress call treatments merely stopgaps. Certainly no one would debate this. Means must be found to reduce populations to supportable levels and to alter habitats so that birds do not enter problem areas. This, however, is not always possible, and so stopgap measures are needed.

Actually, birds chased from one area do not necessarily go to another to be again objectionable. Bruns (1961, 1962) reports that Starlings driven from orchards in Germany usually go to moorlands or other noncultivated places to feed, even possibly learning to avoid cultivated areas if sound treatments are widely used. As he points out, the existence of alternative foods, or areas for roosting, which might not be objectionable to man, may mean that the birds easily leave the places where they are not desired to become economically neutral or even positive. Gramet and Hanoteau (1965) likewise report that, where alternative foods are present, pest birds often move within 24–28 hours after the beginning of sound irradiation to these areas and stay permanently away from the treated areas; where alternative foods are not present, they go shorter distances and continued treatments are necessary. Boudreau (1964, 1965) reports the like for feeding House Finches and Prairie Horned Larks. We (Frings *et al.,* 1955a) found that, at Millheim, Pennsylvania, Starlings that had roosted in town left there for a woodlot outside, where they were not pests. They thus exposed themselves to more rigorous wild conditions, which presumably could reduce their population.

This is not to say that birds may not go to new areas where they are not desired. This is particularly true in cities, where buildings are used as roosts during winter. The birds, if driven from one building, must find some warm place to roost, and so will settle on others nearby. If driven centrifugally from the original roost, they may become scattered enough that their small numbers on any one building are supportable. But these may later rejoin to form a single group. Continued centrifugal driving may be necessary, but this may be too expensive. Actually, the solution here is to drive the birds from the roost before they become firmly established in it, that is, when they first begin to arrive in the autumn. Then they may be driven farther away, or may move southward or out into rural districts in small groups. A case of serious problems arising from moving pest birds from one feeding area to another occurred in Tunisia, where olive orchards were cleared of Starlings by broadcasts of the distress call (Buchmann and Müller, 1957). The birds here moved from olive groves of the larger landowners, who could afford to hire sound trucks, into the holdings of poorer farmers who could not, creating a political problem. Fortunately, however, such aggravated circumstances would probably be unusual.

3. *Expense*

There is little in general that can be said here. Sound projection equipment is not cheap, although it is fairly durable, and the initial cost is thus amortized over long times. Labor costs are determined by the time of attendance needed. Thus, if a man has to be with a sound system, the cost may be rather high. However, in Germany, attendants are used in vineyards. If the system can operate automatically, the cost is much reduced.

In agricultural situations, usually the agriculturist knows something of his potential losses from bird damage. He ought, therefore, be able to judge the economics of acoustical controls. In general, crops with high cash value, e.g., grapes, other fruits, and specialty vegetables, are damaged sufficiently to make acoustical controls economically feasible. Crops with low cash value, e.g., grains and forage, may yield too little per acre to make installations practical. Where human lives are endangered, as at airports, expense is a secondary consideration, although not to be ignored. In general, installation expenses are not so great that the long-continued use of a sound system cannot make it easily feasible; this is much better than the constant expense for pyrotechnics, etc.

Urban roosting problems involve public opinion with respect to supportable expense, often politically debated. Actually, most cities now have sound trucks available and rental or purchase of recordings is a minor item. The hiring of personnel, however, may be major. From personal experience, we can say that public attitude toward this expense is unpredictable. There usually is no obvious economic loss against which to balance the expense, and debates rage about use of public funds for the comfort of a few. To the biologist, it might seem obvious that getting rid of an objectionable roost is desirable and not outside the limits of public interest, but the biologist usually has not been involved with political infighting.

4. *Specificity and Interspecificity*

The earliest report on the use of the distress call for control of Starlings (Frings and Jumber, 1954) emphasizes that the call is specific, that in mixed groups only a given species is affected. Later tests indicated that the situation is much more complex, however, and that the calls might not only be specific, but that local populations of birds might have their own "dialects" (Frings *et al.,* 1958; Frings and Frings, 1959b). This has been supported by fundamental research on bird behavior, such as that of Marler and Tamura (1962) among others. On the other hand, the French workers (Brémond, 1965; Busnel, 1965; Giban, 1962a, 1965; Gramet, 1959, 1961, 1962, 1965) report that their corvids and larids show much interspecificity in response, so that the distress call of one species can be used to move members of another. Hardenberg (1965), in Holland, conversely finds that gulls there show specificity.

Obviously the matter is complex and involves many facets in the inborn and learned behavior of the birds (cf. Thorpe, 1961; Brémond, 1963). Thus, Busnel (1965) reports that mixed flocks of gulls usually show interspecificity, but *Larus marinus* reacts much less to calls of other species, and its distress call is ineffective when broadcast to other species. He believes that possibly its more pelagic and solitary habits have reduced its use of acoustical signals and increased the importance of visual signals (very important in gulls, but little studied, except on breeding grounds). He postulates that interspecificity has three roots—in instinct, in mutual conditioning by learning, and in the fact that the reactogenic properties of the signals of two species are basically similar.

In general, studies on this subject have been incomplete and critical research is badly needed. About all that comes out of the reports is that apparently in some populations of birds (usually where the species is isolated from contact with related forms) communication signals used for alarm and distress are specifically responded to, as are generally reproductive signals (cf. Marler, 1955, 1957). Where, however, the population moves or feeds with individuals of other than its own species, the calls of the various species may be learned interspecifically, inducing similar reactions in all. Certainly also there are similarities in the acoustic structure of the distress calls of related species.

In practical terms, under some circumstances specificity is desirable, e.g., where one wishes to chase only one species and leave all others. In other cases, interspecificity is desirable, e.g., where a mixed population of birds is feeding or resting, and one wishes most economically to chase all. The former is illustrated by urban tree roosts, where one might wish to remove the hordes of Starlings, yet leave robins, etc. The latter is illustrated by the situation in orchards or at airports. If, in this case, one has interspecificity, as in the case of the French crows and gulls, he can use the calls of one species for a time, and if habituation sets in, shift to those of another (Gramet and Hanoteau, 1965). Where, as at the airports in Holland, the birds react only specifically, and the calls of all species must be used serially to drive them all away, this factor of safety does not exist. Luckily in the latter case, habituation has so far not developed.

The French workers seem inclined to believe that interspecificity among related species will be the rule, rather than specificity (Brémond, 1965), thus allowing shifting of calls for increased effectiveness. This remains to be seen. Busnel and Giban (1960) have postulated that, if the effective parameters of the signals could be determined, synthetic signals could be made that might be just as effective, or possibly more effective, than the true ones. This might, however, be a mixed blessing, as Gramet points out, for then variations are not easily possible. This is not necessarily true, however, for variations in detail could be produced by modulating, on the basic signals, various secondary sounds. At present, however, such synthetic signals are not available, and specificity and interspecificity are empirical matters, to be determined for each set of circumstances, even with the same species.

5. *Hazards and Irritations*

Unfortunately, the communication signals used for control of pest birds are audible to man. If communication signals can be used for rodent control, they may be ultrasonic. Distress calls broadcast at high intensities are disturbing to some people, particularly if the broadcasts come in the evening or at night. Generally, the only situation in which this is serious is in treating urban roosts. This, however, is the only source of irritation in the use of these methods for control. There is no fire hazard, as with firecrackers, etc. There is little uncontrolled drift and absolutely no residues, as there are with insecticides.

III. CONCLUSIONS

A variety of visual, mechanical, and acoustical methods for frightening vertebrate pests, chiefly birds, have been used since antiquity—scarecrows, display of dead birds, flashing lights and objects, rockets, and mechanically and pyrotechnically produced loud noises. In some form, all of these are still in wide use. Where times of protection are short or methods can be varied continually, these may keep pest species from crops, resting places, or breeding areas. However, habituation to these stimuli is usually rapid; sooner or later they fail to frighten pests. Modern research has centered mainly on means for retarding habituation or for finding substitutes for methods that fail.

Rodents and other small mammals can hear sounds that are ultrasonic for man, but attempts to use these for control have been few and generally not encouraging; furthermore, these are still essentially frightening noises. Birds probably cannot hear ultrasonic sounds, so these seem to have little promise for bird control. Critical studies with radar beams show no response by birds to these. Rodents, when exposed to certain high frequency sounds, undergo epileptiform seizures; practical use of these for rodent control, however, remains to be developed.

Since 1954, recorded acoustical communication signals of birds, to which the animals respond specifically with complex behavior patterns, have been used for practical control. Distress calls (emitted when birds are restrained) and alarm calls (emitted when birds sight an enemy) have been most studied. Both are repellent, often not directly so, and not simply frightening. Attractive and identifying signals, e.g., food-finding, mobbing, assembly, and courtship calls, may have value for practical control, but have so far not been tested in this regard.

Studies in France and the United States show that high fidelity is quite unnecessary for recognition of the signals by the birds. But tests in Germany show conversely that high fidelity may be needed under some circumstances. The matter needs further study.

Possession of a sound, no matter how effective, does not assure successful pest control. Many factors must be considered before application of sound can be made with hope of success: (1) identification of pest and other species; (2) nature of the problem (roosting, feeding, resting, etc.); (3) size of areas to be protected and size and composition of populations; (4) environmental conditions, such as weather, topography, etc.; (5) behavior patterns of pest and associated species; and (6) factors that could interfere with application of sounds, such as background noises, activities of persons, etc.

With these data, selection and deployment of equipment become problems mainly in sound engineering—the sound must reach the birds at the proper intensity and fidelity. Timing of emissions of sound can, at present, be determined only empirically. Usually the sounds are broadcast as little as possible to avoid habituation. Experience with a species under specific conditions can allow one to set up automated schedules.

Habituation seems to be much less a problem with communication signals than with noises. In fact, some workers have used these signals for years with only increased effectiveness—the animals apparently learn to avoid protected areas. Where habituation does occur, changing the signal usually restores effectiveness. Since most species have a variety of alarm signals, this is usually easy to do.

As with all repellents, distress and alarm calls move the pests, but do not destroy them. In many cases, however, the birds find alternative foods that are not economically important, or seek new roosts that are not objectionable. With communally roosting corvids in France, driving the birds with distress calls from their eggs at night resulted in destruction of the embryos and thus a decrease in population. Removal of roosting birds to wilder situations also may foster increased natural predation.

Communication signals may be specific—even subspecific, with local "dialects"—or may be nonspecific. So far, only specific tests can tell which, for species and populations differ. When a species feeds or roosts with others, they may develop reactions to signals of others; when they remain more or less isolated, they may respond only to their own specific signals. For bird control, specificity or interspecificity may be useful or not, depending upon circumstances. If one wishes to chase only one species and leave others, specificity is desired; if he wishes to clear an area of all birds, interspecificity is advantageous.

Acoustical communication controls are generally reasonably priced, for equipment is relatively permanent and so costs are spread over a long time. Whether its use is justified economically or not must be determined for each case, based on loss or damage figures and degree of alleviation expected. Besides usually allowing precision in manipulation of pest species, broadcasts of acoustical communication signals have no fire or soiling hazards, and leave no troublesome residues.

REFERENCES

Allen, C. H., Frings, H., and Rudnick, I. (1948). *J. Acoust. Soc. Am.* **20,** 62.

Altmann, M. (1952). *Behaviour* **4,** 116.

Andrew, R. J. (1961). *Behaviour* **18,** 161.

Andrieu, A. J. (1965). *In* "Le Problème des Oiseaux sur les Aérodromes," Colloq., Nice, 1963 (R. G. Busnel and J. Giban, eds.), pp. 305–320. Inst. Natl. Recherche Agron., Paris.

Backus, R. H. (1963). *In* "Sharks and Survival" (P. W. Gilbert, ed.), pp. 243–254. Heath, Boston, Massachusetts.

Berlin, C. L. (1963). *J. Speech Hearing Res.* **6,** 359.

Boudreau, G. W. (1960). "Report of Starling Control Test in Arizona." Mimeo., Phoenix, Arizona.

Boudreau, G. W. (1962). "Summary Report of Western House Finch Control Experiments, Delano, California." Mimeo., Phoenix, Arizona.

Boudreau, G. W. (1963) "Report of Bird Control Study at Mayacamas Vineyards, Napa, California." Mimeo., Phoenix, Arizona.

Boudreau, G. W. (1964). "Bird Pests in Arizona Spring Lettuce." Mimeo., Phoenix, Arizona.

Boudreau, G. W. (1965). Personal communication.

Brémond, J. C. (1962). *Angew. Ornithol.* **1,** 49.

Brémond, J. C. (1963). *In* "Acoustic Behaviour of Animals" (R. G. Busnel, ed.), pp. 709–750. Elsevier, Amsterdam.

Brémond, J. C. (1965). *In* "Le Problème des Oiseaux sur les Aérodromes," Colloq., Nice, 1963 (R. G. Busnel and J. Giban, eds.), pp. 233–245. Inst. Natl. Recherche Agron., Paris.

Brough, T. (1965). *In* "Le Problème des Oiseaux sur le Aèrodromes," Colloq. Nice, 1963 (R. G. Busnel and J. Giban, eds.), pp. 279–286. Inst. Natl. Recherche Agron., Paris.

Broughton, W. B. (1963). *In* "Acoustic Behaviour of Animals" (R. G. Busnel, ed.), pp. 824–910. Elsevier, Amsterdam.

Brown, R. G. B. (1962). *In* "Colloque sur les Moyens de Protection Contre les Espèces d'Oiseaux Commettant des Dégâts en Agriculture," Versailles, 1961 (J. Giban, ed.), *Ann. Epiphyties* **13** (No. Hors Ser.), 153.

Brown, R. G. B., and Sugg, R. W. (1961). "Experiments on the Removal of Birds from Airfields by Distress Calls." Ministry of Aviation, London.

Brown, R. G. B., Sugg, R. W., and Brough, T. (1962). "Experiments on the Removal of Birds from Airfields by Distress Calls," Pt. II. Ministry of Aviation, London.

Bruns, H. (1956). *Pflanzenschutz Ber.* **5,** 1.

Bruns, H. (1959a). *Pflanzenschutz Ber.* **11**(5), 79.

Bruns, H. (1959b). *Dritter Jahresber. Staatl. Vogelschutzwarte, Hamburg, April 1959-March 1959.*

Bruns, H. (1959c). *Mitt. Obstbauversuchsringes Alten Landes,* **14,** 140.

Bruns, H. (1959d). *Gesunde Pflanzen* **11,** 151.

Bruns, H. (1960). *Tagesber. Deut. Akad. Landwirtschwiss. Berlin* **30,** 105.

Bruns, H. (1961). *Angew. Ornithol.* **1,** 32.

Bruns, H. (1962). *In* "Colloque sur les Moyens de Protection Contre les Espèces d'Oiseaux Commettant des Dégâts en Agriculture," Versailles, 1961 (J. Giban, ed.), *Ann Epiphyties* **13** (No. Hors Ser.), 77.

Buchmann, E., and Müller, O. (1957). *Z. Obst-Weinbau* **66,** 575.
Büttiker, W. (1962). *In* "Colloque sur les Moyens de Protection Contre les Espèces d'Oiseaux Commettant des Dégâts en Agriculture," Versailles, 1961 (J. Giban, ed.), *Ann. Epiphyties* **13** (No. Hors Ser.), 167.
Burner, C. J., and Moore, H. L. (1953). *U.S. Fish Wildlife Ser. Spec. Sci. Rept. Fisheries* **111.**
Busnel, R. G., ed. (1963a). "Psychophysiologie, Neuropharmacologie, et Biochimie de la Crise Audiogène." C. N. R. S., Paris.
Busnel, R. G., ed. (1963b). "Acoustic Behaviour of Animals." Elsevier, Amsterdam.
Busnel, R. G. (1963c). *In* "Acoustic Behaviour of Animals," (R. G. Busnel, ed.), pp. 69–111. Elsevier, Amsterdam.
Busnel, R. G. (1965). *In* "Le Problème des Oiseaux sur les Aérodromes," Colloq., Nice, 1963 (R. G. Busnel and J. Giban, eds.), pp. 247–258. Inst. Natl. Recherche Agron., Paris.
Busnel, R. G., and Giban, J., eds. (1960). "Colloque sur le Protection Acoustique des Cultures et Autres Moyens d'Effarouchement des Oiseaux," Jouy-en-Josas, 1958. Inst. Natl. Recherche Agron., Paris.
Busnel, R. G., and Giban, J., eds. (1965). "Le Problème des Oiseaux sur le Aérodromes," Colloq., Nice, 1963. Inst. Natl. Recherche Agron., Paris.
Busnel, R. G., Giban, J., Gramet, P., and Pasquinelly, F. (1955). *Compt. Rend.* **241,** 1846.
Busnel, R. G., Giban, J., Gramet, P., and Pasquinelly, F. (1956). *Compt. Rend. Soc. Biol.* **150,** 18.
Busnel, R. G., Giban, J., Gramet, P., Frings, H., Frings, M., and Jumber, J. (1957). *Compt. Rend.* **245,** 105.
Calhoun, J. B. (1963). *U.S. Public Health Ser. Publ.* **1008.**
Cardinell, H. A. (1937). Mich. State Univ. *Agr. Expt. Sta. Circ. Bull.* **160.**
Cooke-Smith, R. A. W. (1965). *In* "Le Problème des Oiseaux sur les Aérodromes," Colloq., Nice, 1963 (R. G. Busnel and J. Giban, eds.), pp. 131–137. Inst. Natl. Recherche Agron., Paris.
Creutz, G. (1956). *Anz. Schaedlingskunde* **29,** 149.
Crook, J. H. (1960). *Behaviour* **16,** 1.
Cummings, M. W., ed. (1964). *Proc. 2nd Vertebrate Pest Control Conf., Anaheim, Calif., 1964.* Univ. Calif. Davis Agr. Exten. Serv., Davis, California.
Danner, P. A., Ackerman, E., and Frings, H. (1954). *J. Acoust. Soc. Am.* **26,** 731.
Dijkgraaf, S. (1963a). *Biol. Rev. Cambridge Phil. Soc.* **38,** 51.
Dijkgraaf, S. (1963b). *Nature* **197,** 93.
Drury, W. H., and Keith, J. A. (1962). *Ibis* **104,** 449.
Ehlgen (1958). *Deut. Weinbau* **13,** 177.
Finger, F. W. (1947). *Psychol. Bull.* **44,** 201.
Frings, H. (1948). *Pests* **16**(4), 9, 44.
Frings, H. (1954). *Proc. 30th Natl. Shade Tree Conf., Atlantic City, N.J., 1954* pp. 108–112.
Frings, H. (1955). *Am. Fruit Grower* **75,** 45.
Frings, H. (1962). *In* "Colloque sur les Moyens de Protection Contre les Espèces d'Oiseaux Commettant des Dégâts en Agriculture," Versailles, 1961 (J. Giban, ed.), *Ann. Epiphyties* **13** (No. Hors Ser.), 87.
Frings, H. (1963). *Am. Fruit Grower* **83**(6), 14, 16, 29.

Frings, H. (1964). *Proc. 2nd Vertebrate Pest Control Conf., Anaheim, Calif., 1964* (M. Cummings, ed.), pp. 50–56. Univ. Calif. Davis Agr. Exten. Serv., Davis, California.

Frings, H., and Boudreau, G. W. (1964). Final Rept., Contract NBY-53155, U.S. Navy, Washington, D.C.

Frings, H., and Cook, B. (1964). *Condor* **66,** 56.

Frings, H., and Frings, M. (1952). *J. Acoust. Soc. Am.* **24,** 163.

Frings, H., and Frings, M. (1957). *J. Wildlife Management* **21,** 91.

Frings, H., and Frings, M. (1959a). *Elepaio* **20,** 6, 14, 23, 30.

Frings, H., and Frings, M. (1959b). *Sci. Am.* **201,** 119, 122, 126, 128, 130.

Frings, H., and Frings, M. (1960). *Stanford Res. Inst. J.* **4,** 11.

Frings, H., and Frings, M. (1963). *Sound* **2,** 39.

Frings, H., and Frings, M. (1964). "Animal Communication." Blaisdell, New York.

Frings, H., and Jumber, J. (1954). *Science* **119,** 318.

Frings, H., and Senkovits, I. (1951). *J. Cellular Comp. Physiol.* **37,** 267.

Frings, H., and Slocum, B. (1958). *Auk* **75,** 99.

Frings, H., Allen, C. H., and Rudnick, I. (1948). *J. Cellular Comp. Physiol.* **31,** 339.

Frings, H., Frings, M., Cox, B., and Peissner, L. (1954). *Anat. Record* **120,** 734.

Frings, H., Jumber, J., and Frings, M. (1955a) *Penn. State Univ. Occasional Papers Dept. Zool. Entomol.* **55-1.**

Frings, H., Frings, M., Cox, B., and Peissner, L. (1955b). *Science* **121,** 340.

Frings, H., Frings, M., Cox, B., and Peissner, L. (1955c). *Wilson Bull.* **67,** 155.

Frings, H., Frings, M., Jumber, J., Busnel, R. G., Giban, J., and Gramet, P. (1958). *Ecology* **39,** 126.

Geiger, R. (1965). "The Climate Near the Ground." Harvard Univ. Press, Cambridge, Massachusetts.

Giban, J. (1962a). *Festschr. Vogelschutzwarte Hessen, Rheinland-Pfalz, Saarland 1937–1962* pp. 52–58.

Giban, J., ed. (1962b). "Colloque sur les Moyens de Protection Contre les Espèces d'Oiseaux Commettant des Dégâts en Agriculture," Versailles, 1961. *Ann. Epiphyties* **13** (No. Hors Ser.).

Giban, J. (1965). *In* "Le Problème des Oiseaux sur les Aérodromes," Colloq., Nice, 1963 (R. G. Busnel and J. Giban, eds.), pp. 223–232. Inst. Natl. Recherche Agron., Paris.

Gramet, P. (1959). *Nature (Paris)* **3286,** 49.

Gramet, P. (1961). *Defense Vegetaux* **15,** 69.

Gramet, P. (1962). *In* "Colloque sur les Moyens de Protection Contre les Espèces d'Oiseaux Commettant des Dégâts en Agriculture," Versailles, 1961 (J. Giban, ed.), *Ann. Epiphyties* **13** (No. Hors Ser.), 111.

Gramet, P. (1965). *In* "Le Problème des Oiseaux sur les Aérodromes," Colloq., Nice, 1963 (R. G. Busnel and J. Giban, eds.), pp. 259–265. Inst. Natl. Recherche Agron., Paris.

Gramet, P., and Hanoteau, J. (1965). *In* "Le Problème des Oiseaux sur les Aérodromes," Colloq., Nice, 1963 (R. G. Busnel and J. Giban, eds.), pp. 267–273. Inst. Natl. Recherche Agron., Paris.

Gross, A., Pfeifer, S., and Keil, W. (1959). *Umschau,* 1959, pp. 105–106.

Haag, O. (1958). *Deut. Weinbau* **31,** 737, 742.

Hamilton, W. J., III. (1962). *Condor* **64,** 390.

Hardenberg, J. D. F. (1960). *In* "Colloque sur le Protection Acoustique des Cultures et Autres Moyens d'Effarouchement des Oiseaux," Jouy-en-Josas, 1958 (R. G. Busnel and J. Giban, eds.), pp. 149–158. Inst. Natl. Recherche, Agron., Paris.

Hardenberg, J. D. F. (1962). *In* "Colloque sur les Moyens de Protection Contre les Espèces d'Oiseaux Commettant des Dégâts en Agriculture," Versailles, 1961 (J. Giban, ed.), *Ann. Epiphyties* **13** (No. Hors Ser.), 145.

Hardenberg, J. D. F. (1965). *In* "Le Problème des Oiseaux sur les Aérodromes," Colloq., Nice, 1963 (R. G. Busnel and J. Giban, eds.), pp. 121–126. Inst. Natl. Recherche Agron., Paris.

Hasse, B. L. (1963). *Ohio J. Sci.* **63,** 145.

Jackson, W. B. (1964). *Pest Control* **32**(9), 11, 50.

Jumber, J. F. (1956). *Auk* **73,** 411.

Kalmbach, E. R. (1928). *U.S. Dept. Agr. Farmers' Bull.* **1571.**

Kalmbach, E. R. (1945). *U.S. Fish Wildlife Serv. Wildlife Leaflet* **172.**

Kalmbach, E. R. (1954). *Pest Control.* **22**(6), 32.

Keil, W. (1962). *In* "Colloque sur les Moyens de Protection Contre les Espèces d'Oiseaux Commettant des Dégâts en Agriculture," Versailles, 1961 (J. Giban, ed.), *Ann. Epiphyties* **13** (No. Hors Ser.), 191.

Keil, W. (1965). *In* "Le Problème des Oiseaux sur les Aérodrome," Colloq., Nice, 1963 (R. G. Busnel and J. Giban, eds.), pp. 287–291. Inst. Natl. Recherche Agron., Paris.

Kleerekoper, H., and Chagnon, E. C. (1954). *J. Fisheries Res. Board Can.* **11,** 130.

Knorr, O. A. (1954). *Wilson Bull.* **66,** 264.

Kuchly, J. (1961). *Defense Vegetaux* **15,** 15.

Kuhring, M. S. (1965a). *In* "Le Problème des Oiseaux sur les Aérodromes," Colloq., Nice, 1963 (R. G. Busnel and J. Giban, eds.), pp. 95–102. Inst. Natl. Recherche Agron., Paris.

Kuhring, M. S. (1965b). *In* "Le Problème des Oiseaux sur les Aérodromes," Colloq., Nice, 1963 (R. G. Busnel and J. Giban, eds.), pp. 301–304. Inst. Natl. Recherche Agron., Paris.

LaFond, E. C. (1961). *Naval Res. Rev.* **April,** p. 15.

Lehmann, A., and Busnel, R. G. (1963). *In* "Acoustic Behaviour of Animals" (R. G. Busnel, ed.), pp. 244–274. Elsevier, Amsterdam.

Little, R. S. (1964). *BioAcoust. Bull.* **4** (1),1.

McAtee, W. L. and Piper, S. E. (1936). *U.S. Dept. Agr. Leaflet* **120.**

Marler, P. (1955). *Nature* **176, 6.**

Marler, P. (1957). *Behaviour* **11,** 13.

Marler, P., and Tamura, M. (1962). *Condor* **64,** 368.

Marsh, B. T., Jackson, W. B., and Beck, J. R. (1962). *Grain Age* **Nov.,** p. 27.

Miller, L. (1952). *Condor* **54,** 89.

Moulton, J. M. (1963). *In* "Acoustic Behaviour of Animals" (R. G. Busnel, ed.), pp. 655–693. Elsevier, Amsterdam.

Müller, O. (1960). *In* Colloque sur la Protection Acoustique des Cultures et Autres Moyens d'Effarouchment des Oiseaux," Jouy-en-Josas, 1958 (R. G. Busnel and J. Giban, eds.), pp. 180–188. Inst. Natl. Recherche Agron, Paris.

Müller, O., and Gerig, L. (1957). *Schweiz. Z. Obst-Weinbau* **66,** 34, 51.

Neff, J. A. (1945). *U.S. Fish Wildlife Serv. Wildlife Leaflet* **268.**

Neff, J. A., and Meanley, B. (1957). Arkansas Univ. (Fayetteville) Agr. Expt. Sta. *Bull.* **584.**

Neff, J. A., and Mitchell, R. T. (1956). *U.S. Fish Wildlife Serv. Wildlife Leaflet* **365.**

Nelson, S. O., and Seubert, J. L. (1964). "Electromagnetic Energy and Sound for Use in Control of Certain Pests," *Proc. 131st Meeting, Am. Assoc. Advance. Sci. Montreal, Canada, 1964.* (Also In "Scientific Aspects of Pest Control." Nat. Acad. Sci., Publ. 1402, 1966).

Pfeifer, S. (1965). *In* "Le Problème des Oiseaux sur le Aérodromes," Colloq., Nice, 1963 (R. G. Busnel and J. Giban, eds.), pp. 275–278. Inst. Natl. Recherche Agron., Paris.

Prosser, C. L., and Brown, F. A., Jr. (1961). "Comparative Animal Physiology," 2nd Ed. Saunders, Philadelphia, Pennsylvania.

Przygodda, W. (1962). *In* "Colloque sur les Moyens de Protection Contre les Espèces d'Oiseaux Commettant des Dégâts en Agriculture," Versailles, 1961 (J. Giban, ed.), *Ann. Epiphyties* **13** (No. Hors Ser.), 13.

Reich, H. (1950). *Mitt. Obstbauversuchsringes Alten Landes.* **5,** 84.

Reich, H. (1955). *Rhein. Monatsschr. Gemuese-Obst-Gartenbau* **43,** 141.

Schilling, H. K., Drumheller, C. E., Nyborg, W. L., and Thorpe, H. A. (1946). *Am. J. Phys.* **14,** 343.

Schilling, H. K., Givens, M. P., Nyborg, W. L., Pielemeier, W. A., and Thorpe, H. A. (1947). *J. Acoust. Soc. Am.* **19,** 222.

Schleidt, W. M. (1961). *Z. Tierpsychol.* **18,** 534.

Schmitt, N. (1957). *Deut. Weinbau* **12,** 701.

Schmitt, N. (1958). *Deut. Weinbau* **13,** 488.

Schmitt, N. (1959a). *Gesunde Pflanzen* **11,** 32.

Schmitt, N. (1959b). *Deut. Weinbau* **14,** 182.

Schmitt, N. (1960a). *In* "Colloque sur la Protection Acoustique des Cultures et Autres Moyens d'Effarouchement des Oiseaux," Jouy-en-Josas, 1958 (R. G. Busnel and J. Giban, eds.), pp. 159–168. Inst. Natl. Recherche Agron., Paris.

Schmitt, N. (1960b). *In* "Colloque sur la Protection Acoustique des Cultures et Autres Moyens d'Effarouchement des Oiseaux," Jouy-en-Josas, 1958 (R. G. Busnel and J. Giban, eds.), pp. 169–176. Inst. Natl. Recherche Agron., Paris.

Schmitt, N. (1960c). *22nd Jahresber. Vogelschutzwarte Hessen, Rheinland-Pfalz, Saarland.* Frankfurt-am-Main.

Schmitt, N. (1962a). *In* "Colloque sur les Moyens de Protection Contre les Espèces d'Oiseaux Commettant des Dégâts en Agriculture," Versailles, 1961 (J. Giban, ed.), *Ann. Epiphyties* **13** (No. Hors Ser.), 57.

Schmitt, N. (1962b). *Festschr. Vogelschutzwarte Hessen, Rheinland-Pfalz, Saarland, 1937–1962* Frankfurt-am-Main, pp. 94–102.

Schwartzkopff, J. (1955a). *Acta 11th Congr. Intern. Ornithol. Basel, Switzerland, 1954* pp. 189–208.

Schwartzkopff, J. (1955b). *Auk* **72,** 340.

Schwartzkopff, J. (1959a). *Vogelwarte* **15,** 194.

Schwartzkopff, J. (1959b). *Orion* **1959,** 884.

Seubert, J. L. (1963). *Angew. Ornithol.* **1,** 163.

Seubert, J. L. (1964). *Proc. 2nd Vertebrate Pest Control Conf., Anaheim, Calif. 1964* (M. W. Cummings, ed.), pp. 150–159. Univ. Calif. Davis Agr. Exten. Serv., Davis, California.

Seubert, J. L. (1965). *In* "Le Problème des Oiseaux sur les Aérodromes," Colloq., Nice, 1963 (R. G. Busnel and J. Giban, eds.), pp. 143–171. Inst. Natl. Recherche Agron., Paris.

Spurlock, E. M. (1962). "Control of Bird-Strike Hazard at Airports," Mimeo., Stanford Res. Inst., Menlo Park, California.

Sugg, R. W. (1965). *In* "Le Problème des Oiseaux sur les Aérodromes," Colloq., Nice, 1963 (R. G. Busnel and J. Giban, eds.), pp. 293–299. Inst. Natl. Recherche Agron., Paris.

Tavolga, W. N. (1960). *In* "Animal Sounds and Communication" (W. E. Lanyon and W. N. Tavolga, eds.), pp. 93–136. Am. Inst. Biol. Sci., Washington, D.C.

Tavolga, W. N. (1965). *Tech. Rep.* NAVTRADEVCEN **1212–1.** U.S. Naval Training Center, Port Washington, New York.

Tembrock, G. (1963). *In* "Acoustic Behaviour of Animals" (R. G. Busnel, ed.), pp. 751–786. Elsevier, Amsterdam.

Thorpe, W. H. (1961). "Bird-song. The biology of vocal communication and expression in birds." Cambridge Univ. Press, London and New York.

Thorpe, W. H., and Griffin, D. R. (1962). *Ibis* **104,** 220.

Tinbergen, N. (1951). "The Study of Instinct" Oxford Univ. Press (Clarendon), London and New York.

U.S. Fish Wildlife Serv. (1948). *Wildlife Leaflet* **254.**

Van Bergeijk, W. A. (1964). *In* "Marine Bio-acoustics" (W. N. Tavolga, ed.), pp. 281–299. Macmillan (Pergamon), New York.

Van der Heyde, J. J. M. (1965). *In* "Le Problème des Oiseaux sur les Aérodromes," Colloq., Nice, 1963 (R. G. Busnel and J. Giban, eds.), pp. 127–130. Inst. Natl. Recherche Agron., Paris.

Wiener, F. M. (1961). *J. Acoust. Soc. Am.* **33,** 1200.

Winn, H. E. (1964). *In* "Marine Bio-acoustics" (W. N. Tavolga, ed.), pp. 213–231. Macmillan (Pergamon), New York.

Wisby, W. J., Richard, J. D., Nelson, D. R., and Gruber, S. H. (1964). *In* "Marine Bio-acoustics" (W. N. Tavolga, ed.), pp. 255–268. Macmillan (Pergamon), New York.

Wright, E. N. (1965). *In* "Le Problème des Oiseaux sur les Aérodromes," Colloq., Nice, 1963 (R. G. Busnel and J. Giban, eds.), pp. 113–119. Inst. Natl. Recherche Agron., Paris.

Wright, R. H. (1964). *Science* **144,** 487.

AUTHOR INDEX

Numbers in italics refer to the pages on which the complete references are listed.

C

H

L

U

V

W

X

Y

Z

SUBJECT INDEX

C

P

Q

R